D1525824

3 Springer Series in Solid-State Sciences
Edited by Hans-Joachim Queisser

Springer Series in Solid-State Sciences

Editors: M. Cardona P. Fulde H.-J. Queisser

Volume 1 **Principles of Magnetic Resonance** 2nd Edition
By C. Slichter

Volume 2 **Introduction to Solid-State Theory**
By O. Madelung

Volume 3 **Dynamical Scattering of X-Rays in Crystals**
By Z. G. Pinsker

Volume 4 **Inelastic Electron Tunneling Spectroscopy**
Editor: T. Wolfram

Z. G. Pinsker

Dynamical Scattering of X-Rays in Crystals

With 124 Figures

Springer-Verlag Berlin Heidelberg New York 1978

Professor Zinovii Grigorievich Pinsker
Shubnikov Institute of Crystallography, Academy of Sciences of the USSR
Leninsky prospekt 59, Moscow B-333, USSR

Editors:

Professor Dr. Manuel Cardona
Professor Dr. Peter Fulde
Professor Dr. Hans-Joachim Queisser

Max-Planck-Institut für Festkörperforschung
Büsnauer Strasse 171, D-7000 Stuttgart 80, Fed. Rep. of Germany

ISBN 3-540-08564-5 Springer-Verlag Berlin Heidelberg New York
ISBN 0-387-08564-5 Springer-Verlag New York Heidelberg Berlin

Offset printing and bookbinding: Zechnersche Buchdruckerei, Speyer.
2153/3130-543210

Dedicated to my wife

Preface

The present book is devoted to one of the latest achievements in optics, namely crystal optics of X-rays. The scientific value of this contribution resides in widening of the crystal-optics visual range to the short-wave region of the electromagnetic spectrum.

It should be particularly emphasized that the scattered radiation parameters are closely allied to the parameters of the crystal medium. As a result, various processes of elastic and inelastic scattering of X-rays are characterized by extremely high sensitivity to deviations from the ideal and idealized crystal structure. This creates (and has already largely produced) probably the most effective refined methods for verifying crystal perfection and, accordingly, detection and thorough investigation of the different types of defects in single-crystal materials. The importance of such methods for various branches of modern technology is quite obvious.

In this book the theory of dynamical scattering of X-rays and the most important experimental methods and results are presented in the most systematic manner possible.

In comparison with the Russian edition of 1974[1], Chapters 4,9,10,11 and 12 are essentially modified, and relevant supplements have been added. Chapter 11 has been revised by F.N. Chukhovskii, and Chapter 12 by A.M. Afanasiev and V.G. Kohn. The author is very grateful to his colleagues for helpful discussions of various problems treated in this book. I am also indebted to E.L. Lapidus-Pinsker for her help in the preparation of the manuscript.

Moscow, October 1977 Z.G. PINSKER

---

[1] Z.G. Pinsker: *Dinamicheskoye rasseyanie rentgenovskikh luchey v ideal'nykh krystallakh* (Nauka, Moscow 1974).

Contents

1. Introduction

(Historical Survey)

The discovery of X-ray diffraction in crystals by LAUE, FRIDRICH and KNIPPING in 1912 [1.1] served as the starting point for the development of scientific research along a number of important lines. We shall discuss just a few of them.

The above discovery convincingly demonstrated the wave properties of X-rays. This, together with the previously established electromagnetic nature of radiation, confirmed the hypothesis that X-rays form the short-wave part of the electromagnetic spectrum.

Further, this discovery was the first and decisive experimental proof of the periodic structure of crystals. In fact, theoretical crystallography had already arrived at this conclusion, mainly as an outcome of the theory of the space groups of symmetry elaborated by FEDOROV [1.2] and SCHOENFLIES [1.3].

From the optics of visible light we know that the radiation of a wavelength of the same order as, and preferably less than, the period of a grating suffers diffraction on periodic objects of the type of optical grating. Thus, the discovery proved that the wavelength of an X-ray must be of the order of interatomic distances. It became clear why the visible light of wavelengths exceeding the crystal lattice periods by about 500 to 1000 times failed to reveal the periodic structure of crystals in diffraction experiments.

At the time LAUE was writing the chapter on diffraction for the Enzyklopädie der Mathematischen Wissenschaften, he found a simple form for the theory of two-dimensional optical gratings. The result of FRIDRICH'S and KNIPPING's experiment inspired him further, and he devised the simplest theory of spatial, or three-dimensional, diffraction and interference. It was referred to as geometrical, or kinematical, theory.

One of the first and best-known lines of investigation, which advanced rapidly as a result of the 1912 discovery, was X-ray analysis of the atomic structure of crystals. The huge amount of experimental data on X-ray structural studies accumulated over more than half a century was one of the most important preconditions for developing solid-state physics and chemistry, on the one hand, and the production, processing, and utilization of many materials of contemporary technology, on the other [1.4].

At the same time it is obvious that the discovery of X-ray diffraction formed the start of a new and extremely interesting era in optics. While in visible-light optics the crystalline medium is regarded as a continuum characterized by anisotropy, X-ray optics must be incomparably closer to the periodic atomic structure. Unfortunately, the spectacular successes of X-ray structural analysis did not contribute much to the development of this new branch of optics. The kinematical scattering theory was later supplemented with highly accurate calculations of intensities and atomic amplitudes, inclusion of the effect of thermal vibrations, the method of determining the phases of structural amplitudes based on anomalous dispersion, and so on. In this form, the theory effectively satisfied the requirements stipulated by the contemporary studies of the atomic structure of crystals.

A characteristic feature of the kinematical theory is that it takes into account only the interaction of each atom with the primary, or refracted, wave in a crystal. It neglects the interaction of an atom with the wave field induced in the crystal by the collective scattering of all the other atoms. In other words, it ignores multiwave scattering and, what is especially important, the interaction of the diffracted waves with the refracted one.

At first glance it might seem that the weakest point in kinematical theory is its neglect of the law of conservation of energy, since the energy of the primary wave passing through the crystal is considered constant though a part of this energy is imparted to the scattered waves.

However, the validity of such a consideration in ordinary experimental conditions was proved and was adequately explained by the negligibly low energy of the diffracted beams in small crystals.

DARWIN [1.5] (see also [1.6]) was the first to notice the schematic nature of the kinematical theory, and he made a successful attempt at more rigorous consideration of the X-ray scattering in crystals. In his theory, multiwave scattering was interpreted, although insufficiently rigorously, as multiple scattering. The reflection of X-rays from a crystal was regarded as a successive passage through and multiple reflection from the parallel planes forming the crystal - an idea suggested by the British school headed by Bragg. Multiple scattering within this plane was neglected. In spite of the half-qualitative character of his treatment, DARWIN obtained the correct form for the maximum of reflection from the perfect crystal, which scatters not only at the Bragg angle, but also within some finite angular interval.

During that period, i.e., 1913-1914, DARWIN, together with Mosley, studied the intensity of crystal-scattered X-rays under Rutherford in Manchester. In verifying the agreement between the experimental intensity and the theoretical value, he found that the former was some tenfold larger.

The explanation for this discrepancy, which was suggested by DARWIN, is of fundamental importance. He proposed an entirely new idea: an imperfect crystal consists of slightly disoriented subunits, or blocks which, however, possess a perfect internal structure, i.e., in contemporary terminolgy it exhibits a mosaic structure (this term was later introduced by Ewald).

When constructing his theory, DARWIN predicted the phenomenon of *primary extinction*, that is total reflection of X-rays by a perfect crystal in a certain region near the reflection maximum and, hence, rapid exponential damping of waves in a crystal. Besides, a mosaic crystal should be expected to exhibit *secondary extinction*, i.e., limited penetration of incident radiation due to gradual reflection from separate blocks properly oriented with respect to the incident beam.

These results greatly influenced subsequent development. First, they introduced considerable improvements in the intensity measurement methods. Secondly, they pointed the way to transition from the rigorous scattering theory to the kinematical one as a theory of scattering in a mosaic crystal. However, the widely used concepts of ideal or nonideal-mosaic crystal are not sufficiently precise and clear. In any case, LAUE, ZACHARIASEN and others later demonstrated that the kinematical equations can be derived from the equations of the perfect crystal for the limiting case of small-crystals. Further, the idea of the block-type, or mosaic, structure of real crystals served as a springboard for the development of an independent line in solid-state physics and for important technological applications of X-ray topography.

The idea that a real crystal is a mosaic structure was doubted by physicists. It was widely believed that no perfect crystals of appreciable size existed, nor could they be created.

The phenomena occurring during the propagation of "short-length" waves in crystals were thoroughly investigated by EWALD [1.7], who had worked as far back as 1911, i.e., even before X-ray diffraction was discovered, on a theory of dispersion of light in a crystal with a regular, periodic structure. This work played an essential role in the formation of those physical ideas which were introduced by EWALD into the dynamical theory of X-ray scattering [1.8].

The fundamental principles of EWALD's theory are expounded in JAMES' book [1.9]. Here we shall mention some of EWALD's remarks [1.4], which explain his approach to the phenomenon of radiation scattering in a crystal.

A typical problem in crystal optics - determination of the refractive index - is considered as an eigenvalue problem similar to calculation of the frequencies of the vibration spectrum of a mechanical system in steady state.

The vibrations of resonators, the initiation and propagation of the elementary waves induced by them through an infinite, unbounded crystal are regarded as proper, rather than forced, vibrations of the system. An essential feature of such an oscillating system is self-consistency. It is manifested in the fact that each resonator vibrates under the action of the wave field formed by the superposition of the elementary waves of all other resonators. In other words, the existence of the wave field requires (or can only be realized with) a connection between the vibrating resonators; the resonator vibrations - a common wave field, the correspondence between the two types of connections are the condition determining the refractive index n. For a given frequency ν, the self-consistency will predetermine the value of λ or the phase velocity q of the optical waves in a crystal, and hence the refractive index of the medium, $n = c/q$.

The role of the phase velocity q as a regulator for ensuring finite amplitude in an (infinite) unbounded crystal is evident. Indeed, the existence of the phase velocity, $q < c$, is equivalent to the statement that elementary waves emerging from some atomic planes and propagating along, say, the direction of the x-axis, will interfere with the subsequent elementary waves with a certain phase shift. Thus, the amplitude of the total wave will be proportional to $(c-q)^{-1}$. If $q = c$ the amplitude of an infinite crystal tends to infinity. Hence, the phase velocity of proper value will determine the "capacity" of a system of resonators to form a wave field of appropriate strength in a "self-consistent" regime.

EWALD's theory, published in 1916-1917, was slowly and reluctantly accepted by physicists and attracted almost no attention even in the postwar period. At that time it was thought that this theory eventually led through a more complicated path, to the same results as DARWIN's theory.

In fact, the content of EWALD's theory, which is given in [1.8], was much richer. Here we shall mention the dispersion of an incident wave within a crystal and the natural transition to multiwave scattering; detailed discussion of the two-wave case and the use of the dispersion surface, which, in principle, enables one to predict the X-ray anomalous penetration in absorbing crystals; symmetrical and asymmetrical reflection; LAUE reflection; and, finally, the "Pendellösung fringe" solution and, hence, the principal interference effects, during scattering in perfect crystals.

In subsequent works EWALD's theory took a somewhat different form.

This theory regarded a crystal as a periodic structure consisting of *point* resonators or dipoles. Under the action of an external wave, the dipoles are excited and start vibrating, and vibration propagates through the crystal. This dipole wave, in turn, generates an electromagnetic wave. The model of point radiator atoms used in the theory was somewhat not in agreement with the concepts which had begun to form in the late 1920's and early 1930's. In X-ray diffraction analysis, use was made of the Fourier method, i.e., analysis of electron density distribution on the basis of the experimental structural amplitudes. This yielded patterns of continuous electron density distribution, whose maxima corresponded to the atom positions.

These results agreed with the new physical concept corresponding to the wave, or quantum mechanics of Louis De Broglie-Heisenberg, which was first given in full in Schrödinger's works in 1926. In 1927-1928, Wentzel, Waller, and especially Hartree and Fock, laid the foundation of the most refined (nonrelativistic) calculations of X-ray scattering amplitudes, using the model of continuous electron density distribution within the atom.

In 1928, BETHE published a fundamental paper [1.10] in which he developed a dynamical theory of the electron scattering by analogy with Ewald's dynamical theory of X-ray scattering. This work appeared after the discovery of electron diffraction in 1927.

In Bethe's theory, continuous triple periodic distribution of the internal crystal potential was taken as the "scattering medium" for electrons.

All this induced LAUE to publish a paper in 1931 [1.11] in which he gives a different substantiation of Ewald's theory, i.e., in place of a structure consisting of point resonators, he introduced a model of continuously distributed electron density with positive charges localized at the centres of atoms. The electric field of the incident wave causes polarization, which is proportional at each point to the strength of the local electric field.

Somewhat later KOHLER (1933-1935) [1.12] published papers where he gave a quantum-mechanical interpretation to the above model utilized in the dynamical theory by Laue in his book [1.13].

The same approach was used by ZACHARIASEN in his book [1.14] in Chapter III dealing with the dynamical theory. Clear and precise presentation and successful choice of the variables and scattering parameters made this chapter very popular. Many theorists and researchers engaged in the field of dynamical scattering of X-rays used it and referred to it.

For some reason, the dynamical theory as formulated by ZACHARIASEN and by LAUE, particularly in the last (1960) edition of his book [1.13] and in the surveys [1.15-17] gained wider recognition than the previous version of Ewald's theory.

Comparing the results of the kinematical and dynamical treatment of X-ray scattering in crystals, from 1920 to 1940 researchers reduced the differences to two points, namely in the directions and the angular width of the diffracted beams and in the magnitudes of integral reflection. As for the interference geometry, during the period indicated, a considerable number of investigations were devoted to the deviation from the Wulff-Bragg equation, measurements of the refractive index by methods borrowed from visible light optics, determination of the universal constants such as the electron charge, and absolute determination of wavelengths and other quantities. These investigations, which were carried out by Parrat, Beardin, Bergen and Davis, Larsson, Beklin, Stenström, Renninger and others, convincingly demonstrated the validity of the dynamical theory equations. At the same time, the results obtained were, in many cases, of a qualitative rather than a quantitative nature.

Experimental verification of the validity of the dynamical theory obtained by measuring the maxima and integral values of reflection in the well-known works of James, Brindley and Wood, Wagner and Kulenkampf and Renninger was of limited and partly qualitative significance [Ref. 1.13, pp. 351-358; Ref. 1.9, Chap. VI].

Two serious difficulties prevented the development of the necessary experimental possibilities for accurate reproduction and measurement of the dynamical scattering parameters. The first difficulty was the need to obtain an X-ray beam which would, in its spectral spread and angular divergence, correspond to the incident plane wave approximation that forms the basis of Darwin's and Ewald's theories. The solution of this problem took more than half a century, including COMPTON's pioneering work [1.18] in 1917 (he used a monochromator crystal); even now the excellent schemes for obtaining such beams, which were used by Renninger, Authier, Bubakova, Hildebrandt and Batterman, and particularly by KOHRA and co-workers [1.19], still need to be improved for one reason or another.

Of great importance is the fact that the task of overcoming the first difficulty is intimately connected with the second one, i.e., the absence of sufficiently perfect crystals.

As regards natural specimens, it was established by Parrat, Beardin et al. as far back as 1932-1933, and later in an important series of investigations by Brogren (1952-1954), that the natural crystals of calcite and quartz include specimens of perfect structure eminently suitable for quantitative verification of the dynamical theory. Artificially grown sodium chloride crystals were investigated by Renninger and showed a more or less sharply defined mosaic structure.

These investigations resulted in convincing proof of the validity of the dynamical theory. It was evident, however, that the equations of this theory, for the most part, namely in interference geometry, largely lead to results differing by negligibly small angles of the order of several angular seconds from the values prescribed by the kinematical theory. In other words, the dynamical theory has a very narrow and specialized field of application. As regards the intensities, whose values differ in some cases by more than an order of magnitude, even this difference is not essential since it pertains to very rare crystal specimens.

Let us now turn to the contemporary period.

Prominent against the general background of scientific and technological progress, which gained momentum particularly after the recess caused by World War II, is the rise and development of the industry dealing with semiconductor materials, and instruments and various devices based on them.

The preconditions for the appearance of this industry consisted of two interconnected factors: the development of semiconductor physics and the elaboration of techniques for producing ultrapure semiconductor materials, particularly of highly perfect single crystals, in the first place Si, Ge, and GaAs. It was further established that semiconducting single crystals possess the necessary properties and perform certain preassigned functions, either with negligibly small, specially introduced additives, or in the ultrapure state, but in any case with the most perfect crystalline structure possible. Naturally, methods of investigating such single crystals began to be developed vigorously. Thus, the requirements of practical work stimulated a keen interest, among other things, in the dynamical scattering of X-rays as one of the principal methods of investigation and control of the degree of perfection of crystals.

The use of single-crystal materials in contemporary technology is not restricted to semiconductors. Quartz single crystals are of great importance in radio engineering, and single crystals of various materials are particularly essential in quantum electronics.

In these circumstances, a substantial role has been played by the intensive theoretical and experimental investigations performed over the last 15 to 20 years and dealing both with dynamical scattering in perfect crystals and scattering in slightly distorted crystals.

An important stage in the development of the dynamical theory was the study of scattering in an absorbing crystal and the derivation of equations for the reflection factor and integral values of reflection and transmission of rays in the Laue case. It is significant that one of the most important effects observed during dynamical scattering in perfect crystals, namely the effect of anomalous penetration in the maximum region, was not predicted theoretically, but discovered purely experimentally in investigations carried out in 1941 and particularly in 1950; it is called the BORRMANN effect in honour of the author of these works [1.20]. At the same time, ZACHARIASEN says nothing about anomalous penetration in his book [1.44] (1944), although the equations given by him, when slightly transformed, unambiguously pointed to the existence of this effect.

In actual fact, LAUE showed in his 1949 paper that the Borrmann effect was an obvious result of the dynamical theory [1.21].

Anomalous penetration of X-rays through a thick absorbing crystal is of great fundamental and practical importance. BORRMANN also devised a pictorial model of physical interpretation of the anomalous penetration effect [1.22]. In the simplest case of the two-wave approximation, each of the two fields exhibits a periodic modulation of the amplitude across the depth of the crystal as a result of interference of the refracted and diffracted waves. The modulation period coincides with that of the lattice, and the minima and maxima (the nodes and antinnodes, respectively) of the two fields are displaced by a half-period. As a result, the photoelectric absorption, which plays the principal part, sharply increases for the field whose maxima lie on the atomic planes and, contrarywise, decreases for the other field.

As well as investigation of absorbing crystal, largely in the 1960's a comprehensive investigation began into the interference effect during dynamical scattering in weakly absorbing crystals. These interference effects, which are of great interest from a purely physical point of view, open up entirely new ways of determining accurately and absolutely the important quantitative parameters of the wave field in the crystal and of the crystal itself. The importance of interference effects is due, in particular, to the fact that they make it possible to measure not only the parameters of perfect crystals, but also the characteristics of various distortions in distorted crystals.

Interference effects were first detected experimentally in 1959 by KATO and LANG when investigating dynamical scattering on the wedge-shaped parts of some crystals [1.23]. In his detailed analysis of the experimental material obtained, KATO showed that one of the fundamental starting points of the dynamical theory in all its forms, namely the incidence of *a plane wave* on a crystal, must be revised. He established that the definite type of interference patterns observed can be formed only if a spherical wave falls on a crystal. In this connection, KATO developed a version of the dynamical theory in the incident spherical-wave approximation [1.24,25].

Later, in 1968-1970, Hildebrandt and Batterman, Authier and Malgrange and other authors experimentally obtained interference effects for an incident plane wave.

It should be noted that similar interference effects had been observed in dynamical scattering of electrons much earlier (beginning with 1939-1940) [1.26,27].

It is remarkable that the interference effects indicated during dynamical scattering were predicted by Ewald in his dynamical theory in 1916 but could not be realized experimentally for about 43 years, in any case for X-rays.

Thus, we have an example of a profound theoretical prediction which was years ahead of the possibilities of physical experiment.

The further study of interference phenomena evidently resulted in the construction and use (from 1965 onwards) of X-ray interferometers by BONSE and HART [1.28]. With these instruments, researchers were for the first time able to detect the interference of coherent X-rays beams outside *the scattering crystal*. Among the relevant interference effects we shall mention the X-ray moiré, which makes it possible to measure the deviations from the periodic structure of crystals with a previously unattainable sensitivity and accuracy. Moreover, X-ray interferometers open up possibilities for ultraprecise metrology.

An interesting and important class of phenomena originally discovered experimentally by RENNINGER in 1937 [1.29] and considerably supplemented later by the experimental works of BORRMANN and HARTWIG [1.30] in 1965 and by other authors is dynamical scattering with an allowance for several interacting waves in the crystal. Here, the character of effects of anomalous penetration in absorbing crystals changes in some cases. The relevant theory was worked out by EWALD and HENO [1.31] and PENNING and POLDER [1.32,33] (and some other authors). Recently AFANASIEV and KOHN succeeded in further developing of the multiwave theory [1.34-37]. When investigating in general form the enhancement effect of the anomalous transmission on transition from the two-wave to the n-wave case, where $n = 3,4 \ldots$, these authors obtained the relevant expression in analytical form. They also elaborated a formalism for direct use of computer calculation of the parameters of the multiwave problem.

Considerable advances have been made in theoretical and experimental investigations into the various mechanisms of absorption of X-rays in perfect crystals.

Mention should be made of the theoretical calculations of Wagenfeld and the thorough measurements of photoelectric absorption in Si and Ge performed by Grimwell and Persson, as well as by HILDEBRANDT [1.38]. Along with this, the classical concepts about the photoelectric absorption as the only absorption mechanism were supplemented by the results of the study into thermal diffuse scattering and Compton scattering. The contribution of these effects, which increases under certain conditions, can be detected and even measured at the existing level of experiment. In their works, AFANASIEV and KAGAN [1.39] and DEDERICHS [1.40] showed that in additon to the Debye-Waller factor (elastic thermal scattering), the inelastic scattering on phonons must be taken into account, and it is derived, not only from general relations, but also from more detailed equations for near-experimental conditions.

Here we shall also mention the investigations of AFANASIEV et al. [1.41] devoted to the theory of thermal diffuse scattering. In these studies they solved, for the first time, the problem of the divergence of the intensity of an inelastically scattered wave in the maximum region and the characteristic dependence of the intensity on the diffraction conditions of the experiment. Noteworthy is the theoretical paper of OHTSUKI [1.42] treating both thermal diffuse and Compton scattering, and particularly the recent experimental work of GIARDINA and MERLINI [1.43] in which they measured the additional effects of X-radiation energy absorption with a sufficient margin of confidence.

Besides the given experimental investigations aimed at rigorous quantitative studies of dynamical scattering, especially in the late 1950's, the *X-ray diffraction topography* of single crystals began to be developed. The methods of topography, which permit direct observation of diffraction patterns of various defects in a given specimen, have acquired an important practical value and are being widely used at present. Interpretation topographs, based on qualitative treatment of dynamical scattering (not always unambiguous), in many cases supplies useful information on the real structure of the objects of investigation and its transformations during service or under treatment. Note that analysis of X-ray topographs has much in common with the deciphering of electron micrographs; this is important because of the considerable successes of the dynamical theory of electron scattering in distorted crystals.

In the last few years, advances have also been made in the quantitative treatment of X-ray topographs in cases where the nature of defects in a crystal can be considered as established. Experimental methods of X-ray topography, have been elaborated by Lang, Authier, Newkirque, Elistratov,Miuskov, and other authors. Detailed calculation of the diffraction patterns of certain typical cases of structural defects have been made by Taupen, Balibar, Authier, Schtolberg and Chukhovsky and Epelboin. Experimental investigations and quantitative treatment of topographs, using the Bragg reflection have largely been carried out by Bonse.

Further progress in elaborate theoretical and experimental methods of studying defects in crystals and the use of these methods in technology is associated with overcoming the fundamental limitation of the classical theory of Ewald-Laue. This limitation primarily applies two main features of this theory. The integral form of its fundamental equations, which reflects the indicated concept of a uniform wave field in a crystal, is unsuitable for describing the physical picture of wave propagation in distorted crystals. Such crystals, even with small defects permitting dynamical scattering, have two regions or two-dimensional areas with continuous or jumpwise changes in the fundamental parameters of the medium - the dielectric constant and polarization. At the same time, the condition of incidence of a plane wave with a front width exceeding the thickness of the scattering crystal plate not only imposes high requirements upon the experiment but also excludes from consideration many very important physical phenomena inherent in the dynamical scattering in crystals, both perfect and imperfect.

A similar situation, as regards the difficulty of using integral equations of the dynamical problem, obtains in the field of electron diffraction. The development of diffraction electron microscopy in connection with investigations into defects in thin lagers induced HOWIE and WHELAN to develop in in 1960-1961 [1.44] a dynamical theory based on the fundamental differential-type equations. Soon after that, TAKAGI [1.45] proposed similar expressions describing the dynamic scattering of X-rays, but, in contrast to Howie and Whelan's equations, they took the form of partial differential equations in two arguments, x and z, in the plane of incidence. This complication is known to imply a larger angular width of X-ray diffraction.

Kato and Lang in 1959, and later Kato in 1961-1968 published the results of their experimental and theoretical investigations on dynamical scattering in crystals caused by the incidence of an X-ray beam which has passed through a system of thin slits. On Kato's suggestion, these conditions were given the name of the incident spherical wave approximation. They can be called an incident wave with a front width in the shape of the δ-function. The experiments were performed by Kato and co-workers on near-perfect crystal specimens, and the above theory may be regarded as referring to the perfect crystal. The calculation of the wave field in the crystal under conditions differing from the incidence of a plane wave was carried out by Kato by expanding in plane waves and then integrating with respect to a variable depending on the angles of incidence. The wave fields corresponding to the separate plane waves are described with the aid of the classical theory. As a result, in addition to the interference effects, which are important for absolute determinations of structure amplitudes, Kato established that the wave field in a crystal is restricted to a triangular region, which was previously known as Borrmann's fan (triangle). This shape of the wave field was predicted by Borrmann in 1955 on the basis of simple physical considerations.

It should be noted that the method indicated, which was used by Kato in calculating the wave field, has been successfully employed in solving a number of dynamical theory problems whose common feature is the use of a slit-shaped diaphragm for the incident beam (the wave front takes the shape of the δ-function).

It is well known that the Ewald-Laue theory considers diffraction both on reflection and on transmission as applied to a semi-infinite crystal bounded either by a single plane or to a plane-parallel crystal bounded by two planes, or to a wage-shaped slice. This can be regarded as yet another limitation of the classical dynamical theory as compared with the kinematical theory, which considers scattering from crystal blocks bounded all around. In two of their papers, BORRMANN and LEHMAN (1963 [1.46] and 1967 [1.47]) discuss dynamical scattering from a crystal bounded all around. In this case both Laue and Bragg reflection takes place. A more rigorous solution of this problem was given in investigations by KATO and co-workers [1.48] with the use of the above method.

Another fundamental problem, which was studied by the same method, is associated with the successive transmission of an incident wave with a narrow wavefront through two crystals. Both scattering crystals have an identical structures and their mutual orientation is rigidly fixed.

KATO et al. [1.49] and AUTHIER [1.50] considered the scattering from a plane defect within the perfect crystal. The pictorial nature of the plane wave method used by them made it possible to trace formation of three interference patterns: from each of the crystal regions, on both sides of the plane defect (interference from a wedgelike plate), and the mutual interference of the two wave fields. It is significant that in the presence of appreciable absorption only this last effect remains. A natural sequel to this work was a thorough (partly experimental) investigation carried out by AUTHIER et al. [1.51] into the interference patterns formed on transmission of an incident wave with a narrow wavefront through two plates. Here, the formation of the three interference patterns indicated can also be traced. This scheme can be regarded as a two-plate interferometer. The authors showed how the intensity distribution varies with the ratio of the thicknesses (and shapes) of the plates and the spacing between them. At a certain definite value of these parameters, one observes a system of parallel bands spaced at intervals which are simply related to the structure amplitude of the reflecting plane (HART and MILNE's paper [1.52]). Problems dealing with consecutive reflection from two crystals, as well as with the Laue-Bragg reflection from a finite crystal, were also investigated with the aid of a generalized theory (see further) in the papers of URAGAMI [1.53-55] and of PINSKER and CHUKHOVSKY [1.56].

While the papers mentioned supplement the Ewald-Laue theory, mainly with respect to scattering in a perfect crystal, another line initiated by the well-known publication of PENNING and POLDER [1.57] and developed further by KATO [1.58] and BONSE [1.59a] is of interest in studying the dynamical wave field in a distorted crystal. In these papers, the authors took the first steps in considering the wave field in a geometrical optics approximation - the ray approximation. As in the optics of the visible range of the electromagnetic scale, in the case of the X-ray diffraction one takes into account the change in the direction and magnitude of the Poynting's vector of the wave field in the crystal under the action of the distorted areas. KATO [1.58] developed this theory in detail for the Laue case, and in BONSE's paper [1.59a], the beam approximation was used to study the defects upon Bragg reflection. Notably, the papers indicated showed that this approach is useful and in accordance with the experimental data referring to the scattering from areas of a crystal with small deformation gradients.

In spite of the successes achieved in the above theories, which considerably supplement the classical Ewald-Laue theory, it was obvious that a more general theory must be elaborated which would make it possible to calculate the wave field in both a perfect crystal and a distorted one, provided that a wave packet with an arbitrary wave front width and a finite angular width is incident on the crystal. The attention of theorists was attracted by the above-mentioned, very brief paper by TAKAGI. The fundamental equations of the dynamical theory proposed by him may be regarded as a generalized form of the old recurrence relations of Darwin's theory. In 1968, SLOBODETSKII et al. [1.60] in the USSR and AUTHIER and SIMON [1.61] in France simultaneously published a solution to the Takagi equations for the Laue case. In the following year, 1969, AFANASIEV and KOHN [1.62] in the USSR and URAGAMI [1.53] in Japan supplied a solution to the same equations for the Bragg case simultaneously, too. All these papers considered X-ray scattering in a perfect crystal. The new dynamical theory can be called a generalized theory based on Takagi-type equations.

Note that the Takagi equations are derived from the Maxwell equations in the quasiclassical approximation. The fundamental equations of the Ewald-Laue theory are also derived from the Maxwell equations, which are, of course, the most general ones. As noted above, however, the form of the Takagi equations makes them more universal when applied to the solution of various problems related to the dynamical theory.

Unfortunately, the mathematical apparatus of the new theory is more complicated. The use of the Riemann function in the Laue case and, further, of the functions generalized by the authors of [1.53-56, 60-62] in relation to the boundary conditions on Bragg reflection is very difficult to analyze qualitatively. Nevertheless, as has been demonstrated by Chukhovsky in [1.63], the G-function with its derivatives used in solving the Takagi equation, both in the Laue and in the Bragg case, which is referred to as the influence function, describes the propagation of a local perturbation of the wave field in a crystal similarly to the function R^{-1} exp(ikR) in light optics. The latter describes a wave induced by a point source according to the Huygens-Fresnel principle.

Over the last five to seven years, the possibilities of the new theory have been demonstrated in solving the following problems: scattering of a wave packet incident upon a perfect crystal in the most general case of Laue-Bragg diffraction [1.54,55]; the moiré theory [1.64]; transition to the geometrical optics approximation during X-ray scattering from defects [1.65]; and also the general treatment of the problem on X-ray dynamical diffraction from a homogeneously bent crystal [1.66]. Note that, as shown in [1.66], the exact solution of the X-ray, confirm the correctness of the geometrical optics theory (in the case of small strain gradients) [1.58]. On the other hand, the exact solution does, in principle, permit one to describe X-ray diffraction for arbitrary distortion of a crystal or in the vicinity of the ray caustics when geometrical optics is a priori inapplicable.

2. Wave Equation and Its Solution for Transparent Infinite Crystal

2.1 Wave Equation and Its Solution

The physical model for the propagation of electromagnetic X-rays, which forms the basis of the theory under review [1.13], is built up as follows. The interaction of X-rays with atomic nuclei is neglected here. In the absence of an external field, a crystal can be assumed neutral. During the transmission of an incident electromagnetic wave, however, negative charges are displaced and, consequently, some polarization is observed.

Thus, under the effect of an external field, a perturbed electron density, and hence an additional Schrödinger current or probability electron current, is developed in the crystal. The electromagnetic wave field in a crystal is associated precisely with the propagation of this perturbation and is described by the methods of electrodynamics with the aid of Maxwell equations.

The physical conditions for the applicability of the theory expounded in this chapter and also for a detailed analysis of the solutions to Laue and Bragg reflections in a transparent and an absorbing crystal, given in Chapters 3 through 8, are as follows.

Incidence of a plane wave and its scattering in a perfect crystal, in which a unified wave field arises. The parameters determining this field, the dielectric constant $\varepsilon(\underline{r})$ or electric susceptibility $\chi(\underline{r})$, are three-dimensional periodic functions with the same periods as those of the crystal lattice. The functions of the angle of deviation from the maximum midpoint (the Bragg angle) do not depend on the coordinates within the crystal-scatterer.

The total perturbed electron density at each point $\rho(\underline{r})$ is obtained by taking the sum of the values pertaining to each electron, as is done in the single-electron approximation. Thus, we have

$$\rho(\underline{r}) = -e\psi(\underline{r})\psi^*(\underline{r}), \qquad (2.1)$$

where e is the electron charge, $\psi(\underline{r})$ is the wave function, and $\psi^*(\underline{r})$ is the corresponding conjugate complex value. Here, $\rho(\underline{r})$ is the number density of the electrons in the crystal averaged with respect to the quantum-mechanical electron state and the statistical distribution of the thermal vibration of the nuclei in the lattice. The Schrödinger current $\underline{J}$ and the dielectric constant ε, related to $\dot{\rho}(\underline{r})$, as well as $\rho(\underline{r})$, are continuous functions of the coordinates. Thus, this differs substantially from the ordinary dielectric constant in the macroscopic theory of dielectrics.

Let us turn to the Maxwell equations in the conventional form:

$$\text{curl } \underline{E} = c^{-1} \frac{\partial \underline{H}}{\partial t} \qquad \text{curl } \underline{H} = c^{-1} \left(\frac{\partial \underline{E}}{\partial t} + 4\pi \underline{J}\right)$$

$$\text{div } \underline{E} = 4\pi\rho \qquad \text{div } \underline{H} = 0 \quad . \tag{2.2}$$

The energy volume density W and the vector of the energy current-density of the electromagnetic field in vacuum S are expressed by the usual equations

$$W = \frac{1}{8\pi} \left(|\underline{E}|^2 + |\underline{H}|^2\right) \quad , \tag{2.3}$$

$$\underline{S} = \frac{c}{4\pi} [\underline{E}\underline{H}] \tag{2.4}$$

The scalar value of the electromagnetic wave intensity is equal to the average value of the modulus of $\underline{S}$ for a sufficiently long period of time as compared with that of the X-ray oscillations ν^{-1}.

The equation of continuity for our perturbed charges can be obtained by taking the divergence curl $\underline{H}$ (2.2); then

$$\frac{d\rho}{dt} + \text{div } \underline{J} = 0 \quad . \tag{2.5}$$

In general, the vectors $\underline{E}$ and $\underline{H}$ are expressed, through the vector potential $\underline{A}$ and the scalar potential φ, as follows:

$$\underline{H} = \text{curl } \underline{A} \tag{2.6}$$

$$\underline{E} = -\frac{1}{c} \frac{\partial \underline{A}}{\partial t} - \text{grad}\varphi \quad . \tag{2.7}$$

Since φ vanishes in a wave field, for $\underline{E}$ we get:

$$\underline{E} = c^{-1} \frac{\partial \underline{A}}{\partial t} \tag{2.8}$$

The displacement of the negative charge due to the electromagnetic wave is described by polarization $\underline{P}$, which here is a continuous function of the coordinates, and therefore by induction we have

$$\underline{D} = \underline{E} + 4\pi\underline{P} \quad , \tag{2.9a}$$

and for the wave field

$$\text{div } \underline{D} = 0 \quad . \tag{2.9b}$$

Obviously

$$\underline{J} = \frac{\partial \underline{P}}{\partial t} \quad . \tag{2.10}$$

Using the same analogy with the field in a dielectric, we can write

$$\underline{D} = \underline{E} + 4\pi\underline{P} = \varepsilon\underline{E}, \qquad \varepsilon = 1 + 4\pi \frac{|\underline{P}|}{|\underline{E}|} \tag{2.11}$$

We shall now derive the wave equation for the transverse wave of the electric displacement. Using (2.9), we can rewrite Maxwell's first equation as

$$\text{curl } (\underline{D}-4\pi\underline{P}) = -c^{-1} \frac{\partial \underline{H}}{\partial t} \quad . \tag{2.12}$$

Applying the well-known equation of vector analysis

$$\text{curlcurl} = \text{grad div} - \Delta \quad , \tag{2.13}$$

we find the curl of (2.12):

$$\Delta\underline{D} = c^{-1}\partial \ (\text{curl } \underline{H})/\partial t = -4\pi \ \text{curlcurl } \underline{P} \quad . \tag{2.14}$$

Now we shall rewrite Maxwell's second equation as

$$\text{curl } \underline{H} = \frac{1}{c}\left(\frac{\partial \underline{E}}{\partial t} + 4\pi\frac{\partial \underline{P}}{\partial t}\right) = \frac{1}{c}\frac{\partial \underline{D}}{\partial t} \quad . \tag{2.15}$$

Hence

$$\Delta\underline{D} - \frac{1}{c^2}\frac{\partial^2 \underline{D}}{\partial t^2} + 4\pi \ \text{curlcurl } \underline{P} = 0 \quad . \tag{2.16}$$

In order to transform this wave equation into a more convenient form, we shall turn again to the value of the current $\underline{J}$. As is known, the general expression for the Schrödinger current, with an allowance for the perturbed electron density, takes the form

$$\underline{J} = \frac{eh}{4\pi im} (\psi^* \mathrm{grad}\psi - \psi \mathrm{grad}\psi^*) - \frac{e^2}{mc^2} \underline{A}\psi\psi^* \quad , \tag{2.17}$$

where the second term describes Thomson scattering. When an electromagnetic wave passes, the first term in the parentheses also changes, but if the frequencies of incident waves differ greatly from the proper frequencies of scattering electrons, this change may be ignored. Another assumption we make is that in the second term we use the quantity ρ, which, according to (2.1), refers to the unperturbed atoms in the lattice. Thus, we have

$$J \approx -\frac{e^2}{mc^2} \underline{A}\psi\psi^* \equiv \frac{e}{mc} \rho\underline{A} \quad . \tag{2.18}$$

Further, assuming for the vector $\underline{A}$

$$\underline{A} = \underline{A}_0 \exp 2\pi i[\nu t - (\underline{K}\underline{r})], \quad \frac{\partial \underline{A}}{\partial t} = 2\pi i\nu\underline{A} \quad . \tag{2.19}$$

and by virtue of (2.8), we rewrite (2.18) as

$$\underline{J} = -\frac{e\rho}{2\pi im\nu} \underline{E} \quad . \tag{2.20}$$

Comparing (2.20) and (2.10) and using functions of the type (2.19) for $\underline{P}$, we obtain

$$\underline{P} = \frac{e\rho}{4\pi^2 m\nu^2} \underline{E} \tag{2.21}$$

and, according to (2.9a),

$$\underline{D} = \left(1 + \frac{e\rho}{4\pi^2 m\nu^2}\right)\underline{E} \quad . \tag{2.22}$$

Hence we get the following equation for the dielectric constant of a medium where the second term describes the mutual effect of scattering on the field in the medium

$$\varepsilon(\underline{r}) = 1 + \frac{e\rho(\underline{r})}{\pi m\nu^2} \quad . \tag{2.23}$$

Now we shall introduce a new scalar function of the lattice χ (see (2.9a))

$$4\pi\underline{P} \approx \chi\underline{D}, \quad \chi = 1 - \varepsilon^{-1} \quad . \tag{2.24}$$

Assuming that $\underline{D}$ is also a function of the type (2.19), i.e.,

$$\frac{1}{c^2}\,\frac{\partial^2\underline{D}}{\partial t^2} = \frac{4\pi^2\nu^2}{c^2}\,\underline{D} = 4\pi^2|\underline{K}|^2\underline{D}, \quad |\underline{K}| = \frac{1}{\lambda} \quad , \tag{2.25}$$

we obtain the transformed wave equation for induction

$$\Delta\underline{D} + 4\pi^2|\underline{K}|^2\underline{D} + \text{curlcurl}\,(\chi\underline{D}) = 0 \quad . \tag{2.26}$$

χ accurate within the factor $(4\pi)^{-1}$ is the polarizability and can be calculated in the following manner.

Let N denote the number of electrons per unit volume.
In this case from (2.23) ($\rho = \bar{\rho} = -|e|N$) we have

$$\varepsilon = 1 + \frac{e\rho}{\pi m\nu^2} = 1 - \frac{e^2N\lambda^2}{mc^2\pi} \quad . \tag{2.27}$$

Since $e^2/mc^2 = 2.8 \cdot 10^{-13}$ cm, $\lambda^2 \approx 10^{-16}$ cm^2 and $N \approx 10^{23} - 10^{25}$ cm^{-3}, the second term on the right in (2.27) is of the order $< 10^{-4}$. Hence, we can expand χ in (2.24) into a power series of $e\rho/\pi m\nu^2$

$$\chi = \frac{e\rho}{\pi m\nu^2} - \left(\frac{e\rho}{\pi m\nu^2}\right)^2 + \ldots \tag{2.28}$$

and restrict ourselves, without impairing the accuracy, to the first term, assuming

$$\chi = \frac{e\rho}{\pi m\nu^2} < 0 \quad . \tag{2.29}$$

As electric susceptibility χ is a continuous periodic function of the coordinates, it can be expanded as a Fourier series of the form

$$\chi = \sum_m \chi_m \exp[-2\pi i(\underline{h}_m \underline{r})] \quad , \tag{2.30}$$

where $\underline{h}_m$ is the radius vector of the reciprocal lattice:

$$\underline{h}_m = m_1\underline{a}_1 + m_2\underline{a}_2 + m_3\underline{a}_3 \quad . \tag{2.31}$$

Summation is performed over all the vectors of the reciprocal lattice. The coefficients of expansion (2.30) are expressed by the equations

$$\chi_m = \Omega^{-1} \int_{cell} \chi \exp[2\pi i(\underline{h}_m \underline{r})]\, d\tau \quad , \tag{2.32}$$

$$\chi_o = \Omega^{-1} \int_{cell} \chi d\tau \quad . \tag{2.33}$$

The subscripts m may be either positive or negative, whereas χ_m and $\chi_{\bar{m}}$ are conjugate complex values; χ_o, i.e., the average of χ over the entire lattice, is real.

Hereafter we shall express χ_m and χ_o as a function of the structure amplitudes F_m:

$$\chi_m = - \frac{e^2}{\pi m \nu^2 \Omega} \int_{cell} - \frac{\rho}{e} \exp 2\pi i(\underline{h}_m \underline{r}) d\tau = - \frac{e^2}{mc^2} \frac{\lambda^2}{\pi\Omega} F_m \tag{2.34}$$

$$\chi_o = - \frac{e^2 F_o}{\pi m \nu^2 \Omega} = - \frac{e^2}{mc^2} \frac{\lambda^2}{\pi} N \quad . \tag{2.35}$$

It is clear that $\chi_o > \chi_m$, because the scattering power of an atom decreases with the angle of deviation from the direction of the primary beam; in the case of χ_o, summation with respect to the unit cell is always carried out at phases that are equal for all atoms.

The solution of the wave equation (2.26) of the dynamical problem is the Bloch wave:

$$\underline{D} = \exp \left\{2\pi i[\nu t - (\underline{k}_o \underline{r})]\right\} \sum_m \underline{D}_m \exp[-\, 2\pi i(\underline{h}_m \underline{r})] \quad . \tag{2.36}$$

The factors $\underline{D}_m$ are complex vectors. The expression (2.36) can be regarded as a plane wave with a wave vector $\underline{k}_o$ and periodically varying amplitude; at identical points of different unit cells $|\underline{D}|^2$ has the same value.

Another interpretation of the Bloch wave is based on the condition

$$\underline{k}_m = \underline{k}_o + \underline{h}_m \quad , \tag{2.37}$$

hence

$$\underline{D} = \exp(2\pi i\nu t) \sum_m \underline{D}_m \exp[-2\pi i(\underline{k}_m\underline{r})] \quad . \tag{2.38}$$

In this case the Bloch wave describes a wave field consisting of an infinite number of plane waves with wave vectors $\underline{k}_m$. Consequently, (2.38) describes the multiwave solution of the dynamical theory which was mentioned in the Introduction. It has also been indicated that the two-wave approximation is extremely important for X-ray scattering; we will dwell on this in Section 2.2.

In order to verify the solution proposed, the last terms in (2.16) and (2.26) need to be investigated more thoroughly and transformed. For this purpose, we shall represent the polarization expression as a triple Fourier series

$$\underline{P} = \exp 2\pi i\nu t \sum_m \underline{P}_m \exp[-2\pi i(\underline{k}_m\underline{r})] \quad . \tag{2.39}$$

On the other hand, comparing the last terms and using the expansions (2.30) and (2.36), we obtain

$$4\pi\underline{P} \approx \chi\underline{D} = \exp\left\{2\pi i[\nu t - (\underline{k}_o\underline{r})]\right\} \cdot \sum_q \sum_n \chi_{m-n} \underline{D}_n \exp[-2\pi i(\underline{h}_q+\underline{h}_n,\underline{r})] . \tag{2.40}$$

Since in the expansion (2.38) the summation limits are extended to infinity, the expansion remains valid if, for instance, all the subscripts m are increased by the same value n, and then the subscript o becomes p. Further, it is obvious that the vector

$$\underline{h}_{q+n} = \underline{h}_q + \underline{h}_n \tag{2.41}$$

is also a vector of the reciprocal lattice. The expression (2.40) can therefore be rewritten differently if we put m = q + n:

$$\chi\underline{D} = \exp\left\{2\pi i[\nu t - (\underline{k}_o\underline{r})]\right\} \sum_q \sum_n \chi_{m-n} \underline{D}_n \exp[-2\pi i(\underline{h}_m\underline{r})] \quad , \tag{2.42}$$

$$\chi \underline{D} = \exp 2\pi i\nu t \sum_m \sum_n \chi_{m-n} \underline{D}_n \exp[- 2\pi i(\underline{k}_m\underline{r})] \quad . \tag{2.43}$$

Thus, we obtain the coefficients of the Fourier expansion (2.39)).

$$\underline{P}_m = (4\pi)^{-1} \sum_n \chi_{m-n} \underline{D}_n \quad . \tag{2.44}$$

Now, in order to substitute the values of the terms contained in the wave equation (2.16) and to establish the conditions in which the solutions (2.36) and (2.38) can be applied, we differentiate:

$$\begin{aligned} \Delta\underline{D} &= -4\pi^2 \exp 2\pi i\nu t \sum_m k_m^2 \underline{D}_m \exp[- 2\pi i(\underline{k}_m\underline{r})] = \\ &= - 4\pi^2 \exp \left\{2\pi i[\nu t - (\underline{k}_o\underline{r})]\right\} \sum_m k_m^2 \underline{D}_m \exp[- 2\pi i(\underline{h}_m\underline{r})] \quad . \end{aligned} \tag{2.45}$$

Further (see (2.39))

$$\begin{aligned} \text{curl}\underline{P} &= -2\pi i \exp 2\pi i\nu t \sum_m [\underline{k}_m\underline{P}_m] \exp[-2\pi i(\underline{k}_m\underline{r})], \\ \text{curlcurl } \underline{P} &= -4\pi^2 \exp 2\pi i\nu t \sum_m [\underline{k}_m[\underline{k}_m\underline{P}_m]] \exp[-2\pi i(\underline{k}_m\underline{r})] \quad . \end{aligned} \tag{2.46}$$

We transform the triple vector product in the right-hand side of (2.46) according to the usual rule:

$$[\underline{A}[\underline{B}\underline{C}]] = \underline{B}(\underline{A}\underline{C}) - \underline{C}(\underline{A}\underline{B}) \tag{2.47a}$$

and get

$$[\underline{k}_m[\underline{k}_m\underline{P}_m]] = \underline{k}_m(\underline{k}_m\underline{P}_m) - \underline{P}_m\underline{k}_m^2 \quad . \tag{2.47b}$$

It is obvious that in the first term on the right we have a term $\underline{P}_m$ parallel to $\underline{k}_m$; this is of no interest to us in describing the transverse waves; the second term contains a component $\underline{P}_m$ perpendicular to $\underline{k}_m$. We shall denote it by $\underline{P}_{m[m]}$.

In (2.16), substituting the new values of all the three terms from (2.25), (2.45), and (2.46), we obtain an equation which must be satisfied identically, i.e., for each m separately. The condition for vanishing of coefficient of each component oscillation (or mode)

$$-4\pi^2 k_m^2 \underline{D}_m + 4\pi^2 \underline{K}^2 \underline{D}_m + 4\pi^2 \underline{P}_{m[m]} k_m^2 = 0, \quad (k_m^2 - K^2)\, D_m / k_m^2 = 4\pi\, \underline{P}_{m[m]} , \qquad (2.48)$$

or, by virtue of (2.44),

$$\frac{k_m^2 - K^2}{k_m^2} \underline{D}_m = \sum_n \chi_{m-n}\, D_{n[m]} \quad . \qquad (2.49)$$

The component $\underline{D}_{n[m]}$ is the electric displacement of the waves n normal to the wave vector of the given wave $\underline{k}_m$.

Equation (2.49) is called the fundamental equations of the dynamical theory.

Incidentally, (2.2), (2.9a), and (2.15) may be used to locate the mutual positions of important vectors of the wave field.

Indeed, from (2.2) and from the Fourier representation of the magnetic field strength

$$\underline{H} = \exp 2\pi i\nu t \sum_m \underline{H}_m \exp[-2\pi i(\underline{k}_m \underline{r})] \qquad (2.50)$$

we obtain

$$\operatorname{div} \underline{H} = -2\pi i \exp(2\pi i\nu t) \sum_m (\underline{k}_m \underline{H}_m) \exp[-2\pi i(\underline{k}_m \underline{r})] = 0. \qquad (2.51)$$

Obviously it is an identity, i.e., all $(\underline{k}_m \underline{H}_m)$ vanish and, hence, the magnetic field strengths of the separate waves are normal to their wave vectors.

In a similar way, we can prove the normality of the electric displacement vectors $\underline{D}_m$ to the wave vectors, which is obvious from (2.49).

Finally, from (2.15) it follows that

$$-2\pi i \exp(2\pi i\nu t) \sum_m [\underline{k}_m \underline{H}_m] \exp[-2\pi i(\underline{k}_m \underline{r})]$$

$$= 2\pi i K \exp(2\pi i\nu t) \sum_m \underline{D}_m \exp[-2\pi i(\underline{k}_m \underline{r})] \quad ;$$

hence

$$\underline{D}_m = -K^{-1} [\underline{k}_m \underline{H}_m] \quad . \qquad (2.52)$$

Thus, in "plane" waves m, the vectors $\underline{H}_m$, $\underline{D}_m$, and $\underline{k}_m$ form a right-handed orthogonal system precisely as in electromagnetic optics of visible light of anisotropic media [2.1].

It is only the waves of electric displacement $\underline{D}$ that are purely transverse. The analogy with optics also applies to the electric field vector $\underline{E}$, which, being normal to $\underline{H}$, lies in a plane passing through $\underline{k}$ and $\underline{D}$. This follows from Maxwell's first equation (2.2), which gives

$$\underline{H}_m = \underline{K}^{-1}\,[\underline{k}_m\underline{E}_m] \quad , \tag{2.53}$$

as well from (2.9) and (2.24) which give $\underline{E}_m$ in the form

$$\underline{E}_m = \underline{D}_m - \sum_n \chi_{m-n}\,\underline{D}_n \quad . \tag{2.54}$$

At the same time it should be emphasized that the representation of the Bloch wave in the form of plane waves is only an approximation. As can be seen from (2.37) and (2.49), "plane" waves are generated only in a combination, forming a unified wave field. Moreover, the dynamical theory propounded here is not sufficiently rigorous in some other respects as well. The use of the electric displacement $\underline{D}$ in place of the electric field $\underline{E}$ is not fully substantiated because in this theory $\underline{D}$ is a function of the coordinates, rather than a macroscopic parameter.[1]

From these remarks it follows that the solution (2.38) of the wave equation in a crystal, as the system of fundamental equations (2.49), describes a wave field consisting of an infinitely large number of separate plane waves. Obviously a solution in this form cannot be calculated in practice and hence cannot be compared with the experiment. But the experiment indicates the nature of the approximate solution, which makes it possible to make comprehensive calculations and develop the corresponding theory. In order to formulate this approximation, we shall turn our attention to the diagram of the reflection sphere in a reciprocal lattice which is familiar in the kinematical theory (Fig.2.1).

---

[1] Incidentally, as was shown by WAGENFELD [2.2], Ewald's use of the electric vector $\underline{E}$ to describe the wave field in a crystal is also based on an appoximation: $\underline{E} \approx \underline{D}$ = curlcurl $\underline{Z}$ ($\underline{Z}$ is the Hertz vector). This approximation is no better substantiated than the Laue approximation (2.24): $4\pi\,\underline{P} = \chi\underline{E} \approx \chi\underline{D}$. In numerical calculations of the parameters of the dynamical theory, the difference due to the use of $\underline{D}$ or $\underline{E}$ is a value of an order not greater than 10^{-3}%.

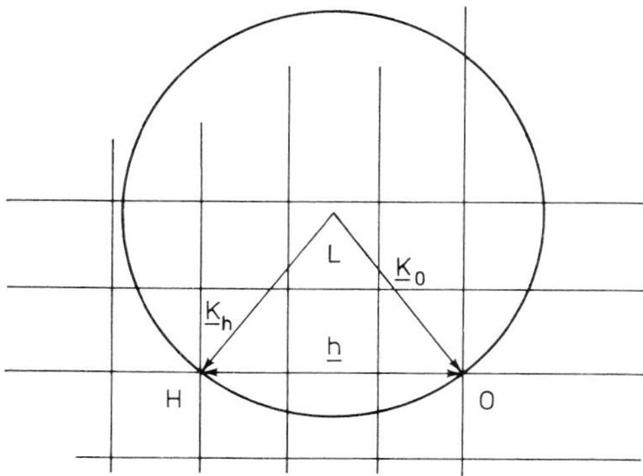

Fig. 2.1 Sphere of reflection in reciprocal space (kinematical approximation

This diagram implies that at least two points of the reciprocal lattice lie precisely on a sphere of radius $\underline{K}$ equal to λ^{-1}, the wave vector in vacuum. One of them corresponds to the origin of the coordinates (point O), the node or reflection (000), i.e., to the incident wave. The other corresponds to the point or reflection hkl (point H). In the dynamic theory this pattern is somewhat different. If the radius of the sphere is equal to the wave vector in vacuum $\underline{K}$, the origin must be at a distance of $\underline{k}_m \neq \underline{K}$ from the sphere centre L, because the refractive index of X-rays is slightly different from unity. For the formation of reflections in the dynamical theory we do not need such a strong condition as the Wolff-Bragg equation to be fulfilled. For any given experimental conditions, i.e., for a given position of the sphere centre and of the sphere itself in the reciprocal lattice, any node, in principle, may correspond to certain reflection, and the wave vector of this reflected wave $\underline{k}_m$, after being replaced in the fundamental equation (2.49), yields the vector $\underline{D}_m$ as a function of all $\underline{D}_{n[m]}$. It should, however, be taken into consideration that the sum of the right-hand side of all the equations of the type (2.49) should be approximately constant, and almost all the nodes of the reciprocal lattice are spaced at a distance from the centre which is large and very large as compared with $\underline{K}$, i.e., in general, $\underline{K} \ll \underline{k}_m$ and hence $(k_m^2-K^2)/k_m^2 \approx 1$. For a point located close to the sphere, however, this ratio will be very small. Consequently, the corresponding wave $\underline{D}_m$ will, on the contrary, be very large as compared with all other $\underline{D}_n$.

In other words, ours is a two-wave approximation: one of the waves may be regarded as the incident wave (more precisely a refracted one), and the other, as the reflected one.

The distance of the points in the reciprocal lattice from the sphere is given according to the excitation error (Resonanzfehler)

$$\varepsilon_m = \frac{(|\underline{k}_m| - |\underline{K}|)}{|\underline{K}|} \quad . \tag{2.55}$$

In the dynamical theory the construction of the dispersion surface in a reciprocal space is more important than the construction of the sphere. In a general case, if we consider the system of fundamental equations (2.49) as a system of equations with p unknown vectors $\underline{D}$ to be solved, then it obviously decomposes into 3p scalar equations. Since the components $\underline{D}_m$, parallel to the respective wave vectors $\underline{k}_m$, do not form transverse electromagnetic waves, we have only 2p equations. For a nontrivial solution to exist, the determinant of this system must vanish. The equation thus obtained will describe the dispersion surface.

For the variable in this equation we can use $\varepsilon_{m=o}$ from (2.55), or another quantity related to the geometry of the dispersion surface. The highest power of this variable, and hence the degree of the equation, will be 2p. Using (2.55), from the roots found for the variable $\varepsilon_{m=o}$ it is possible to derive the respective wave vectors $\underline{k}_o^{(i)}$, $\underline{k}_h^{(i)}$ and, thus, $\underline{D}_o^{(i)}$ and $\underline{D}_h^{(i)}$, where $i = 1,2,3 \ldots p$. The following mental construction explains the foregoing with the aid of a dispersion surface image.

Let us construct, in a reciprocal space, vectors $\underline{k}_o^{(i)}$ converging at the origin of the reciprocal lattice. Their directions and magnitudes will be determined by the angle of incidence and the refractive index, which differs for different i (see further (3.21)). The starting points of these vectors will be termed the excitation points (after Laue), or the tie points (after Ewald). By changing the angle of incidence near the Bragg angle, we can slightly oscillate the wave vectors without displacing their terminal points from the origin of the coordinates. Their starting points will describe surfaces, which form a 2p-sheet dispersion surface. Each sheet of this surface should be regarded as the locus of the excitation points. From one of the points on sheet No.i, if we draw, vectors $\underline{k}_h^{(i)}$ to all the points of the reciprocal lattice (h = 1,2, ...), this set, together with the initial vector $\underline{k}_o^{(i)}$, will belong to a certain wave field or mode of the dynamic problem. In all there will be 2p such fields, according to the number of scalar equations in our fundamental system (2.49). The term excitation point reflects the effect of wave field generation in the crystal under the influence of an external vacuum wave for a given angle of incidence.

In considering a wave field in an infinite crystal, however, we cannot, in principle, single out from the set of waves that particular one which corresponds to the refracted wave $\underline{k}_o^i$. Oscillations in an infinite crystal are regarded as proper, not constrained. The tie point [1.4] corresponds to this concept. In the general case, these wave vectors may be complex

$$\underline{k}_o^{(i)} = \underline{k}_{or}^{(i)} + i\underline{k}_{oi}^{(i)} \quad . \tag{2.56}$$

It is known that the complex value of a wave vector leads to the appearance, in the wave expression, of the damping factor

$$\exp[-2\pi i(\underline{k}_o^{(i)}\underline{r})] = \exp[-2\pi i(\underline{k}_{or}^{(i)}\underline{r})]\exp[2\pi(\underline{k}_{oi}^{(i)}\underline{r})] \quad . \tag{2.57}$$

In the expression of intensity we obtain the square of this factor

$$\exp[-2\pi i(\underline{k}_{or}^{(i)} + i\underline{k}_{oi}^{(i)},\underline{r}]\exp[+2\pi i(\underline{k}_{or}^{(i)}\underline{r}-i\underline{k}_{oi}^{(i)},\underline{r})] = \exp 4\pi(\underline{k}_{oi}^{(i)}\underline{r}), \tag{2.58}$$

$$|\underline{k}_{oi}^{(i)}| = -k_{oiz} < 0 \quad .$$

Our fundamental equation (2.49) is written for vectors of the electric displacement $\underline{D}_m$. Of course, by simple transformation it can be represented as the equation for magnetic vectors $\underline{H}_m$. Let us multiply (2.52) vectorially by $\underline{k}_m$. Thus, on the right-hand side we have

$$[\underline{k}_m[\underline{k}_m\underline{H}_m]] = k_m^2\underline{H}_m \quad ,$$

and the equation will take the form

$$\underline{H}_m = \frac{K}{k_m^2}[\underline{k}_m\underline{D}_m] \quad . \tag{2.59}$$

Now multiply the vector equation (2.49) by $\underline{k}_m K$. According to (2.59), we will obtain $(k_m^2-K^2)\underline{H}_m$ on the left-hand side and $\Sigma\chi_{m-n}K(\underline{k}_m\underline{D}_n)$ on the right-hand (the subscript [m] can be omitted because here we have a vector product which includes only the component $\underline{D}_n$ normal to $\underline{k}_m$). In this vector product, replacing $\underline{D}_n$ by its value from (2.52) and then replacing the triple vector product by the difference of the scalar products, we get

$$(k_m^2 - K^2)\underline{H}_m = \sum_n \chi_{m-n} [\underline{H}_n(\underline{k}_m\underline{k}_n) - \underline{k}_n(\underline{k}_m\underline{H}_n)] \quad . \tag{2.60}$$

This is the fundamental equation for the magnetic field vectors.

2.2 Two-Wave Approximation. Dispersion Surface

Before proceeding to a detailed description of the theory for two strong waves, we shall consider scattering in the absence of regular reflection, i.e., when the angle of incidence of the primary wave is quite different from the Bragg angle. In this case the system (2.49) degenerates to a single equation, the subscript m vanishes, and so does n. The wave vector for this case of transmission of the primary wave without reflection will be denoted by k. The equation takes the form $(k^2-K^2)k^{-2} - \chi_0 = 0$.

Since the magnitude of χ_0 is very small, we neglect χ_0^2 and transform the equation obtained as

$$k^2(1-\chi_0) = K^2, \quad |\underline{k}| \approx \frac{K}{(1 - 1/2\ \chi_0)} \approx K(1 + 1/2\ \chi_0) \quad . \tag{2.61}$$

We shall introduce the refractive index

$$n = \frac{k}{K} = 1 + 1/2\ \chi_0 = \left(1 - \frac{e^2N}{2\pi m\nu^2}\right) < 1 \tag{2.62}$$

This value of n coincides with the X-ray refractive index calculated by the classical dispersion theory. As is generally known, the experimental verification of this value for frequencies higher than the K-absorption edge is accurate to several percent. This confirms the applicability of the assumptions and the physical model underlying the theory under review.

Obviously, in this case the dispersion surface degenerates to the sphere of radius

$$|\underline{k}| = K(1 + 1/2\ \chi_0) < K \quad . \tag{2.62a}$$

We shall now pass on to the two-wave approximation. If, by continuously varying the angle of incidence of the primary wave on the crystal surface, we find ourselves in the maximum region in the vicinity of the Bragg angle, a Bloch wave arises in the crystal, which can be represented, in this case, as a set of four plane waves - two refracted and two reflected ones -

because i=1,2. This applies to each of the two states of polarization of our electromagnetic waves.

For the sake of a more accurate and comprehensive description of the scattering process we shall introduce a plane of reflection passing through the vectors $\underline{k}_o^{(i)}$ and $\underline{k}_h^{(i)}$ of the waves indicated in the crystal.

We shall discuss the simplest scheme, most widely used experimentally, in which the plane of reflection coincides with the plane of incidence.

In those cases, where the frequencies of the incident waves are high compared with the K-absorption edge, electric susceptibilities χ_o and χ_h are true scalars, and the plane under consideration will be the symmetry plane of the scattering process.

Let us now turn to (2.49). In our two-wave approximation, we must put $m = 0$ in the first equation and $m = h$ in the second; n on the right-hand sides of both equations successively takes the values 0 and h. As regards the vectors $\underline{D}_m$, more precisely $\underline{D}_o$ and $\underline{D}_h$, we shall decompose them, and the vectors $\underline{H}_o$ and $\underline{H}_h$, along two directions: normal and parallel to the plane of reflection. Thus, in the subsequent text we will deal with scalar equations. For oscillations of the induction vectors perpendicular to the plane of reflection (σ-polarization) we obtain

$$\frac{k_o^2 - K^2}{k_o^2} D_o = \chi_o D_o + \chi_{\bar{h}} D_h \quad , \quad \frac{k_h^2 - K^2}{k_h^2} D_h = \chi_h D_o + \chi_o D_h \quad . \tag{2.63}$$

Waves with the other state (π-polarization) can be described by the fundamental equations for magnetic vectors, namely for the components of these vectors normal to the initial plane. We shall make use of (2.60).

In this case m takes the values 0 and h, and so does n. It is easy to see that the second term within the brackets on the right-hand side of (2.60) vanishes, while the scalar product before H_h is equal to $\cos 2\vartheta$ where ϑ is the Bragg angle. Thus, the system takes the form

$$\frac{k_o^2 - K^2}{k_o^2} H_o = \chi_o H_o + \chi_{\bar{h}} \cos 2\vartheta H_h \quad ,$$

$$\frac{k_o^2 - K^2}{k_h^2} H_h = \chi_h \cos 2\vartheta H_o + \chi_o H_h \quad . \tag{2.64}$$

$$(H_{o,h} \equiv H_{o,h}^{\sigma})$$

It is clear that π-polarization can be described with the aid of equations for the components D_o and D_h lying in the initial plane; the equations are obtained by replacing H_o and H_h by D_o^π and D_h^π, as is seen from Fig. 2.2.

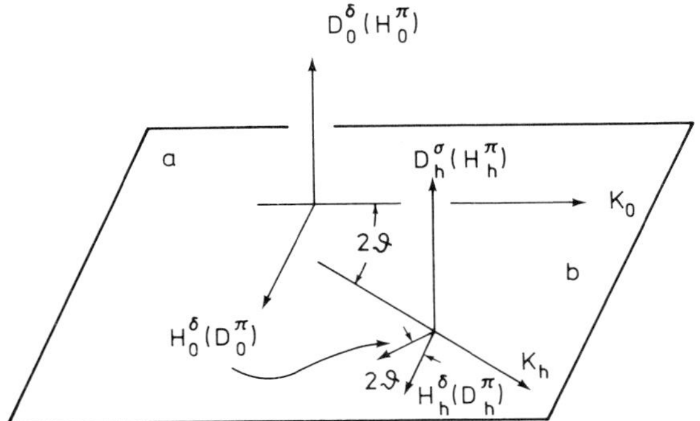

Fig. 2.2 Mutual positions of the induction vectors in two states of polarization: a) σ-polarization; the induction vectors are perpendicular to the plane of reflection and are mutually parallel; b) π-polarization; the induction vectors lie on a plane parallel to the plane of reflection

Note that while the components $\underline{D}_o^\sigma$ and $\underline{D}_h^\sigma$, $\underline{H}_o^\pi$ and $H_{h,\pi}^\pi$ are pairwise parallel, this is not the case with $\underline{D}_o^\pi$ and D_h^π .

From now on we shall deal mainly with D_o^σ and D_h^σ which are normal to the plane of reflection (σ-polarization). We shall treat these quantities (as well as D_o^π and D_h^π) as scalars, since their directions have been determined unambiguously.

We shall now transform the system of equations (2.63). For this purpose, we shall assume, to a sufficient approximation, that

$$k_o^2 - K^2 - k_o^2\chi_o \approx k_o^2 - K^2(1 + 1/2\ \chi_o)^2 = k_o^2 - k^2 \quad , \tag{2.65}$$

$$k_h^2 - K^2 - k_h^2\chi_o \approx k_h^2 - k^2 \quad , \tag{2.66}$$

$$\frac{k_o^2}{2K} \approx \frac{k_h^2}{2K} \approx 1/2\ K, \quad k_o + k \approx k_h + k \approx 2K \tag{2.67}$$

and rewrite (2.63):

$$(k_o-k)D_o - 1/2\ K\chi_{\bar{h}}D_h = 0, \quad - 1/2\ K\chi_h D_o + (k_h-k)D_h = 0 \quad . \tag{2.68}$$

The determinant of this system, being zero,

$$\Delta = \begin{vmatrix} k_o - k & - 1/2\ K\chi_{\bar{h}} \\ - 1/2\ K\chi_h & k_h - k \end{vmatrix} = 0 \quad , \tag{2.69}$$

gives the equation of a dispersion surface in reciprocal space. This surface is (in the adopted approximation) a two-sheet surface of revolution (a hyperbolic cylinder) about the axis OH, where O and H are the respective nodes of the reciprocal lattice. A section of this surface by the plane reflection is shown in Fig.2.3 (hyperbolas $S^{(1)}$, $S^{(2)}$).

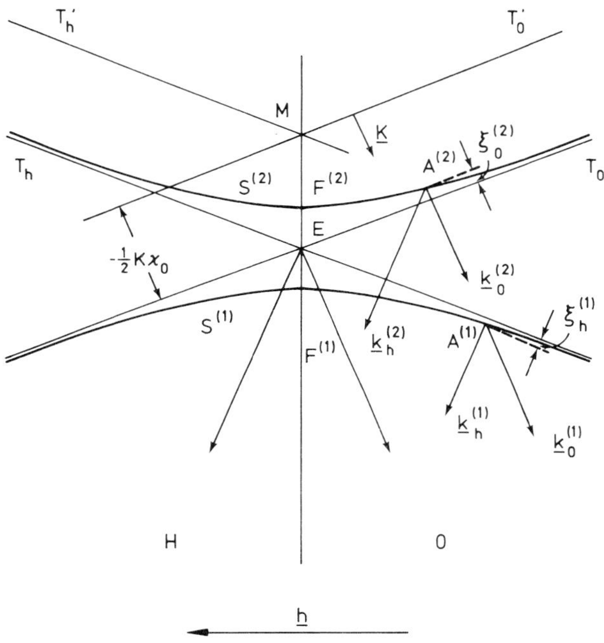

Fig. 2.3 Dispersion surface in a reciprocal space in two-wave approximation (infinite crystal)

With the chosen scale, this figure is too small to include the points O and H, and, of course, the part of the section opposite with respect to the OH axis. Describing spheres of radius K from the points O and H, in our section we obtain lines T_0' and T_h', respectively, and their point of intersection M. Because of the great distance between the points O and H, within a small angular interval the lines T and T' may be regarded as straight lines.

It is clear that the point M, spaced at a distance of $K = \lambda^{-1}$ from the points O and H (λ is the wavelength in vacuum) corresponds to the centre of the sphere of propagation in the kinematical theory, i.e., to reflection with an accurate value of the Bragg angle and a constant wave-length.

Point E may also be considered as the centre of a sphere constructed for reflection with an accurate value of the Bragg angle, but with a value of the wave vector $\underline{k}$ which includes the correction for reflection.[2] Finally, the points $F^{(1)}$ and $F^{(2)}$ on the two branches of the hyperbola already correspond to the dynamical reflection with a particular value of the angle of reflection, namely the Bragg angle. These points are spaced from points O and H at $\underline{k}_o^{(1)} = \underline{k}_h^{(1)}$ and $\underline{k}_o^{(2)} = \underline{k}_h^{(2)}$, respectively. In this particular case, the wave vectors with the subscripts o and h are equal.

The dispersion surface (a hyperbola in our section) is the locus of the excitation points. Therefore, in the case of dynamical reflection at angles slightly differing from the Bragg angle, the displacement of the point $F^{(1)}$ or $F^{(2)}$, or both, will correspond along the curves $S^{(1)}$ and $S^{(2)}$. Let the points $A^{(1)}$ and $A^{(2)}$ be the excitation points. Join them to the points O and H. Introduce segments $\xi_o^{(i)}$ and $\xi_h^{(i)}$, which have the following values (see Fig.2.3):

$$\xi_o^{(i)} = k_o^{(i)} - k, \quad \xi_h^{(i)} = k_h^{(i)} - k, \quad i = 1,2 \quad . \tag{2.70}$$

The quantities $\xi_o^{(i)}$ and $\xi_h^{(i)}$ will be positive if they form acute angles with the positive direction of the vectors $\underline{k}$ and $\underline{k}^{(i)}$. It should be recalled that the vectors $\underline{K}$, $\underline{k}$, and $\underline{k}^{(i)}$ are directed from the points M,E, or $A^{(i)}$ to O and H. Thus, we have

$$\xi_o^{(i)} = \mp|\xi_o^{(i)}|, \quad \xi_h^{(i)} = \mp|\xi_h^{(i)}| \quad . \tag{2.71}$$

Now we can rewrite the dispersion surface equation (2.69)

$$\Delta = \begin{vmatrix} \xi_o & -1/2\, K\chi_{\bar{h}} \\ -1/2\, K\chi_h & \xi_h \end{vmatrix} = 0 \quad , \tag{2.72}$$

$$\xi_o\xi_h = \frac{1}{4} K^2\chi_h\chi_{\bar{h}} \quad . \tag{2.73}$$

---

[2] By Ewald's suggestion (see [1.4]), point M is sometimes called the Laue (La) point, because it corresponds to the terminus of the vector $\underline{K}$, used in the kinematical theory when describing the wave field in a crystal. For point E, Ewald proposed the term the Lorentz (Lo) point, because it appears when the refraction at the vacuum-crystal boundary is taken into account. It is generally known that (2.62) can be derived from the Lorenz-Lorentz equation in the dispersion theory from the condition $(\omega_o-\omega) \approx -\omega$.

It is obvious that if we write an equation of the type (2.68) for the other state of polarization with magnetic vectors normal to the initial plane, according to (2.64), we have

$$\xi_o \xi_h = \frac{1}{4} K^2 (\cos 2\vartheta)^2 \chi_h \chi_{\bar{h}} \quad . \tag{2.73a}$$

Therefore, we shall henceforth introduce into our equations a factor C, which has the following meaning:

$$C = \begin{cases} 1 \text{ - for } \sigma\text{-polarization} \\ |\cos 2\vartheta| \text{ - for } \pi\text{-polarization} \end{cases} \quad . \tag{2.74}$$

Accordingly, the dispersion surface equation takes the form

$$\xi_o \xi_h = \frac{1}{4} K^2 C^2 \chi_h \chi_{\bar{h}} \quad . \tag{2.75}$$

Eliminating k_o from (2.55) and (2.70) and using (2.61), we obtain the relation between ε_o and ξ_o:

$$\xi_o = K(\varepsilon_o - 1/2\ \chi_o) \quad . \tag{2.76}$$

It should be recalled that χ_o is an essentially negative value.

The dispersion surface equation may be regarded as the equation of a hyperbola on the plane of reflection.

The values ξ_F drawn from the points $F^{(i)}$ on the real diameter of hyperbola have the magnitudes

$$(\xi_o^{(i)})_F = (\xi_h^{(i)})_F = \frac{KC}{2} \sqrt{\chi_h \chi_{\bar{h}}} \quad . \tag{2.77}$$

If we proceed from the equation of hyperbola about the axes

$$\frac{x^2}{a^2} - \frac{y^2}{b^2} = 1 \quad , \tag{2.78}$$

the magnitude of the real semiaxis

$$a = \frac{\xi_F}{\cos\vartheta} = \frac{KC}{2\cos\vartheta} \sqrt{\chi_h \chi_{\bar{h}}} \tag{2.79}$$

and the imaginary one

$$b = \frac{KC}{2\sin\vartheta} \sqrt{\chi_h \chi_{\bar{h}}} \tag{2.80}$$

and the equation of hyperbola (2.78) will take the form

$$x^2 \cos^2\vartheta - y^2 \sin^2\vartheta = \frac{K^2C^2\chi_h\chi_{\bar{h}}}{4} = \xi_o\xi_h \quad . \tag{2.81}$$

From (2.79) it is clear that for electric displacement vector oscillations perpendicular to the plane of reflection, the real axis 2a is $(\cos 2\vartheta)^{-1}$ times larger than for the π-state of polarization (Fig.2.4).

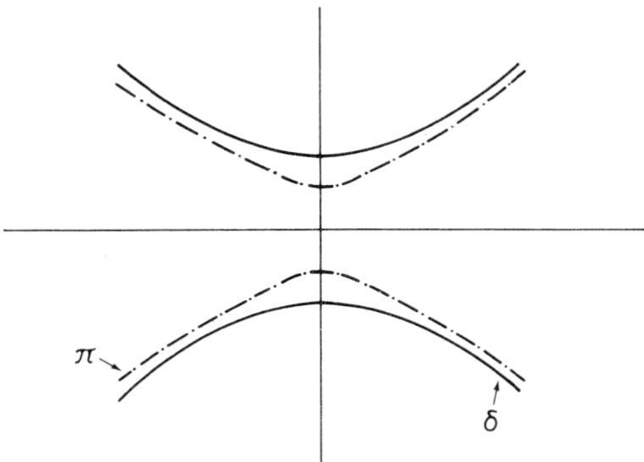

Fig. 2.4 Dispersion surfaces for σ- and π-polarizations

Now we shall introduce values $c^{(i)}$ expressing the ratio $D_h^{(i)}/D_o^{(i)}$. From (2.68) and (2.70) it follows that

$$c^{(i)} = \frac{D_h^{(i)}}{D_o^{(i)}} = \frac{2\xi_o^{(i)}}{K\chi_{\bar{h}}C} = \frac{K\chi_h C}{2\xi_h^{(i)}} = \sqrt{\frac{|\xi_o^{(i)}|}{|\xi_h^{(i)}|}} \quad . \tag{2.82}$$

These expressions determine the ratios of the amplitudes in the wave fields within the crystal. Using the last ratio, which represents $c^{(i)}$ as a function of the variables $\xi^{(i)}$, we can express the change in $c^{(i)}$ with continuous change in the angle of incidence inside the maximum region.

Evidently the entire maximum region corresponds to the range of angles from one side of the dispersion curves (and surfaces, respectively) to the other.

In the lower branch, as the negative deviation from the angle ϑ (on the left) changes to positive ones (on the right), the value of the ratio $c^{(1)} = D_h^{(1)}/D_o^{(1)}$ increases from ~ 0.01 to 100 (at the right-hand side of the diagram), turning into unity at point $F^{(1)}$ (Fig.2.3). For $c^{(2)}$, i.e., for the upper branch, we obtain a reverse picture, i.e., a decrease in the ratio $D_h^{(2)}/D_o^{(2)}$ in going from left to right. These features of the functions (2.82), however, are not so characteristic and essential as the ratio of the absolute values of $D_h^{(i)}$ and $D_o^{(i)}$ (see (3.22,23)).

It should be recalled that in the two-wave approximation, to which the diagram of Fig.2.3 is applicable, only two waves, D_o and D_h, among the infinite number of waves of the dynamical problem are assumed sufficiently strong. Therefore, for incidence angles close to the boundaries of the maximum, i.e., to the boundaries of the dispersion curves, when one of the waves, namely D_h, becomes weak, the two-wave approximation and this diagram cease to be valid. Hence, one should not adjust (2.82) to the conditions corresponding to the edges of Fig.2.3. Outside the maximum region we find ourselves in conditions corresponding to a single wave which propagates in the crystal without reflection. Its direction coincides with that of an incident wave in vacuum with an accuracy up to that of the refractive index. Thus, in these cases we have a transition from two waves in a crystal to a single one. It is clear that, from the point of view of this transition, the dispersion surface is an intermediate region of intersection of two spheres of propagation (Fig.2.5).

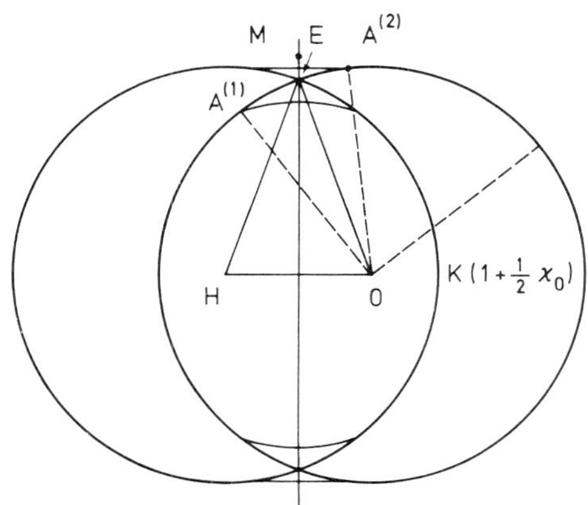

Fig. 2.5 Dispersion surface - region of intersection of two spheres of propagation (after JAMES [1.16])

With the help of (2.82) and Fig.2.3 we can establish the relative phases of the induction waves of the two fields in each state of polarization. As already mentioned, χ_h is a complex quantity. Let

$$\chi_h = |\chi_h| \exp(i\eta_h), \quad \chi_{\bar{h}} = |\chi_h| \exp(-i\eta_h) \quad , \tag{2.83}$$

hence[3]

$$\chi_h\chi_{\bar{h}} = |\chi_h|^2 \ , \quad \sqrt{\frac{\chi_h}{\chi_{\bar{h}}}} = \exp i\eta_h \quad ,$$

$$\left|\sqrt{\frac{\chi_h}{\chi_{\bar{h}}}}\right|^2 = \sqrt{\frac{\chi_h}{\chi_{\bar{h}}}}\sqrt{\frac{\chi_h^*}{\chi_{\bar{h}}^*}} = 1 \quad , \tag{2.84a}$$

$$\chi_h = \chi_{\bar{h}}^*, \quad \chi_h^* = \chi_{\bar{h}} \quad . \tag{2.84b}$$

In the particular case of a centrosymmetrical crystal

$$\eta_h = \pi \quad , \quad \exp i\eta_h = -1 \quad |\chi_h/\chi_{\bar{h}}| = 1 \quad .$$

In (2.82) the signs of $c^{(i)}$, determining the phase ratios of the oscillations D_o and D_h, will depend on the signs of both ξ_o and χ_h. Since for a given η_h all χ_h become negative, $c^{(1)} < 0$ and $c^{(2)} > 0$, i.e., $D_h^{(1)}$ oscillates in phase with $D_o^{(1)}$, and $D_h^{(2)}$, in counterphase with $D_o^{(2)}$.

Thus, in the dynamical theory of the two-wave approximation we have the formation inside a crystal, in the maximum region, of four wave fields (two fields for each state of polarization), each field consisting of two waves D_o and D_h, which makes a total of eight waves.

We must make another very important reservation here. Fig.2.3 and the relevant analysis refer to the transmission of the X-rays, i.e., to the Laue case. A systematic discussion of the Bragg case will be given in Chapters 7 and 8.

[3] These relations refer to a particular case of transparent crystals for which the atomic scattering amplitudes can be considered as real. Otherwise, as will be shown in Chapter 4, in absorbing crystals only the pairs of χ_{hr} and $\chi_{\bar{h}r}$, χ_{hi} and $\chi_{\bar{h}i}$ will be conjugate complex values

3. Transmission of X-Rays Through a Transparent Crystal Plate. Laue Reflection

3.1 Wave Fields Inside a Crystal

3.1.1 Semi-infinite Crystal. Connection with Experimental Conditions. Refraction Effect

So far we have been discussing the propagation of Bloch waves in an infinite crystal. This entails indeterminateness in the localization of the real excitation centres on the dispersion surface and, hence, the impossibility of determining the acting wave vectors $\underline{k}_o^{(i)}$ and $\underline{k}_h^{(i)}$. The indeterminateness is removed if we take account of the vacuum-crystal boundary and the wave incident on this boundary represented by a wave vector $\underline{K}_o^{(a)}$ with an angle of incidence ψ_o and a reflecting plane with an angle φ relative to the boundary or the entrance surface.

In theoretical consideration of the crystal-vacuum boundary, because of the small (of the order of the lattice periods) wavelength of the X-rays, the real structure of the surface, with its atomic steps, oxide or adsorbed films and other microscopic deviations from the perfect state, there is every possibility for considerable distortions creeping into the scheme of the boundary conditions. But the problem is solved by experiment. The scheme using the mathematical crystal-vacuum boundary agrees fairly well with the experimental data available. Thus, we pass over from an infinite crystal to a semi-infinite crystal.

The wave field in a crystal is induced by an incident wave, and the wave vector $\underline{k}_o$ is a direct continuation of the incident wave vector $\underline{K}_o^{(a)}$ in the crystal. In this case the phase velocities along the boundary, and hence the tangential components of the wave vectors, should be equal.

It can be shown that if, in the three Laue conditions for the formation of maxima of three-dimensional interference, we write two conditions for the vectors lying in the plane of the entrance surface, these conditions will be fulfilled in the dynamical theory as well. This follows from the continuity of the tangential components of the wave vectors at the boundary. As regards the third condition, in the kinematical theory it is fulfilled only for a definite angle of incidence, and the more precisely, the larger is the scattering crystal. On the contrary, in the dynamical theory the third condition is not so strict, even for an infinitely large scattering crystal.

Into the scheme of Fig.2.3 we shall now introduce parameters describing the boundary conditions. The line PP' in Fig.3.1 shows the position of a (unit) normal to the boundary. The vector of the normal is assumed to be directed into the crystal. Since the direction of the wave vector $\underline{K}_0^{(a)}$ of the incident wave in vacuum corresponds to a segment drawn from point P on the straight line T_0' to the point 0 of the reciprocal lattice, the angle P'PO, which we shall denote by ψ_0, will be the angle of incidence.

Because of the large distance between the points 0 and H, the lines converging at one of these points in Figs.2.3 and 3.1 are represented as parallel lines. The wave vectors of the waves in the crystal $\underline{k}_0^{(i)}$ and $\underline{k}_h^{(i)}$ are drawn from the points $A^{(i)}$ to the points 0 and H.

For the tangential component wave vectors at the crystal-vacuum boundary to be equal it is necessary that the vectors $\underline{K}_0^{(a)}$ and $\underline{k}_0^{(i)}$ in Fig.3.1 be supported on a common normal $\underline{n}_0$ lying on PP'. This can be written in the form of vector equations

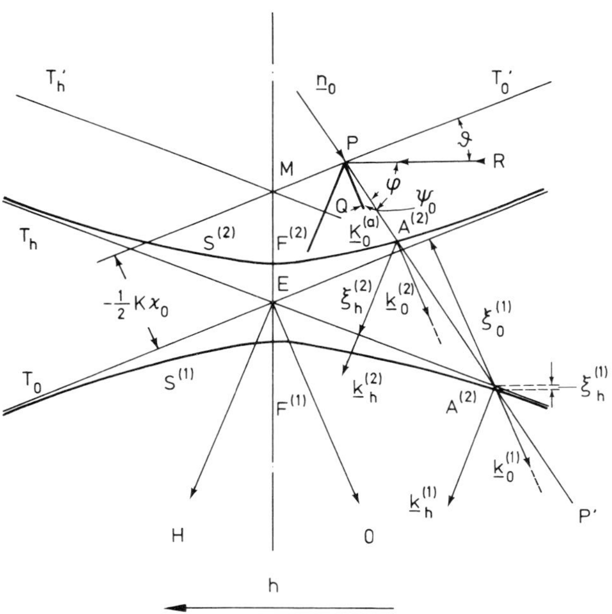

Fig. 3.1 Dispersion surface in a reciprocal space (semi-infinite, crystal)

$$\underline{K}_o^{(a)} = K\delta^{(i)}\underline{n}_o + \underline{k}_o^{(i)}, \quad \underline{k}_o^{(i)} = \underline{K}_o^{(a)} - K\delta^{(i)}\underline{n}_o \quad , \tag{3.1}$$

where $K\delta^{(i)}\underline{n}_o$ are represented by the segments $PA^{(i)}$ directed away from the point P. In Figs.2.3 and 3.1 and, accordingly, in the schemes referring to real space, we shall consider the plane of reflection from the positive side facing the vector $[\underline{k}_h\underline{k}_o]$. The angle φ between the reflecting plane ME and the entrance surface in Fig.3.1 is represented by the angle between the straight line $\overline{RP} \parallel \overline{OH}$ ($\overline{OH} = \underline{h}$) and by the normal $\underline{n}$. This angle may vary from $-\pi/2$ to $+\pi/2$ and will be assumed positive if the vector product $[\underline{h}\underline{n}_o]$ is directed along the same side as $[\underline{k}_h\underline{k}_o]$. The angle of incidence ψ_o may also vary from $-\pi/2$ to $+\pi/2$ and will be assumed positive if the vector product $[\underline{K}^{(a)}\underline{n}_o]$ is directed along the same side, i.e., is (in Fig.3.1) directed towards the reader.

Thus, the slope of the normal $\underline{n}_o$ with respect to the line T_o' is determined by the angles ϑ and φ.

As regards the position of the normal on T_o', it is determined by the angle of incidence. If the angle of incidence ψ_{ok} corresponds to the kinematical theory, i.e., to the Bragg angle for reflection from the given plane, the normal will pass through the point M, with which the point P coincides. From Fig.3.1 it follows that

$$\psi_{ok} = \frac{\pi}{2} - \varphi - \vartheta \quad . \tag{3.2}$$

As the angle of incidence deviates from this value, the normal $\underline{n}_o$ shifts along T_o' to the left or right of the point M. The increment in ψ_{ok}, which we shall denote by η, can be represented by the angle MOP in Fig.3.1.

In the subsequent discussion we shall assume

$$\gamma_o = \cos\psi_o = \cos(\psi_{ok} + \eta) \approx \cos\psi_{ok} \tag{3.3}$$

and similarly

$$\gamma_h = \cos\psi_h = \cos(\psi_{hk} + \eta) \approx \cos\psi_{hk} \quad . \tag{3.4}$$

For the increment η we get

$$\eta = \psi_o - \frac{\pi}{2} + \varphi + \vartheta \quad . \tag{3.5}$$

When a sphere of radius K is drawn from point P as the centre, it will pass through the point O, but not through the point H. The distance of the sphere from the point H along the radius is equal to the segment $\overline{PQ}$. Obviously,

$$\overline{PQ} = K\alpha = k\eta \sin 2\vartheta \quad . \tag{3.6}$$

These definitions apply completely only to reflections hkl with positive indices. In the case of reflection from the other side of the same planes, i.e., with reflections $\bar{h}\bar{k}\bar{l}$, in order to preserve the condition of the former treatment with the vector product $[\underline{k}_h\underline{k}_o]$ facing the reader, the drawing must be reversed.

Figures 3.2a and b illustrate the diagrams of rays in real space.

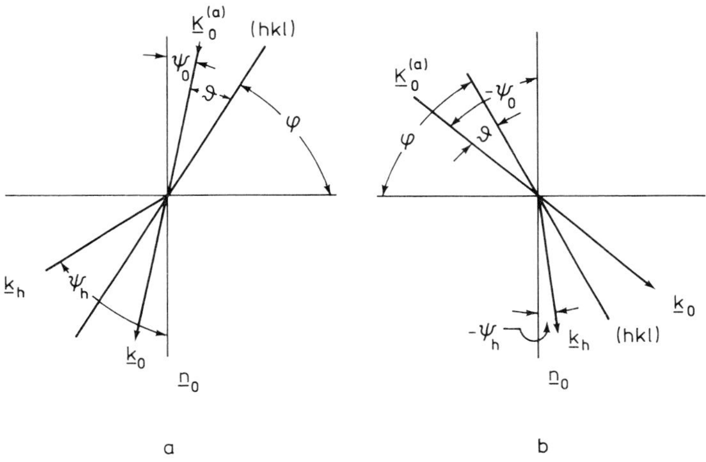

Fig. 3.2 Diagrams reflections: a) hkl, b) $\bar{h}\bar{k}\bar{l}$

For reflections with negative indices, the equality (3.2) takes the form $\psi_o = -\varphi + \vartheta$; the sign of ψ_o is assumed negative as the vector products $[\underline{k}_h\underline{k}_o]$ and $[\underline{K}_o^{(a)}\underline{n}_o]$ for diagram Fig.3.2b are antiparallel.

According to (2.84), for transparent crystals the transmission power coefficient T and reflection power coefficient R are the same for planes with indices of both signs. This is not, however, true for absorbing crystals.

An important result of drawing the normal PP' in our diagram is the formation of two excitation points $A^{(1)}$ and $A^{(2)}$ on the dispersion curves.

Now we have to determine the values of accommodation (Anpassung) $\delta^{(i)}$

and then calculate coefficients $c^{(i)}$ inside the crystal as a function of the experimental parameters. For this purpose it is convenient to write the moduli of the vectors $|K\delta^{(i)}\underline{n}_o|$ in terms of the segments $\xi_o^{(i)}$ and $\xi_h^{(i)}$. According to the definition (2.71) they are positive for the second field and negative for the first. From Fig.3.1 we get

$$\begin{aligned} \overline{PA}^{(i)} &= |K\delta^{(i)}\underline{n}_o| = \gamma_o^{-1}\left(-\xi_o^{(i)} - \frac{1}{2}K\chi_o\right) = \\ &= \gamma_h^{-1}\left(-\xi_h^{(i)} - \frac{1}{2}K\chi_o + K\alpha\right) \quad . \end{aligned} \tag{3.7}$$

We solve these equations together with (2.75) and obtain a quadratic equation for $\xi_o^{(i)}$:

$$\xi_o^{(i)2} + \frac{K}{2}\frac{\gamma_o}{\gamma_h}\left[2\alpha - \chi_o\left(1-\frac{\gamma_h}{\gamma_o}\right)\right]\xi_o^{(i)} - \frac{1}{4}K^2\chi_h\chi_{\bar{h}}C^2\frac{\gamma_o}{\gamma_h} = 0 \quad . \tag{3.8}$$

Introduce a new angular variable

$$\beta = 2\alpha - \chi_o\left(1-\frac{\gamma_h}{\gamma_o}\right) . \tag{3.9}$$

In this case the solution of (3.8) for $\xi_o^{(i)}$ will be

$$\xi_o^{(i)} = \frac{K}{4}\frac{\gamma_o}{\gamma_h}\beta \mp \sqrt{\frac{K^2}{16}\frac{\gamma_o^2}{\gamma_h^2}\beta^2 + \frac{K^2}{4}\chi_h\chi_{\bar{h}}C^2\frac{\gamma_o}{\gamma_h}} \quad . \tag{3.10}$$

In considering the radicals of (3.10), two cases should be distinguished: the Laue case, i.e., diffraction on transmission, when $\gamma_h > 0$, and the Bragg case, i.e., diffraction on reflection, when $\gamma_h < 0$. In the Laue case the second term under the radical in (3.10) has a plus sign, and the radical is real. Considering further the expression (3.10) as a whole, we note that with a positive sign in front of the radical $\xi_o^{(i)} > 0$ and with a negative sign, $\xi_o^{(i)} < 0$. By virtue of the conditons (2.71), we obtain, in (3.10), $\xi_o^{(1)}$ with a negative sign in front of the radical and $\xi_o^{(2)}$, with a positive sign.

We can now write the expressions for $c^{(i)}$ inside the crystal as a function of the experimental parameters. To do this, it will suffice to use (2.82) and take $\xi_o^{(i)}$ from (3.10). We have

$$c^{(i)} = -\frac{\beta \pm \sqrt{\beta^2 + 4\chi_h\chi_{\bar{h}}C^2(\gamma_h/\gamma_o)}}{2\chi_{\bar{h}}(\gamma_h/\gamma_o)C} \tag{3.11}$$

or

$$\left(\frac{\gamma_h}{\gamma_o}\right)^{1/2} \frac{D_h^{(i)}}{D_o^{(i)}} = - \frac{\beta \pm \sqrt{\beta^2 + 4\chi_h\chi_{\bar{h}}\, C^2(\gamma_h/\gamma_o)}}{2\chi_{\bar{h}}(\gamma_h/\gamma_o)^{1/2}C} \tag{3.12}$$

With an eye to what follows, we introduce new angular coordinates

$$y = \sinh v = \frac{\beta}{2C(\chi_h\chi_{\bar{h}})^{1/2}(\gamma_h/\gamma_o)^{1/2}} \quad , \tag{3.13}$$

where we have, for the hyperbolic functions of v, the familiar relations

$$\cos v = \sqrt{1 + \sinh^2 v}, \ \sinh v \pm \cosh v = \pm \exp(\pm v).$$

Using the new variables, we obtain for $|\xi_o^{(i)}|$ and $|\xi_h^{(i)}|$

$$\begin{aligned} |\xi_o^{(i)}| &= m\,(\sqrt{1+y^2} \pm y) = m \exp(\pm v) \quad , \\ |\xi_h^{(i)}| &= m\,\frac{\gamma_h}{\gamma_o}(\sqrt{1+y^2} \mp y) = m\,\frac{\gamma_h}{\gamma_o} \cdot \exp(\mp v), \quad m = \frac{KC|\chi_h|}{2}\sqrt{\frac{\gamma_o}{\gamma_h}} \quad . \end{aligned} \tag{3.14}$$

We rewrite the expression (3.12)

$$\sqrt{\frac{\gamma_h}{\gamma_o}}\,\frac{D_h^{(i)}}{D_o^{(i)}} = - \sqrt{\frac{\chi_h}{\chi_{\bar{h}}}}\,(y \pm \sqrt{1+y^2}) \tag{3.15}$$

$$\sqrt{\frac{\gamma_h}{\gamma_o}}\,\frac{D_h^{(i)}}{D_o^{(i)}} = \mp \sqrt{\frac{\chi_h}{\chi_{\bar{h}}}}\,\exp(\pm v) \tag{3.16}$$

or, with an allowance for (2.83,84),

$$\sqrt{\frac{\gamma_h}{\gamma_o}}\,\frac{D_h^{(i)}}{D_1^{(i)}} = -(y \pm \sqrt{1+y^2})\,\exp(i\eta_h) = \mp \exp(i\eta_h \pm v) \quad . \tag{3.17}$$

It is obvious that the positive sign of the ratio $D_h^{(i)}/D_o^{(i)}$ indicates the coincidence of the phases of these amplitudes and refers to the first field, while the negative sign means that the respective amplitudes in the second field are in counterphase (see Sec.2.2).

The midpoint of the maximum corresponds to the values $\beta = y = v = 0$. In this case

$$\gamma_h^{1/2}|D_h^{(i)}| = \gamma_o^{1/2}|D_o^{(i)}| \tag{3.18}$$

and for intensities

$$\gamma_h|D_h^{(i)}|^2 = \gamma_o|D_o^{(i)}|^2 \quad , \tag{3.19}$$

i.e., the per-second energy flows are equal, since the presence of factors $\gamma_h = \cos\psi_h$ and $\gamma_o = \cos\psi_o$ is due to the change in the cross sections on reflection.

The amplitude reflection coefficients $c^{(i)}$ are given in (3.12) and (3.15) as a function of the magnitudes β, y or v, which, in turn, depend on η. The ambiguity in the values of these angular functions in the general case for a given value of η, is highly essential.

Close inspection of (3.6), (3.9), and (3.13) makes it possible to given a graphic geometrical interpretation and reveal the specific refraction effect which is observed in the Laue case.

Equation (2.62), which determines the value of the refractive index for a wave incident at angles which lie beyond any reflection maximum, corresponds to a decrease by $(1/2)K\chi_o$ in the modulus of the wave vector of the vacuum wave $\underline{K}$ (see (Fig.2.3 or 3.1). When the same wave falls within the maximum, a wave field characterized by dispersion is formed. Denoting, quite formally, the refractive indices for the transmitted and diffracted waves in the crystal by $n_o^{(i)}$ and $n_h^{(i)}$, we can write the following expressions for the wave vectors and $n^{(i)}$:

$$\begin{aligned} k_o^{(i)} &= K(1+1/2\chi_o) + \xi_o^{(i)} \quad , \quad n_o^{(i)} = 1 + 1/2\,\chi_o + (\xi_o^{(i)}/K) \quad , \\ k_h^{(i)} &= K(1+1/2\chi_o) + \xi_h^{(i)} \quad , \quad n_h^{(i)} = 1 + 1/2\,\chi_o + (\xi_h^{(i)}/K) \quad . \end{aligned} \tag{3.20}$$

The magnitude $n_h^{(i)}$ may be regarded as the refractive index on the transition of reflected waves from the crystal into vacuum. In contrast to n, the values $n_o^{(i)}$ and $n_h^{(i)}$ may be either more than unity or less. On the other hand, the total refraction effect in the case of a plane-parallel plate can be defined by

$$\xi_h^{(i)} - \xi_o^{(i)} \frac{\gamma_h}{\gamma_o} = \frac{K\beta}{2} \quad . \tag{3.21}$$

The right-hand side of this expression is obtained by substituting $\xi_h^{(i)}$ and $\xi_o^{(i)}$ into the left-hand side from (3.14) and by further replacing y with its value in (3.13).

The left-hand side of (3.21) is the difference between the projections of the segments $\xi_h^{(i)}$ and $\xi_o^{(i)}$ onto $\underline{n}_o$, which is further projected onto the direction of $\underline{k}_h$ practically coinciding with that of $\underline{K}_h^{(d)}$, the vector of the diffracted wave in the vacuum.

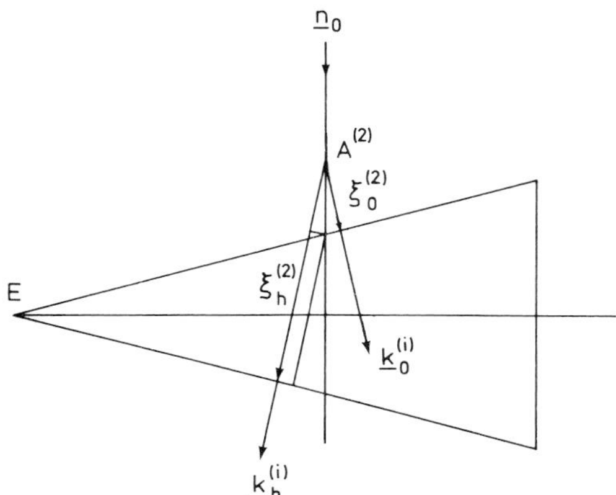

Fig. 3.3 Derivation of (3.22)

From the geometrical interpretation of the magnitude of β (or $K\beta/2$), we find that (as demonstrated by Fig.3.3) for symmetrical reflection (the segment $\overline{AL}$ in Fig.3.3 is equal to $\overline{PQ}$ in Fig.3.1)

$$\left(\xi_h^{(i)} - \xi_o^{(i)} \frac{\gamma_h}{\gamma_o}\right)_s = K\eta \sin 2\vartheta \quad ; \tag{3.22}$$

this is in full agreement with (3.9) and (3.21) for $\gamma_h = \gamma_o$. The general case of an asymmetrical reflection is illustrated in Fig.3.4.

Projecting the segment $\overline{N^I N^{II}}$ successively onto $\overline{MR} \parallel \underline{n}_o$ and then $\overline{MR}$ onto $\overline{N^{III} N^V}$, we find that

$$\overline{N^{IV} N^V} = \overline{FF''} = -\frac{1}{2} K\chi_o\left(1 - \frac{\gamma_h}{\gamma_o}\right)$$

is added to the segment $\overline{F''C} = \overline{PQ} = \eta \sin 2\vartheta$, so that

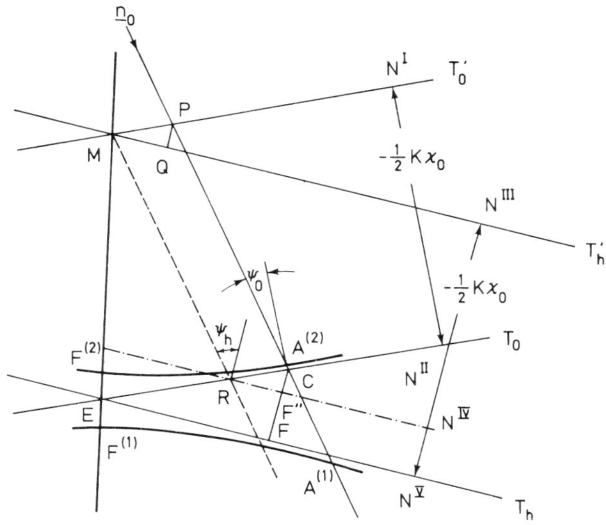

Fig. 3.4 Consideration of the refraction effect for an asymmetrical reflection (Laue case)

$$\overline{F''C} = K\eta \sin 2\vartheta - \frac{1}{2} K\chi_0 \left(1 - \frac{\gamma_h}{\gamma_0}\right) = \frac{K\beta}{2} \quad . \tag{3.23}$$

The construction of the difference (3.23) in the general case is shown in Fig.3.5, where the immediate vicinity of the point $A^{(i)}$ is given.

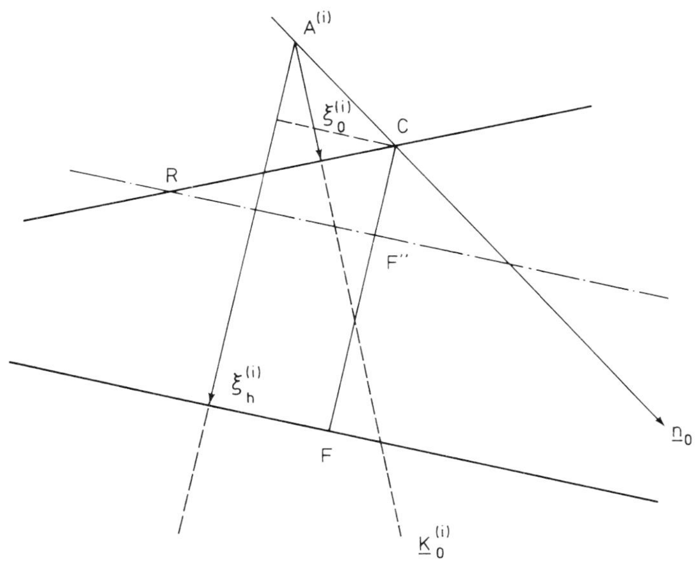

Fig. 3.5 Derivation of (3.21) in the general case

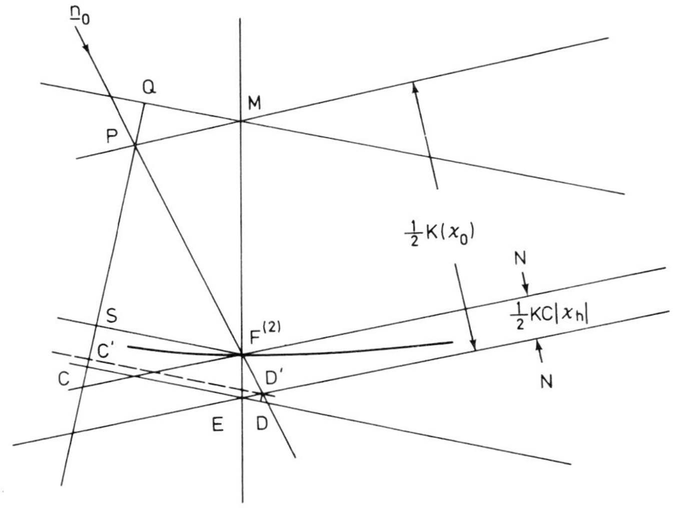

a

b

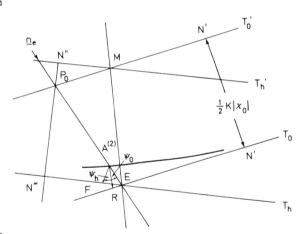

Fig. 3.6 a and b Derivation of (3.24) and (3.25)

Figure 3.6 shows typical particular cases of the position of the normal, i.e., the values of η and of one of the parameters β, y, $\dot{v}$.
Thus, in Fig.3.6a the normal intersects the upper branch of the dispersion hyperbola at the point $F^{(2)}$. It can easily be seen that in this DD', which is generally equal to 1/2K, takes on a particular value:

$$\overline{DD'} = \frac{1}{2} K|\chi \;|\; 1-\left(\frac{\gamma_h}{\gamma_o}\right) \quad ,$$

$$\overline{PQ} = K\eta \sin 2\vartheta = \frac{1}{2} K(|\chi_o| - |\chi_h|)\left(\frac{\gamma_h}{\gamma_o} - 1\right) < 0 \quad . \tag{3.24}$$

The negative value of $\overline{QP}$ is due to the displacement of the normal to the left of point M. Fig.3.6b shows the case of incidence of a vacuum wave at $\beta = y = v = 0$. Here, the segment

$$\overline{N^{III}P_o} = -K\eta \sin 2\vartheta = \frac{1}{2} K|\chi_o|\left(1- \frac{\gamma_h}{\gamma_o}\right) \quad . \tag{3.25}$$

Direct determination of the refraction effect is given by η expressed, say, as a function of y, because the results of the theory, in particular, the reflection and transmission curves, are given as a function of y. From (3.13) we obtain

$$\eta = ay + b, \quad a = \frac{C|\chi_h|(\gamma_h/\gamma_o)^{1/2}}{\sin 2\vartheta} \quad ,$$
$$b = \frac{|\chi_o|[(\gamma_h/\gamma_o) - 1]}{2 \sin 2\vartheta} \quad . \tag{3.26}$$

In a symmetrical reflection $\eta = ay$ or, in seconds of arc,

$$\eta''_s = \frac{C|\chi_h|y}{\sin 2\vartheta} \; \frac{1}{4.85 \cdot 10^{-6}} \quad . \tag{3.27}$$

Since C and $\sin 2\vartheta$ are values of the order of unity and $|\chi_h| \approx 10^{-5}$-10^{-6}, for a symmetrical reflection the value of y is close to η''_s and has the same sign.

In the general case of an asymmetrical reflection, according to (3.26), both the signs and the ratio of y and η may differ, depending on the ratio b/a. For small glancing angles, i.e., for large ψ_o, the value of b may appreciably exceed a; this expresses the displacement of the maximum relative to the Bragg angle due to refraction.

3.1.2 Wave Amplitudes; Pendulum Solution. Extinction. Quasi-standing Waves

Now we shall pass on to the boundary conditions for amplitudes. These conditions differ substantially in two respects from those used in the electromagnetic theory of light. In the first place, bacause of the negligible difference between the refractive index and unity in the case of X-rays we can restrict ourselves to the condition of continuity of the component induction vectors. Secondly, for the same reason we can neglect the wave which is a mirror reflection back into vaccum, as it follows directly from the corresponding well-known Fresnel formula.

At the same time it is known that in the case of glancing incidence of the X-rays at angles $(\pi/2)-\psi_0$ of the order of the critical angles for total external reflection, the reflection coefficient takes on values which cannot be neglected. Therefore, in these special conditions the corresponding calculation for dynamical scattering has to be made. The case of small glancing angles has been attracting ever-increasing interest over the last few years in connection with the construction of spectrometers with an asymmetrical Bragg reflection. It should be stressed that in such devices the glancing angles of incidence $(\pi/2)-\psi_0$ exceed $1\text{-}2^0$, which permits one to neglect the reflected wave.

From the formulation of the boundary conditions, we find that here we are dealing with a particular case of a plane-parallel crystal plate onto whose entrance surface a plane monochromatic wave falls at an angle of $\psi_{ok} \pm \eta$. The plate is assumed nonabsorbing, and its thickness is small compared with the front of the incident wave. In this case the wave fields in the crystal remain unseparated.

Thus, the conditions at the entrance surface take the form of the following equations:

$$D_o^{(1)} + D_o^{(2)} = D_o^{(a)} \quad , \tag{3.28}$$

$$D_h^{(1)} + D_h^{(2)} = 0 \quad , \tag{3.29}$$

if the point of origin for wave propagation is chosen somewhere on this surface. Equation (3.29) indicates that here we have the Laue case, because in the Bragg case the diffracted wave re-emerges through the same surface.

Solving (3.28), (3.29), (3.17), and (3.18) simultaneously, we obtain the amplitudes $D_h^{(i)}$ and $D_o^{(i)}$ within the crystal in terms of the amplitudes $D_o^{(a)}$

of the incident vacuum wave:

$$D_o^{(i)} = \frac{1}{2}\left(1 \mp \frac{y}{(1+y^2)^{1/2}}\right) D_o^{(a)} = \frac{\exp(\mp v)}{2 \cosh v} D_o^{(a)} \quad , \tag{3.30}$$

$$\sqrt{\gamma_h} D_h^{(i)} = \pm \frac{\exp i\eta_h}{(1+y^2)^{1/2}}, \quad \sqrt{\gamma_o} D_o^{(a)} = \pm \frac{\exp i\eta_h}{2 \cosh v} \sqrt{\gamma_o} D_o^{(a)} . \tag{3.31}$$

In the particular case of scattering in a centre-symmetrical crystal

$$\sqrt{\gamma_h} D_h^{(i)} = \pm \frac{1}{(1+y^2)^{1/2}} \quad \sqrt{\gamma_o}\, D_o^{(a)} = \pm \frac{1}{2 \cosh v} \sqrt{\gamma_o}\, D_o^{(a)} \quad . \tag{3.32}$$

Considering the wave field within a crystal, one should, in the first place, distinguish two cases. If the incident wave is not polarized, the two states of polarization will be independent oscillations in the sense that there will be no constant phase relationships between them. The interconnection between the two states of polarization for the plane-polarized incident wave is obvious.

As for the relationship between the oscillations of two wave fields with identical states of polarization, they will be mutually coherent because of the boundary conditions (3.28) and (2.29) relating them to each other. If the condition of nonseparation of the fields within the crystal is also preserved, the total wave field for oscillations with identical types of polarization can be described by a pair of expressions of the kind

$$\begin{aligned} &\sum_i D_o^{(i)} \exp 2\pi i[\nu t - (\underline{k}_o^{(i)} \underline{r})] \quad , \\ &\sum_i D_h^{(i)} \exp 2\pi i[\nu t - (\underline{k}_h^{(i)} \underline{r})] \quad , \end{aligned} \tag{3.33}$$

the first of which refers to the field of the refracted wave and the second, to that of the diffracted wave in the crystal.

These expressions can be transformed as follows:

$$\sum_i D_o^{(i)} \exp 2\pi i[\nu t - (\underline{k}_o^{(i)} \underline{r})] =$$

$$= \exp 2\pi i[\nu t - (\underline{k}_o \underline{r})] \quad \sum_i D_o^{(i)} \exp(\pm\pi i \Delta k z) \quad , \tag{3.34}$$

$$\sum_i D_h^{(i)} \exp 2\pi i[\nu t - (\underline{k}_h^{(i)}\underline{r})] = $$
$$= \exp 2\pi i[\nu t - (\underline{k}_h \underline{r})] \sum_i D_h^{(i)} \exp(\pm\pi i \Delta kz) \quad , \tag{3.34}$$

where

$$\underline{k}_o = \underline{k}_o^{(1)} + \frac{\Delta \underline{k}}{2} \quad , \quad \underline{k}_h = \underline{k}_h^{(1)} + \frac{\Delta \underline{k}}{2} \quad ,$$
$$\Delta \underline{k} = \underline{k}_o^{(2)} - \underline{k}_o^{(1)} = \underline{k}_h^{(2)} - \underline{k}_h^{(1)} \quad . \tag{3.35}$$

In calculating the modulus $|\Delta \underline{k}|$ we can use the relation

$$|\Delta \underline{k}| = \frac{1}{\gamma_o} [|\xi_o^{(1)}| + |\xi_o^{(2)}|] = \frac{1}{\gamma_h} [|\xi_h^{(1)}| + |\xi_h^{(2)}|] \quad . \tag{3.36}$$

The values of ξ can be taken from (3.14), using the angular variables y and v according to (3.13). We finally get

$$|\Delta \underline{k}| = \frac{KC|\chi_h|}{\sqrt{\gamma_o \gamma_h}} = \sqrt{1 + y^2} = \frac{KC|\chi_h|}{(\gamma_o \gamma_h)^{1/2}} \cosh v \quad . \tag{3.37}$$

We begin the analysis of the wave field in a transparent crystal by considering the contribution to the maxima from the refracted and diffracted waves of each of the fields. This can be done on the basis of (3.30) and (3.32) with the aid of Figs. 3.1 and 3.7.

As the positive value of the angular function y or v increases, i.e., it is displaced along the dispersion surface to the right or left of the line ME, the point $A^{(2)}$ approaches the asymptote T_o and, as can easily be seen from (3.30), $D_o^{(2)}$ increases, approaching the value $D_o^{(a)}$.

Hence, this is true for the positive sign in these equations, when

$$\lim_{y\to\infty} \frac{y}{(1+y^2)^{1/2}} = 1, \quad \lim_{v\to\infty} \frac{\exp v}{2 \cosh v} = 1 \quad .$$

Conversely, under these conditions $D_o^{(1)}$ tends to zero. It is easy to see that in the region of negative values of y and v, i.e., to the left of the line ME in Fig.3.1, the sign of change in the amplitudes $D_o^{(2)}$ and $D_o^{(1)}$

will be opposite. In other words, in the region of the positive values of y and v, the second field of the transmitted wave predominates, whereas in the region of negative y and v, the first field predominates.

As distinct from the two amplitudes of the refracted wave, the moduli of both amplitudes of the diffracted wave remain equal throughout the entire maximum region according to (3.31); thus, the maximum will be symmetrical about points y = 0 and v = 0 for both fields.

At the edges of the maximum $D_h^{(i)}$ vanish as one of $D_o^{(i)}$; we observe the phenomenon of propagation in the crystal of a single wave undergoing refraction at the entrance surface. At the midpoint of the maximum, for $\beta = y = v = 0$, we obtain

$$D_o^{(1)} = D_o^{(2)} = \frac{1}{2} D_o^{(a)} \quad , \quad D_h^{(1)} = D_h^{(2)} = \left(\frac{\gamma_o}{\gamma_h}\right)^{1/2} \frac{1}{2} D_o^{(a)} \quad . \tag{3.38}$$

Finally, for a symmetrical reflection from a plane normal to the entrance surface

$$D_h^{(1)} = D_h^{(2)} = \frac{1}{2} D_o^{(a)} \quad . \tag{3.39}$$

These results are illustrated by Fig.3.7, which is constructed for reflection 220 Si, at $\vartheta = 8^o 28'$; $(\gamma_o/\gamma_h)^{1/2} = 1.23$, Ag K$\alpha$, radiation. Here absorption was ignored.

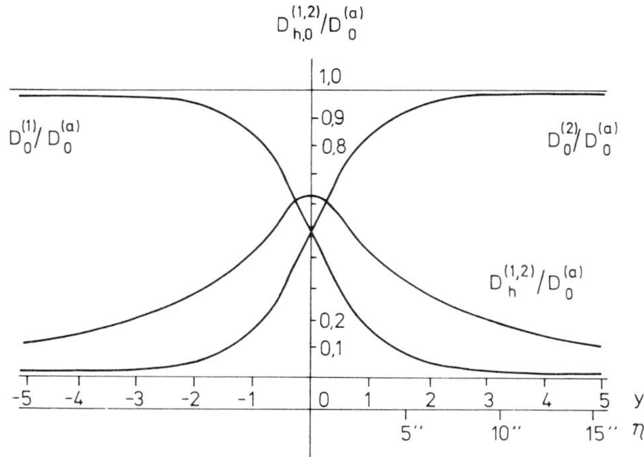

Fig. 3.7 Variations of the values of amplitude reflection coefficients $|D_o^{(i)}/D_o^{(a)}|$ and $|D_o^{(i)}/D_o^{(a)}|$ inside the crystal; C = I (see text)

The space distribution of maxima of the wave field in the crystal can be determined by analysing (3.34). If we use the values $D_o^{(i)}$ and $D_h^{(i)}$ from (3.30) and (3.31), and switch from exponential to trigonometric functions in the summation in (3.34), we obtain the following expressions:

$$D_o^{(a)} \left[\cos\pi\Delta kz - i\frac{y}{(1+y^2)^{1/2}}\sin\pi\Delta kz\right] \exp 2\pi i[\nu t - (\underline{k}_o\underline{r})] \quad , \tag{3.40}$$

$$D_o^{(a)} \left[-i\frac{1}{(1+y_r^2)^{1/2}}\sin\pi\Delta kz\right] \exp 2\pi i[\nu t - (\underline{k}_h\underline{r})] \quad . \tag{3.41}$$

A simpler form, convenient for qualitative physical analysis, is acquired by these expressions if we pass over from the general case to the point $y = 0$,

$$D_o^{(a)} = (\cos\pi\Delta kz) \exp 2\pi i[\nu t - (\underline{k}_o\underline{r})] \quad , \tag{3.42}$$

$$\begin{aligned} &D_o^{(a)} (-i \sin\pi\Delta kz) \exp 2\pi i[\nu t - (k_o r)] = \\ &= D_o^{(a)} (\sin\pi\Delta kz) \exp\left\{2\pi i[\nu t - (\underline{k}_o\underline{r})] - i\frac{\pi}{2}\right\} \quad , \end{aligned} \tag{3.43}$$

where the value $-i = \exp(-i\pi/2)$ is transferred from the amplitude to the phase factor, and $\underline{k}_h$ is replaced by $\underline{k}_o$.

According to (3.35), the vector $\Delta\underline{k}$ is antiparallel to the vector of the normal $\underline{n}_o$. Its modulus is equal to the distance between the two excitation points on the branches of the hyperbola. In the particular case of a symmetrical reflection, for $y = 0$, it transforms into the vector

$$\Delta\underline{k}_o = \frac{1}{(1+y_r^2)^{1/2}} \Delta\underline{k} \quad , \tag{3.44}$$

whose modulus is equal to the real diameter of the dispersion hyperbola (see (2.79)):

$$|\Delta\underline{k}_o| = \frac{KC|\chi_h|}{\cos\vartheta} = 2a \quad . \tag{3.45}$$

From the expressions (3.42) and (3.43), we find that they describe oscillations with amplitudes (in brackets) which are periodic functions of the depth of penetration of z-radiation into the crystal. In other words, the amplitudes have constant values on planes parallel to the entrance boundary.

For $y = 0$ the maxima of the wave function of the diffracted wave (3.43) will lie at the distances

$$z_n = \frac{2n + 1}{2\Delta k_o} \quad . \tag{3.46}$$

For $n = 0$ the first maximum is found on a plane at a depth equal to half the characteristic value of the extinction distance:

$$\tau_o = |\Delta \underline{k}_o|^{-1} \quad . \tag{3.47}$$

The magnitude τ_o is the distance between two adjoining planes with maximum values of the wave function of the diffracted wave. According to (3.42), the distance between the maxima of a transmitted wave is the same. One family of planes is displaced relative to the other by half the extinction distance $(2\Delta \underline{k}_o)^{-1}$.

Incidentally, this structure of the wave field in a crystal resembles, but does not completely coincide with, a system of standing waves. The difference is due to the dependence of the exponents in (3.42) and (3.43) on the coordinates x,y, and time. Hence, for each given value of z, these equations describe a wave running in the plane (xy) with an amplitude depending on z.

An essential feature of such a wave field is periodic "pumping" of energy form the transmitted wave to the diffracted one and back as the radiation penetrates the crystal and also (which is very important) in the stationary wave field induced by the incident vacuum wave. This remarkable physical phenomenon, which is detected with the aid of the appropriate interference experiments, was first predicted by EWALD on the basis of his dynamical theory. By analogy with the interaction of compound and oscillating pendulums, which results in a periodic oscillation energy exchange between them, EWALD termed this phenomenon a "Pendellösung fringe" ≡ pendulum solution of the dynamical problem. This solution also includes the formation of secondary maxima, which will be considered below in connection with the analysis and numerical evaluations of the expressions for the intensity of transmitted and diffracted waves in a vacuum (~T and R).

We shall now give numerical estimates of the extinction distances. For $y = 0$ the extinction distance

$$\tau_o = \frac{\lambda \cos\vartheta}{C|\chi_h|} = \frac{\pi mc^2}{e^2} \frac{\cos\vartheta}{\lambda} \frac{\Omega}{|F_h|} \quad . \tag{3.48}$$

Since the pendulum solution is considered in this chapter for the case of a transparent crystal, of greatest interest in numerical evaluations for comparison with the experiment is the scattering of short-wave radiation in crystals consisting of atoms of light and medium elements. Table 3.1 lists examples referring to the scattering of Mo $K\alpha_1$, Cu $K\alpha_1$, Ag $K\alpha_1$ radiations in silicon, germanium, and quartz crystals.

It should be remembered that the above values of extinction depths refer to the case $y = 0$, which, in a symmetrical reflection, corresponds to the angle ϑ. With a deviation from ϑ, according to (3.44), the extinction distance decreases according to the law $\tau = \tau_0\,(1+y^2)^{-1/2}$ which, at $y = 3$, corresponds to $\tau \approx 0.32\,\tau_0$.

In connection with the foregoing analysis of the expressions for the wave fields in a crystal (3.33), (3.34) and (3.40-43) we must make a reservation which is of special interest for scattering in an absorbing crystal (see Chap.4). Each separate wave field can evidently be described with the aid of an expression of the type (2.26) suitable for the two-wave approximation:

Table 3.1. Values of extinction distances τ_0 (in μm) for scattering of Ag $K\alpha_1$, Mo $K\alpha_1$, and Cu $K\alpha_1$ radiations by silicon, quartz, and germanium crystals. Symmetrical reflection. $C = 1$, $y = 0$

Material	Reflection	Radiation		
		Ag $K\alpha_1$	Mo $K\alpha_1$	Cu $K\alpha_1$
Silicon	111	54.1	42.4	18.4
	220	46.5	36.6	15.4
	422	62.4	48.2	16.4
	333	92.8	71.5	23.2
	444	88.7	66.9	6.43
Quartz	1010	144.2	114.0	
	1120	126.6	100	44
Germanium	111	23.6	18.6	8.64
	220	19.25	15.2	7.0
	422	26.2	20.7	8.1
	333	39.8	30.6	11.2
	444	36.8	29.1	5.8

$$D^{(i)} = \exp 2\pi i[\nu t - (\underline{k}_0\underline{r})] \left\{D_0^{(i)} + D_h^{(i)} \exp[-2\pi i(\underline{h}\underline{r})]\right\} \quad . \qquad (3.49)$$

This equation can be interpreted as the equation of a wave with an amplitude (in brackets) which is a periodic function of the coordinates. In distinction to similar problems for an arbitrary medium, where the geometrical parameters of the wave field are determined by the oscillation wavelengths, here the oscillations adapt themselves to the lattice period. Indeed, in this equation the amplitude will take extreme values when the scalar product ($\underline{hr}$) is integral. Such values correspond to the atomic planes and n planes lying between them, where the whole number n is the ratio of the interference indices to the Miller indices. Whether amplitude maxima or minima will be observed on these planes depends on the relationship between the signs or phases of $D_o^{(i)}$ and $D_h^{(i)}$.

For the first field, the signs of $D_o^{(1)}$ and $D_h^{(1)}$ coincide, and thus we observe the maxima on atomic planes. Conversely, for the second field we observe minima on these planes, while there will be maxima between them when ($\underline{hr}$)equals to -1.

This structure of each of the two fields is of considerable importance in the phenomena of transmission of the X-rays in absorbing crystals. Since the main mechanism of true X-ray absorption is the photoelectric effect, absorption is largely localized on the atoms and, hence, on the atomic planes. The field which has maxima on the atomic planes will be absorbed more intensively.

This structure of each separate field does not manifest itself in nonabsorbing and weakly absorbing crystals. The interactions between the transmitted ($D_o^{(i)}$) and diffracted ($D_h^{(i)}$) waves of both fields become more substantial.

3.2 Transmission and Reflection Coefficients. Analysis of Pendulum Solution in the Case of Plane-Parallel Plate

We shall now calculate the wave functions of waves entering the vacuum from the lower, or exit, crystal surface, and the transmission and reflection coefficients.

We rewrite the condition (3.29) for the value of $D_h^{(i)}$ on the entrance surface

$$c^{(1)}D_o^{(1)} + c^{(2)}D_o^{(2)} = 0 \quad , \tag{3.50}$$

where $c^{(i)}$ are given in (3.11).[1]

[1] In their investigations [3.1,2] carried out in 1971-1972, a group of authors studied Bragg and Laue reflections arising at very small glancing angles of incidence or reflection, of the order of minutes of arc. In theoretical considerations of such experiments, the authors take into account, in the boundary conditions of the dynamical problem, the mirror-reflected wave on the entrance face. As a result they obtain a modified form of the dynamical theory. Their experiments are in agreement with the relations derived.

The conditions on the exit surface must include the phase factors corresponding to the propagation of the waves across a plate of thickness t (point of origin on the entrance surface)

$$D_o^{(1)} \exp(-2\pi i t k_{oz}^{(1)}) + D_o^{(2)} \exp(-2\pi i t k_{oz}^{(2)}) = \\ = D_o^{(d)} \exp(-2\pi i t K_{oz}^{(d)}) \quad , \tag{3.50a}$$

$$D_h^{(1)} \exp(-2\pi i t k_{hz}^{(1)}) + D_o^{(2)} \exp(-2\pi i t k_{hz}^{(2)}) = \\ = D_h^{(d)} \exp(-2\pi i t K_{hz}^{(d)}) \quad . \tag{3.50b}$$

Solving (3.50) and (3.28) simultaneously, we get

$$D_o^{(1)} = \frac{c^{(2)}}{c^{(2)} - c^{(1)}} D_o^{(a)}, \qquad D_o^{(2)} = -\frac{c^{(1)}}{c^{(2)} - c^{(1)}} D_o^{(a)} \quad , \tag{3.51}$$

$$D_h^{(1)} = c^{(1)} D_o^{(1)} = \frac{c^{(1)}c^{(2)}}{c^{(2)} - c^{(1)}} D_o^{(a)} \ , \quad D_h^{(2)} = -\frac{c^{(1)}c^{(2)}}{c^{(2)} - c^{(1)}} D_o^{(a)}. \tag{3.52}$$

Substituting the values of $D_o^{(i)}$ and $D_h^{(i)}$ in (3.50), we obtain

$$D_o^{(d)} \exp(-2\pi i t K_{oz}^{(d)}) = \frac{c^{(2)}}{c^{(2)} - c^{(1)}} D_o^{(a)} \exp(-2\pi i t k_{oz}^{(1)}) - \\ - \frac{c^{(1)}}{c^{(2)} - c^{(1)}} D_o^{(a)} \exp(-2\pi i t k_{oz}^{(2)}) \quad , \tag{3.53}$$

$$D_h^{(d)} \exp(-2\pi i t K_{hz}^{(d)}) = \frac{c^{(1)}c^{(2)}}{c^{(2)} - c^{(1)}} D_o^{(a)} \exp(-2\pi i t k_{hz}^{(1)}) + \\ - \frac{c^{(1)}c^{(2)}}{c^{(1)} - c^{(2)}} D_o^{(a)} \exp(-2\pi i t k_{hz}^{(2)}) \quad . \tag{3.54}$$

Since (see (3.1)) $k_{oz}^{(i)} = K_{oz}^{(a)} - K\delta^{(i)}$, and in the case of a plane-parallel plate

$$\underline{K}_o^{(d)} = \underline{K}_o^{(a)} \quad , \tag{3.55}$$

we rewrite (3.53):

$$D_o^{(d)} = D_o^{(a)} \frac{1}{c^{(2)} - c^{(1)}} \exp[\pi i t K(\delta^{(1)}+\delta^{(2)})] \quad .$$

$$\cdot \quad c^{(2)}\Big\{\exp[\pi i t K(\delta^{(1)}-\delta^{(2)})] - c^{(1)} \exp[-\pi i t K(\delta^{(1)}-\delta^{(2)})]\Big\} . \qquad (3.56)$$

A similar transformation can be made in (3.54) because

$$K_{hz}^{(d)} - k_{hz}^{(i)} = K_o^{(a)} - k_{oz}^{(i)} = \delta^{(i)} \quad . \qquad (3.57)$$

From (3.7) and (3.10) follows

$$K(\delta^{(1)}+\delta^{(2)}) = \frac{K}{2\gamma_h}\beta - \frac{K\chi_o}{\gamma_o} \quad , \quad K(\delta^{(1)}-\delta^{(2)}) = \frac{K}{2\gamma_h} W;$$

$$W = \sqrt{\beta^2 + 4C^2 \frac{\gamma_h}{\gamma_o} \chi_h\chi_{\bar{h}}} = 2C \sqrt{\frac{\gamma_h}{\gamma_o}} \sqrt{\chi_h\chi_{\bar{h}}} \sqrt{1 + y^2} \quad . \qquad (3.58)$$

Passing over from the wave functions to the values of the coefficients of transmission T and reflection R, we multiply the right-hand sides of (3.56) and of the similar expression for $D_h^{(d)}$ by the respective conjugate complex expressions. Complex values are those values of $c^{(i)}$ which contain complex fractions $\chi_h/\chi_{\bar{h}}$. The products $\chi_h\chi_{\bar{h}}$ appearing in (3.11) and (3.13) are real values, as well as y and v.

For the transmission coefficient we have

$$T = \frac{|D_o^{(d)}|^2}{|D_o^{(a)}|^2} = \frac{1}{|c^{(2)} - c^{(1)}|^2} [|c^{(2)}|^2 + |c^{(1)}|^2 - c^{(1)}c^{(2)*} \cdot$$

$$\cdot \exp(-i2\alpha) - c^{(1)*}c^{(2)} \exp i2\alpha] \quad ,$$

or, switching from exponential to trigonometric functions,

$$T = \frac{1}{|c^{(2)} - c^{(1)}|^2} [|c^{(2)}|^2 + |c^{(1)}|^2 + 2c^{(1)}c^{(2)*} \cos 2\alpha] \quad . \qquad (3.59)$$

In like manner, we obtain for the reflection coefficient

$$R = \frac{\gamma_h}{\gamma_o} \frac{|c^{(1)}|^2 |c^{(2)}|^2}{|c^{(2)} - c^{(1)}|^2} (2-2 \cos 2\alpha) \quad , \tag{3.60}$$

$$\alpha = \pi t K(\delta^{(1)}-\delta^{(2)}) = \pi t \frac{KC}{\sqrt{\gamma_o \gamma_h}} \sqrt{\chi_h \chi_{\bar{h}}} \sqrt{1 + y^2} = A\sqrt{1 + y^2} \quad . \tag{3.61}$$

And, finally, using (3.17), we get explicit expressions for T and R as a function of y and v:

$$T(y) = \frac{|D_o^{(d)}|^2}{|D_o^{(a)}|^2} = \frac{1}{1 + y^2} (y^2 + \cos^2 A \sqrt{1+y^2}) \quad , \tag{3.62}$$

$$R(y) = \frac{\gamma_h}{\gamma_o} \frac{|D_h^{(d)}|^2}{|D_o^{(a)}|^2} = \frac{\sin^2 A \sqrt{1 + y^2}}{1 + y^2} \tag{3.63}$$

$$T(v) = \frac{\cosh^2 v - \sin^2 (A \cosh v)}{\cosh^2 v} = 1 - R(v) \quad , \tag{3.64}$$

$$R(v) = \frac{\sin^2 (A \cosh v)}{\cosh^2 v} \tag{3.65}$$

$$A = \pi t \frac{KC}{(\gamma_o \gamma_h)^{1/2}} \sqrt{\chi_h \chi_{\bar{h}}} \quad . \tag{3.66}$$

Considering (3.64) and (3.65), we note, first of all, that the sum of any respective pair of values T and R for any given value of the angular argument y or v is equal to unity. This is obviously in agreement with our idea of scattering in a transparent crystal in which the energy of the incident wave is divided into the energy of the transmitted and reflected waves. Eqs. (3.62,65) represent the pendulum solution. Indeed, these expressions are, via the variable A, periodic functions of the crystal thickness t. T and R are complementary in that for any given value of t definite values of y will correspond to the maximum of T and minimum of R, and conversely. Thus, for $t = (2n+1) \tau_1$ (n-integer) at the midpoint of the maximum, i.e., at $y = 0$, the reflection factor R reaches its maximum value, unity. For values $t = n\tau$, at the midpoint of the maximum $R = 0$ and $T = 1$. Similar periodic variation in the values of R and T with thickness is observed at other, non-zero, values of y and v. Note that at values $y \neq 0$ the maximum values of R no longer reach unity.

It is obvious that the pendulum solution of the equations obtained will manifest itself especially pictorially in the case of scattering from a wedge-shaped plate. Here, the periodic variation indicated in the values of T and R will be observed directly on the sites of the exit face corresponding to the critical values of t. Besides (and this is especially important), the intervals between the maxima of the two values increase in certain

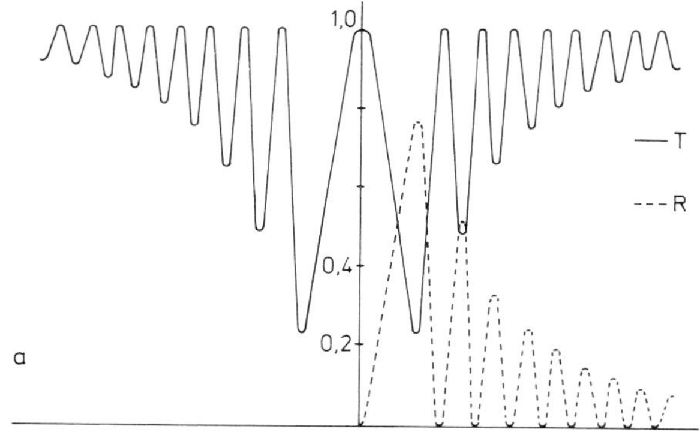

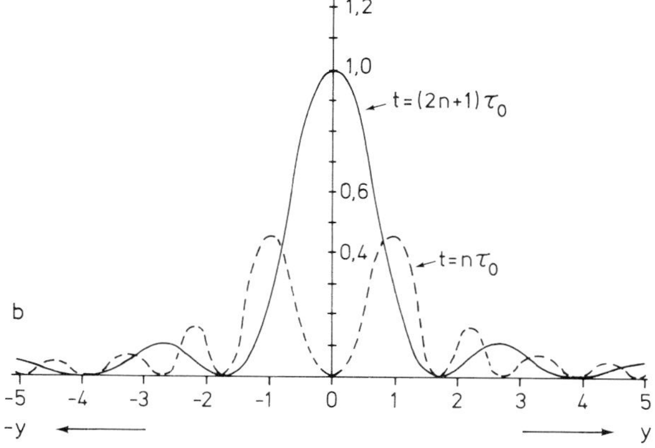

Fig. 3.8 a and b Variations in values of T and R in the maximum region in the case of transparent crystal: a) R and T curves correspond to $t = n\tau_0$; b) R curves at various values of crystal thickness t; subsidiary maxima of pendulum solution are shown (after ZACHARIASEN [1.14])

experimental setups such as $\sim (\sin\mu)^{-1}$ or $(\tan\mu)^{-1}$, where μ is the wedge angle. The corresponding geometrical theory is outlined below.

Apart from their dependence on thickness, the coefficients R and T are also quasi-periodic functions of the angular variables y and v. The presence of $(1+y^2)$ in (3.63,65) leads to a monotonic drop of the comb and a decrease in oscillation period on the R and T curves (Fig.3.8).

At large values of the angular variables, R vanishes and T becomes unity. Thus, the pendulum solution for R and T as functions of the angle of incidence leads to the formation of a fine structure or subsidiary maxima.

The angular distribution of the subsidiary maxima of the reflection power coefficient R is determined from the condition

$$\frac{\pi t C|\chi_h|}{\lambda(\gamma_o\gamma_h)^{1/2}} \sqrt{1+y_h^2} = \frac{2n+1}{2}\pi \tag{3.67}$$

or

$$y_h^2 = \left(\frac{2n+1}{2}\right)^2 \frac{\lambda^2\gamma_o\gamma_h}{C^2|\chi_h|^2 t^2} - 1 \quad . \tag{3.68}$$

Since for typical experimental conditions

$$\frac{\lambda^2\gamma_o\gamma_h}{C^2|\chi_h|^2 t^2} \equiv \varphi\,(|\chi_h|,t) \ll 1 \quad , \tag{3.69}$$

for the second and subsequent subsidiary maxima $(2n-1/2)^2 \geq 25/4$ the unity in (3.68) can be neglected without impairing the accuracy, and in this case the distance between the subsidiary maxima will be

$$\Delta y \approx \frac{\lambda(\gamma_o\gamma_h)^{1/2}}{tC|\chi_h|} = \frac{\tau_o}{t} \quad . \tag{3.70}$$

For symmetrical reflection (3.68) and (3.70) enable us to represent

$$\varphi\,(|\chi_h|,t) = \varphi_1(t) - \varphi_2(\chi_h) \quad , \tag{3.71}$$

which is obviously important for experimental verification of the theory.

Indeed, for symmetrical reflection (see (3.27))

$$\eta = -\frac{C|\chi_h|y}{\sin 2\vartheta}$$

and, passing over from the angular function y to the argument in (3.68), we obtain

$$\eta_h^2 = \frac{2n+1}{2}^2 \frac{d_h^2}{t^2} - \frac{C^2|\chi_h|^2}{\sin^2 2\vartheta} \quad . \tag{3.72}$$

For the maxima following the first one we can again use (3.70), which is transformed as follows for symmetrical reflection

$$\Delta\eta_s'' = \frac{d_h}{t \cdot 4.85 \cdot 10^{-6}} \quad . \tag{3.73}$$

Thus, comparing the measurements of the angular position of the first maximum (3.72) and the subsequent ones (3.73), one can determine separately both the thickness of a given crystal plate and the value of χ_h or the structure amplitude F_h of a given reflection [3.3].

In the case of an asymmetrical reflection, (3.70) can be transformed thus:

$$\Delta\eta = \frac{d_h \gamma_h}{t \cos\vartheta} = \Delta\eta_s \frac{\cos\psi_h}{\cos\vartheta} \quad . \tag{3.74}$$

For numerical evaluations of the fine structure of the maxima due to the pendulum solution we apply (3.73) to reflections 220 and 333 from single-crystal Si slices. If we assign an angular interval of about 1", the necessary plate thickness will be 40 and 22 μm, respectively. In order to obtain a large interval, still thinner plates must be used, which involves great experimental difficulties.

Comparing the two forms of the Pendellösung fringe solution -- bands of equal thickness and subsidiary maxima -- we will mention some interesting differences between these phenomena [3.3a]. First of all we shall turn our attention to the possible separate determination of the plate thickness t and the value of χ_h in the case of subsidiary maxima, according to (3.68-73). In the case of bands of equal thickness, (3.44) includes a parameter y which, taken separately, does not always lend itself to accurate determination. It should further be mentioned that, in principle, the two types of maxima differ in number. In the case of bands of equal thickness, the number of bands cannot exceed the value t/τ_0. At the same time, it follows from (3.68) and (3.70) that there is no limit to the number of subsidiary maxima.

Finally, it is of interest to consider the model representing the pendulum solution as the effect of interference of waves from two image sources.

Image sources are used in visible light optics for interpretation of classical interference experiments. In contrast to optical schemes, however, where image sources are introduced for calculating the interaction of waves in space, here use is made of excitation points $A^{(i)}$ (see, for instance, Fig.3.1) in reciprocal space. Ignoring the difference between the amplitudes of $D_o^{(i)}$ and $D_h^{(i)}$ and choosing the direction of the vector $\underline{K}_o^{(a)}$ (which is close to the directions of $\underline{k}_o^{(1)}$ and $\underline{k}_o^{(2)}$ (Fig.3.9a)) as the direction of propagation of the waves $D_o^{(i)}$ interferring in the crystal, we write down the phase difference of the waves $D_o^{(i)}$:

$$2\pi|\Delta\underline{k}|\gamma_o = 2\pi \frac{\gamma_o}{\tau} \quad . \tag{3.75}$$

This phase difference corresponds to the first extinction distance in the crystal. Using (3.37), we write out the value of the difference in the path of propagation of two interferring waves which corresponds to the n-th extinction distance, or the n-th band (maximum):

$$\Delta l_n = \frac{1}{2\pi/\lambda} 2\pi \frac{\gamma_o}{n\tau} = \frac{C|\chi_h|}{n} \frac{\gamma_o}{\gamma_h} \sqrt{1 + y^2} = \frac{\Delta l_1}{h} \quad , \tag{3.76}$$

where Δl_1 is the path difference corresponding to the first band. Thus, as the number of the band of equal thickness increases, the path difference in the interferring waves decreases. A different picture is observed during the formation of secondary maxima. In this case (3.75) corresponds to the phase difference of the interferring waves for a certain definite subsidiary maximum, i.e., a definite position of the normals 1-4 in Fig.3.9b or a definite value of $y_n(n=1,2,\ldots)$.

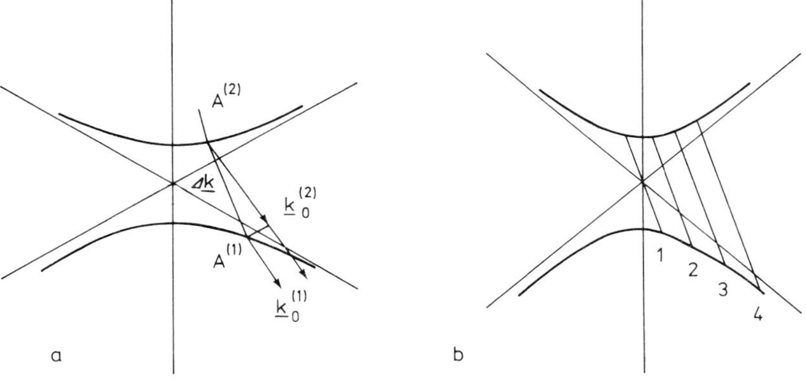

Fig. 3.9 a and b Path difference of waves in crystal: a) during formation of bands of equal thickness; b) during formation of subsidiary maxima

The zero maximum (y=0) evidently corresponds to the position of the normal 1 in the same figure. It can be seen that the path difference corresponding to the n-th maximum,

$$\Delta l_n = C|\chi_h| \frac{\gamma_o}{\gamma_h} \sqrt{1 + y_n^2} \tag{3.77}$$

increases with the number of the maximum.

A further analysis of the effects of the pendulum solution associated, in particular, with the experimental determination of the structure amplitudes, is given for the incident spherical wave approximation in Chapters 6 and 9, and for the incident-plane-wave approximation, in Chapter 9.

From the foregoing it is clear that experimental observation and measurement of the effects associated with the pendulum solution is a difficult task. It has in fact been performed only very recently, as shown in Chapter 9. Therefore, until now, use was mainly made of the expressions (3.62-65) in a form corresponding to the available measurement techniques. Here we imply the averaging out of functions R and T over t and y. In order to obtain $\bar{R}$ it will suffice, for instance, to replace the value of the sine squared in (3.63) by its average value over the full oscillation period. The average value of $\bar{T}$ will be obtained by deducting $\bar{R}$ from unity:

$$\bar{R} = \frac{1}{2(1+y^2)} \quad , \; \bar{T} = 1 - \frac{1}{[2(1+y^2)]} \quad . \tag{3.78}$$

The curves for R and T are given in Fig.3.10.

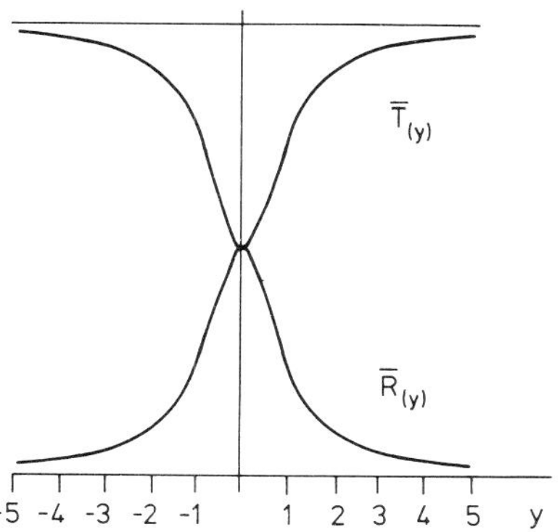

Fig. 3.10 Profiles of maxima of averaged functions $\bar{T}$ and $\bar{R}$ in a transparent crystal (after JAMES [1.16])

The half-width of the maximum w or its width at values $\bar{R} = 1/2\ \bar{R}_{max}$ is determined from the condition 2y = 2. Passing over again to absolute values of η, we obtain (see (3.27))

$$w(\eta) = 2\eta_{1/2} = \frac{2C|\chi_h|(\gamma_h/\gamma_o)^{1/2}}{\sin 2\vartheta} \quad . \tag{3.79}$$

If we vary the wavelength of the radiation used, i.e., if we assume that a definite interval $\lambda = \lambda - \lambda_B$ corresponds to a given value of $\Delta\eta$, where λ_B is the wavelength precisely corresponding to the Bragg angle ϑ, we can express the half-width of the maximum in wavelength intervals:

$$2\Delta\lambda_s = \frac{\lambda_B C|\chi_h|(\gamma_h/\gamma_o)^{1/2}}{\sin^2\vartheta} \quad . \tag{3.80}$$

We now pass on to determining the integral reflection

$$R_i^y = \int_{-\infty}^{\infty} \frac{\sin^2 A\sqrt{1 + y^2}}{1 + y^2}\, dy \quad . \tag{3.81a}$$

This integral is calculated by differentiating the integrand with respect to the parameter A:

$$\int_0 \frac{\sin (2A \sin\varphi)}{\sin\varphi}\, d\varphi \quad ,$$

to which the right-hand side in (3.81a) can be reduced.

In what follows we use the definition

$$\frac{\pi}{2} J_o(x) = \int_0 \cos(x \sin\varphi) d\varphi \quad ,$$

and the result has the form:

$$R_i^y = \frac{\pi}{2} \int_0^{2A} J_o(x) dx = \pi \sum_{n=0}^{\infty} J_{2n+1}\ (2A) \quad . \tag{3.81b}$$

Here, J_o and J_n, as usual, denote the Bessel functions of the zero and, accordingly, the n-th order.

For numerical calculation of integrated power reflection from (3.81b), one can use either the asymptotic properties of the series

$$\sum_{n=0}^{\infty} J_{2n+1}(2A) = \begin{cases} A \ (A \ll 1) & , \\ \frac{1}{2} \ (A \gg 1) & , \end{cases} \tag{3.81c}$$

or, according to De MARCO and WEISS [3.4], the values of the integral $\int_0^{2A} J_o(x)dx$ for $A < 5$ given in Appendix A. For $A > 5$, sufficiently accurate results are yielded by the function

$$\int_0^{2A} J_o(x)dx = 1 - \frac{1}{\sqrt{2A}} [P(A) \cos(2A+\tfrac{\pi}{4}) + Q(A) \sin(2A+\tfrac{\pi}{4})] ,$$

$$P(A) = 0{,}7979 - \frac{0.2010}{A^2} + \frac{0.4575}{A^4} , \tag{3.81d}$$

$$Q(A) = \frac{0.2493}{A} - \frac{0.2586}{A^3} + \frac{1.0332}{A^5} .$$

The curve of variation in the function $R_i^y(A)$ is given in Fig.3.11. At small values of A, which is proportional to the crystal thickness t, a monotonic increase is observed in the integral reflection, this being in agreement with the kinematical theory. With a further increase in A the extinction effect is revealed, namely periodic changes in integral

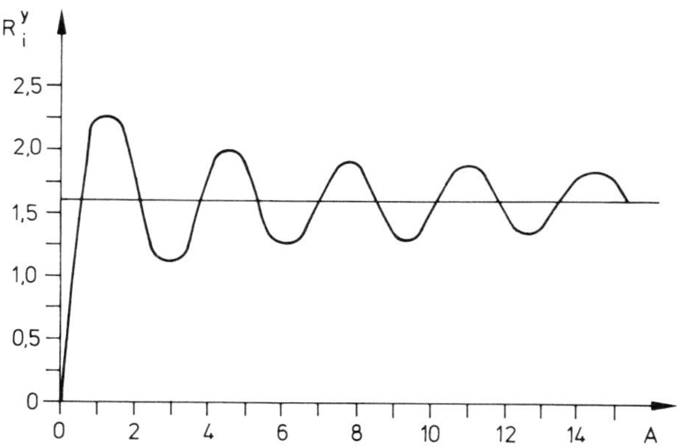

Fig. 3.11 Curve of variation in integrated reflection $R_i^y(A)$ (see also (3.66) after ZACHARIASEN [1.14])

reflection, with a gradual decrease in the oscillation amplitude with respect to the average value $\pi/2$. It is easy to see that the same average value $\pi/2$ is obtained on integrating the value of $\bar{R}$ from (3.78):

$$\int_{-\infty}^{\infty} \bar{R}\, dy = \frac{1}{2} \int_{-\infty}^{\infty} (1+y^2)^{-1}\, dy = \frac{\pi}{2} \quad . \tag{3.81e}$$

It should also be kept in mind that for large thicknesses t of the crystal slice the transparent-crystal approximation is inapplicable, and (4.124-126) must be used. Numerical evaluation of the critical thicknesses for this transition can be performed on the basis of the theory explained in Chapter 4.

Adopting the indicated average value of integrated power reflection, we will write out the following final values:

$$R_i^y = \frac{\pi}{2} \quad , \quad R_i^\eta = \frac{C|\chi_h|}{\sin 2\vartheta} \sqrt{\frac{\gamma_h}{\gamma_o}}\, \frac{\pi}{2} \quad . \tag{3.82a}$$

In the case of incident nonpolarized radiation

$$R_i^\eta = \frac{|\chi_h|}{\sin 2\vartheta} \sqrt{\frac{\gamma_h}{\gamma_o}}\, \frac{\pi}{2}\, \frac{1 + |\cos 2\vartheta|}{2} \quad . \tag{3.82b}$$

For an asymmetrical reflection the reflection maxima of the rocking curve R (and, accordingly, minima of the rocking curve T) are displaced from the positions corresponding to the kinematical theory. The shift of the maxima is determined from (3.26) at y = 0.

$$\eta_o = \frac{|\chi_o| \left[\left(\frac{\gamma_h}{\gamma_o}\right) - 1\right]}{2 \sin 2\vartheta} \tag{3.83}$$

Using (2.62) and introducing

$$\delta_D = 1 - n = - 1/2\chi_o \quad , \tag{3.84}$$

we rewrite (3.83)

$$\eta_o = - \frac{\delta_D}{\sin 2\vartheta} \left(1- \frac{\gamma_h}{\gamma_o}\right) = \frac{\chi_o}{2 \sin 2\vartheta} \left(1- \frac{\gamma_h}{\gamma_o}\right) \quad . \tag{3.85}$$

Thus, the shift effect is not associated with the dynamical scattering proper (is independent of χ_h) and is determined by the specific refraction (considered in Sec.3.1) on passing through a plane-parallel slice. It is interesting to note that in switching from hkl and $\bar{h}\bar{k}\bar{l}$ the shift value changes. Indeed, the change in the signs of the reflection coefficients is equivalent to transition from reflection from the other side of the same system of planes. Here, the values of γ_0 and γ_h are interchanged.

It should be kept in mind that an increase in η means an increase in ψ_0, i.e., a decrease in the glancing angle of incidence.

Finally, we note that in an asymmetrical diffraction the angular half-width of the maxima $w = 2\eta_{1/2}$ of reflections with the opposite of the indices is different. Therefore, if in a symmetrical reflection the angular position and the half-width of the maxima R_h and $R_{\bar{h}}$ are completely identical, in an asymmetrical reflection these maxima acquire a different angular half-width and diverge from the point corresponding to $\eta = 0$.

As for numerical evaluations, the effects described are, generally speaking, insignificant, i.e., of the order of an angular second, which, however, can be detected by modern recording instruments.

3.3 Transmission Through a Wedge-Shaped Plate

So far we have been discussing the transmission of an X-ray wave through a plane-parallel plate. If the incident wave front is quite wide as compared with the plate thickness and, hence, the wave fields within the crystal remain unseparated, then one common wave $D_0^{(d)}$ and one wave $D_h^{(d)}$ emerge from the crystal and enter into the vacuum.

It is easy to show that if the crystal plate is not plane but wedge shaped, then we may expect that the waves entering the vacuum might diverge as a result of different wave fields within the crystal. Here, at the exit surface, the waves on both sides of the boundary will be related, not by (3.29), but by boundary conditions for each field separately. The same holds true of the continuity condition for the tangential components of the wave vectors (3.1).

We shall now consider the transmission of a nonpolarized X-ray wave through a wedge-shaped crystal plate whose maximum thickness is far less than the wave front of the incident vacuum wave. The boundary conditions at the entrance surface, and the wave field within the crystal, which is formed according to these boundary conditions, remain unchanged and, hence, are described in accordance with the analysis already discussed. Now, however, for each state of polarization, four waves will emerge from the crystal into the vacuum.

The amplitudes and wave vectors of the incident wave in the vacuum and of the four waves emerging from the crystal will be denoted in this section as follows:

$$D_o^{(a)},\ D_{od}^{(1)},\ D_{od}^{(2)},\ D_{hd}^{(1)},\ D_{hd}^{(2)},$$
$$\underline{K}_{oa},\ \underline{K}_{od}^{(1)},\ \underline{K}_{od}^{(2)},\ \underline{K}_{hd}^{(1)},\ \underline{K}_{hd}^{(2)} \quad . \tag{3.86}$$

Now we shall derive an expression for the wave vectors in the vacuum, for instance $\underline{K}_{hd}^{(i)}$. To do this, we shall use the boundary conditions for the wave vectors, first at the entrance, and then at the exit surfaces. Turning to (3.1), we add the vector h to both sides of the equation:

$$\underline{k}_h^{(i)} = \underline{K}_{oa} + \underline{h} - K\delta^{(i)}\underline{n}_o \quad . \tag{3.87}$$

By virtue of (3.7) and (3.14), we get

$$\underline{k}_h^{(i)} = \underline{K}_{oa} + \underline{h} \mp \frac{\underline{n}_o}{\gamma_h}\frac{KC}{2}|\chi_h|\sqrt{\frac{\gamma_h}{\gamma_o}} \cdot (\sqrt{1+y^2}+y) \mp \frac{\underline{n}_o}{\gamma_h}\frac{K\chi_o}{2}\frac{\gamma_h}{\gamma_o} - P \quad . \tag{3.88}$$

To formulate the boundary conditions at the exit surface, we shall use the diagram shown in Fig.3.12.

The straight lines $\overline{A^{(i)}P^{(i)}}$ are drawn through the excitation points $A^{(i)}$ parallel to the normal $\underline{n}_d$ to the exit surface. The normal is directed from the crystal into the vacuum. The boundary conditions at the exit surface for each of the vectors in the vacuum can be written in the following way:

$$\underline{K}_{hd}^{(i)} = \underline{k}_h^{(i)} + \overline{KP^{(i)}A^{(i)}}\ \underline{n}_d \quad , \tag{3.89}$$

where, by analogy with (3.7), we have

$$\overline{KP^{(i)}A^{(i)}}\ \underline{n}_d = (-\xi_h^{(i)} - 1/2K\chi_o + K\alpha_i)\frac{\underline{n}_d}{\gamma_h^{(i)}} \quad , \quad K\alpha_i = \overline{P^{(i)}Q^{(i)}} \quad . \tag{3.90}$$

After substituting the values of $k_h^{(i)}$ in (3.89) from (3.88) and then using (3.14) and (3.90), we obtain the final expression for $K_{hd}^{(i)}$

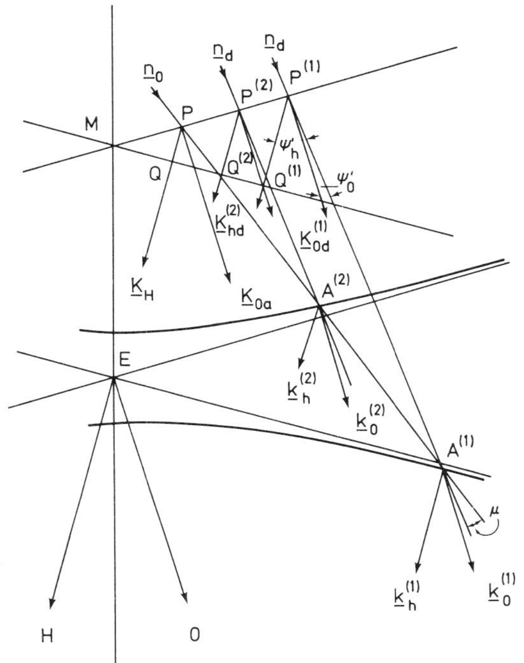

Fig. 3.12 Dispersion surface in a reciprocal space for a wedge-shaped plate

$$\underline{K}_{hd}^{(1)} = \underline{K}_{oa} + \underline{h} - m_1 \left[\frac{\underline{n}_o}{\gamma_h} (\sqrt{1+y^2}+y) - \frac{\underline{n}_d}{\gamma_h^{(1)}} (\sqrt{1+y^2}-y)\right] + \\ + \frac{K\chi_o}{2} \left(\frac{\underline{n}_o}{\gamma_h} - \frac{\underline{n}_d}{\gamma_h^{(1)}}\right) - (\underline{p}-\underline{p}_1) \quad . \tag{3.91}$$

$$\underline{K}_{hd}^{(2)} = \underline{K}_{oa} + \underline{h} + m_1 \left[\frac{\underline{n}_o}{\gamma_h} (\sqrt{1+y^2}-y) - \right. \\ \left. - \frac{\underline{n}_d}{\gamma_h^{(2)}} (\sqrt{1+y^2}+y)\right] + \frac{K\chi_o}{2} \left(\frac{\underline{n}_o}{\gamma_h} - \frac{\underline{n}_d}{\gamma_h^{(2)}}\right) + (\underline{p}+\underline{p}_2) \quad , \tag{3.92}$$

$$m_1 = m \frac{\gamma_h}{\gamma_o} \ , \quad \underline{p} = \frac{\underline{n}_o}{\gamma_h} K\alpha \ , \quad \underline{p}_i = \frac{\underline{n}_d}{\gamma_h^{(i)}} K\alpha_i , \ i = 1,2 \quad .$$

We shall now calculate the angles between the pairs of wave vectors, i.e., x_o between $\underline{K}_{od}^{(i)}$ and x_h between $\underline{K}_{hd}^{(i)}$. For this purpose, turn to Fig.3.13 which shows the corresponding details of Fig.3.12 on a larger scale.

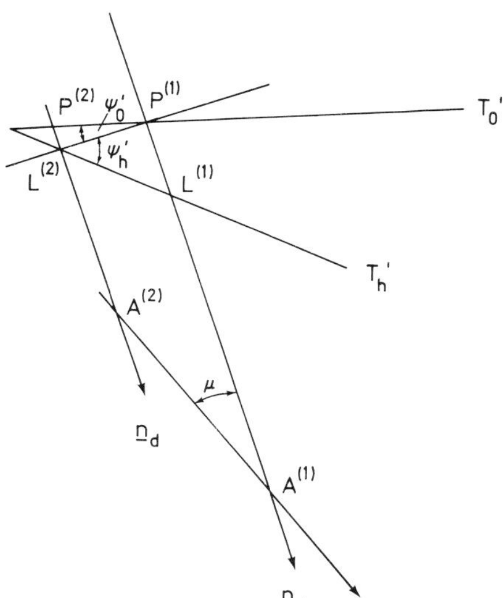

Fig. 3.13 For the derivation of equations for the angles x_0 and x_h

The angle $x_o \approx K^{-1}P^{(1)}P^{(2)}$, where the segment $\overline{P^{(1)}P^{(2)}}$ is approximately normal to the vetors $\underline{K}_{od}^{(i)}$. Similarly, the angle $x_h \approx K^{-1}Q^{(1)}Q^{(2)}$, where the segment $\overline{Q^{(1)}Q^{(2)}}$ is almost normal to $\underline{K}_{hd}^{(i)}$.

According to (3.35), the vector $\overline{A^{(2)}A^{(1)}}$, which represents the vector $K\delta^{(1)}\underline{n}_o - K\delta^{(2)}\underline{n}_o$, is at the same time the vector difference $\underline{k}_o^{(2)}-\underline{k}_o^{(1)}$ or $\underline{k}_h^{(2)}-\underline{k}_h^{(1)}$. For the boundary conditions for wave vectors at the exit surface it follows that

$$[\underline{k}^{(2)} - \underline{k}_o^{(1)}]_{tan} = [\underline{k}_h^{(2)} - \underline{k}_h^{(1)}]_{tan} =$$
$$=[\underline{K}_{od}^{(2)} - \underline{K}_{od}^{(1)}]_{tan} = [\underline{K}_{hd}^{(2)} - \underline{K}_{hd}^{(1)}]_{tan} \quad . \tag{3.93}$$

That validity of these equalities is clear from Fig.3.13, where the segments $\overline{A^{(2)}A^{(1)}}$, $\overline{P^{(2)}P^{(1)}}$, $\overline{Q^{(2)}Q^{(1)}}$, which represent

$$|\Delta\underline{k}_o^{(i)}| \text{ and } |\Delta\underline{k}_h^{(1)}|, \ |\Delta\underline{K}_{od}|, \ |\Delta\underline{K}_{hd}|,$$

respectively, have equal projections on the line $\overline{P^{(1)}Q^{(2)}}$. This line, normal to the vector $\underline{n}_d$, is parallel to the exit surface.

From (3.37) using the value $|\Delta \underline{k}| = \overline{A^{(2)}A^{(1)}}$ from Fig.3.13, we find that

$$\overline{P^{(1)}Q^{(2)}} = \overline{A^{(2)}A^{(1)}} \sin\mu = \frac{KC|\chi_h|}{(\gamma_o\gamma_h)^{1/2}} \sqrt{1+y^2} \sin\mu \quad . \tag{3.94}$$

In order to determine the approximate values of the angles x_o and x_h we shall introduce the average values ψ_o' for angles

$$\angle[\underline{k}_o^{(1)}\underline{n}_d] \approx \angle[\underline{K}_{od}^{(1)}\underline{n}_d], \quad \angle[\underline{k}_o^{(2)}\underline{n}_d] \approx \angle[\underline{K}_{od}^{(2)}\underline{n}_d] \quad , \tag{3.95}$$

ψ_h' for angles

$$\angle[\underline{k}_h^{(1)}\underline{n}_d] \approx \angle[\underline{K}_{hd}^{(1)}\underline{n}_d], \quad \angle[\underline{k}_h^{(2)}\underline{n}_d] \approx \angle[\underline{K}_{hd}^{(2)}\underline{n}_d] \tag{3.96}$$

and the symbols

$$\cos\psi_o' \equiv \gamma_o', \quad \cos\psi_h' \equiv \gamma_h' \quad . \tag{3.97}$$

Thus, for these angles we obtain

$$x \approx \frac{1}{K}\,\overline{P^{(2)}P^{(1)}} = \frac{1}{K\gamma_o'}\,\overline{Q^{(2)}P^{(1)}} = \frac{C|\chi_h|}{(\gamma_o\gamma_h)^{1/2}} \sqrt{1+y^2}\,\frac{\sin\mu}{\gamma_o'} \quad , \tag{3.98}$$

$$x_h \approx \frac{1}{K}\,\overline{Q^{(2)}P^{(1)}} = \frac{C|\chi_h|}{(\gamma_o\gamma_h)^{1/2}} \sqrt{1+y^2}\,\frac{\sin\mu}{\gamma_h'} \quad . \tag{3.99}$$

Using the expressions (2.34), (2.35), and (2.62), we can write

$$|\chi_h| = |\chi_o|\,\frac{|F_h|}{N\Omega} = 2(1-n)\,\frac{|F_h|}{F_o} \quad ; \tag{3.100}$$

hence

$$x_o = P\,\frac{\sin\mu}{\gamma_o'} \quad , \quad x_h = P\,\frac{\sin\mu}{\gamma_h'} \quad , \tag{3.101}$$

$$P = \frac{2C(1-n)}{(\gamma_o\gamma_h)^{1/2}}\,\frac{|F_h|}{F_o}\,\sqrt{1+y^2} \quad .$$

These formulas can be used to calculate the angles of diversion of the waves $\underline{K}_{od}^{(i)}$ and $\underline{K}_{hd}^{(i)}$ when X-ray waves pass through a wedge-like plate.

Of great interest, from an experimental point of view, is the observation and calculation of interference in the weakly diverging pairs $\underline{K}_{ho}^{(i)}$ and $\underline{K}_{hd}^{(i)}$. This interference applies to bands of equal thickness, i.e., the phase difference corresponding to the periodic changes in intensity observed is related to $\Delta\underline{k}$ according to (3.37)

The exact period of the interference pattern is defined by $(\Delta\underline{K}_{hd})^{-1}$, i.e., by subtracting (3.91) from (3.92). Putting

$$\gamma_h^{(1)} \approx \gamma_h^{(2)} \approx \gamma_h', \quad \alpha_1 \approx \alpha_2 \quad , \tag{3.102}$$

we obtain

$$\Lambda_h = (|\Delta\underline{K}_{hd}|)^{-1} = \left[\lambda\ C|\chi_h| \sqrt{\frac{\gamma_h}{\gamma_o}} \sqrt{1+y^2}\left(\left|\frac{\underline{n}_o}{\gamma_h} - \frac{\underline{n}_d}{\gamma_h'}\right|\right)\right]^{-1} \quad . \tag{3.103}$$

The expression for Λ_h in explicit form is obviously determined by the vector difference in the parentheses on the right-hand side of (3.103). This difference is determined by one of the following possible experimental schemes [1.23]: 1) the wedge edge in perpendicular to the plane in which the normals $\underline{n}_o$ and $\underline{n}_d$ to the entrance and exit surfaces lie; 2) the wedge edge is inclined to the above plane. We will discuss only the first case here. The experimental schemes will differ in the ratios of the angles φ and φ' between the reflecting plane and the boundary planes. Finally, two ray paths are possible in each scheme, which differ in the mutual directions of the vectors:

$$(\underline{k}_h\underline{k}_o) \ , \ (\underline{n}_o\underline{n}_d) \quad . \tag{3.104}$$

This condition is equivalent to reflections from the two opposite sides of a reflecting plane. The schemes are shown in Figs.3.14 and 3.15.

Derivation of (3.101) in an explicit form, i.e., calculation of the modulus of the vector difference on the right-hand side of (3.103), is facilitated by Fig.3.16.

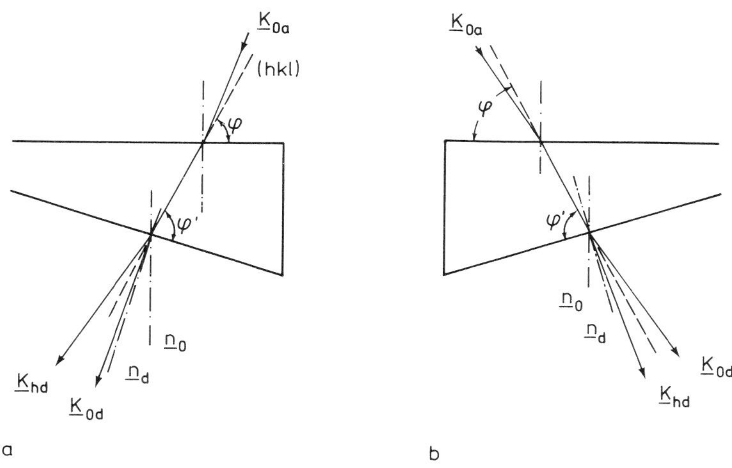

Fig. 3.14 a and b Schemes of transmission through a wedge-shaped plate: $\psi_h' = \psi_h - \mu$ a) $\psi_h = \psi_o + 2\vartheta$ reflections hkl; b) $\psi_h = \psi_o - 2\vartheta$ reflections $\bar{h}\bar{k}\bar{l}$

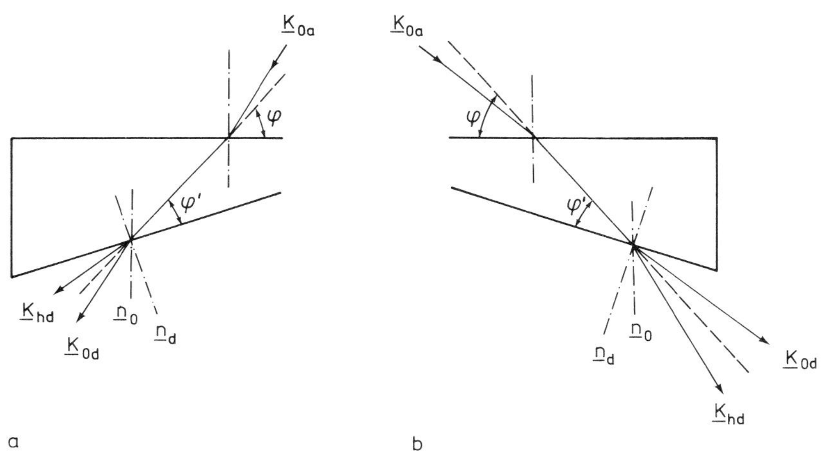

Fig. 3.15 a and b Schemes of transmission through a wedge-shaped plate: $\psi_h' = \psi_h + \mu$ a) $\psi_h = \psi_o + 2\vartheta$ reflections hkl; b) $\psi_h = \psi_o - 2\vartheta$ reflections $\bar{h}\bar{k}\bar{l}$

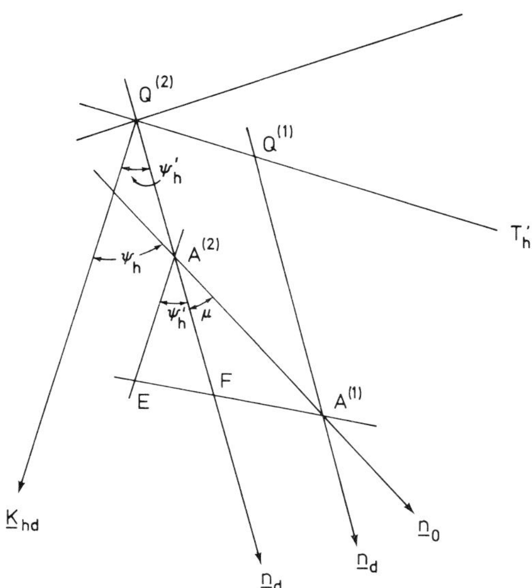

Fig. 3.16 For the derivation of (3.106)

Case $\psi_h = \psi'_h + \mu$.
In this figure:

$$\overline{A^{(1)}A^{(2)}} = |\Delta \underline{k}|, \ \overline{A^{(2)}F} = |\Delta \underline{k}| \frac{\gamma_h}{\gamma'_h} \quad ,$$

$$\overline{A^{(1)}F} = \overline{Q^{(2)}Q^{(1)}} = |\Delta \underline{K}_{hd}|, \quad \psi_h = \psi'_h + \mu \quad .$$

The value of the segment $\overline{A^{(1)}F}$ is determined from the triangle $A^{(2)}A^{(1)}F$:

$$\overline{A^{(1)}F} = |\Delta \underline{k}| \sqrt{1 + \frac{\gamma_h^2}{\gamma_h'^2} - 2\frac{\gamma_h}{\gamma'_h}\cos\mu} \quad ; \tag{3.105}$$

hence, substituting the value of $|\Delta \underline{k}|$ from (3.37), we have

$$\Lambda_h = (|\Delta \underline{K}_{hd}|)^{-1} = \lambda \gamma'_h \sqrt{\gamma_o \gamma_h} \ \cdot \\ \cdot \left(C|\chi_h| \sqrt{1 + y^2} \sqrt{\gamma_h^2 + \gamma_h'^2 - 2\gamma_h \gamma'_h \cos\mu} \right)^{-1} \tag{3.106}$$

After substituting γ'_h for γ_h in the numerator of (3.106) we obtain Λ for the case $\psi'_h = \psi_h + \mu$. This expression is simplified in certain cases of the schemes (3.104) and that of Fig.3.14b for $\psi_h \approx \mu$, Fig.3.15b for $\psi'_h \approx \mu$.

For the schemes shown in Fig.3.14b we shall take $\gamma_h \approx \cos\mu$ and select a reflection with $\gamma_h' \approx 1$, hence

$$\Lambda_h \approx \lambda \sqrt{\gamma_o \gamma_h}\left(C|\chi_h| \sqrt{1 + y^2} \sin\mu\right)^{-1} \qquad (3.107)$$

On the other hand, for the schemes in Fig.3.15b we shall take $\gamma_h' \approx \cos\mu$ and select a reflection with $\gamma_h \approx 1$; as a result

$$\Lambda_h \approx \lambda \sqrt{\gamma_o \gamma_h}\left(C|\chi_h| \sqrt{1 + y^2} \tan\mu\right)^{-1} \quad . \qquad (3.108)$$

For small μ (3.107) and (3.108) are almost equivalent. Finally, by virtue of (3.99), from (3.108) we obtain

$$\Lambda_h = \lambda x_h^{-1} \quad . \qquad (3.109)$$

Interference in a wedge plate will be discussed in Chapter 6 in connection with the spherical-incident-wave approximation and also in Chapter 9 when dealing with experimental verification of the dynamical theory. In particular, we will consider the effect of modulation of the period Λ_h due to the beats of two oscillations corresponding to two states of polarization and, hence, to the two values of C in (3.106-108).

For considering the propagation of X-rays through a crystal slice, in this chapter we proceeded from a model which included two important conditions: the incident wave is plane monochromatic, and the width of its front is large compared with the plate thickness. In many cases, however, the situation is different. The incident radiation is actually represented by a bundle of wave vectors in an angular interval exceeding the angular width of the dynamical maximum, and the span of the front of the incident radiation at the entrance surface is much less than the plate thickness. A detailed discussion of the wave field in a crystal and of the vacuum waves emerging through the exit surface in this incidence scheme is given in Chapter 6.

Here we will restrict ourselves to some relevant remarks.

In Sections 3.1-3.3 we considered the conditions in which both the fields in the crystal overlap. In Section 3.3 we dealt with a wave field in a wedge-shaped plate, and the boundary conditions at the exit surface were formulated for each field separately, although the wave fields within the plate remained unseparated. This resulted from the fact that the front of the incident vacuum wave was assumed wide as compared with the plate thickness. A different situation will be observed during the incidence of a wave with a narrow

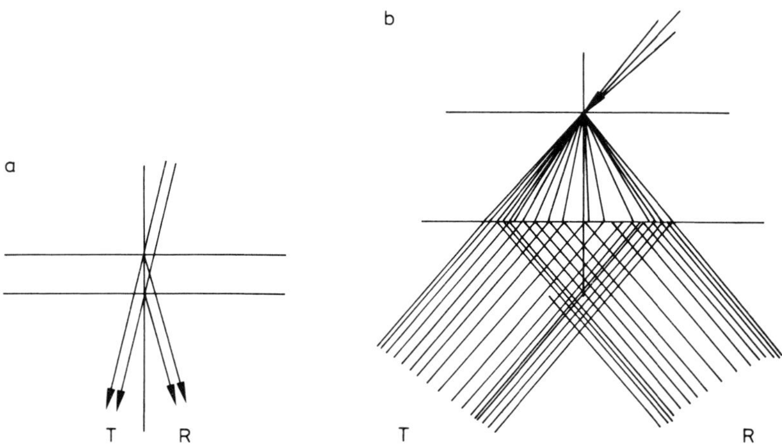

Fig. 3.17 a and b Approximate schemes of the Laue reflection under conditions other than for the case of incidence of a plane wave with a wide wave front. See text for details

front, for instance, through a slit. If one could, in this case, ensure parallelism of the wave vectors across the slit, i.e., assume, as before, a strictly definite angle of incidence, separation of two wave fields within the crystal would be expected; waves with a different state of polarization would separate in each of them. At the exit surface of the crystal, each field must make its own contribution to both the transmitted and diffracted wave (see Fig.3.17a). The separation of fields inside a thick crystal results from the fact that the propagation of X-ray energy in a crystal, as in the case of crystal optics of visible light, takes place along the directions of Poynting's vectors, which do not coincide with those of the wave vectors. These problems are treated in detail in Chapter 5.

In reality, however, the use of slits leads to the formation of a ray fan. As a matter of fact, a beam of rays from an X-ray tube has a similar structure. In both cases, it is the spherical wave, rather than the plane wave that is the best approximation, as will be demonstrated in Chapter 6. Here we will draw the reader's attention to the fact that since the incident beam contains all the directions or all the angles of incidence simultaneously, at least within the range of maximum, a wide fan of rays or directions of propagation of X-rays is formed in the crystal (Fig.3.17b).

At the same time, the pictorial models shown in Fig.3.17 must be regarded as particular cases of incidence at a crystal of a wave with an arbitrary front width which corresponds to the more general concept of a wave packet. This concept is discussed in the generalized theory on the basis of the Takagi equations and is briefly described in Chapter 11 and also, partly, in Chapters 7 and 10.

4. X-Ray Scattering in Absorbing Crystal. Laue Reflection

The scattering theory described in Chapters 2 and 3 refers to a transparent crystal without absorption. This approximation is reasonable and is in good agreement with the experimental data, provided the following conditions are met. The crystal scatterer must be sufficiently thin and contain no atoms of heavy elements, and the scattered radiation must have a small wavelength compared with those corresponding to the absorption edges of the elements forming the crystal. A typical example is the scattering of Ag $K\alpha$ and Mo $K\alpha$ radiation in thin slices of a silicon crystal on which the effects associated with the Pendellösung fringe solution were studied (see Chaps. 3,6,9). It is obvious that transitions to the wide range of phenomena accompanying the transmission of X-rays of any wavelength through thicker crystals means that the various elastic and inelastic scattering channels must be taken into account. The most important among them are elastic, coherent scattering, and photoelectric absorption.

We shall first consider the interaction of radiation with separate atoms.

4.1 Atomic Scattering and Absorption

Elastic channels of atomic scattering include, as the basic value, the atomic amplitude f_0, representing elastic coherent scattering from the electron shell of the atom, which is independent of the wavelength of the incident radiation. An allowance for resonance effects or dispersion adjusts the value of f by adding the dispersion correction Δ_f, whose modulus and sign depend on the difference $\lambda - \lambda_K$, where λ is the wavelength of the incident radiation, and λ_K is the wavelength for the absorption discontinuity of the given element.

While dispersion can either increase or decrease the atomic amplitude f_0, another effect - inelastic diffuse scattering due to thermal oscillations - only decreases this amplitude. The thermal disturbance of the lattice symmetry corresponding to temperature 0 K is covered by the familiar Debye-Waller factor.

We shall now pass on to a brief description of the important inelastic scattering channels. This is primarily photoelectric absorption μ^{Ph}, whose contribution is related to the well-known absorption curve as a function of λ, or frequency, and is characterized by discontinuities or bands for the K-,L-,M- and subsequent shells of the atom. If we consider $\mu^{Ph}(\lambda)$, the values of this parameter will obviously decrease for atoms of light elements and radiations of a wavelength either much less than λ_K (for the K-shell), or corresponding to a small displacement towards the larger λ, to the right of the absorption band edge. Here we imply the values of the normal, linear coefficient of absorption μ.

In most cases, the conditions of dynamical scattering are such that the parameters of other mechanisms of inelastic scattering are at least two orders less than the values of μ^{Ph}. Nevertheless, as will be explained in more detail later on, two such effects should be taken into consideration even at the present level of experiment. These are the Compton effect and inelastic scattering from phonons. Thus, the total value of the normal, linear coefficient of absorption can be written as:

$$\mu = \mu^{Ph} + \mu^{CS} + \mu^{TDS} \quad . \tag{4.1}$$

In this section we consider scattering and absorption on separate atoms, which presupposes coherent relations between the waves and additivity of the absorption terms only within a single atom. When passing over to dynamical scattering from a crystal, however, account must be taken of the phase relations between the radiation scattered from different atoms; this, as is well known, is actually done within a single unit cell with the aid of the expression for the structure amplitude. Considering phenomenologically the absorption effect as due to the phase shift by π, include in the value of the scattering amplitude the terms corresponding to absorption, using the complex form of the structure amplitude and, hence, of polarisation:

$$\chi = \chi_r + i\chi_i \quad . \tag{4.2}$$

Accordingly, for the atomic amplitude we write:

$$f = (f_0^{T=0} + \Delta f)\exp(-M) + if_i \tag{4.3}$$

where f_i is determined by the absorption effects. The explicit form of this function, as well as the temperature dependence, is considered below.

The numerical values of the magnitudes indicated, including the Debye temperature θ_D, are given in some publications, such as the International Tables [4.1]. However, these values, which are sufficiently accurate for use in X-ray analysis, do not meet the requirements of modern-precision experiment in the field of dynamical scattering in near-perfect crystals. The following discussion deals with an analysis of the data available.

As regards the value of $f_0^{T=0}$, the most accurate values of this atomic amplitude are, by present standards, those calculated by the Hartree-Fock method (below we use the abbreviation HF), which has become available due to advances in computer techniques. When using the HF method, each electron is assumed to move in "a self-consistent field" induced by the nucleus with all the other electrons, the exchange interaction of the electrons in the atom being taken into account. The fundamentals of the HF method are explained in [2.1] (see also papers by CLEMENTI [4.2] and DOWSON [4.3,4]). The problem of including relativistic corrections in calculation of atomic amplitudes was discussed by CROMER [4.5], who pointed out the inconsiderable influence of the relativistic effect for elements with atomic numbers of less than $Z < 40$.

Later, however (in 1967-1968), a paper by CROMER (and also that of CROMER and WABER) [4.5] was contested and in 1968 DOYLE and TURNER published a paper [4.6] containing tables of atomic amplitudes calculated on the basis of relativistic wave functions according to HF, and then RHF.

A comparison of the data given in [4.6] and [4.3] for Si shows the differences within 0.01 to 0.3 percent for a set of reflections up to 555. In this range $\sin\vartheta/\lambda$ RHF > HFCL (HFCL according to CLEMENTI [4.2]). For Ge, the divergences $f_{RHF} - f_{HFCL}$ increase gradually from ~ 0.1 percent at $\sin\vartheta/\lambda = 0.15$ Å^{-1} to ~ 0.75 percent at $\sin\vartheta/\lambda = 0.9$ Å^{-1} ([4.6] and [4.3]).

However, the above-mentioned calculation methods referred to "free" atoms, whereas the electron density of atoms in crystals, and above all the valent electrons, are redistributed as a result of interaction with their closest neighbours (chemical bond). Theoretical consideration of such perturbation of electron shells in crystals is a separate problem in its own right. For us, it is the effect of this perturbation of f_0 ($\sin\vartheta/\lambda$) that is essential here. Without dwelling on the earlier literature, we shall mention a number of publications by DOWSON [4.3,4,7], who refers to the pioneer work of KONOBEJEVSKII [4.8] and also to the surveys in [4.9]. DOWSON's analysis consists in introducing into the standard equations for structure amplitudes F_{hr} parameters of the type F_c, f_a, T_c, T_a, which are, respectively, symmetrical and asymmetrical components of the atomic amplitudes, as well as similar components of the temperature factors. The asymmetrical parameters indicated are related to certain harmonics of the series expansions of electric density functions in the theory of von der Lage and Bethe.

Because of the above-mentioned complexity of theoretical calculation of problems related to the electron density distribution, Dowson's theory is actually of greatest interest in interpreting experimental data.

It is obvious that the electron density perturbation indicated refers mainly to valence shells and therefore decreases both with the increasing atomic number Z and with the transition to remote reflections of high order (increase in $\sin\vartheta/\lambda$).

It is clear from the foregoing that, along with the theoretical values of the atomic amplitudes f_0, experimental data are quite essential, particularly for light- and medium-weight elements at small $\sin\vartheta/\lambda$ and for the other elements over the entire reflection range.

Passing on to dispersion corrections, we note that, at present, published works mainly use the values of Δf calculated by CROMER [4.10]. This calculation is based on semiclassical conceptions of the scattering atom as a set of oscillators, which are comprehensively described in JAMES' book [1.9]. In the quantum theory, to the oscillator number or strength g_s with a given, natural (circular) frequency there correspond the probabilities of the respective transitions. We can write

$$\Delta f = \sum_s \int_{\omega_s}^{\infty} \frac{\omega^2 (dg/d\omega)_s}{\omega_i - \omega} d\omega \quad , \tag{4.4}$$

where $dg/d\omega$ is the oscillator density, and ω_i is the frequency of the incident radiation. Determining the oscillator strengths g_s, which can be related to the absorption coefficient μ, involves considerable difficulty in calculation

$$g_s = \frac{mc}{2\pi^2 e^2} \cdot \frac{\omega_s}{n - 1} \mu(\omega_s) \quad . \tag{4.5}$$

The function $\mu(\omega_s)$ cannot be determined accurately enough from experiment, which fact is emphasized by the author of [4.10]. The tables of Δf values given in the paper indicated cover elements with numbers from 10 to 98 for conventional radiations (see also [4.11,12]).

The imaginary part of the atomic amplitude in (4.3) stands for the absorption of a set of inelastic scattering mechanisms. The connection between the f_i values and the linear coefficient of absorption μ is established in a very simple way by using the refractive index n, which appears in (2.62).

This expression refers to the incidence of a vacuum wave beyond the maximum. Expressing the electron density ρ as

$$\rho = eN_e = eNf_{o,o} \tag{4.6}$$

where N_e is the number of electrons, N is the number of atoms per unit volume, and $f_{o,o}$ is the atomic amplitude of scattering in the direction of the incident wave, we rewrite (2.62):

$$n = 1 - \frac{e^2}{2mc^2}\frac{\lambda^2}{\pi} N\,(f_{or,o}+if_{oi,o}) = n_r + in_i \quad . \tag{4.7}$$

This expression reflects the above-mentioned nonconversion to unity of the imaginary part of the phase factor in transition to intensities (see (2.58)):

$$\exp 4\pi\,(Kn_i,r) = \exp(-\mu r) \quad . \tag{4.8}$$

We now introduce the value of atomic absorption μ_a, which has a dimension of cm^2 and is also called the atomic absorption cross section:

$$\mu = \Omega^{-1}\sum_j \mu_{a,j}; \quad \mu_a = \frac{e^2}{mc^2}\,2\lambda f_{oi,o} \quad ; \qquad f_{oi,o} \equiv f_{oi} \quad . \tag{4.9}$$

In the subsequent text, when considering scattering in the maximum region, we shall use the symbol f_{oi} for the imaginary part of the atomic amplitude of the refracted wave and, respectively, f_{hi} for a diffracted wave.

The relations between the values of f_{oi} and μ given in (4.7-9) refer to normal absorption. In transition to dynamic scattering, n_i values of the type (3.20) or the imaginary parts of complex accomodations (Anpassung) $\delta_i^{(i)}$ (see (3.7)) must be used. In the equations for T and R, additional terms appear in the attenuation factor as a result of interference (see (4.99)). For the second field, we observe anomalous transmission and the corresponding anomalous coefficient of absorption because of the abrupt decrease in the normal coefficient of absorption:

$$\mu_{an} = \mu_{an}^{Ph} + \mu_{an}^{CS} + \mu_{an}^{TDS} \quad . \tag{4.10}$$

When passing on to shortwave radiation and light elements, the relative contribution of μ^{Ph} to (4.10) decreases by 25 to 35 percent as compared with (4.1) [1.43].

Theoretical calculations of μ^{Ph} have been made in a number of investigations. We shall mention Hönl's method described and developed to the numerical calculation stage by JAMES [1.9] and used by many authors. Among the more recent works we point out Wagenfeld's calculations, whose main results are given here together with the numerical calculations from the most recent publication [1.38]. Wagenfeld used the hydrogen-like approximation. In other words, the real atom is replaced by a one-electron model, and the force field of all the inner electrons is taken into account by the so-called Slater's screening constant. Choosing the x-axis in the direction of propagation of the scattered wave, and the y-axis parallel, respectively, to the vector $\underline{D}_\sigma$ or $\underline{D}_\pi$, we can write the following expression.

$$\mu_a = \frac{4\pi e^2\hbar^2}{m^2c^2K}\sum_\rho\sum_{\rho'}\left|\langle\rho'|\,\frac{\partial}{\partial y}\ \exp iKx|\rho\rangle\right|^2 \quad ; \quad K = \frac{2\pi}{\lambda} \quad . \tag{4.11}$$

In this expression, summation is performed over quantum numbers $\rho = n,\ell,m$ of the ground state and $\rho' = n',\ell',m'$ of the excited state of the atom. $n' = (Z-s)/ia_0k$, where Z is the atomic number, s is Slater's shielding constant, and k is the modulus of the wave vector of knocked-out electrons.

Through expansion of the exponent in (4.11) the problem is approximated by a series corresponding to the superposition of scattering by a dipole, quadrupole, octupole, and so on, i.e., to a definite angular distribution of the knocked-out electrons. The series indicated has the form:

$$\exp iKx = 1 + iKx - \frac{1}{2}K^2x^2 \ldots \quad . \tag{4.12}$$

Denoting the quantum numbers of the orbital angular momenta in the initial state by ℓ and in the excited state by ℓ', we use the familiar transition rules:
for a dipole $\ell \to \ell' = \ell \pm 1$; for a quadrupole $\ell \to \ell' = \ell + 2$, $\ell' = \ell$;
for an octupole $\ell \to \ell' = \ell + 1$, $\ell \pm 3$; ($\ell' = \ell$ is forbidden).
Further we introduce, in the notation of absorption cross sections, the symbol of the electron shell as a subscript and multipole as a superscript. The corresponding equations, given by Wagenfeld, have the form

$$\mu^D_{a,K} = \frac{2^7}{3}\pi r_0\lambda\left(\frac{\lambda}{\lambda_K}\right)^3\frac{\exp(-4z_1/\cot z_1)}{1-\exp(-2\pi z_1)} \quad ;$$
$$\mu^Q_{a,K} = \frac{2}{5}\mu^D_{a,K}(\lambda_c/\lambda)(4-3\lambda/\lambda_K);\ \mu^{D,0}_{a,K} = \frac{2}{5}\mu^D_{a,K}(\lambda_c/\lambda)(1-2\lambda/\lambda_K) \quad . \tag{4.13}$$

Here:

λ_K is related to the hydrogen-like proper value of E_K:

$$\lambda_K = [(Z-s_K)^2 R_\infty]^{-1} \quad ; \quad z_1 = [\lambda/(\lambda_K-\lambda)]^{1/2} \quad . \tag{4.14}$$

$R_\infty = 2\pi^2 mc^4/h^3\, c = 109737.3$ is the Rydberg constant;
$r_o = e^2/mc^2 = 2.8179 \cdot 10^{-13}$ cm; $\lambda_c = h/mc = 24.263 \cdot 10^{-11}$ cm is the Compton wavelength; $s_K = 0.30$ is Slater's shielding constant for the K-shell; Z is the atomic number, and λ_K is the wavelength of the radiation absorbed. The symbols D, Q and D, O denote, respectively, dipole, quadrupole, and mixed dipole-octupole absorption. The total atomic absorption cross section on the K-shell is expressed as the sum of these components.

The total absorption cross section on the L-shell is expressed by the relation [4.13]

$$\mu^D_{a,L} + \mu^Q_{a,L} + \mu^{D,O}_{a,L} = \mu_{a,2s} + \mu_{a,2p} = \frac{2^{11}}{3}\pi r_o \left(\frac{\lambda}{\lambda_L}\right)^2 \cdot$$

$$\cdot\left(\left(1+3\frac{\lambda}{\lambda_L}\right)\left\{1 + 2\frac{\lambda}{\lambda_c}\left[1 - 2\frac{\lambda}{\lambda_L} + \frac{4}{5}\left(\frac{\lambda}{\lambda_L}\right)^2\right]\right\} + \right. \tag{4.15}$$

$$\left. + \left(\frac{\lambda}{\lambda_L}\right)^2 \left\{3 + 8\frac{\lambda}{\lambda_L} - \frac{2}{5}\frac{\lambda}{\lambda_c}\left[25 + 56\frac{\lambda}{\lambda_L} - 52\left(\frac{\lambda}{\lambda_L}\right)^2\right]\right\}\right)\frac{\exp(-4z_2/\cot\frac{1}{2}z_2)}{1-\exp(-2\pi z_2)} .$$

$$\lambda_L = 4(Z-s_L)^2 R_\infty \quad , \quad z_2 = 2[\lambda/(\lambda_L-\lambda)]^{1/2} \quad . \tag{4.16}$$

Similar expression are given in [4.13] for the atomic absorption cross sections on the M- and N-shells; the contribution of absorption on these levels to the total value of μ_a for light and medium elements is negligibly small.

In numerical calculation of μ_a from the above expressions, of great importance, along with the exact values of the wavelengths of the absorbed radiation and the edge of the absorption band ($\lambda_K, \lambda_L, \lambda_M$), is the adopted value of Slater's screening constants, which ensures the applicability of the hydrogen-like approximation. Ref. [1.38] includes a table of these values obtained from spectroscopic data.

Using the method described for Ge, for absorbed radiation wavelengths lying on both sides of the K-absorption edge of this element (λ_K=1.1166 Å), we shall find it impossible to use (4.13). Indeed, the value E = 13.67 keV

Table 4.1 Slater's screening constants and absorption edge energies: E_k is the exact value; E_{1s} is the value adopted in the hydrogen-like approximation

	Slater screening constants				Absorption edges (keV)	
	1s	2s,2p	3s	3p,3d	E_k	E_{1s}
Si	0.30	3.25	8.20	-	1.839	2.554
Ge	0.30	3.41	8.62	12.86	11.10	13.67

corresponds to $\lambda_K = 0.9068$ Å, which leads, for radiations with $\lambda > \lambda_K$, to imaginary values of z, according to (4.14). The authors of [1.38] obviate this difficulty by neglecting the exponent in the denominator of the expression for z and using the moduli of the values $(\lambda_K-\lambda)$ in the numerator of the fraction in (4.13). Nevertheless, for the radiation indicated we observe a greater discrepancy between theory and experiment (by ~ 6 percent for μ).

It is further obvious that a change from μ_a to the linear coefficient of absorption μ^{Ph} is achieved with the aid of (4.9); in contrast to μ_a, μ has a dimension of cm^{-1}.

The authors of [1.38] give the values of $\mu_a(\equiv\sigma)$ of the separate shells and components according to their multipole calculated in line with the above-described theory, as well as the total values of photoelectric absorption μ^{Ph} for Si and Ge and 12 different characteristic radiations. Moreover, they give the ratios μ^Q/μ essential for subsequent analysis of the parameters of dynamical scattering in absorbing crystals.

The theory of Compton scattering in the dynamical regime was developed by SANO et al. [4.14]. They give graphs of the numerical values of the contribution of Compton scattering to the normal and minimal (for y_r=0) coefficients of absorption, as a function of two parameters - the temperature and the wavelength λ of the incident radiation. The plots have been calculated for a number of reflections from Al, Si, Ge, Ag, and Pb. As might be expected, the contribution of Compton scattering increases with the decreasing serial number of the element and wavelength λ. This contribution decreases only slightly when the temperature increases. These results are partly rendered by the following approximate expression for μ_{an}^{CS} obtained by GIARDINA and MERLINI [1.43]:

$$\mu_{an}^{CS} \approx \frac{8\pi N r_0^2}{3} \frac{\lambda^2}{(\lambda-\lambda_c)^2} [Z - f_h \exp(-M)] \tag{4.17}$$

where r_o is the classical radius of the electron, and λ_c is the Compton wavelength. The authors of [1.43] calculated, according to (4.17), the values of μ^{CS} and μ_{an}^{CS} for Si and Ag Kα and Mo Kα radiations. They are equal to about 0.5 and 0.2 percent of μ^{Ph} and 67 and 32 percent of μ_{an}^{Ph}, respectively, for Ag Kα and Mo Kα radiations. μ_{an} are related to the temperature of 295 K and reflection (220).

According to the graphs of [4.14], respective values for Ge and for Mo Kα radiation are $\sim$ 0.2 and $\sim$ 2.5 percent.

Passing on to thermal diffusion scattering, we must emphasize the complexity of the corresponding theory and the difficulty of finding reliable numerical values for the contribution of TDS to the coefficients of absorption.

The most rigorous and comprehensive consideration of the contribution of TDS to the attenuation of the refracted and diffracted waves is given in the theory of AFANASIEV and KAGAN [1.39]. They analyse the two components of TDS - coherent and noncoherent inelastic scattering from phonons.

If one takes into account the nature of the temperature dependence of different contributions to absorption weakening the wave field in a crystal, the various scattering mechanisms, or the total value of the coefficients μ and μ_{an} can be subdivided into two groups of the type

$$\mu_{an} = \mu_{an,o} + \mu_{an,T} \quad . \tag{4.18}$$

The temperature dependence of the first term, which includes photoelectric absorption and Compton scattering, is determined by the Debye-Waller factor, while that of the second, due to inelastic scattering from phonons, is of a more complex nature. Accurate calculation of the values of $\mu_{an,T}$ requires knowledge of the phonon spectrum of the crystal and, in particular, of the vectors of polarization and normal oscillations throughout the range of wave vector determination. In a number of cases of dynamical scattering important from a practical point of view, however, it was possible to write relatively simple approximation equations for the corresponding numerical calculations. These calculations use only the values of the atomic amplitudes, of the Debye-Waller factors, and the elastic constants of the crystal. In this connection we shall mention the investigations of EFIMOV [4.15], KOHN [4.16], and CHEZZI et al. [4.17].

KOHN [4.16] gives graphs of the temperature dependence of the values of $\mu_{an,T}$ normalized to unity at T = 0 for Si, Ge, Al, Ni, Cu and InSb. These values increase linearly with the temperature. At the same time, as shown by another graph given by the same author, the ratio $\mu_{an,T}/\mu_{an,0}$ increases up to ~ 200 K, after which it remains approximately constant. The numerical estimates of this ratio given in [4.15-17] differ considerably. Some remarks made by KOHN as to the favourable conditions for the observation of the contribution of TDS to anomalous absorption, for instance for reflection of Cu Kα from crystals of Ni,Cu, of Mo Kα from Zr and some others, need verification by staging new, more precise experiments.

Over the last few years, the linear coefficients of absorption μ for Si and Ge have in several cases been determined quite precisely. Here we shall mention the work of GRIMVALL and PERSSON [4.18], whose numerical results are given in [4.19] in a form convenient for use, then the above-cited work of three authors [1.38], and also a number of experimental data given in the paper of GIARDINA and MERLINI [1.43], to which we shall revert later.

The accompanying Table 4.2 for Ge mainly uses experimental determinations made in [1.38] and the theoretically calculated values of $\mu, \mu^{D} + \mu^{D0}$, and μ^{Q}, which are also given in [1.38]. Note that experimental values obtained in [1.38,4.18] for Ge diverge by 0.8 percent for most radiations, except those of Fe Kα,Co Kβ, and Co Kα, for which the difference is ~ 4 percent. For Si, the paper [1.43] is used along with [1.38].

Table 4.2 Experimental values of the linear coefficient of absorption μ, theoretical values of the coefficients of photoabsorption $\mu^{Ph}, \mu^{D} + \mu^{D,0}$, μ^{Q}, and ratios μ^{Q}/μ^{Ph} for Si and Ge in cm^{-1}

Characteristic radiation	λ[Å]	μ_{exp}	μ^{Ph}	$\mu^{D}+\mu^{D,0}$	μ^{Q}	μ^{Q}/μ^{Ph}
		Si				
Ag $K\alpha_1$	0.558	7.39	6.92	6.51	0.41	0.59
Mo $K\alpha_1$	0.708	14.64	14.16	13.56	0.66	0.046
Cu $K\alpha_1$	1.538	144	142.4	139.7	2.72	0.019
Fe $K\alpha_1$	1.932	275	276.5	272.6	3.91	0.014
Cr $K\alpha_1$	2.285		447.3	442.3	4.99	0.013
		Ge				
Ag $K\alpha_1$	0.558	168	170.5	163.9	6.65	0.039
Mo $K\alpha_1$	0.708	319	325.3	317.0	8.25	0.025
Cu $K\alpha_1$	1.538	353	334.2	326.5	7.65	0.023
Fe $K\alpha_1$	1.932	677	645.8	634.3	11.39	0.018
Cr $K\alpha_1$	2.285	1070	1045.8	1030.8	15.03	0.014

In comparing the experimental and theoretical values of μ, one should remember that the latter take into account only the contribution of photoelectric absorption (μ^{Ph}). From this point of view the data referring to Ge are still insufficiently accurate, because for Ag Kα, and Mo Kα, for which the contribution of other absorption mechanisms must be more pronounced, $\mu_{th}^{Ph} > \mu_{exp}$ by 1-2 percent. The considerable excess of μ_{exp} over μ_{th}^{Ph} for Ge is evidently due to the above-mentioned limitation of the hydrogen-like approximation. For Si, the relations indicated correspond to those expected. If we use the average values of μ_{exp} and μ_{th} given in [1.43,38], we can see that for Mo Kα_1 and Ag Kα_1 radiations the values of $\mu_{exp} - \mu_{th}^{Ph}$ are about an order higher than the deviations from the average values.

4.2 Complex Form of Dynamical Scattering Parameters [1.16,4.20]

Let us now turn to the dynamical parameters of scattering in ideal absorbing crystals.

In this case the values of polarizability and the coefficients of its Fourier expansion become complex:

$$\chi = \chi_r + i\chi_i \quad , \quad \chi_o = \chi_{or} + i\chi_{oi} \quad , \quad \chi_h = \chi_{hr} + i\chi_{hi} \quad . \tag{4.19}$$

The wave vectors of waves in the crystal also become complex.

When a vacuum wave falls beyond the reflection maxima range, the imaginary part of the wave vector will be related only to the zero component of the Fourier expansion χ, namely to χ_o (see (2.61)):

$$k = \underline{K}\left(1+\frac{1}{2}\chi_{or}+i\frac{1}{2}\chi_{oi}\right) \quad , \tag{4.20}$$

$$\underline{K}_i = \frac{1}{2}\,\underline{K}\chi_{oi} \quad . \tag{4.21}$$

On conversion to intensities, the phase factor for such a wave in a crystal transforms according to (2.58).

If the angle of incidence is inside some maximum, the wave vectors of the wave in the crystal $\underline{k}_o^{(i)}$ and $\underline{k}_h^{(i)}$ become complex through $\underline{k}_o^{(i)}$ or $\underline{k}_h^{(i)}$. In accordance with (3.1) we can write

$$\underline{k}_o^{(i)} = \underline{K}_o^{(a)} - K\delta_r^{(i)}\underline{n}_o - iK\delta_i^{(i)}\underline{n}_o \quad . \tag{4.22}$$

A similar relation can be written for the vector $k_h^{(i)}$; as follows from (3.7), the vector $\underline{k}_h^i$ also becomes complex because of the function $\delta^{(i)}$. Thus

$$\underline{k}_h^{(i)} = K\delta_i^{(i)} \underline{n}_o = \underline{k}_{oi}^{(i)} \quad . \tag{4.23}$$

Therefore, in contrast to $\underline{k}_{or}$ and $\underline{k}_{hr}$, the vectors $\underline{k}_{oi}$ and $\underline{k}_{hi}$ are directed along the normal $\underline{n}_o$ towards the entrance surface. Comparing (4.8), (4.20) and (4.21), we can write

$$\chi_{oi} = -\frac{\mu}{2\pi K} \quad . \tag{4.24}$$

We shall now turn to (4.19) and use the analogy between the amplitudes of scattering f_r anf F_{hr} and the absorption f_i and F_{hi}.

For the real part of χ and χ_h we can write the obvious expressions

$$\chi_r^{at} = \frac{e^2}{mc^2} \frac{\lambda^2}{\pi V} \int_{at} (-\frac{\rho}{e}) \exp 2\pi ik\,(s-s_{o,r})\, d\tau = -\frac{e^2}{mc^2} \frac{\lambda^2}{\pi V} f_r \tag{4.25}$$

(where f_r must include both the dispersion correction and the Deby-Waller factor),

$$\chi_{hr} = \frac{1}{\Omega} \int_{cell} \chi_r \exp 2\pi i\,(\underline{r}\underline{h})\, d\tau = \frac{e^2}{mc^2} \frac{\lambda^2}{\pi\Omega} F_{hr} \quad . \tag{4.26}$$

At h = 0 we obtain the expression (2.35). Similarly to (4.25) we can write for χ_{hi} the following expression by formally introducing the imaginary atomic amplitude f_i:

$$\chi_i^{at} = -\frac{e^2}{mc^2} \frac{\lambda^2}{\pi V} f_i \quad , \tag{4.27}$$

$$\chi_{oi} = -\frac{e^2}{mc^2} \frac{\lambda^2}{\pi} \frac{1}{\Omega} \sum_{cell} f_{oi} = -\frac{\mu}{2\pi K} < 0^{1} \tag{4.28}$$

On the other hand, from (4.9) we obtain

$$f_{oi} = \frac{mc^2}{e^2} \frac{1}{2\lambda} \mu_a \quad , \quad f_{hi} = \frac{mc^2}{e^2} \frac{1}{2\lambda} \mu_a^{(h)} \quad . \tag{4.29}$$

---

[1] The conditions: $\chi_{hi} < 0$ and $\chi_{oi} < 0$ correspond to the adopted form of the exponent in the wave function.

When a vacuum wave falls beyond the maximum, there is only normal absorption (defined by the linear coefficient μ) during which the vector nature of X-radiation does not reveal itself. When a regular reflection is formed, not only the two standard states of polarization (as in elastic scattering) become essential, but also the various components of expansion (4.12). Along with the value μ_a the atomic interference absorption cross section $\mu_a^{(h)}$ must be introduced, which is different for the two states of polarization [4.20]:

$$f_{hi}^{\sigma} = \frac{mc^2}{e^2}\frac{1}{2\lambda}\mu_a^{(h)\sigma} \quad , \quad f_{hi}^{\pi} = \frac{mc^2}{e^2}\frac{1}{2\lambda}\mu_a^{(h)\pi} \quad . \tag{4.30}$$

$$\mu_a^{(h)\sigma} = \mu_a^{D} + \mu_a^{D,0} + \mu_a^{Q}|\cos 2\vartheta| \tag{4.31}$$

$$\mu_a^{(h)\pi} = (\mu_a^{D}+\mu_a^{D,0})\ |\cos 2\vartheta| + \mu_a^{Q}\ |\cos 4\vartheta| \quad . \tag{4.32}$$

Switching to the coefficient of expansion of electric susceptibility similar to (4.28), and taking into consideration the structure and temperature factors, we can write

$$\chi_{hi} = -\ r\sum \mu_a^{(h)}\ \exp 2\pi i(\underline{h}\underline{r})\ \exp(-M) \quad , \tag{4.33}$$

$$r = (2\pi K\Omega)^{-1} \quad . \tag{4.34}$$

Thus, both here and in the following sections, for the coeffcients T and R we will have two independent series of expressions for the two standard states of polarization. In the subsequent text the subscripts σ and π are largely omitted. Replacing the exponential function by trigonometrical ones and assuming that the temperature factor is included in the values of f and $\mu_a^{(h)}$, we write down the following formal relations. For reflection hkl (j is the atomic number in the cell):

$$\chi_h = (\chi_{hr}+i\chi_{hi}) = -\Gamma\left[\sum_{cell}(f_{rj}+if_{ij})(\cos\varphi_j+i\ \sin\varphi_j)\right] \quad , \tag{4.35}$$

$$\chi_{hr} = -\Gamma \sum_{cell} (f_{rj} \cos\varphi_j + i f_{rj} \sin\varphi_j) \quad , \tag{4.36}$$

$$\chi_{hi} = -\Gamma \sum_{cell} (f_{ij} \cos\varphi_j + i f_{ij} \sin\varphi_j) =$$

$$\doteq -r \sum_{cell} (\mu_{aj}^{(h)} \cos\varphi_j + i\mu_{aj}^{(h)} \sin\varphi_j) \quad . \tag{4.37}$$

For reflection $\bar{h}\bar{k}\bar{l}$:

$$\chi_{\bar{h}} = \chi_{\bar{h}r} + i\chi_{\bar{h}i} = -\Gamma \sum_{cell} (f_{rj} \cos\varphi_j - i f_{rj} \sin\varphi_j) -$$

$$- ir \sum_{cell} (\mu_{aj}^{(h)} \cos\varphi_j - i\mu_{aj}^{(h)} \sin\varphi_j) \quad , \tag{4.38}$$

$$\Gamma = \frac{e^2}{mc^2} \frac{\lambda^2}{\pi\Omega} \quad , \quad \varphi_j = 2\pi(hx_j + ky_j + lz_j) \quad . \tag{4.39}$$

It is obvious that the values χ_{or}, and χ_{oi}, which do not contain phase factors, are real, while χ_{hr} and χ_{hi} are complex. Further, from (4.38) and (4.37) it follows that

$$\chi^*_{hr} = \chi_{\bar{h}r} \quad , \quad \chi^*_{hi} = \chi_{\bar{h}i} \quad . \tag{4.40}$$

In contrast to the parameters of a transparent crystal, in the absence of a centre of symmetry,

$$\chi^*_h \neq \chi_{\bar{h}} \quad . \tag{4.41}$$

The form of writing the complex values χ_{hr} and χ_{hi} used here is compared with another form:

$$\chi_{hr} = |\chi_{hr}| \exp i\eta_h \quad , \quad \chi_{\bar{h}r} = |\chi_{hr}| \exp(-i\eta_h) \quad , \tag{4.42}$$

$$\chi_{hi} = |\chi_{hi}| \exp i\omega_h \quad , \quad \chi_{\bar{h}i} = |\chi_{hi}| \exp(-i\omega_h) \quad . \tag{4.43}$$

A comparsion of (4.35), (4.36), and (4.30) enables one to determine the phases η_h and ω_h (the subscript j is omitted)

$$\cos\eta_h = \frac{-1}{|F_{hr}|} \sum f_r \cos\varphi \quad , \quad \sin\eta_h = \frac{-1}{|F_{hr}|} \sum f_r \sin\varphi \quad , \tag{4.44}$$

$$\cos\omega_h = \frac{-1}{|F_{hi}|} \sum f_i \cos\varphi = \frac{-1}{|\sum\mu_a^{(h)}|} \sum\mu_a^{(h)} \cos\varphi \quad , \tag{4.45}$$

$$\sin\omega_h = \frac{-1}{|F_{hi}|} \sum f_i \sin\varphi = \frac{-1}{|\sum\mu_a^{(h)}|} \sum\mu_a^{(h)} \sin\varphi \quad , \tag{4.46}$$

$$|F_{hr}| = \sqrt{(\sum f_r \cos\varphi)^2 + (\sum f_r \sin\varphi)^2} , \tag{4.47}$$

$$|F_{hi}| = \sqrt{(\sum f_i \cos\varphi)^2 + (\sum f_i \sin\varphi)^2} , \tag{4.48}$$

$$|\sum\mu_a^{(h)}| = \sqrt{(\sum\mu_a^{(h)} \cos\varphi)^2 + (\sum \mu_a^{(h)} \sin\varphi)^2} \quad . \tag{4.49}$$

It is noteworthy that $|F_{hr}|$ and $|F_{hi}|$ are dimensionless values, and according to (4.49), $|\sum \mu_a^{(h)}|$ has the dimension of μ_a, i.e., [cm^2].

We shall now consider the values of $\chi_h\chi_{\bar{h}}$ and $|\chi_h/\chi_{\bar{h}}|$, which will appear in the equations for T and R. As has already been indicated (see (4.39)), in absorbing crystals $\chi_{\bar{h}}$ and χ_h are not conjugate complex values. Using (4.37-40), we obtain

$$\chi_h\chi_{\bar{h}} = (\chi_{hr}+i\chi_{hi})(\chi_{\bar{h}r}+i\chi_{\bar{h}i}) \equiv (\chi_{hr}+i\chi_{hi})(\chi^*_{\bar{h}r}+i\chi^*_{\bar{h}i}) \tag{4.50}$$

On the basis of (4.42)

$$\begin{aligned} \chi_h\chi_{\bar{h}} = {} & |\chi_{hr}|^2 - |\chi_{hi}|^2 + i|\chi_{hr}||\chi_{hi}| \exp(-i\nu_h) + \\ & + i|\chi_{hr}||\chi_{hi}| \exp(+i\nu_h) = \Phi_h + i\Psi_h \quad , \end{aligned} \tag{4.51}$$

$$\Phi_h = |\chi_{hr}|^2 - |\chi_{hi}|^2 \; ; \quad \Psi_h = 2|\chi_{hr}||\chi_{hi}| \cos\nu_h \; ; \quad \nu_h = \eta_h - \omega_h \quad . \tag{4.52}$$

Further,

$$\left|\frac{\chi_h}{\chi_{\bar{h}}}\right| = \frac{\sqrt{(\chi_{hr}+i\chi_{hi})(\chi^*_{hr}-i\chi^*_{hi})}}{\sqrt{(\chi_{\bar{h}r}+i\chi_{\bar{h}i})(\chi^*_{\bar{h}r}-i\chi^*_{\bar{h}i})}} = \frac{\sqrt{(\chi_{hr}+i\chi_{hi})(\chi_{hr}-i\chi_{\bar{h}i})}}{\sqrt{(\chi_{\bar{h}r}+i\chi_{\bar{h}i})(\chi_{hr}-i\chi_{hi})}} \approx 1 + 2\left|\frac{\chi_{hi}}{\chi_{hr}}\right| \sin\nu_h \tag{4.53}$$

and similarly,

$$\left|\frac{\chi_{\bar{h}}}{\chi_h}\right| \approx 1 - 2 \left|\frac{\chi_{hi}}{\chi_{hr}}\right| \sin\nu_h \quad . \tag{4.54}$$

In deriving (4.53), the values $|\chi_{hi}/\chi_{hr}|^2$ were discarded as values of the second order of smallness. According to (4.44-46),

$$\cos\nu_h = \left[|F_{hr}||\sum \mu_a^{(h)}|\right]^{-1} \Delta_1 \quad , \quad \Delta_1 = \begin{vmatrix} \sum f_r \cos\varphi & \sum \mu_a^{(h)} \sin\varphi \\ -\sum f_r \sin\varphi & \sum \mu_a^{(h)} \cos\varphi \end{vmatrix} , \tag{4.55}$$

$$\sin\nu_h = \left[|F_{hr}||\sum \mu_a^{(h)}|\right]^{-1} \Delta \quad , \quad \Delta = \begin{vmatrix} \sum f_r \sin\varphi & \sum f_r \cos\varphi \\ \sum \mu_a^{(h)} \sin\varphi & \sum \mu_a^{(h)} \cos\varphi \end{vmatrix} . \tag{4.56}$$

Note that the quantities $\cos\nu_h$ and $\sin\nu_h$ are dimensionless, as they are actually supposed to be. For centro-symmetrical structures, η_h and ω_h turn into zero or π, and then $\cos\nu_h = \pm 1$, and $\sin\nu_h = 0$.

Making use of these relations, we transform (4.53,54) as follows:

$$\left|\frac{\chi_h}{\chi_{\bar{h}}}\right| \approx 1 + \left(\lambda \frac{e^2}{mc^2}\right)^{-1} \frac{\Delta}{|F_{hr}|^2} \quad , \quad \left|\frac{\chi_{\bar{h}}}{\chi_h}\right| \approx 1 - \left(\lambda \frac{e^2}{mc^2}\right)^{-1} \frac{\Delta}{|F_{hr}|^2} \quad . \tag{4.57}$$

Thus, the ratio between moduli of the two relations is determined by the sign of the determinant Δ or, according to (4.50,51) by the sign of $\sin\nu_h$.

As well as the linear coefficient of absorption μ, which refers to any direction in the crystal, below we use the linear coefficient of absorption in the direction of the normal to the entrance surface:

$$\sigma_o = + \frac{2\pi K|\chi_{oi}|}{\gamma_o} = \frac{\mu}{\gamma_o} > 0 \quad . \tag{4.58}$$

Finally, for the complex angular function β we can write

$$\beta = \beta_r + i\beta_i \quad , \quad \beta_r = 2\eta \sin 2\vartheta + |\chi_{or}| \left(1 - \frac{\gamma_h}{\gamma_o}\right) \quad ,$$
$$\beta_i = |\chi_{oi}| \left(1 - \frac{\gamma_h}{\gamma_o}\right) \quad . \tag{4.59}$$

Tables 4.2-4 list the values of exp(-M), χ_{or}, χ_{oi}, χ_{hr}, and χ_{hi} for a number of reflections of Ag $K\alpha_1$, Mo $K\alpha_1$, and Cu $K\alpha_1$ radiations from Si and Ge crystals. Both theoretical calculations and experimental results were used in obtaining these data.

The values of $f_0 + \Delta f$ necessary for calculating χ_{hr} were borrowed from various sources.

In addition to the above-indicated values of f according to HF-Clementi [72], we used new experimental results. Highly reliable and accurate measurements of the atomic amplitudes scattering from silicon crystals have been taken by ALDRED and HART [4.21] with the aid of Pendellösung fringe bands. In order to obtain these images, an X-ray beam which had passed through a narrow slit was directed at a wedge-shaped crystal. The theory and techniques for taking such measurements, first suggested by KATO, are described in Chapter 6. The accuracy of the measurements taken at near-room temperature (~ 293 K) in Ag $K\alpha_1$ and Mo $K\alpha_1$ rays, which we used in compiling Table 4.3, is estimated by the authors at ±0.1 percent. However, since the data obtained correspond to $(f_0+\Delta f)$ exp(-M), if f_0 is referred to 0 K, the conversion of the measured values to f_0, and also to the values $(f_0+\Delta f)$ exp(-M) for the Cu $K\alpha_1$ radiation evidently entails an increase in the error indicated.

Nevertheless, a comparison of these converted absolute values with those of f_0 obtained by the HF-Clementi method is interesting. This comparison reveals a discrepancy of ~ 1.8 percent for 111 and 311 to ~ 0.8 percent for 220, 400, and 331 for the first five reflections. These discrepancies undoubtedly reflect distortions in the electron density distribution of the valency shells of the atoms in the crystals.

Note that measurements of f values for certain reflections were taken during other investigations, using both the same method and other methods: measurements of the interference bands formed on beam transmission through a plane-parallel crystal with a narrow slit; of intensities of scattering from a powder-like specimen; and of the parameters of the reflection curves obtained from a highly monochromatized incident wave with negligible divergence. In some cases the divergence of different measurements is close to the above-mentioned measurement error in [4.21], i.e., +0.2 percent and less.

For atomic amplitudes of scattering from Ge crystals, reliable experimental data are more limited. As already indicated above, proceeding from general considerations one can restrict oneself to non-HF amplitude values (experimental or theoretical) only for the first five reflections. The most comprehensive measurements (of atomic amplitudes for nine reflections) were taken by MATSUSHITA and KOHRA [4.22]. A comparison of their results with

Table 4.3 Values of Debye-Waller temperature factors exp (-M) for reflection from Si and Ge crystals at temperature of 293 K. θ_D^{Ge} = 293 K; B = 0.556 Å^2. θ_D^{Si} = 532 K; B = 0.461 Å^2

hkl →	111	220	311	400	331	422	333	440	444	555
Si	0.989	0.969	0.958	0.939	0.928	0.910	0.900	0.882	0.829	0.746
Ge	0.987	0.966	0.953	0.933	0.921	0.901	0.890	0.871	0.812	0.722

Table 4.4 Values of the coefficients of the real parts of the Fourier expansion of polarizability $|\chi_{hr}|$ 10^6

	hkl	000	111	220	311	400	331	422	333	440	444	555
Si	Cu $K\alpha_1$	15.07	8.117	9.128	5.920	7.634	5.197	6.734	4.486	5.650	4.575	2.224
	Mo $K\alpha_1$	3.162	1.698	1.901	1.234	1.588	1.081	1.392	0.931	1.214	0.944	0.455
	Ag $K\alpha_1$	1.966	1.054	1.179	0.764	0.984	0.670	0.860	0.576	0.751	0.584	0.281
Ge	Cu $K\alpha_1$	28.73	17.26	20.30	13.14	16.64	11.16	14.18	9.480	12.21	9.330	4.555
	Mo $K\alpha_1$	6.391	3.866	4.595	2.983	3.80	2.554	3.27	2.20	2.845	2.222	1.120
	Ag $K\alpha_1$	3.994	2.452	2.93	1.89	2.42	1.622	2.078	1.394	1.810	1.41	0.725

the theoretical values of RHF [4.6] show divergences for the first five reflections exceeding similar divergences in the case of Si. This is physically unacceptable, because the relative contribution of the valency electrons to the scattering by Ge atoms should be substantially less. At the same time, the measurement errors indicated by the authors of [4.22] - 0.25 to 0.6 percent - actually increase in connection with the adopted values of the dispersion correction and the temperature factor. Therefore we have corrected the values of f_0 for the first five reflections in such a way that the deviations from the RHF values are less than the respective values for Si, and adopted for the other five reflections the theoretical values of f_0 according to RHF.

Measurements of atomic scattering amplitudes for Ge (reflections 111 and 220) have also been taken by some other authors [4.23-27]. The spread of the values obtained may reach 2.5 percent.

Turning to estimations of errors arising on certain corrections of the values of f obtained by accurate theoretical or experimental methods, we emphasize that for the dispersion correction Δf this is a very difficult task. As for the temperature factors exp(-M) or, more precisely, the Debye temperature θ_D, these values have been analyzed in a number of recent investigations. Here we shall mention the work of BATTERMAN and CHIPMAN [4.28] and, in particular, the most recent papers [1.38,4.3,4] already quoted, as well as those by LUDEWIG [4.29] and GIARDINA and MERLINI [1.43]. The temperature variation for Ge and Si has obviously been calculated with the utmost reliability in the theoretical paper of SALTER [4.30]. LUDEWIG has demonstrated that the difficulty of determining the temperature parameters experimentally is due to the need to take into account the contributions of TDS and CS to absorption.

In our calculations, we assumed, for Si, θ_D = 532 K and B = 0.461 $Å^2$. For Ge, θ_D = 293 K (see [4.29]) makes it possible, in particular, to avoid inaccurate interpolation procedure in determining $\Phi(x)$ and, accordingly, B = 0.556. Note that switching from the indicated values of the parameters for Ge to θ_D = 290 K leads to a divergence of the values $(f_0 + \Delta f)$ exp(-M) by $\sim$ 0.45 percent for reflection 444 (Table 4.3).

Now we pass on to the values of χ_{hi} listed in Tables 4.5a and 4.5b. It can be seen from (4.31-33) that theoretical calculation of χ_{hi}^{σ} requires a knowledge of the values of μ^D (or $\mu^D + \mu^{D,0}$) and μ^Q. It should be kept in mind that there is a considerable difference in the variation of the values χ_{hi}^{σ} and χ_{hi}^{π} throughout the series of reflections from 111 to 555. So, according to (4.31) and (4.33) for the face centered lattices of the elements, the main value determining absorption (μ_a^D) is multiplied only by the structure factor and exp(-M); then for χ_{hi}^{σ} we have a small, gradual decrease for

Table 4.5a Values of the coefficients of the imaginary parts of the Fourier expansion of polarizability $\chi_{oi} \cdot 10^7$ and $\chi_{hi}^{\sigma} \cdot 10^7$

	hkl →	000	111	220	311	400	331	422	333	440	444	555
Si	Cu $K\alpha_1$	3.523	2.456	3.398	2.366	3.269	2.279	3.148	2.195	3.033	2.812	1.754
	Mo $K\alpha_1$	0.165	0.1150	0.1589	0.1109	0.1536	0.1078	0.148	0.103	0.143	0.138	0.085
	Ag $K\alpha_1$	0.065	0.0453	0.0625	0.0456	0.0608	0.0424	0.058	0.041	0.0519	0.053	0.033
Ge	Cu $K\alpha_1$	8.637	6.012	8.282	5.769	7.938	5.535	7.601	5.295	7.299	6.702	
	Mo $K\alpha_1$	3.593	2.449	3.462	2.418	3.337	2.332	3.215	2.243	3.099	2.881	1.810
	Ag $K\alpha_1$	1.493	1.041	1.437	1.004	1.380	0.9675	1.333	0.931	1.282	1.196	0.752

Table 4.5b Values of the coefficients of the imaginary parts of the Fourier expansion of polarizability $\chi_{hi}^{\pi} \cdot 10^7$

	hkl	111	220	311	400	331	422	333	440	444	555
Si	Cu $K\alpha_1$	2.140	2.265	0.826	1.190	0.549	0.186	0.217	0.877	2.640	
	Mo $K\alpha_1$	0.1085	0.1430	0.0990	0.1270	0.0566	0.1125	0.064	0.0864	0.078	0.032
	Ag $K\alpha_1$	0.0423	0.0568	0.0382	0.0520	0.0354	0.0463	0.0278	0.0423	0.0360	0.019
Ge	Cu $K\alpha_1$	5.020	5.440	3.250	3.175	1.633	0.979	0.1185	1.347	5.04	
	Mo $K\alpha_1$	2.482	3.292	2.255	2.950	2.002	2.650	1.792	2.340	2.335	0.835
	Ag $K\alpha_1$	1.035	1.398	0.962	1.292	0.882	1.188	0.817	1.097	0.647	0.480

even and odd reflections (separately) with an increase in indices. In contrast to this, according to (4.32), for reflections with a small value of cos 2ϑ, we shall observe a sharp decrease in the values of χ^{π}_{hi}, which will, generally speaking, be less than χ^{σ}_{hi}.

On the other hand, the values of π-components in χ_{hi} are appreciable only at low absorption, because for scattering in strongly absorbing specimens the component χ^{π}_{hi} reduces almost completely to zero. The strong absorption of the π-component of the incident radiation makes it particularly important that the parameter

$$\varepsilon^{\sigma} = \varepsilon^{\sigma}_{o} \exp(-M) = |\chi^{\sigma}_{hi}|/|\chi_{oi}| \tag{4.60}$$

be determined as accurately as possible. This parameter appears in the expressions for the coefficients R and T and their integrated values in the case of an absorbing crystal. In this connection we shall mention the method of integral characteristics elaborated by ELISTRATOV and EFIMOV [4.15]. (That Paper also includes references to previous investigations). According to this method, the value of anomalous absorptions μ_{an} is determined from the slope of the line $\ell n\, R_i t$. The method was also applied, in a modified form, in [1.43]. The authors used the graph ($\ell n\, R_i$ + 1/2 ℓn t) as a function of t in determining μ_{an} for reflection (220) Si in Mo $K\alpha_1$ and Ag $K\alpha_1$ rays. The authors of [1.38] determined the values of ε^{σ} systematically and experimentally for many reflections of Ge in Fe $K\alpha_1$, Cu $K\alpha_1$, Mo $K\alpha_1$, and Ag $K\alpha_1$ ryas. The authors divided the expression for $\ell n\, R_i$ into three parts: A, which is independent of μ, and two terms $\mu t/\gamma_0$ and $F(\varepsilon) = (h - 1/2\ \ell n\ h)$, where $h = \mu\varepsilon t/\gamma_0$. The procedure consisted in measuring the values of R_i for a set of Ge slices of different thicknesses and taking such values of ε^{σ} that the value of A remained constant. The accuracy of determination of ε^{σ} in this work was the highest for Cu $K\alpha_1$ and Mo $K\alpha_1$ radiation and for strong reflections. The minimum error for 220 Cu $K\alpha_1$, or Mo $K\alpha_1$) was 0.1 percent, and the maximum one, up to ~ 10 percent. Note that the value of μ_{an} for 220 Cu $K\alpha_1$ calculated from the data of this work coincides, within experimental error, with that obtained by EFIMOV [4.15].

The values of the parameters χ^{σ}_{hi} were calculated from the experimental values of ε^{σ}, when these were not available they were calculated from (4.31,33,34) in which the theoretical ratios μ^{Q}/μ for photoabsorption were used. Finally, the parameters χ^{π}_{hi} were calculated from the theoretical values (also for photoabsorption) of $\mu^{D} + \mu^{D,0}$ and μ^{Q}, according to (4.33) and (4.34).

The structure factor

$$F_h = \sum_{cell} f_i \exp 2\pi i(\underline{h}\underline{r})$$

for Ge and Si becomes unity for even indices and to $1/\sqrt{2}$ for odd ones. The values of μ^D, $\mu^{D,0}$ and μ^Q are taken from Table 4.2.

4.3 Derivation of Exact Formulae for Transmission (T) and Reflection (R) Coefficient in the Case of an Absorbing Crystal

Here we consider the case of unseparated fields, which corresponds to the incidence of a vacuum wave with a wave front exceeding the thickness of the crystal slice. The derivation of the relevant formulae is similar to the one used in the case of a transparent crystal and is given in Section 3.3. The calculation is more complicated, however, because of the complex nature of the variables $c^{(i)}$, $\delta^{(i)}$, and W.

The exact formulae for T and R referring to a general case of scattering in a crystal without a centre of symmetry and with any ratio between the moduli χ_{hr} and χ_{hi} were derived by ZIELINSKA-ROHOSINSKA [4.31]. Our presentation is very close to the derivation given by this author, but we have retained the meaning of the symbols χ_{hr} and χ_{hi}, which we introduced previously following Laue.

We proceed from (3.11) for $c^{(i)}$:

$$c^{(i)} = - \frac{\beta \pm W}{2C(\gamma_h/\gamma_o)\chi_{\bar{h}}} \quad , \tag{4.61}$$

$$\beta \pm W = (\beta_r \pm W_r) + i(\beta_i \pm W_i) \quad , \tag{4.62}$$

where W is determined from (3.58). Multiplying the numerator and denominator of the fraction by $\chi^*_{\bar{h}}$, we get

$$c^{(i)} = - \frac{(\chi_{\bar{h}r} - i\chi_{\bar{h}i})[(\beta_r \pm W_r) + i(\beta_i \pm W_i)]}{2C(\gamma_h/\gamma_o)\ |\chi_{\bar{h}}|^2} \ , \quad i = 1,2 \quad . \tag{4.63}$$

Replacing the denominator on the right-hand side of (4.63) by the letter p, we write out, in explicit form, the four values of $c_r^{(i)}$, $c_i^{(i)}$

$$
\begin{aligned}
c_r^{(1)} &= -\, p^{-1}\, [\chi_{\bar{h}r}(\beta_r+W_r) + \chi_{\bar{h}i}(\beta_i+W_i)] \quad ,\\
c_i^{(1)} &= -\, p^{-1}\, [\chi_{\bar{h}r}(\beta_i+W_i) - \chi_{\bar{h}i}(\beta_r+W_r)] \quad ,\\
c_r^{(2)} &= -\, p^{-1}\, [\chi_{\bar{h}r}(\beta_r-W_r) + \chi_{\bar{h}i}(\beta_i-W_i)] \quad ,\\
c_i^{(2)} &= -\, p^{-1}\, [\chi_{\bar{h}r}(\beta_i-W_i) - \chi_{\bar{h}i}(\beta_r-W_r)] \quad .
\end{aligned} \tag{4.64}
$$

The exact expression for the quantities W_r and W_i can be obtained by applying the rule of taking the root of a complex number:

$$
W_r = \left[\frac{1}{2}(\sqrt{a^2+b^2}+a)\right]^{1/2} > 0 \quad , \quad W_i = \left[\frac{1}{2}(\sqrt{a^2+b^2}-a)\right]^{1/2} > 0 \quad , \tag{4.65}
$$

$$
a = \beta_r^2 - \beta_i^2 + 4C^2 \frac{\gamma_h}{\gamma_o} \Phi_h \quad , \quad b = 2\beta_r\beta_i + aC^2 \frac{\gamma_h}{\gamma_o} \Psi_h \quad . \tag{4.66}
$$

Now we shall give the explicit expressions for $[\delta^{(1)} + \delta^{(2)}]$ and $[\delta^{(1)} - \delta^{(2)}]$ which are complex values. From (3.10) follows

$$
\frac{1}{\gamma_o} \xi_o^{(i)} = - \frac{K}{4\gamma_h} [(\beta_r \pm W_r) + i(\beta_i \pm W_i)] \quad , \tag{4.67}
$$

and then (see (3.7))

$$
\delta_r^{(i)} = \frac{\beta_r}{4\gamma_h} - \frac{\chi_{or}}{2\gamma_o} \pm \frac{1}{4\gamma_h} W_r \quad , \quad \delta_i^{(i)} = \frac{\beta_i}{4\gamma_h} - \frac{\chi_{oi}}{2\gamma_h} \pm \frac{1}{4\gamma_h} W_i \quad . \tag{4.68}
$$

Substituting these values in the sums and differences of $\delta^{(i)}$:

$$
\begin{aligned}
\delta^{(1)} + \delta^{(2)} &= (\delta_r^{(1)}+\delta_r^{(2)}) + i(\delta_i^{(1)}+\delta_i^{(2)})\\
\delta^{(1)} - \delta^{(2)} &= (\delta_r^{(1)}-\delta_r^{(2)}) + i(\delta_i^{(1)}-\delta_i^{(2)})
\end{aligned} \tag{4.69}
$$

we obtain, for the real and imaginary components,

$$
\begin{aligned}
\delta_r^{(1)} + \delta_r^{(2)} &= \frac{\beta_r}{2\gamma_h} - \frac{\chi_{or}}{2\gamma_o} \quad ,\\
\delta_i^{(1)} + \delta_i^{(2)} &= \frac{\beta_i}{2\gamma_h} - \frac{\chi_{oi}}{\gamma_o} = - \frac{\chi_{oi}}{2} \left(\frac{1}{\gamma_h} + \frac{1}{\gamma_o}\right) \quad ,
\end{aligned} \tag{4.70}
$$

$$\delta_r^{(1)} - \delta_r^{(2)} = \frac{W_r}{2\gamma_h} \quad , \quad \delta_i^{(1)} - \delta_i^{(2)} = \frac{W_i}{2\gamma_h} \quad . \tag{4.71}$$

It should be also taken into account that

$$\begin{aligned} i\pi Kt[(\delta_r^{(1)}+\delta_r^{(2)}) + i(\delta_i^{(1)}+\delta_i^{(2)})] = \pi Kt \left(\frac{1}{\gamma_o}+\frac{1}{\gamma_h}\right) \frac{\chi_{oi}}{2} + \\ + i\pi Kt\left(\frac{\beta_r}{2\gamma_h} - \frac{\chi_{or}}{\gamma_o}\right) \quad , \end{aligned} \tag{4.72}$$

$$i\pi Kt[(\delta_r^{(1)}-\delta_r^{(2)}) + i(\delta_i^{(1)}-\delta_i^{(2)})] = - \pi Kt \frac{W_i}{2\gamma_h} + i\pi Kt \frac{W_r}{2\gamma_h} \quad . \tag{4.73}$$

Now, repeating the derivation given in (3.3), we rewrite (3.53) and (3.54), using (3.56) for the wave functions of a refracted and diffracted wave as applied to a crystal with absorption

$$\begin{aligned} D_o^{(d)} = D_o^{(a)} \frac{1}{(c_r^{(2)}-c_r^{(1)}) + i(c_i^{(2)}-c_i^{(1)})} \cdot \exp\left[\pi Kt \frac{\chi_{oi}}{2}\left(\frac{1}{\gamma_o}+\frac{1}{\gamma_h}\right)\right] \\ \cdot \exp\left[i\pi Kt\left(\frac{\beta_r}{2\gamma_h} - \frac{\chi_{or}}{\gamma_o}\right)\right]\left[(c_r^{(2)}+ic_i^{(2)}) \exp\left(-i\pi Kt \frac{W_i}{2\gamma_h}\right)\right. \\ \cdot \exp\left(i\pi Kt \frac{W_r}{2\gamma_h}\right) - (c_r^{(1)}+ic_i^{(1)}) \exp\left(\pi Kt \frac{W_i}{2\gamma_h}\right) \\ \left.\cdot \exp\left(-i\pi Kt \frac{W_r}{2\gamma_h}\right)\right] \quad . \end{aligned} \tag{4.74}$$

$$\begin{aligned} D_h^{(d)} = D_o^{(a)} \frac{(c_r^{(1)}+ic_i^{(1)})(c_r^{(2)}+ic_i^{(2)})}{(c_r^{(2)}-c_r^{(1)}) + i(c_i^{(2)}-c_i^{(1)})} \exp\left[\pi Kt \frac{\chi_{oi}}{2}\left(\frac{1}{\gamma_o}+\frac{1}{\gamma_h}\right)\right] \cdot \\ \cdot \exp\left[i\pi Kt\left(\frac{\beta_r}{2\gamma_h} - \frac{\chi_{or}}{\gamma_o}\right)\right]\left[\exp\left(-\pi Kt \frac{W_i}{2\gamma_h}\right) \cdot\right. \\ \left.\cdot \exp\left(i\pi Kt \frac{W_r}{2\gamma_h}\right) - \exp\left(\pi Kt \frac{W_i}{2\gamma_h}\right) \exp\left(-i\pi Kt \frac{W_r}{2\gamma_h}\right)\right] \quad . \end{aligned} \tag{4.75}$$

Passing on to the quantities T and R, we multiply the above equations by the conjugate complex values. Here, we use (4.28) and apply trigonometric functions in interference terms.

The exact equations for T and R have the following form:

$$T = \left|\frac{D_o^{(d)}}{D_o^{(a)}}\right|^2 = \frac{\exp[-(\mu/2)(\gamma_o^{-1}+\gamma_h^{-1})t]}{|c_r^{(1)} - c_r^{(2)}|^2 + |c_i^{(1)} - c_i^{(2)}|^2} \left[|c^{(2)}|^2 \exp\left(-\pi Kt\frac{W_i}{\gamma_h}\right) + \right.$$

$$+ |c^{(1)}|^2 \exp\left(\pi Kt \frac{W_i}{\gamma_h}\right) + 2(c_r^{(1)}c_r^{(2)*}+c_i^{(1)}c_i^{(2)*}) \cos\left(\pi Kt \frac{W_r}{\gamma_h}\right)$$

$$\left. + 2[c_i^{(1)*}c_r^{(2)}-c_r^{(1)}c_i^{(2)*}] \sin\left(\pi Kt \frac{W_r}{\gamma_h}\right)\right] \quad , \tag{4.76}$$

$$R = \frac{|c^{(1)}|^2|c^{(2)}|^2}{|c_r^{(1)}-c_r^{(2)}|^2 + |c_i^{(1)}-c_i^{(2)}|^2} \exp\left[-\frac{\mu}{2}\left(\frac{1}{\gamma_o}+\frac{1}{\gamma_h}\right)t\right] \cdot$$

$$\cdot \left[\exp\left(-\pi Kt \frac{W_i}{\gamma_h}\right) + \exp\left(\pi Kt \frac{W_i}{\gamma_h}\right) - 2\cos\left(\pi Kt \frac{W_r}{\gamma_h}\right)\right] \quad . \tag{4.77}$$

Despite the rather complicated form of these equations and the difficulty of analysing them qualitatively, they clearly reveal one remarkable feature of the phenomenon of transmission of X-rays through perfect absorbing crystals namely the Borrmann effect.

Indeed, when inspecting these equations it can be seen that the terms in the braces succesively describe the waves belonging to the first and second fields and their interferences. The absorption is represented by the common factor standing in front of the braces and by the additional factors belonging to each of the two fields. Here, the additional factor at the first term in the braces, which refers to the first field, has the sign of the exponent, which coincides with that of the exponent of the common factor in front of the braces, whereas the additional factor at the second term has the opposite sign. Hence, if we refer to the common multiplier as the "arithmetical mean" of the absorption of the transmitted and diffracted waves in the crystal,

$$-\sigma_m = -\left(\frac{\mu}{2}\right)(\gamma_o^{-1}+\gamma_h^{-1}) \quad , \tag{4.78}$$

then, during the incidence of the vacuum wave, additional, or interference, absorption arises in the maximum region:

$$\mp\pi Kt \frac{W_i}{\gamma_h} \quad . \tag{4.79}$$

A characteristic feature of it is the difference in sign, which leads to an increase in the absorption of one field and to a substantial decrease in the absorption of the other. To put it differently, by X-ray scattering in an absorbing crystal we observe anomalous absorption of one field and anomalous penetration of X-rays into the other. Where thickness t of the crystal slice is sufficient, the transmitted and diffracted waves in the vacuum will be formed by only one of the fields in the crystal.

4.4 Derivation of Approximate Equations for Transmission Coefficient T and Reflection Coefficient R

At present, the principal materials used for quantitative experimental investigation of dynamical scattering are single crystals of Si, GaAs, quartz, calcite, Ge, Cu, and some other substances. These single crystals mainly consist of atoms of light and medium elements of the Periodic System. We believe that by scattering in crystals with radiations with a wavelength λ, which is not very close to the wavelength for the absorption edge on the K- or L-shells of the scattering atoms, the following conditions obtains:

$$|\chi_{hi}| \ll |\chi_{hr}| , \qquad |\chi_{oi}| \ll |\chi_{or}| , \qquad |\Psi_h| \ll |\Phi_h| \quad . \tag{4.80}$$

Moreover, interaction of incident X-radiation with atoms in such crystals (compared with scattering from atoms of elements with a high atomic number) is more amenable to the approximate methods of calculation of the absorption parameters described in Section 4.1. As has been shown by ZACHARIASEN [1.14] and LAUE [1.13], the use of these conditions and neglect of the squares of the imaginary components in the presence of the squares of the real components $|\chi_{or}|$ and $|\chi_{hr}|$ makes it possible to derive simpler equations for T and R. Such equations are convenient for detailed qualitative analysis of the phenomenon and actually yield sufficiently accurate results in quantitative comparison with the experiment. The approximate equations can be derived with the aid of simple transformations.

As angular variables we shall use the quantities y and v which should be regarded here as complex functions. On the basis of (3.13) and (4.59), and the conditions (4.80) we can write for y

$$y = y_r + iy_i = \frac{\beta_r + i\beta_i}{2C\left[\frac{\gamma_h}{\gamma_o}(\Phi_h + i\Psi_h)\right]^{1/2}} \approx \frac{\beta_r + i\beta_i}{2C\left(\frac{\gamma_h}{\gamma_o}\Phi_h\right)^{1/2}}\left(1 - i\frac{\Psi_h}{2\Phi_h}\right) \quad , \tag{4.81}$$

$$y_r \approx \frac{\beta_r}{2C\left(\frac{\gamma_h}{\gamma_o}\Phi_h\right)^{1/2}} \quad , \quad y_i \approx \frac{\beta_i - (\beta_r\Psi_h/2\Phi_h)}{2C\left(\frac{\gamma_h}{\gamma_o}\Phi_h\right)^{1/2}} \quad .$$

The value of the imaginary part of the angular variable v is obtained with the aid of the equality

$$\sinh v = \sinh v_r \cos v_i + i\cosh v_r \sin v_i \quad . \tag{4.82}$$

On the basis of the approximation

$$\cos v_i \approx 1, \quad \sin v_i \approx v_i \tag{4.83}$$

we can write

$$(\sinh v)_r \approx \sinh v_r \quad , \quad v_i \approx \sin v_i = \frac{(\sinh v)_i}{\cosh v_r} \quad . \tag{4.84}$$

Hence, using (4.81) and (3.13), we obtain

$$v_i \approx \frac{y_i}{(1+y_r^2)^{1/2}} = \frac{\beta_i - (\beta_r\Psi_h/2\Phi_h)}{2C\left(\frac{\gamma_h}{\gamma_o}\Phi_h\right)^{1/2}(1+y_r^2)^{1/2}} =$$

$$= \left[\frac{\beta_i}{2C\left(\frac{\gamma_h}{\gamma_o}\Phi_h\right)^{1/2}} - \frac{y_r\Psi_h}{2\Phi_h}\right]\frac{1}{(1+y_r^2)^{1/2}} \quad . \tag{4.85}$$

For the values of W_r and W_i we use approximate values, which are obtained from the exact ones by the following simple transformation. Neglecting the value of β_i^2 in (4.66), compared with β_r^2 (see (4.59)), we get for W_r'

$$W_r \approx W_r' = \sqrt{\beta_r^2 + 4C^2 \frac{\gamma_h}{\gamma_o} \Phi_h} \quad . \tag{4.86}$$

The approximate expression for W_i can be obtained by multiplying and dividing the second equation (4.65) by W_r and replacing W_r by W_r' in the denominator:

$$W_i \approx W_i' = \frac{b}{2W_r'} = \frac{\beta_r\beta_i + 2C^2(\gamma_h/\gamma_o)\,\Psi_h}{\sqrt{\beta_r^2 + 4C^2(\gamma_h/\gamma_o)\,\Phi_h}} \quad . \tag{4.87}$$

The approximate expressions for T and R can be derived by using the method described in Section 3.3, as has been done above for the exact values of the coefficients indicated, but using approximate values of the functions appearing in these expressions.

For the values of $c^{(i)}$ we first use (3.16) in the following form

$$\begin{aligned} c^{(i)} &= \mp \sqrt{\frac{\gamma_o}{\gamma_h}} \sqrt{\frac{\chi_h}{\chi_{\bar{h}}}} \exp[\pm(v_r + iv_i)] \quad , \\ c^{(i)*} &= \mp \sqrt{\frac{\gamma_o}{\gamma_h}} \sqrt{\frac{\chi_h}{\chi_{\bar{h}}}} \exp[\pm v_r - iv_i)] \quad . \end{aligned} \tag{4.88}$$

The squares of the moduli of the values $[c^{(2)} - c^{(1)}]$ and $c^{(2)}c^{(1)}$ can be written thus (see (4.83)):

$$\begin{aligned} |(c^{(2)} - c^{(1)})|^2 &= \frac{\gamma_o}{\gamma_h} \left| \frac{\chi_h}{\chi_{\bar{h}}} \right| [\exp(2v_r) + \exp(-2v_r) + 2\cos 2v_i] \approx \\ &\approx \frac{\gamma_o}{\gamma_h} \frac{|\chi_h|}{|\chi_{\bar{h}}|} 4\cosh^2 v_r \quad , \end{aligned} \tag{4.89}$$

$$|c^{(1)}c^{(2)}|^2 = \left(\frac{\gamma_o}{\gamma_h} \left|\frac{\chi_h}{\chi_{\bar{h}}}\right|\right)^2 \quad . \tag{4.90}$$

Further, the exponents $\pi iKt[\delta^{(1)} + \delta^{(2)}]$ and $\pi iKt[\delta^{(1)} - \delta^{(2)}]$ are transformed according to (4.72) and (4.73). The first terms on the right-hand side of these expressions, when multiplied by the conjugate complex values, will appear in the exponents of the absorption factors (with replacement of W_r and W_i by W_r' and W_i'). The second term from the right in (4.72) is cancelled out, as well as the second term from the right in (4.73), at values

$|c^{(i)}|^2$. As regards the terms $c^{(1)}c^{(2)*}$ and $c^{(2)}c^{(1)*}$, here we obtain

$$c^{(1)}c^{(2)*}\exp\left(i\pi Kt\frac{W_r'}{2\gamma_h}\right)+c^{(1)*}c^{(2)}\exp\left(-i\pi Kt\frac{W_r'}{2\gamma_h}\right)=$$
$$=2\frac{\gamma_o}{\gamma_h}\left|\frac{\chi_h}{\chi_{\bar{h}}}\right|\cos\left(\pi Kt\frac{W_r'}{\gamma_h}-2v_i\right)\quad . \tag{4.91}$$

In accordance with (4.52) and (4.80) we write

$$\Phi_h\approx|\chi_{hr}|^2\ ,\qquad \Psi_h=2|\chi_{hr}|\,|\chi_{hi}|\cos\nu_h\quad . \tag{4.92}$$

In this case, we get from the above expressions, for interference absorption,

$$\pi K\frac{W_i'}{\gamma_n}=\pi K\chi_{oi}\left(\frac{1}{\gamma_o}-\frac{1}{\gamma_h}\right)\frac{\beta_r}{\left(\beta_r^2+4C^2\frac{\gamma_h}{\gamma_o}\Phi_h\right)^{1/2}}+$$
$$+\frac{2\pi KC|\chi_{hi}|\cos\nu_h}{(\gamma_o\gamma_h)^{1/2}(1+y_r)^{1/2}}=\sigma_h'+\sigma_h\quad . \tag{4.93}$$

Thus, in addition to the mean coefficient of absorption σ_m, whose value was given in (4.78), two additional coefficients appear in reflection:

$$\sigma_h'=-\frac{\mu}{2}\left(\gamma_o^{-1}-\gamma_h^{-1}\right)\tanh v_r\qquad \sigma_h=\frac{2\pi KC|\chi_{hi}|\cos\nu_h}{(\gamma_o\gamma_h)^{1/2}\cosh v_r}\quad . \tag{4.94}$$

Finally,

$$\pi Kt\frac{W_r'}{\gamma_h}=\frac{2\pi KCt|\chi_{hr}|}{(\gamma_o\gamma_h)^{1/2}}\cosh v_r=2A\cosh v_r=2\alpha\quad . \tag{4.95}$$

Here, A has the same value (3.66) as for a transparent crystal (taking $\chi_{hr}=\chi_h$).

Thus, the exact equations (4.76) and (4.77) for R and T are transformed into the approximate Laue equations, which have the following form (we retain the previous notation for T and R):

$$T(v) = \frac{\exp(-\sigma_m t)}{2\cosh^2 v_r}\{\cosh[2v_r + (\sigma_h+\sigma_h')t] + \cos(2\alpha-2v_i)\} \quad . \tag{4.96a}$$

In contrast to T, the reflection coefficient R will, in a general case of a crystal without a centre of symmetry, be different for reflections hkl and $\bar{h}\bar{k}\bar{l}$. Replacing $|\chi_h|/|\chi_{\bar{h}}|$ and $|\chi_{\bar{h}}|/|\chi_h|$ by their values according to (4.53) or (4.54), we obtain, for the two reflection factors,

$$R_{\pm h}(v) = \frac{\exp(-\sigma_m t)}{2\cosh^2 v_r}\{\cosh[\sigma_h+\sigma_h')t] - \cos 2\alpha\}\left(1\pm 2\left|\frac{\chi_{hi}}{\chi_{hr}}\right| \sin v_h\right) \quad .\tag{4.96b}$$

Another form of writing the expressions for T and R as a function of the angular variable y is

$$T(y) = \frac{\exp(-\sigma_m t)}{4(1+y_r^2)}\left\{(y_r-\sqrt{1+y_r^2})^2 \exp[-(\sigma_h+\sigma_h')t] + \right.$$
$$\left. + (y_r+\sqrt{1+y_r^2})^2 \exp[(\sigma_h+\sigma_h')t] + 2\cos\left(2\alpha-\frac{2y_i}{\sqrt{1+y_r^2}}\right)\right\} \quad , \tag{4.97a}$$

$$R_{\pm h}(y) = \frac{\exp(-\sigma_m t)}{2(1+y_r^2)}\{\cosh[(\sigma_h+\sigma_h')t] - \cos 2\alpha\}\left(1\pm 2\left|\frac{\chi_{hi}}{\chi_{hr}}\right| \sin v_h\right) \quad . \tag{4.97b}$$

It should be remembered that the above equations (4.96,97) for the transmission coefficient T and the reflection coefficient R should be applied separately for each of the two standard polarization states. Use is made of the respective values of C in (4.81) for y_r and (4.95) for A, as well as the values χ_{hi}^{σ} or χ_{hi}^{π}. The total value of T and R is obtained by adding up the values for both states of polarization. When superimposing the components of the curves T(y) and R(y), it should be kept in mind that $y_i^{\sigma} \neq y_i^{\pi}$.

Now we shall also give the expressions for T and R (in our notation) used by ZACHARIASEN in his book [1.14]. These expressions are easy to obtain with the aid of simple transformations from (4.97):

$$T(y) = \exp(-\sigma_m t) \left\{ 1 - \frac{1}{1+y_r^2} \sin^2\alpha + \frac{1+2y_r^2}{1+y_r^2} \right. \cdot$$

$$\left. \cdot \sinh^2 [(\sigma_h+\sigma_h')\tfrac{t}{2}] + \frac{y_r}{\sqrt{1+y_r^2}} \sinh [(\sigma_h+\sigma_h')t] \right\} , \tag{4.98a}$$

$$R(y) = \frac{\exp(-\sigma_m t)}{1+y_r^2} \{\sin^2\alpha + \sinh^2[(\sigma_h+\sigma_h')\tfrac{t}{2}]\} \cdot$$

$$\cdot \left(1 \pm 2\left|\frac{\chi_{hi}}{\chi_{hr}}\right| \sin \nu_h\right) . \tag{4.98b}$$

In Laue equations of the type (4.96a) or (4.97a) the braces contain successive terms referring to the first and second fields and to interference between them; Zachariasen's equations have first the terms referring to a transparent crystal and then those referring to an absorbing one. Note that (4.98a) does not take into account the additional phase $2v_i = 2y_i/(1+y_r^2)^{1/2}$. It is obvious that the Laue equations demonstrate the Borrmann effect more clearly, in explicit form. A pictorial, physical interpretation of this effect is given in Chapter 3 in connection with the analysis of (3.49).

4.5 Analysis of Approximate Equations for the Transmission Coefficient T and the Reflection Coefficient R

Equations (4.96) and (4.97), as well as (4.76) and (4.77) clearly indicate the interference absorption, which is different for the two fields. The total value of the coefficient of absorption in the direction of the normal to the entrance surface is equal to

$$\sigma = \sigma_m \pm (\sigma_h+\sigma_h') . \tag{4.99}$$

It will be the same for the refracted and diffracted waves since, according to (4.20) and (4.23), this value is determined by the imaginary part of the wave vector $\underline{k}_{oi}^{(i)}$.

At the same time the ratio of the coefficients T and R is essentially determined by the experimental conditions. The most important values are the products $\sigma_m t$ (for symmetrical reflection μt), the condition $\varphi = \pi/2$ (symmetrical reflection) or $\varphi \neq \pi/2$ (asymmetrical reflection) and, finally, the presence or absence of a centre of symmetry in the scattering crystal. One should also keep in mind the remarks made in Section 4.1 as to the inapplicability of the foregoing theory to scattering by crystals which consist of atoms of heavy elements, and to radiations with a wavelength close to the absorption edge. Turning to (4.99), we shall note that interference absorption proper, depending on χ_{hi}, is represented by the quantity σ_h, which has a maximum at $y_r = 0$. As regards σ_h', this quantity is, according to (4.94), determined similarly to σ_m, by the linear coefficient of absorption μ and depends on the difference between the angles of incidence ψ_0 and the reflection ψ_h, vanishing to zero in a symmetrical reflection.

The presence of oscillating terms in the equations for T and R means that at small values of the product $\mu_m t$ (up to ~ 2) subsidiary maxima of the pendulum solution are observed in absorbing, as well as in transparent crystals. The presence of the additional phase $2v_i$ in the cosine term of the equation for T leads to certain peculiarities of scattering in an absorbing crystal. In the first place, the subsidiary maxima of T are slightly displaced relative to those on the R curve, which disturbs the rigorous complementarity of the two curves, typical of scattering in a transparent crystal. Secondly, since v_i depends on y_r to the first power, the T curve, which contains subsidiary maxima, acquires a slight additional asymmetry with respect to the midpoint.

With transition to larger values μt, the anomalous absorption of one of the fields increases sharply, and this results in the weakening and then complete disappearance of the subsidiary maxima. This is due to the increased "intrinsic" absorption of the interference portion (the cosine term).

We shall now proceed to a more detailed analysis of the equations for R and T.

4.5.1 Symmetrical Reflection [1.16,4.20,32]

In the case of symmetrical reflection, the scattering parameters take the following particular values. The additional subscript means that the given parameter refers to the symmetrical case:

$$\gamma_{os} = \gamma_{hs} = \cos\vartheta \quad , \quad \beta_{rs} = 2\eta\sin 2\vartheta \quad , \quad \beta_{is} = 0 \quad ,$$

$$\sigma'_{hs} = 0 \quad , \quad \sigma_{ms} = \frac{\mu}{\cos\vartheta} = \sigma_o \quad , \tag{4.100}$$

$$\sigma_{hs} = \frac{2\pi KC|\chi_{hi}|}{\cos\vartheta} \, \frac{\cos\nu_h}{(1+y_{rs}^2)^{1/2}} \quad , \quad y_{rs} = \frac{\eta\sin 2\vartheta}{C|\chi_{hr}|} \quad , \tag{4.101}$$

$$y_{is} = -y_{rs}\left|\frac{\chi_{hi}}{\chi_{hr}}\right| \cos\nu_h \quad , \quad v_{is} = -\frac{y_{is}}{(1+y_{rs}^2)^{1/2}} \quad , \tag{4.102}$$

$$\alpha_s = A_s\sqrt{1+y_{rs}^2} \quad , \quad A_s = \frac{\pi KCt}{\cos\vartheta} \quad . \tag{4.103}$$

Let us write out the expressions for the transmission and reflection coefficients:

$$T_s = \frac{1}{4(1+y_{rs}^2)}\left\{(y_{rs}-\sqrt{1+y_{rs}^2})^2 \exp\left[-\frac{\mu t}{\cos\vartheta}\left(1+\frac{\varepsilon\cos\nu_h}{(1+y_{rs}^2)^{1/2}}\right)\right] + \right.$$
$$+ (y_{rs}+\sqrt{1+y_{rs}^2})^2 \exp\left[-\frac{\mu t}{\cos\vartheta}\left(1-\frac{\varepsilon\cos\nu_h}{(1+y_{rs}^2)^{1/2}}\right)\right] + \tag{4.104}$$
$$\left. + 2\exp\left(-\frac{\mu t}{\cos\vartheta}\right)\cos(2\alpha_s-2v_{is})\right\} \quad ,$$

$$R_s = \frac{1\pm 2\left|\frac{\chi_{hi}}{\chi_{hr}}\right|\sin\nu_h}{4(1+y_{rs}^2)}\left\{\exp\left[-\frac{\mu t}{\cos\vartheta}\left(1+\frac{\varepsilon\cos\nu_h}{(1+y_{rs}^2)^{1/2}}\right)\right]\right.$$
$$\left. + \exp\left[-\frac{\mu t}{\cos\vartheta}\left(1-\frac{\varepsilon\cos\nu_h}{(1+y_{rs}^2)^{1/2}}\right)\right] - 2\exp\left(-\frac{\mu t}{\cos\vartheta}\right)\cos 2\alpha_s\right\} \quad . \tag{4.105}$$

We shall now follow up the deviations in the shape of T_s and R_s curves with an increase in the product μt. This increase may be due either to the increase in the thickness of the crystal slice at a fixed value of μ for the radiation used, or to the transition to other, softer radiations, whose absorption in a given crystal increases and occurs on other electron shells of the atom. Experiment shows (see Chap.9) that the two cases exhibit far-reaching similarity.

The change in the shape of the T_s and R_s curves is demonstrated in Figs. 4.1 and 4.2 [1.16,4.32]. At small values of μt the curves correspond to the values of R and T without the cosine term; N is the intensity of the beam which has passed through a slice of a given thickness outside the maximum.

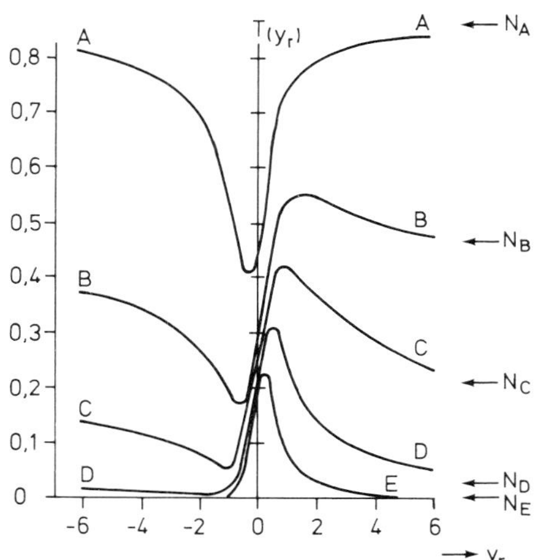

Fig. 4.1 Maxima of transmission coefficient T for absorbing crystal in the case of the symmetrical diffraction series of increasing values of μt(Cu $K\alpha_1$, 200, NaCl, μ=160 cm^{-1}). N_A - N_E indicate the normal transmission factors. μt_A = 0.16; μt_B = 0.8; μt_C = 1.6; μt_D = 3.2; μt_E = 6.4 (after JAMES [1.16])

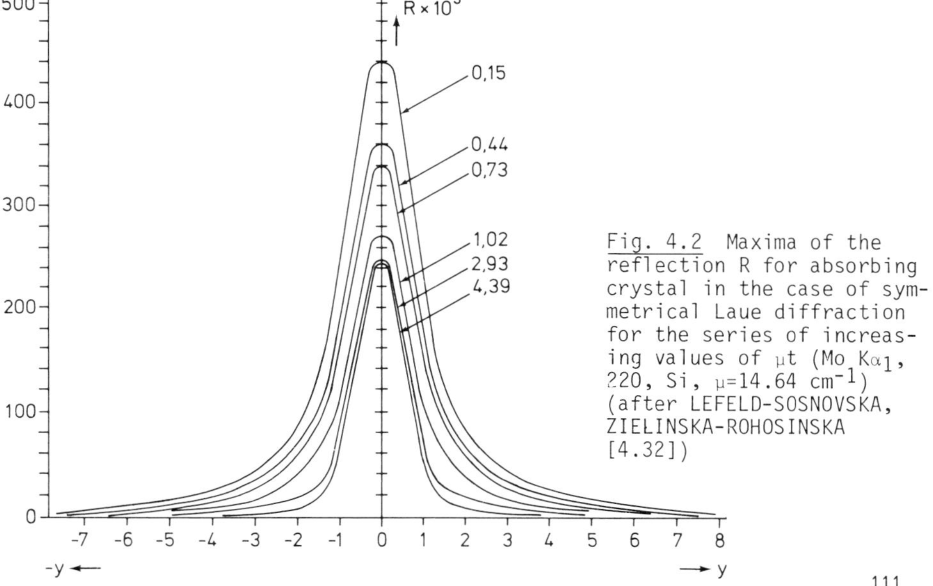

Fig. 4.2 Maxima of the reflection R for absorbing crystal in the case of symmetrical Laue diffraction for the series of increasing values of μt (Mo $K\alpha_1$, 220, Si, μ=14.64 cm^{-1}) (after LEFELD-SOSNOVSKA, ZIELINSKA-ROHOSINSKA [4.32])

As for the transmission curve, at small $\mu t \approx 0.1$-0.2, it shows a characteristic minimum, which is slightly displaced from the zero point. Further, with an increase in μt, the asymmetry of the T curve becomes considerable, the curve is "distorted", and a maximum is gradually formed.

These changes are easy to explain with the aid of a qualitative analysis of (4.104). The curve asymmetry is due to the difference in the attenuation factors for the first and second fields, which are represented, respectively, by the first and second components in the braces. It can be said that at small μt the initial transmission curve characterizes anomalous absorption compared to the "background" or to the value of $\mu t/\cos\vartheta$ valid outside the maximum and denoted by the straight lines N in Fig.4.1. This anomalous absorption is slightly stronger for the first field, which is effective at $y_{rs} < 0$. An increase in μt enhances anomalous penetration, which takes place only for the second field, i.e., to the right of the axis $y = 0$.

At medium values of μt, the T-curves describe the superposition of the functions $[y_{rs} - (1+y_{rs}^2)^{1/2}]^2/4(1+y_{rs}^2)$ on the left and $[y_{rs} + (1+y_{rs})^{1/2}]^2/4(1+y_{rs}^2)$ on the right, which change slowly with the variation of $|y_{rs}|$, and the attenuation factor of the second field.

In distinction to the transmission curves, the reflection curves (Fig.4.2) retain their symmetrical shape and position relative to the axis $y_{rs} = 0$ over the entire range of μt. This is due to the fact that the functions $D_h^{(1)}$ and $D_h^{(2)}$ are symmetrical about this axis.

Now consider the region of approximation of a thick crystal with a symmetrical reflection. At values of $\mu t \approx 5$-6 and more, the first field and also the interference portion weaken to such an extent that the size and shape of the T- and R-curves are determined almost solely by the second field. Here we approach the thick-crystal approximation.

The equations for T and R are written as

$$T_s^e \approx \frac{(y_{rs} + \sqrt{1+y_{rs}^2})^2}{4(1+y_{rs}^2)} \exp\left[-\frac{\mu t}{\cos\vartheta}\left(1 - \frac{\varepsilon\cos\nu_h}{(1+y_{rs}^2)^{1/2}}\right)\right] , \quad (4.106)$$

$$R_s^e \approx \frac{1 \pm 2\left|\frac{\chi_{hi}}{\chi_{hr}}\right| \sin\nu_h}{4(1+y_{rs}^2)} \exp\left[-\frac{\mu t}{\cos\vartheta}\left(1 - \frac{\varepsilon\cos\nu_h}{(1+y_{rs}^2)^{1/2}}\right)\right] . \quad (4.107)$$

Since in transmission through thick crystals the π-component is almost absorbed, the thick-crystal approximation is applicable only to the σ-component.

With increasing μt, the above-mentioned superposition of the functions for T_s^e increasingly tends to the predominance of the exponential function

$$\exp\left[-\mu t\left(1-\frac{\varepsilon\cos\nu_h}{(1+y_{rs}^2)^{1/2}}\right)\right] \quad . \tag{4.108}$$

As can be seen from the lower curve of T_s in Fig.4.1, which corresponds to $\mu t = 6.4$, the maximum of T is determined by the function (4.108). With a further increase in μt, the maximum of T, which decreases and shifts closer to the $y_r = 0$, coincides in position and shape with the maximum of R at values $\mu t \approx 40$-60. The characteristic parameter for the transition region, and especially for the region of thick-crystal approximation, is the value of displacement of the point T_{max} relative to the axis $y_r = 0$. This value is determined from the cubic equation

$$\frac{\partial T}{\partial y_{rs}} = y_{rs}^3 - \alpha y_{rs}^2 + y_{rs} - p^{-1} = 0 \quad , \quad \alpha = \frac{4+p^2}{4p} \quad , \quad p = \mu t\varepsilon \quad . \tag{4.109}$$

Eq. (4.109) has an exact solution.

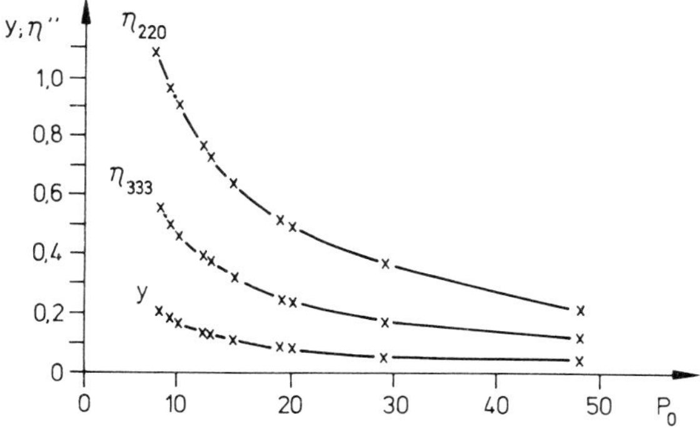

Fig. 4.3 Shifts of maxima of T on the y_r scale and in seconds of arc as a function of $p = \mu t\varepsilon$ for reflections 220 and 311 of $CuK\alpha_1$ radiation from Ge

Figure 4.3 gives the values of displacement y_{rs} and also the displacement of the point T_{max} in seconds of arc for the particular cases of reflection 220 and 311 of Cu $K\alpha_1$ radiation from Ge at increasing values of μt.

Another important parameter of the T_s^e and R_s^e curves is their half-width, which is approximately the same for scattering in sufficiently thick crystals.

The value of the half-width $2y_{1/2}$ or $w = 2\eta_{1/2}$ can be determined from the equation

$$\mu t \varepsilon \log e \left(1 - \frac{1}{(1+y_{rs}^2)^{1/2}}\right) = \log \frac{2}{(1+y_{rs}^2)} \quad . \tag{4.110}$$

The values of $y_{1/2}$ for $\mu t = 3$ are based on inclusion of two wave fields in the crystal (Table 4.6).

Table 4.6 Values of $2y_{1/2}$ and of half-width $w = 2\eta''_{1/2}$ for symmetrical reflection curves R of Cu $K\alpha_1$ radiation from Ge

	220 (ε = 0.96)			333 (ε = 0.61)		
$\mu t\varepsilon$	μt	$2y_{1/2}$	w"	μt	$2y_{1/2}$	w"
0	0	2	10.03	0	2	3.77
3	3.13	1.124	6.61	4.92	1.325	2.50
5	5.21	0.924	5.46	8.20	0.790	1.49
10	10.42	0.686	4.05	16.39	0.576	1.09
15	15.63	0.568	3.35	24.59	0.476	0.896
20	20.8	0.494	2.92	32.8	0.415	0.782
25	26.0	0.462	2.72	41.0	0.372	0.701
30	31.25	0.407	2.40	49.2	0.340	0.641
50	52.1	0.317	1.87	82.0	0.270	0.509

The value of the absorption coefficient in (4.106) and (4.107) in the simplest case of a crystal with a centre of symmetry and σ-polarization has a minimum value

$$\sigma_{min} = \frac{\mu}{\cos\vartheta}(1-\varepsilon) = \frac{\mu}{\cos\vartheta}\left(1 - \left|\frac{\chi_{hi}}{\chi_{oi}}\right|\right) \quad .$$

Comparing the series of reflections of a definite type with increasing $\sin\vartheta/\lambda$ for a given crystal and a given radiation, we shall obviously obtain a (qualitative) similarity for the ratios $|\chi_{hi}|/|\chi_{oi}|$ and $|\chi_{hr}|/|\chi_{or}|$. In the first case the drop is due to the effect of the temperature factor, and in in the second, to the decrease in function $f(\sin\vartheta/\lambda)$ as well. Therefore, for reflections of a high order the similarity will be less pronounced.

At the same time, the ratios χ_{hr}/χ_{or}, according to (2.62) and (2.77), and also according to Fig.3.1, are equal to the ratios of the real semi-diameter of the dispersion hyperbola to the value 1/2 K χ_o, i.e., they are the larger, the closer the point of intersection of the hyperbola with the axis ME (Fig. 3.1) is to the Laue point M.

Thus, from Fig.3.1, the values of interference (and total) absorption, i.e., the Borrmann effect, can be established and compared.

This method, pointed out by EWALD in 1958 [4.33], can be called the Ewald criterion. In the three-wave case it is also essential, as will be shown in Chapter 12.

4.5.2 Asymmetrical Reflection

Asymmetrical reflection [4.20] reveals a number of interesting features, which are, to a considerable extent, formally connected with the effect of σ'_h. These features are more pronounced when the asymmetry of the diffraction arrangement is sharply defined. Compare the following expressions of the values σ_h and σ'_h:

$$\sigma_h = \frac{2\pi KC|\chi_{hi}|}{\sqrt{\gamma_o\gamma_n}} \frac{\cos\nu_h}{\sqrt{1+y_r^2}} \quad , \quad \sigma'_h = -\frac{2\pi K|\chi_{oi}|}{\sqrt{\gamma_o\gamma_h}} \frac{y_r}{\sqrt{1+y_r^2}} q \quad ,$$

$$q = \frac{\gamma_h - \gamma_o}{2\sqrt{\gamma_o\gamma_h}} \quad . \tag{4.111}$$

First of all, turning to Figs.3.2a and b we note that for reflections with the positive indices hkl, $\gamma_o > \gamma_h$ and for the reflections $\bar{h}\bar{k}\bar{l}$, $\gamma_o < \gamma_h$. In the first case, at $y_r > 0$, σ'_h becomes positive and, hence, it increases the anomalous penetration. The value of σ'_h can be formally added to the interference absorption $\mu_{int}^{as} = \sigma_h + \sigma'_h$.

This sum increases within certain limits with increasing y_r and reaches a maximum at $y_r = q/\varepsilon$, which corresponds to the expression $\partial/\partial y_r\ (\sigma_h+\sigma'_h) = 0$. When passing on to the reflections $\bar{h}\bar{k}\bar{l}$, μ_{int}^{as} increases with $-|y_r|$.

The curves of variation in parameters in transition to negative indices are obtained by symmetrical reflection about the ordinate axis. The new origin of the coordinates y_r is rotated by an angle of 2ϑ relative to the old one.

Figures 4.4a and b illustrate these effects for two cases: (Cu $K\alpha_1$ on Ge) of moderate asymmetrical reflection ± 220 at $\varphi = 60^o$ and $q = \pm 0.2476$ and extreme reflection ± 333 at $\varphi = 55^o 10'$ and $q = \pm 0.9713$.

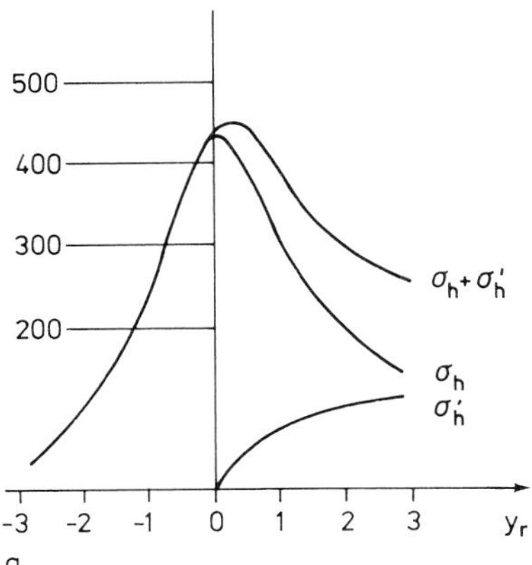

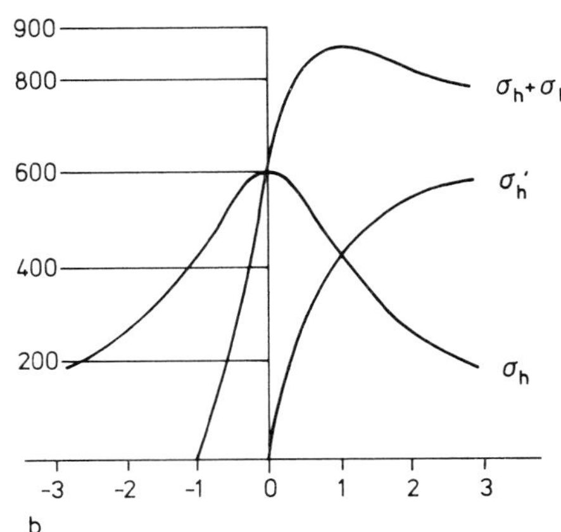

Fig. 4.4 Variations in the absorption parameters σ_h, σ_h' and $(\sigma_h+\sigma_h')$ as a function of y_r; CuKα_1 on Ge. a) moderate asymmetrical reflection 220 at $\varphi = 60^0$ and $q = -0.2476$; b) extreme asymmetrical reflection 333 at $\varphi = 55^0 10''$ and $q = -0.9713$

The increase in μ_{int}^{as} indicated does not lead to values $\mu_{eff}^{as} < \mu_{min}$, where $\mu_{min} = \mu(1-\varepsilon)/\cos\vartheta$ corresponds to minimum absorption in symmetrical reflection from a thick crystal, for $y_r = 0$. However, at μ values from 1 to 10, in symmetrical reflection (depending on the value of ε) we observe a slight shift of the transmission rocking curve from the point $y_r = 0$, as shown in Fig.4.1. To the values of $T_{max}^{y_r \neq 0}$ near $T_{max}^{y_r=0}$ $\mu^s = \mu[1 - \varepsilon/(1+y_r^2)^{1/2}]$ at $y_r \approx 0.4 - 0.1$, corresponds. For this reason the values of μ_{eff}^{as} may exceed μ_{eff}^s. The contribution of the first field, $\mu(1+\varepsilon/\sqrt{1+y_r^2})$ is negligibly small under these conditions.

In asymmetrical reflection, two more essential effects arise. In the first place, the absorption of the first field sharply increases and, hence, the lower boundary of the thick-approximation region decreases:

$$R^e \approx A \exp[-(\sigma_m - \sigma_h - \sigma_h')t] \quad . \quad T^e \approx B \cdot \exp[-(\sigma_m - \sigma_h - \sigma_h')t] \quad . \tag{4.112}$$

$$A = \frac{(1 \pm 2\left|\frac{\chi_{hi}}{\chi_{hr}}\right| \sin\nu_h}{4(1+y_{rs}^2)} \qquad B = \frac{(y_{rs} + \sqrt{1+y_{rs}^2})^2}{4(1+y_{rs}^2)}$$

Secondly, the shift of the maximum of μ_{eff}^{as} on the y_r scale leads to an increase in the amplitude B, according to (4.112), and as a result the transmission coefficients and, in particular, their maximum values $T^{as}_{max} > T^{s}_{max}$. The ratios of these values vary between 1.3 and 2 for the reflections indicated, ±220 and ±333.

In narrower ranges of the values of μt, small shifts of the values of R^{as}_{max} and of their excess over R^{s}_{max} also occur.

It is further obvious that if for reflections hkl the displacement of μ^{as}_{int} to the right on the y_r scale leads to an increase in amplitude, for reflections $\bar{h}\bar{k}\bar{l}$ the displacement of μ^{as}_{int} to the left, in the direction of negative y_r, leads to a sharp decrease in amplitudes, which follows from Fig. 3.7. In the region of negative values of $y \equiv y_r$ the contribution of the second field to the values of T rapidly drops by two or three orders. This is not the case with reflection coefficients R, whose maxima acquire an asymmetrical shape and retain their mirror-symmetry when shifting left and right along the y_r-axis. Here, in the case of centro-symmetric crystals, $R_{hkl} = R_{\bar{h}\bar{k}\bar{l}}$, and:

$$\int R_h(y)dy = \int R_{\bar{h}}(y)dy \quad . \tag{4.113}$$

Figs.4.5a,b, and c show curves of asymmetrical reflection hkl for the above-indicated cases. The asymmetry of the maxima increases with the values of $\sigma_m t$, and, respectively, μt.

One of the most interesting features of asymmetrical reflection is the difference in the half-width of the R_{hkl} and $R_{\bar{h}\bar{k}\bar{l}}$ curves, which is revealed in the change-over from the y_r scale to $\eta = \Delta\vartheta$. According to (3.26), in which y should be replaced by y_r in the case of an absorbing crystal, the switch-over indicated is due to the different values of the coefficients a and b for reflections with opposite indices. This leads to different positions of η_{max} corresponding to R_{max} and to a different half-width $w'' = a\Delta y_{1/2}/4.85\cdot 10^{-6}$. It is easy to see that the w" ratio

$$(w''_{hkl}/w''_{\bar{h}\bar{k}\bar{l}}) = \gamma_h/\gamma_o$$

where γ_h and γ_o are the cosines of the angles of reflection and incidence for reflections hkl. Since the values of R are independent of the signs of their reflection coefficients, in asymmetrical reflection we observe a more or less considerable increase in the integrated reflection in the case of negative coefficients (see Sec.4.7).

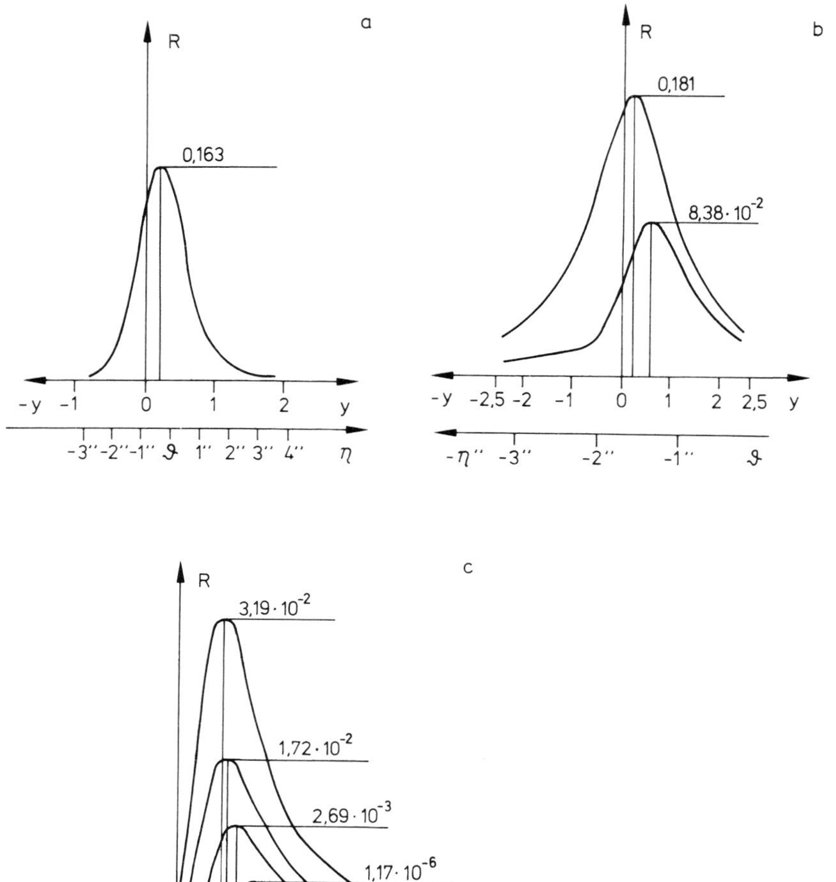

Fig. 4.5a-c The curves of the asymmetrical reflection hkl Cu $K\alpha_1$ from Ge. a) Moderate asymmetrical reflection (see caption to Fig.4.4a); b,c) Extreme asymmetrical reflection (see Fig.4.4b) at increasing values of μt. When passing on to reflections $\bar{h}\bar{k}\bar{l}$, do the same as for Fig.4.4

It is interesting to note that the transition hkl → $\bar{h}\bar{k}\bar{l}$ is accompanied by an increase in the glancing exit angle of the reflected beam $\pi/2 - \psi_h$; for the reflection ±333 indicated this angle increases from ~ 10° to ≈ 100°. The transmission curves are given in Figs.4.6a-d.

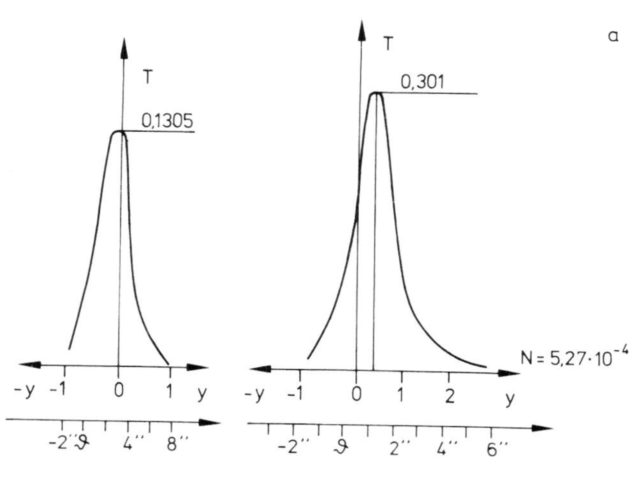

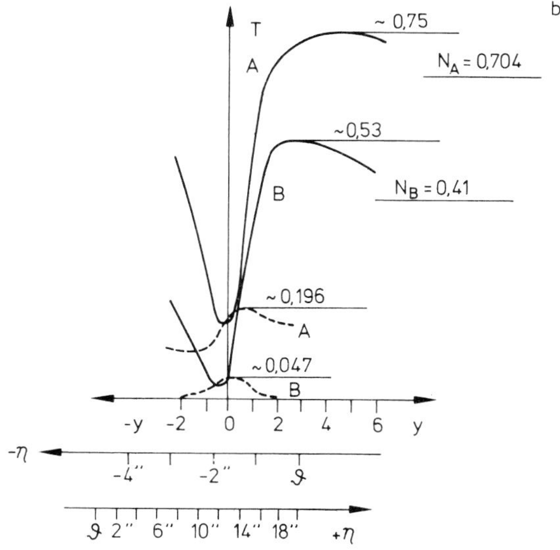

Fig. 4.6a and b

(legend for Fig. 4.6 a-d see p. 120)

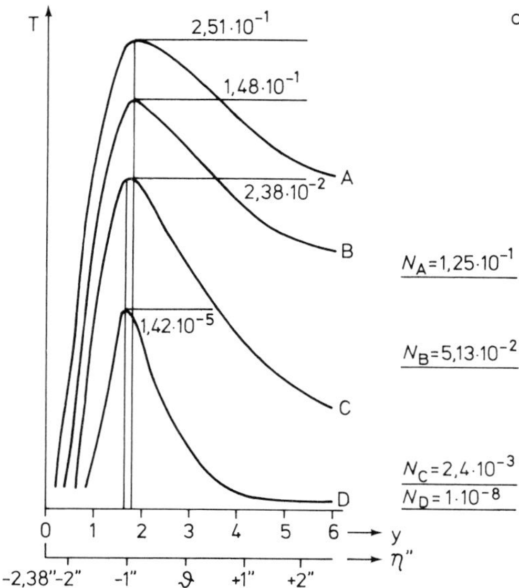

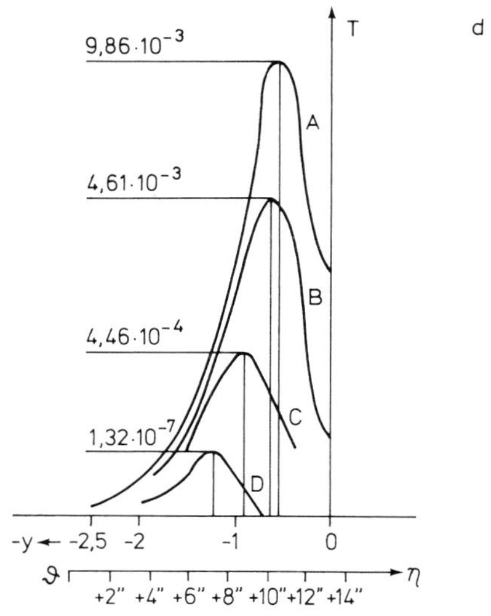

Fig. 4.6a-d Transmission curves in asymmetrical recording; Cu $K\alpha_1$, from Ge; lines N_A, N_B, N_C, N_D show the level of normal absorption outside the maximum. a) Moderate asymmetrical transmission: from the right 220; from the left $\bar{2}\bar{2}0$. (see Fig.4.4a).b) Extreme asymmetrical transmission ±333 (see Fig.4.4b); continuous lines reflections 333; dotted line $\bar{3}\bar{3}\bar{3}$. A: $\mu t \approx 0.89$; $\sigma_m t = 3$. c) Transmission curves +333. A: $\mu t \approx 2.08$; $\sigma_m t = 7$. B: $\mu t \approx 2.97$; $\sigma_m t = 10$. C: $\mu t \approx 5.94$; $\sigma_m t = 20$. D: $\mu t \approx 18.43$; $\sigma_m t = 62$. d) Transmission curves $\bar{3}\bar{3}\bar{3}$: A: $\mu t \approx 2.08$; B: $\mu t \approx 2.97$; C: $\mu t \approx 5.94$; D: $\mu t \approx 18.43$

The R curves, while preserving their asymmetrical shape, will be mirror-symmetrical about the ordinate axis (take into account rotation by an angle of 2ϑ). When switching to angle scale $\eta = \Delta\vartheta$, however, R_{hkl} curves become greater in ratio γ_0/γ_h, in their values for hkl reflections (see text).

The increasing values in the sequence of $\sigma_m t$ and μt in Figs.4.5b and c coincide with the ones in Figs.4.6b,c,d.

Values of T_{max} should be compared with values of N_A, N_B, N_C, N_D in Fig.4.6c for equal values of μt. The effect of anomalous penetration is absent in 4.6d. There is a striking difference between the curves for reflection with opposite signs of the indices. Here, in the case of the reflections hkl, as well as for the values of R when switching to the η scale, some broadening of the T curves will be observed. The value of $|D_0^{(2)}|^2$ is very small. Fig.4.6 depicts T curves with the corresponding numerical values of the main parameters.

4.6 Integrated Values of Reflection R_i and Transmission T_i in the Case of Absorbing Crystal [4.34]

The quantities R_i and T_i are the most important scattering parameters, along with the coefficients R and T, especially in scattering by an absorbing crystal. Their arrangement is a simpler experimental problem than obtaining complete reflection rocking curves, and it has been attracting the attention of many researchers.

The functions R_i and T_i for some particular cases are investigated in ZACHARIASEN's monograph [1.114] and HIRSCH's paper [4.35]. However, a more complete solution of the problem of calculating these integrated values and an analysis of the results are contained in the works of RAMACHANDRAN [4.36] and KATO [4.37]. Useful discussion, particularly with reference to thick crystals, which also takes into account the periodic components, is given by KOV'EV et al. [4.38]. In the present text, we mainly follow the calculation method described in KATO's paper, making slight changes in the values of the variables and successive inclusion of the periodic components, and deriving general expressions for the integrated quantities R_i and T_i. At the same time, the derivations given here refer to cases where $\chi_{hi} \ll \chi_{hr}$, since the Zachariasen-Laue expressions (4.96) and (4.97) are integrated. We further note that if the integrated reflection R_i is calculated quite accurately (to put it more precisely, if the corresponding integrals are proper ones), when calculating T_i we have to deal with improper integrals, for which, however, the main value can be found. The divergence of the integrals indicated reflects the specific features of the integrand at the boundary of the maximum.

Thus, by definition (see (4.81))

$$R_i^y = \int_{-\infty}^{\infty} R(y_r)dy_r \quad , \quad R_i^\eta = \frac{\left(C\frac{\gamma_h}{\gamma_o}\Phi_h\right)^{1/2}}{\sin 2\vartheta} R_i^y \quad , \tag{4.114a}$$

$$T_i^y = \int_{-\infty}^{\infty} [T(y_r) - \exp(-\sigma_o t)]dy_r \quad , \quad T_i^\eta = \frac{\left(C\frac{\gamma_h}{\gamma_o}\Phi_h\right)^{1/2}}{\sin 2\vartheta} T_i^y \quad . \tag{4.114b}$$

Accordingly, using (4.98b), we write the following expression for integrated reflection

$$R_i^y = \int_{-\infty}^{\infty} R(y_r)dy_r = \left|\frac{\chi_h}{\chi_{\bar{h}}}\right| \exp(-\sigma_m t)(W+V) \quad , \tag{4.115a}$$

where

$$W = \int_{-\infty}^{\infty} \frac{\sin^2 A\sqrt{1+y_r^2}}{1+y_r^2} dy_r \quad , \quad V = \int_{-\infty}^{\infty} \frac{1}{1+y_r^2} \sinh[(\sigma_h+\sigma_h')\tfrac{t}{2}]dy_r \quad . \tag{4.115b}$$

The values of A, σ_h, σ_h' are determined from (4.94) and (4.95). The integral W, in accordance with the analysis made in Section 3.2, is expressed as follows:

$$W = \pi \sum_{n=0}^{n} J_{2n+1}(2A) = \begin{cases} \pi A \quad , & A \ll 1 \quad , \\ \frac{\pi}{2} \quad , & A \gg 1 \quad . \end{cases} \tag{4.116}$$

As regards the hyperbolic integrand function in (4.115b) for V, its argument is also common for the hyperbolic functions, which are to be integrated when calculating T_i (see further). In order to reduce the corresponding integrals to the standard form, the following substitutions are made:

$$h = \frac{\mu t}{2\gamma_o\gamma_h} S \quad , \quad \tan\beta = \frac{2\varepsilon\sqrt{\gamma_o\gamma_h}\cos\nu_h}{\gamma_o - \gamma_h} \quad ,$$

$$\cos\beta = \frac{\gamma_o - \gamma_h}{S} = \frac{-q}{\sqrt{\varepsilon^2\cos^2\nu_h + q^2}} \quad , \tag{4.117}$$

$$S = \left|2\sqrt{\gamma_o\gamma_h(\varepsilon^2\cos^2\nu_h + q^2)}\right| \quad , \quad \varepsilon = \left|\frac{\chi_{hi}}{\chi_{oi}}\right| \quad .$$

Here, the "asymmetry factor" q is given by (4.111). Further,

$$h\cos\beta = \frac{\mu t q}{(\gamma_o\gamma_h)^{1/2}} = \sigma_h' t \coth\nu_r \quad , \quad h\sin\beta = \sigma_h t\cosh\nu_r \quad ,$$

$$y_r = \sinh\nu_r = -\cot\theta \quad . \tag{4.118}$$

In this case the given argument of the hyperbolic functions is transformed in the following manner:

$$(\sigma_h+\sigma_h')t = \frac{2\pi K|\chi_{hi}|t\cos\nu_h}{(\gamma_o\gamma_h)^{1/2}(1+y_r^2)^{1/2}}\left[1 + \frac{y_r(\gamma_o-\gamma_h)}{2\varepsilon(\gamma_o\gamma_h)^{1/2}\cos\nu_h}\right]$$

$$= h\sin\beta\sin\theta(1-\cot\beta\cot\theta) = h\cos(\theta+\beta) \quad . \tag{4.119}$$

Passing on from $\sinh^2(\sigma_h+\sigma_h')t/2$ to $\cosh[(\sigma_h+\sigma_h')t]$ and from the integration variable y_r to the variable θ, we obtain

$$V = 1/2\int_0^\pi \{\cosh[h\cos(\theta+\beta)] - 1\}d\theta \tag{4.120}$$

This integral is solved with the aid of Bessel functions of the imaginary argument $I_n(h)$. The general expression, and also the functions of the zero and first orders, have the form

$$I_n(h) = \sum_{k=0}^{\infty} \frac{1}{k!(n+k)!} \left(\frac{h}{2}\right)^{n+2k} \quad ,$$

$$I_o(h) = 1 + \frac{h^2}{4} + \frac{h^4}{2^2 \cdot 16} + \frac{h^6}{6^2 \cdot 64} + \frac{h^8}{24^2 \cdot 256} + \frac{h^{10}}{120^2 \cdot 1024} + \ldots \quad , \tag{4.121}$$

$$I_1(h) = \frac{h}{2} + \frac{h^3}{2 \cdot 8} + \frac{h^5}{12 \cdot 32} + \frac{h^7}{144 \cdot 128} + \frac{h^9}{24 \cdot 120 \cdot 512} + \frac{h^{11}}{120 \cdot 720 \cdot 1024} + \ldots \quad .$$

When passing on to thick crystals, use is made of asymptotic expressions for $I_n(h)$, which show good convergence at sufficiently large values of h. We shall give such expressions for zero- and first-order functions:

$$I_o(h) \approx F(h) \sum_i A_i h^{-i} \quad , \quad I_1(h) \approx F(h) \sum_i A_i' h^{-i} \quad , \tag{4.122a}$$

$$\sum_i A_i h^{-i} = 1 + \frac{1}{8h} + \frac{9}{128h^2} + \frac{75}{16 \cdot 64h^3} + \frac{75 \cdot 49}{8 \cdot 64^2 \cdot h^4} + \frac{15.49.81}{64^3 h^5} + \ldots$$

$$\sum_i A_i' h^{-i} = 1 - \frac{3}{8h} - \frac{15}{128h^2} - \frac{105}{16 \cdot 64h^3} - \frac{63 \cdot 75}{8 \cdot 64^2 \cdot h^4} - \frac{21.35.99}{64^3 \cdot h^5} + \ldots \quad , \tag{4.122b}$$

$$F(h) = \frac{\exp h}{\sqrt{2\pi h}} \quad .$$

Using the functions indicated [4.39], we directly obtain for the integral (4.120)

$$V = \frac{\pi}{2} \left[I_o(h) - 1 \right] \tag{4.123}$$

and for the total value for integrated reflection power:

$$R_i^{y_r} = \left(1 \pm 2 \left|\frac{\chi_{hi}}{\chi_{hr}}\right| \sin\nu_h \right) \exp(-\sigma_m t) Q \quad ,$$

$$Q = \frac{\pi}{2} \left[2 \sum_{n=0}^{n} J_{2n+1}(2A) + I_o(h) - 1 \right] \quad . \tag{4.124}$$

The integral reflection for a centro-symmetrical crystal

$$R_{ic}^{y_r} = \exp(-\sigma_m t)Q \quad . \tag{4.125}$$

A simpler expression, which is often in good agreement with the experiment, can be obtained if we proceed from the quantity $\bar{R}$. In this case we use the transition (see (4.116))

$$\pi \sum J_{2n+1}(2A) \to \pi/2 \quad ;$$

hence

$$R_i^{y_r} \approx \frac{\pi}{2}\left(1 \pm 2\left|\frac{\chi_{hi}}{\chi_{hr}}\right| \sin\nu_h\right) \exp(-\sigma_m t) I_0(h) \quad . \tag{4.126}$$

We shall now proceed to calculate the integrated transmission. In the first place, we note that since integration is performed between the limits $-\infty$ and $+\infty$, it is obvious that in (4.114b), $\exp(-\sigma_0 t)$ represents radiation transmitted through the crystal outside the maximum region.

To facilitate calculation, we evaluate the integral of the sum ($T_i^{y_r}$ + $R_{ic}^{y_r}$). Using (4.97) and (4.98), we can write (see (4.119))

$$\begin{aligned} T_i + R_{ic} = \exp(-\sigma_m t) \int_{-\infty}^{\infty} \Bigg\{ & \frac{2y_r^2 + 1}{2(y_r^2+1)} \cosh[h \cos(\theta+\beta)] + \\ & + \frac{y_r(y_r^2+1)^{1/2}}{2(y_r^2+1)} \sinh [h \cos(\theta+\beta)] + \frac{\cosh[h \cos(\theta+\beta)]}{2(y_r^2+1)} + \\ & + \frac{1}{2(y_r^2+1)} \times [\cos(2\alpha-2v_i) - \cos 2\alpha] - \exp\left[-\frac{\mu t}{2}\left(\frac{1}{\gamma_0} - \frac{1}{\gamma_h}\right)\right]\Bigg\} dy_r \quad . \end{aligned} \tag{4.127}$$

Using (4.85) and taking into account the approximation (4.83), we obtain

$$\frac{1}{2(y_r^2+1)} [\cos(2\alpha-2v_i) - \cos 2\alpha] \approx \frac{2v_i \sin 2\alpha}{2(y_r^2+1)} \quad ;$$

$$\int \frac{v_i \sin 2\alpha}{1 + y_r^2}\, dy_r = \int \frac{\beta_i \sin 2\alpha \; dy_r}{2C\left(\frac{\gamma_h}{\gamma_o}\,\Phi_h\right)^{1/2}\,(y_r^2+1)^{3/2}} \quad ,$$

since

$$\int_{-\infty}^{\infty} \frac{y_r \Psi_h \sin 2\alpha}{2\Phi_h (y_r^2+1)^{1/2}}\, dy_r = 0$$

(y_r makes the integrand function odd relative to the zero point). As a result of obvious transformations, (4.127) takes the following form:

$$T_i + R_{ic} = \exp(-\sigma_m t)(Y_1+Y_2) \quad ,$$

$$Y_2 = \frac{|\chi_{oi}|\left(1-\frac{\gamma_h}{\gamma_o}\right)}{2C\left(\frac{\gamma_h}{\gamma_o}\,\Phi_h\right)^{1/2}} \int_{-\infty}^{\infty} \frac{\sin 2\alpha}{(y_r^2+1)^{3/2}}\, dy_r \quad ,$$

$$Y_1 = \int_{-\infty}^{\infty} \Bigg\{ \cosh[h \cos(\theta+\beta)] +$$

$$+ \frac{y_r(y_r^2+1)^{1/2}}{y^2 + 1} \sinh[h \cos(\theta+\beta)] - \exp(h \cos\beta) \Bigg\}\, dy_r \quad . \tag{4.128}$$

Turning to the integral Y_1, we note that it is an improper one. Indeed, with increasing y_r the coefficient at sinh tends to unity. Further, the sum (cosh + sinh) takes the value $\exp[h \cos(\theta+\beta)]$. Inspecting (4.118) and (4.119) we note that with a further increase in y_r, $\exp[h \cos(\theta+\beta)] \to$ $[\exp(h \cos\beta + h \sin\beta/y_r)$.

As a result, the integrand acquires the following asymptotic value:

$$\exp(h \cos\beta) \left[\frac{h \sin\beta}{y^2} + \frac{1}{2!}\left(\frac{h \sin\beta}{y_r}\right)^2 + \ldots\right] \quad . \tag{4.129}$$

Thus, this integral is a divergent one, and in what follows we shall determine its main value.

Subdivide Y_1 into two parts

$$Y_1' = \int_0^{\pi} \{\cosh[h \cos(\theta+\beta)] - \cosh(h \cos\beta)\} \sin^{-2}\theta d\theta \quad , \tag{4.130}$$

$$Y_1'' = \int_0^{\pi} \{\cos\theta \sinh[h \cos(\theta+\beta)] - \sinh(h \cos\beta)\} \sin^{-2}\theta d\theta \quad . \tag{4.131}$$

In calculating these integrals, use is made of the well-known series expansion in Bessel functions of the imaginary argument [4.39]:

$$\begin{aligned} \cosh(h \cos\varphi) &= I_o(h) + 2I_2(h) \cos 2\varphi + 2I_4(h) \cos 4\varphi + \dots \\ &\dots + 2I_{2n}(h) \cos 2n\varphi \quad , \end{aligned} \tag{4.132}$$

$$\begin{aligned} \sinh(h \cos\varphi) &= 2I_1(h) \cos\varphi + 2I_3(h) \cos 3\varphi + 2I_5(h) \cos 5\varphi + \dots \\ &\dots + 2I_{2n+1}(h) \cos(2n+1)\varphi \quad . \end{aligned} \tag{4.133}$$

The integral Y_1' is transformed with the aid of the expansion (4.132):

$$Y_1' = 2 \sum_{n=0}^{\infty} I_{2n}(h) \int_0^{\pi} [\cos 2n(\theta+\beta) - \cos 2n\beta]\sin^{-2}\theta d\theta \quad , \tag{4.134}$$

and its main value corresponds to the limiting transition

$$\lim_{\varepsilon\to 0} \int_{\varepsilon}^{\pi-\varepsilon} [\cos 2n(\theta+\beta) - \cos 2n\beta]\sin^{-2}\theta d\theta \quad . \tag{4.135}$$

Dividing the range of integration in (4.134) into two parts, we obtain the desired value of this integral

$$\int_0^{\pi/2} [\cos 2n(\theta+\beta) + \cos 2n(\theta-\beta) - 2 \cos n\beta]\sin^{-2}\theta d\theta =$$

$$= - 4 \cos 2n\beta \int_0^{\pi/2} \frac{\sin^2 n\theta}{\sin^2\theta} d\theta = -2n\pi \cos 2n\beta \quad . \tag{4.136}$$

This last integral is a standard one (it is calculated with the aid of integral sums). Thus,

$$Y_1' = -2\pi \sum_{n=0}^{\infty} 2n \cos 2n\beta I_{2n}(h) \quad . \tag{4.137}$$

Similarly, we get for the second integral of (4.131):

$$Y_1'' = -2\pi \sum_{n=0}^{\infty} (2n+1) \cos[(2n+1)\beta]I_{2n+1}(h) \quad . \tag{4.138}$$

The total value of the integral Y_1 is

$$Y_1 = -2\pi \sum_{m=1}^{\infty} [m \cos(m\beta)I_m(h)] \quad . \tag{4.139}$$

Another expression for the integral Y_1, which is used subsequently in an analysis of the equations obtained with reference to certain individual cases, can be obtained by using recurrence equations for functions $I_n(h)$ [4.40]:

$$mI_m(h) = \frac{h}{2}[I_{m-1}(h) - I_{m+1}(h)] \quad , \tag{4.140}$$

hence

$$Y_1' = -\pi hI_1(h) + 2\pi h \sin\beta \sum_{n=0}^{\infty} \sin[(2n+1)\beta]I_{2n}(h) \quad , \tag{4.141}$$

$$Y_1'' = -\pi h \cos\beta I_o(h) + 2\pi h \sin\beta \sum_{n=1}^{\infty} \sin(2n\beta) I_{2n}(h) \quad , \tag{4.142}$$

$$Y_1 = -\pi h[I_1(h) + I_o(h)\cos\beta] + 2\pi h \sin\beta P \quad , \tag{4.143}$$

$$P = \sum_{m=0}^{\infty} I_m(h) \sin m\beta \quad . \tag{4.144}$$

The following expression can also be obtained for P in the general case on hand. Proceeding from another recurrence equation for $I_n(h)$

$$\frac{d}{dh} I_n(h) = \frac{1}{2}[I_{n-1}(h) + I_{n+1}(h)] \quad , \tag{4.145}$$

we obtain the differential equation for P

$$\frac{\partial P}{\partial h} = \frac{1}{2} I_o \sin\beta + P \cos\beta \quad . \tag{4.146}$$

At the initial condition P = 0 for h = 0 we have

$$P = \frac{1}{2} \sin\beta \exp(h \cos\beta) \int_0^h I_o(x) \exp(-x \cos\beta) dx \quad . \tag{4.147}$$

Let us calculate the integral Y_2, which we rewrite as

$$Y_2 = \left|\frac{x_{hi}}{x_{hr}}\right| \cot\beta \cos\nu_h Y_2' \quad , \quad Y_2' = \int_{-\infty}^{+\infty} \frac{\sin 2A(1+y_r^2)^{1/2}}{(1+y_r^2)^{1/2}} dy_r \quad . \tag{4.148}$$

Passing on to the angular variable V_r

$$Y_2' = \int_{-\infty}^{+\infty} \frac{\sin(2A \cosh v_r)}{\cosh^2 v_r} dv_r \tag{4.149}$$

and applying the series expansion in function $I_n(h)$

$$\sin(2A \cosh v_r) = 2I_1(2A) \cosh v_r + 2I_3(2A) \cosh(3v_r) + \ldots \tag{4.150}$$

we transform Y_2' in the integral series

$$Y'_2 = 4I_1(2A) \int_0^\infty \frac{dv_r}{\cosh v_r} + 4I_3(2A) \int_0^\infty \frac{\cosh(3v_r)}{\cosh^2 v_r} dv_r + + 4I_5(2A) \int_0^\infty \frac{\cosh(5v_r)}{\cosh^2 v_r} dv_r + \dots \quad . \tag{4.151}$$

These integrals are standard. Their solutions have the form

$$\int_0^\infty \frac{dv_r}{\cosh v_r} = \frac{\pi}{2} \quad , \quad \int_0^\infty \frac{\cosh(\ell v_r)}{\cosh^2 v_r} dv_r = B(1+\tfrac{\ell}{2},\ 1-\tfrac{\ell}{2}) \quad . \tag{4.152}$$

The functions B are calculated with the aid of infinite products

$$B(x,y) = \prod_{k=0}^{\infty} \frac{k(x+y+k)}{(x+k)(y+k)} \quad , \tag{4.153}$$

which are negligibly small at reasonable values of the parameters.

Thus,

$$Y_2 = \left|\frac{\chi_{hi}}{\chi_{hr}}\right| \cot\beta[2\pi I_1(2A) + 2\sum_{n=1}^{\infty} I_1(2A)\ B(1+\tfrac{\ell}{2},1-\tfrac{\ell}{2})] \approx \approx \left|\frac{\chi_{hi}}{\chi_{hr}}\right| \cot\beta\ 2\pi I_1(2A)\cos\nu_h \tag{4.154}$$

The above calculation leads to the following two expressions for integrated transmission power,

$$T_{i,1}^{y_r} = \exp(-\sigma_m t)(Y_1+Y_2) - R_{ic} = \pi \exp(-\sigma_m t)[M(h) + N(A)] \quad ,$$

$$T_{i,2}^{y_r} = \pi \exp(-\sigma_m t)[M'(h) + N(A) + \tfrac{1}{2}] \quad . \tag{4.155}$$

Here, the functions M and M' of h are the result of integration of the monotonic components R and T, while the function N of A refers to integration of the periodic components. Below we give the explicit expressions for these functions:

$$M(h) = \frac{1}{2}[1 - I_o(h)] - 2 \sum_{m=1} m \cos(m\beta)\ I_m(h) \quad ,$$

$$M'(h) = 2Ph \sin\beta - hI_1(h) - \frac{1}{2}(1+2h \cos\beta)\ I_o(h) \quad , \tag{4.156}$$

$$N(A) = 2\left|\frac{\chi_{hi}}{\chi_{hr}}\right| \cot\beta \cos\nu_h + I_1(2A) - \sum_{n=1}^{\infty} J_{2n+1}(2A) \quad . \tag{4.157}$$

The introduction of the two different expressions for T_i, as well as of the the different expressions for the functions P, is dictated by the need to use, in certain different cases (thin and thick crystals, symmetrical and asymmetrical reflections), such functions as will show the best convergence in certain limiting conditions.

4.7 Analysis of Expressions for Integrated Values R_i and T_i as Applied to Important Particular Cases

Transform the general expressions previously obtained to a form more convenient both for qualitative analysis in individual cases and for numerical evaluations of the convergence of the series used.

1) Scattering in a thin crystal. In this case it is convenient to use the exact expressions for the functions $I_n(h)$ and $I_1(2A)$, similar to those given in (4.121), and investigate them in the range of argument values from A, $h \ll 1$ to A, $h \approx 1\text{-}5$. Applying the relevant condition in (4.116) and also the expressions (4.114), (4.121), and (4.124), for the integrated reflection we have

$$R_i^n \approx \frac{\left(C \frac{\gamma_h}{\gamma_o} \Phi_h\right)^{1/2}}{\sin 2\vartheta} \left(1 \pm 2\left|\frac{\chi_{hi}}{\chi_{hr}}\right| \sin\nu_h\right) \exp(-\sigma_m t) \quad .$$

$$\cdot\ \frac{\pi}{2}\left(2A + \frac{h^2}{4} + \frac{h^4}{64} + \frac{h^6}{2304} + \ldots\right) \quad . \tag{4.158}$$

To calculate the integrated transmission it is convenient to use the function M(h) according to (4.155) and (4.156). Using a general expression (4.121) to calculate the functions $I_n(h)$, we get, for $n > 1$,

$$T_i^\eta = \frac{C\left(\frac{\gamma_h}{\gamma_o}\right)\Phi_h^{1/2}}{\sin 2\vartheta} 2\pi \exp(-\sigma_m t)\left[-\frac{h}{2}\cos\beta - \frac{h^2}{16}(4\cos 2\beta+1) - \right.$$

$$- \frac{h^3}{16}(\cos\beta+\cos 3\beta) - \frac{h^4}{256}(2.67\cos 4\beta+6\cos 2\beta+1) - \tag{4.159}$$

$$-\frac{h^5}{384}(1.5\cos 3\beta+\cos\beta) - \frac{h^6}{64\cdot 144}(4.8\cos 4\beta+6\cos 2\beta+1) +$$

$$\left. + \frac{\chi_{hi}}{\chi_{hr}}\cot\beta\left(A+\frac{A^3}{2}+\frac{A^5}{12}+\frac{A^7}{144}+\dots\right) - \frac{A}{2}\right] .$$

These expressions are simplified considerably when we pass on to symmetrical reflection, and

$$\gamma_o = \gamma_h = \cos\vartheta \quad , \quad \sigma_m = \sigma_o \quad , \quad \cos\beta = \cot\beta = \cos(2n+1)\beta = 0$$

$$(n=1,3,5\dots) \quad , \tag{4.160}$$

$$\cos 2n\beta = -1 \quad , \quad \cos 4n\beta = 1 \quad , \quad h_s = \frac{\mu t\varepsilon\cos\nu_h}{\cos\vartheta} .$$

The expressions (4.158) and (4.159) take the form

$$R_{is}^\eta \approx \frac{C\Phi_h^{1/2}}{\sin 2\vartheta}\frac{\pi}{2}\left(1\pm 2\left|\frac{\chi_{hi}}{\chi_{hr}}\right|\sin\nu_h\right)\exp(-\sigma_o t)\cdot$$

$$\cdot\left(2A + \frac{h_s^2}{4} + \frac{h_s^4}{64} + \frac{h_s^6}{2304} + \dots\right) , \tag{4.161}$$

$$T_{is}^\eta \approx \frac{C\Phi_h^{1/2}}{\sin 2\vartheta} 2\pi \exp(-\sigma_o t)\left(-\frac{A}{2} + \frac{h_s^2}{5.33} + \frac{h_s^4}{110} + \frac{h_s^6}{4600} + \dots\right) . \tag{4.162}$$

Comparing (4.158) and (4.159) for an asymmetrical arrangement with (4.161) and (4.162) for a symmetrical one, we note that the former can be used within a smaller range of μt. This is due to the fact that $h \approx 2\text{-}2.5\ \mu t$, while $h_s \approx \mu t$. Although at small μt the phenomena should be similar to scattering in transparent crystals which are characterized, say, by the negative value of T_i in (4.159) and (4.162) for h and h_s less than unity, we are nevertheless able to note features inherent in absorbing crystals. First of all, we ought to mention the difference in the values of R_i for hkl and $\bar{h}\bar{k}\bar{l}$ reflections because $\sin\nu_h \neq 0$. The value of integrated transmission T_i in asymmetrical arrangement differs for opposite signs of the indices because of the change in the signs at $\cos \ell\beta$ in (4.159); this happens both in the absence and in the presence of a centre of symmetry in the scattering crystal. Moreover, the values of T_i and R_i differ for reflections with opposite indices because of the inversion of the fraction γ_h/γ_o.

The expression (4.162) very pictorially reflects the transition from negative values for $h_s < 1$ to positive ones for $h_s > 1$, which corresponds to the transition to the intermediate region between the thin- and thick-crystal approximations. Although the convergence of the series appearing in the expressions under review deteriorates when the values of h and A increase above ~ 5, these equations can still be used for approximate calculation of the integral values in the intermediate range.

In the limit, at $h \approx 0$, the relations obtained should coincide with the equations for the transparent crystal. Taking into account the fact that in this case μ should be equal to zero, we get, for instance, from (4.161) and (4.162):

$$R_{is}^{y_r} \approx \pi A \quad , \quad T_{is}^{y_r} = -R_{is}^{y_r} \quad . \tag{4.163}$$

2) Scattering in the range intermediate between the thin- and thick-crystal approximations. At values of h up to 10, integral reflections can be determined, in accordance with (4.124), by using the expansion (3.81) for calculating

$$\int_0^{2A} J_o\,(x)dx$$

and the table for the values of $I_o(h) = J_o(ih)$ [4.40].

3) Scattering by a thick crystal. Symmetrical reflection. For the thick-crystal approximation, use is made of the asymptotic expressions for the functions $I_o(h)$ and $I_1(h)$. We shall first consider symmetrical reflection, which is the subject of most of the existing experimental investigations. Since the asymptotic expressions (4.122a) have a common multiplier $\exp h/(2\pi h)^{1/2}$, in the equations for R_i and T_i

$$\exp[-(\sigma_o t-h_s)] = \exp\left[-\frac{\mu t}{\cos\vartheta}(1-\varepsilon\cos\nu_h)\right] \quad . \tag{4.164}$$

For integrated reflection, we obtain (see (4.126))

$$R^{\eta}_{is} \approx \frac{\sqrt{\pi}}{2\sqrt{2}}\frac{C|\chi_{hr}|}{\sin 2\vartheta}\frac{\exp\left[-\frac{\mu t}{\cos\vartheta}(1-\varepsilon\cos\nu_h)\right]}{(\mu t\varepsilon\cos\nu_h)^{1/2}}\sqrt{\cos\vartheta} \quad . \tag{4.165}$$

$$\cdot\left(1\pm 2\left|\frac{\chi_{hi}}{\chi_{hr}}\right|\sin\nu_h\right)\left(1+\frac{1}{8h_s}+\frac{9}{128h_s^2}+\dots\right) \quad .$$

Note the good divergence of the series in (4.165). Even for $h_s = 15$, i.e., in the range intermediate between the thin- and thick-crystal approximations, the third term containing h_s^{-2} will equal only 0.03 percent of unity. In order to calculate the integrated transmission it is necessary to use T_{i2} from (4.155). Here, the function $M'(h)$ is given in (4.156), and the function $N(A)$ takes the asymptotic value for symmetrical reflection, equal to -1/2. As regards the quantity P,

$$P_s = \frac{1}{2}\int_0^h I_o(x)dx \approx \frac{\exp(h_s)}{\sqrt{2\pi h}}\sum_{k=0}^{\infty}\frac{(2k-1)!!}{(2h_s)^k}\sum_{m=0}^{m=k}\frac{(2m-1)!!}{m!\,4m} \quad . \tag{4.166}$$

Further using the asymptotic expressions for $I_o(h)$ and $I_1(h)$, we obtain

$$T^{\eta}_{is} \approx \frac{\sqrt{\pi}}{2\sqrt{2}}\frac{C|\chi_{hr}|}{\sin 2\vartheta}\frac{\exp\left[-\frac{\mu t}{\cos\vartheta}(1-\varepsilon\cos\nu_h)\right]}{(\mu t\varepsilon\cos\nu_h)^{1/2}}\sqrt{\cos\vartheta} \quad . \tag{4.167}$$

$$\cdot\left(1+\frac{2.125}{h_s}+\frac{5.320}{h_s^2}+\frac{18.590}{h_s^3}+\frac{83.587}{h_s^4}+\dots\right) \quad .$$

Obviously, when applying the above equations for R_{is} and T_{is} to scattering in crystals with a centre of symmetry, one should put $\cos\nu_h = 1$, $\sin\nu_h = 0$.

The presence of the factor $\cos\nu_h$ in the exponent of the attenuation factor in scattering in noncentro-symmetrical crystals considerably reduces the effect of anomalous transmission, compared to that in crystals with a centre of symmetry. The reduction is particularly great for strong reflections, i.e., for large values of ε. Thus, at $\varepsilon = 0.96$, $\cos\nu_h = 0.95$, and $\mu t = 25$:

$$\exp[-\mu t(1-\varepsilon)] = 0.368 \quad \exp[-\mu t(1-\cos\nu_h\varepsilon)] = 0.107 \quad . \tag{4.168}$$

An important scattering parameter is the ratio of integrated transmission and reflection. In experimental investigations this ratio may serve as an approximate criterion of the degree of perfection of the scattering crystal.

The convergence of the series for T_i (4.167) is worse than that of the series (4.165) for integrated reflection. Thus, at $\mu t = 30$, the use of only the first term in (4.165), i.e., of unity, leads to an error of ~ 0.5 percent. The same value of the error requires the use of the first terms in (4.167), including the term with h_s^{-2}. With the above accuracy, the ratio

$$Q_s = \frac{T_{is}}{R_{is}} \approx \left(1 \pm 2\left|\frac{\chi_{hi}}{\chi_{hr}}\right| \sin\nu_h\right)^{-1} \left(1 + \frac{2.125}{h_s} + \frac{5.320}{h_s^2} + \frac{18.520}{h_s^3}\right) \tag{4.169}$$

for a centro-symmetrical crystal is at least 1.057 at $h_s \approx \mu t = 40$ and about 1.04 at $\mu t = 50$. In the case of a crystal without a centre of symmetry, when the factor in the front of the series in (4.169) needs to be taken into account, the value of Q for reflections with negative indices may prove to be less than unity.

4) Scattering by a thick crystal. Asymmetrical reflection. In this case, too, we shall evidently use asymptotic expressions for the functions $I_n(h)$. For the exponent of the absorption factor we obtain

$$\sigma_m t - h = \frac{\mu t}{2\gamma_o\gamma_h} [(\gamma_o+\gamma_h) - S] \quad . \tag{4.170}$$

For integrated reflection we get an equation differing but little from (4.165),

$$R_i^{\eta} \approx \frac{\sqrt{\pi}}{2\sqrt{2}} \frac{C|\chi_{hr}|}{\sin 2\vartheta} \frac{\exp\left\{-\frac{\mu t}{2\gamma_o\gamma_h}[(\gamma_o+\gamma_h) - S]\right\}}{(\mu t S)^{1/2}} \sqrt{2\gamma_o\gamma_h} \quad ,$$

$$\cdot \left(1 \pm 2\left|\frac{\chi_{hi}}{\chi_{hr}}\right| \sin\nu_h\right)\left(1 + \frac{0.125}{h} + \frac{0.703}{h^2} + \dots\right)\sqrt{\frac{\gamma_h}{\gamma_o}} \quad . \tag{4.171}$$

In order to obtain the expression for T_i we have to derive an asymptotic equation for the function P in general form. To this end, we introduce a new variable

$$g(h) = 2\sin^{-1}\beta\sqrt{2\pi h}\exp(-h)P \quad . \tag{4.172}$$

Differentiating (4.172) with respect to h and using (4.147), we get the following differential equation for g(h):

$$\frac{\partial g}{\partial h} + g[1 - \cos\beta - (2h)^{-1}] = \sqrt{2\pi h}\exp(-h)I_o(h) \quad . \tag{4.173}$$

We now pass on to large h (see (4.122))

$$\sqrt{2\pi h}\, I_o(h)\exp(-h) = A_o + \frac{A_1}{h} + \frac{A_2}{h^2} + \dots \quad . \tag{4.174}$$

Assume that the asymptotic expression for the function g has a similar form:

$$g \approx B_o + \frac{B_1}{h} + \frac{B_2}{h^2} + \dots \quad . \tag{4.175}$$

In this case, substituting this value of g, and also the right-hand side of (4.174), in (4.173) and equating the coefficients with equal h, we get

$$B_o = \frac{A_o}{1 - \cos\beta} \quad , \quad B_i = \frac{A_i + \left(i - \frac{1}{2}\right)B_{i-1}}{1 - \cos\beta} \quad (i \geq 1) \quad . \tag{4.176}$$

Thus, the general solution of the differential equation (4.173) for g will be:

$$g \approx M\sqrt{2\pi h}\exp[-(1-\cos\beta)h] + B_o + (B_1/h) + (B_2/h^2) + \dots \quad . \tag{4.177}$$

The integration constant M should be equal to zero, since P for large $h(1-\cos\beta)$ has the value

$$P \approx \frac{\sin\beta}{2} \frac{\exp h}{\sqrt{2\pi h}} \frac{1}{1-\cos\beta} \quad , \tag{4.178}$$

and, hence, the desired asymptotic expression for P has the form

$$P \approx \frac{\sin\beta}{2} \frac{\exp h}{\sqrt{2\pi h}} \sum_{i=0} B_i h^{-i} \quad . \tag{4.179}$$

Note that in the particular case of symmetrical reflection $\cos\beta = 0$ and the coefficients B_i take the values of the coefficients of the series (4.166).

Using the solution for $T_{i,2}$ in (4.155) and $M'(h)$ in (4.156), neglecting the small value of the first term in the expression for N(A), and taking into account that at large A $2 \sum_n J_{2n+1}(2A) - 1 = 0$ we obtain

$$T_i^\eta = \frac{\pi}{2} \frac{C|\chi_{hr}|}{\sin 2\vartheta} \exp\left[-\frac{\mu t}{2\gamma_o\gamma_h}(\gamma_o+\gamma_h)\right] 2M'(h) \sqrt{\frac{\gamma_h}{\gamma_o}} \tag{4.180}$$

$$2M'(h) \approx \frac{2\exp h}{(2\pi h)^{1/2}} \left[h(1-\cos^2\beta)\sum_{i=0}^{i} B_i h^{-i} - 2h\sum_i A_i' h^{-i} - \right.$$
$$\left. - \frac{1}{2}(1+2h\cos\beta)\sum_i A_i h^{-i}\right] \quad . \tag{4.181}$$

Substituting the values of the coefficients B_i from (4.176), A_i and A_i' from (4.122), we get for integrated transmission

$$T_i^\eta = \frac{\sqrt{\pi}}{2\sqrt{2}} \sqrt{\frac{\gamma_h}{\gamma_o}\frac{C|\chi_{hr}|}{\sin 2\vartheta}} \frac{\exp\left\{-\frac{\mu t}{2\gamma_o\gamma_h}[(\gamma_o+\gamma_h) - S]\right\}}{(\mu t S)^{1/2}} \sqrt{2\gamma_o\gamma_h} \sum_i L_i h^{-i} \quad . \tag{4.182}$$

The general expression for the coefficients L_i is of the form

$$L_i = 2(1+\cos\beta)(i+\tfrac{1}{2}B_i) + 2(|A'_{i+1}|+A_{i+1}) - A_i \quad . \tag{4.183}$$

Now we give the expressions for the six coefficients L_i in explicit form[2]:

$$L_0 = \frac{1 + \cos\beta}{1 - \cos\beta} \quad ; \quad L_1 = \frac{3}{2} \frac{1 + \cos\beta}{(1-\cos\beta)^2} \left[1 + \frac{1}{4}(1-\cos\beta)\right] + \frac{1}{4} \quad ;$$

$$L_2 = \frac{3 \cdot 5}{4} \frac{1 + \cos\beta}{(1-\cos\beta)^3} \left[1 + \frac{1}{4}(1-\cos\beta) + \frac{3}{32}(1-\cos\beta)^2\right] + \frac{9}{32} \quad ;$$

$$L_3 = \frac{3 \cdot 5 \cdot 7}{8} \frac{1 + \cos\beta}{(1-\cos\beta)^4} \left[1 + \frac{1}{4}(1-\cos\beta) + \frac{3}{32}(1-\cos\beta)^2 + \right.$$

$$\left. + \frac{5}{128}(1-\cos\beta)^3\right] + \frac{3^2 \cdot 5^2}{8^3} \quad ; \tag{4.184}$$

$$L_4 = \frac{3 \cdot 5 \cdot 7 \cdot 9}{16} \frac{1 + \cos\beta}{(1-\cos\beta)^5} \left[1 + \frac{1}{4}(1-\cos\beta) + \frac{3}{32}(1-\cos\beta)^2 + \right.$$

$$\left. + \frac{5}{128}(1-\cos\beta)^3 + \frac{5 \cdot 7}{32 \cdot 64}(1-\cos\beta)^4\right] + \frac{3 \cdot 5^2 \cdot 7^2}{64}$$

$$L_5 = \frac{3^3 \cdot 5 \cdot 7 \cdot 11}{32} \frac{1 + \cos\beta}{(1-\cos\beta)^6} \left[1 + \frac{1}{4}(1-\cos\beta) + \frac{3}{32}(1-\cos\beta)^2 + \right.$$

$$\left. + \frac{5}{128}(1-\cos\beta)^3 + \frac{5 \cdot 7}{32 \cdot 64}(1-\cos\beta)^4 + \frac{3^2 \cdot 7}{2 \cdot 64^2}(1-\cos\beta)^5\right] + \frac{49 \cdot 75 \cdot 81}{32 \cdot 64^2}$$

Let us turn to the qualitative analysis of the relations obtained for the values R_i and T_i in asymmetrical arrangement. First of all, from a comparison of the L_i and A_i series it follows that R_i can, generally speaking, be calculated, with greater accuracy because of the better convergence of the A_i series. This is illustrated by the curves of R and T in Figs.4.5 and 4.6. Then, the most important feature of asymmetrical reflection is the difference between the values of R_i and T_i for reflections with opposite signs of the indices. As far as integrated reflection is concerned, this difference is due to the factor γ_h/γ_o, which leads only to an increase in the half-width of the maxima on interchanging the numerator and the denominator. If we proceed from symmetrical reflection and decrease the angle φ in the range of its positive values in accordance with Figs.3.2a and b, the integral reflection

[2] The coefficients L_i are related to the coefficients D_i of the corresponding expansion in KATO's paper [4.37] by the expression $L_i = 2D_i - A_i$.

R_i for $\bar{h}\bar{k}\bar{l}$ will, to an increasing degree, exceed the same value for hkl. The ratio $R^{\eta}_{i,hkl}/R^{\eta}_{i,\bar{h}\bar{k}\bar{l}}$ is independent of μt and equal γ_o/γ_h if the latter refer to reflection hkl. Eq.(4.171) can be used even at $\mu t = 10$ and $h = \mu t(S/2\gamma_o\gamma_h) = 25$, which corresponds to a moderate asymmetrical reflection ($q \approx 0.25$) at $\varepsilon = 0.96$ or to an extreme reflection ($q \approx 0.97$), $\varepsilon = 0.61$. If we take into account the three terms in the series, the termination error will be of the order of hundredths of one percent. The values of integrated transmission T_i can be calculated with sufficient accuracy only in the case of moderate asymmetrical reflection, which corresponds to values $\cos\beta$ less than ~ 0.5, according to (4.117). An example of this, which we have used (see Fig.4.6), are reflections 220 at $\varphi = 60^o$. The series

$$\sum_i L_i h^{-i} = 1.667 + 4.2085\ h^{-1} + 15.900\ h^{-2} + 65.61\ h^{-3} + 394.07\ h^{-4} + \dots \qquad (4.185)$$

even at $h = 20(\mu t \approx 15.7)$, yields an error of fractions of one percent.

In the case of extreme asymmetrical reflection (+333 at $\varphi = 55^o10'$) it follows from Fig.4.6 that accurate calculation of the values T_i is impossible; since in this case $\cos\beta = 0.8468$, the series $\sum_i L_i h^{-i}$ can be used only within six terms, and even for $h = 42$ (μt=18.6) the termination error equals a few percent.

Noteworthy is the effect of the reduction in $T_{i(\bar{h}\bar{k}\bar{l})}$ as compared with the same value for reflection with positive indices. This reduction is not great for moderate asymmetrical reflection and increases when changing to extreme reflection 333, growing with μt (see Fig.4.6). The reduction in asymmetrical integrated transmission in the case of negative indices is opposite to the effect of reduction in asymmetrical reflection, which takes place for positive indices, as noted above. The reduction in $T_{i,\bar{h}\bar{k}\bar{l}}$, compared to $T_{i,hkl}$, is the result of the superposition of the two opposite effects. A change in the sign of $\cos\beta$ leads to a relative increase in $T_{i,hkl}$, while inversion of the fraction $(\gamma_h/\gamma_o)^{1/2}$ leads to an increase in $T_{i,\bar{h}\bar{k}\bar{l}}$. It can be seen from (4.117) that an increase in the modulus $|\cos\beta|$ is possible only with an increasing ratio γ_o/γ_h.

Keeping in mind the good convergence of the series in (4.171) for R_i, the value of the ratio of integrated transmission to integrated reflection in the asymmetrical case can be determined from the equation

$$Q = \frac{T_i}{R_i} \approx \left(1 \pm 2\left|\frac{\chi_{hi}}{\chi_{hr}}\right| \sin\nu_h\right)^{-1} \cdot \sum_i L_i h^{-i} \quad . \tag{4.186}$$

For scattering in a transparent crystal, transition to π-polarizations in the plane of reflection leads to a reduction in the coefficient of reflection and integrated reflection, but these parameters have the same order of magnitude as for σ-polarizations. This follows directly, for instance, from (3.82b) for integrated reflection.

For reflections 220 and 333 of Mo $K\alpha_1$ radiation from a thin Si slice:

$$R^{220}_{i,\pi} = 0.93\, R^{220}_{i,\sigma} \quad , \quad R^{333}_{i,\pi} = 0.77\, R^{333}_{i,\sigma} \quad .$$

Passing on to scattering in an absorbing crystal, we note that in the case of a thin crystal integral reflection power $R_{i,\pi} \sim |\cos 2\vartheta|\, R_{i,\sigma}$ (4.158,161) for any arrangement. A similar dependence obtains for T_i.

In a thick crystal the propagation of waves with π-polarizations obeys other laws, and in practice such waves hardly ever show anomalous penetration. This results from the fact that the function h (4.117) appears in the exponent at absorption factors $(\sigma_m t - h)$ and $(\sigma_0 t - h_s)$ (see (4.165,167,171,182)).

In the case of symmetrical reflection the effect of the polarization factor is easier to trace. For strong reflections, when $(1-\varepsilon) \ll 1$, the absorption factors for σ- and π-polarizations differ drastically. A typical case is the reflection 220 of Cu $K\alpha_1$ radiation from Ge. Here, $\varepsilon = 0.96$ and, for instance, at $\mu t = 20$ the absorption factor for σ-polarization is equal to 0.45 and for π-polarization, $3 \cdot 4 \cdot 10^6$, i.e., the radiation transmitted through the crystal is almost completely plane-polarized.

For weaker reflections, as well as for asymmetrical ones, the difference between the attenuation factors is not so great.

Another remark refers to the so-called Friedel law, which plays an important part in the X-ray structure analysis based on the kinematical theory of scattering. The Friedel law states that the reflection intensity is insensitive to the presence or absence of a centre of symmetry in the structure. It is particularly important to indicate that here we deal with reflections from planes which are not parallel to the polar directions existing in a non-centro-symmetrical crystal.

It follows from the text of Chapters 3 and 4 that one can speak of the fulfillment or violation of the Friedel law in dynamical scattering with the following reservation. First, in place of the intensity, in dynamical scattering a comparison should be between reflection curves and integrated reflection R_i.

Secondly, as has been shown by a number of authors, in multi-wave scattering the Friedel law for integrated reflection powers is inapplicable to scattering both in a transparent and an absorbing crystal.

However, as is clear from the foregoing, in the two-wave approximation, scattering in a transparent crystal obeys the Friedel law, whereas for scattering in an absorbing crystal this law is not valid due to the presence of the factor $\chi_h/\chi_{\bar{h}} = [1 \pm 2(\chi_{hi}/\chi_{hr}) \sin\nu_h]$ in the equations for integrated reflection R_i. It should be emphasized that this factor acquires values differing appreciably from unity only for scattering of radiations with a wavelength close to the absorption edge by crystals containing heavy atoms. Thus, for instance, in the scattering of radiations with a wavelength close to the absorption edge of a metal atom by crystals of CdS with a sphalerite structure, the values of the factor indicated will differ greatly, according to (4.56). Indeed, the value

$$\Delta = (\sum_{\text{cell}} f_r \sin\varphi)(\sum_{\text{cell}} \mu_a^{(h)} \cos\varphi) - (\sum_{\text{cell}} \mu_a^{(h)} \sin\varphi)(\sum_{\text{cell}} f_r \cos\varphi) \tag{4.187}$$

will be determined by the sums connected with the sines if we choose the description of the structure in which Cd atoms are located at the centres of the octants. In this case, using a radiation of the above-mentioned wavelength, we obtain reduction in f_r of Cd due to the value and negative sign of the corrections Δf. On the other hand, the absorption $(\sum \mu_a^{(h)} \sin\varphi)$ increases in this case. The ratio (see (4.57))

$$\frac{1+\alpha}{1-\alpha} = \frac{\lambda \frac{e^2}{mc^2} + \Delta/|F_{hr}|^2}{\lambda \frac{e^2}{mc^2} - \Delta/|F_{hr}|^2} \quad ; \quad \alpha = 2\left|\frac{\chi_{hi}}{\chi_{hr}}\right| \sin\nu_h = \Delta\left(\lambda \frac{e^2}{mc^2}\right)^{-1} |F_{hr}|^2 \tag{4.188}$$

may be as high as 2 [Ref.1.14,pp.131,138].

5. Poynting's Vectors and the Propagation of X-Ray Wave Energy

5.1 Averaged Poynting's Vector in the General Case

The theory outlined describing the generation and propagation of wave fields in a crystal cannot be complete without a consideration of Poynting's vectors, along which the wave field energy propagates. As in the case of visible light, the directions of Poynting's vectors in X-ray fields in a crystal do not coincide with the wave vector $\underline{k}_o^{(i)}$ or $\underline{k}_h^{(i)}$.

The relations determining Poynting's vectors and their paths can be taken into account in considering the case when the front of the incident wave is small as compared with the crystal thickness, or scattering of X-rays in distorted crystals.

The density vector $\underline{S}$ of energy current, represented by the expression

$$\underline{S} = \frac{c}{4\pi} [\underline{E}\underline{H}] \quad , \tag{5.1}$$

is used in this theory in a slightly modified form. In the general case of absorbing crystals, the electric and magnetic vectors are complex quantities of the type (2.50).

First of all we note that the directions of vector $\underline{S}$ and its modulus vary in time with a frequency corresponding to the frequencies of X-ray oscillations; they vary, moreover, within the unit cell of the crystal, identically in all the cells. Therefore, for comparison with the experiment, the value of $\underline{S}$ should be averaged out over the oscillation period ν^{-1} and the lattice period, along the normal direction to the entrance surface.

We shall refer our electromagnetic wave to a rectangular coordinate system. During the propagation of the wave along the positive direction of the x-axis, the average value of the vector product for one oscillation period will be

$$\overline{[\underline{E}\underline{H}]} = \overline{E_y H_z} = \nu \int_0^{1/\nu} E_y H_z dt \quad . \tag{5.2}$$

The result of this first averaging can be written as

$$\underline{S}^t = \frac{c}{8\pi} \mathrm{Re}[\underline{E}\underline{H}] \quad , \tag{5.3}$$

where the additional factor 1/2 is the average value of the cosine squared. Further it is convenient to switch from the electric field vector $\underline{E}$ to the displacement vector $\underline{D}$, because the dielectric constant of the perturbation wave field is close to unity. Note that the vector $\underline{D}$ forms in the crystal a right-handed triplet of vectors together with $\underline{H}$ and $\underline{S}$. Thus, for the time-average value of Poynting's vector we can write

$$\underline{S}^t \approx \frac{c}{8\pi} \mathrm{Re}[\underline{D}\underline{H}^*] \quad . \tag{5.4}$$

To obtain the expression for the vector $\underline{S}$ which would meet our requirements, we must substitute, in (5.4), the Bloch solutions (2.36) and (2.50) for the induction and magnetic field vectors. It is found that the vector $\underline{S}^t$ in the crystal is the sum of the components referring to the separate fields. As noted above, fields belonging to one and the same state of polarization and interrelated through common conditions at the vacuum-crystal interface are mutually coherent, as well as fields referring to different states of polarization when the incident wave is plane-polarized.

Now write out the expressions for $\underline{D}$ and $\underline{H}$:

$$\underline{D} = \sum_i \exp 2\pi i[vt - (\underline{k}_o^{(i)} r)] \sum_m \underline{D}_m \exp[-2\pi i(\underline{h}_m \underline{r})] \quad , \tag{5.5}$$

$$\underline{H} = \sum_j \exp 2\pi i[vt - (\underline{k}_o^{(j)} \underline{r})] \sum_n \underline{H}_n \exp[-2\pi i(\underline{h}_n \underline{r})] \quad . \tag{5.6}$$

Equations (5.5) and (5.6) are written in a general case of multiwave scattering. The first (external) sum is taken over the excitation points on all the sheets of a multi-sheet dispersion surface, and the second sum, over all the possible reflection indices. Accordingly, the vector product in (5.4) will have the form

$$[\underline{D}\underline{H}^*] = \sum_i \sum_j \exp[2\pi i(\underline{k}_o^{(j)*} - \underline{k}_o^{(i)}, \underline{r})] \cdot$$

$$\cdot \sum_m \sum_n [\underline{D}_m^{(i)} \underline{H}_n^{(j)*}] \exp[2\pi i(\underline{h}_n - \underline{h}_m, \underline{r})] \quad . \tag{5.7}$$

Considering energy propagation in an absorbing crystal, we use (2.58) and (4.37) to calculate the vector of the difference $\underline{k}_o^{(j)*} - \underline{k}_o^{(i)}$. Thus, we have

$$(\underline{k}_o^{(j)*} - \underline{k}_o^{(i)}, \underline{r}) = (\Delta\underline{k}^{(ij)}, \underline{n}z) + (i/4\pi)(\sigma^{(i)} + \sigma^{(j)})z \quad , \tag{5.8}$$

where $\Delta\underline{k}^{(ij)}$ stands for a vector referring to the difference between the wave vectors of any pair of excitation points on two different sheets of the common dispersion surface; $\sigma^{(i)}$ and $\sigma^{(j)}$ denote the full values of the absorption factors for the waves associated with the points indicated.

Now considering the value of the vector product (5.7) and taking into account (5.8), we note that the values of σz slowly change with depth z and are almost constant on segments of the order of the lattice period. The spread of the value $|\Delta\underline{k}^{(ij)}|$ is very small. By contrast,

$$\exp[2\pi i(\underline{h}_n - \underline{h}_m, \underline{r})] = \exp 2\pi i(\underline{h}_q \underline{r})$$

is a function rapidly changing inside the cell. The average value of this magnitude within the unit cell vanishes to zero at all values of q except q = 0, when this factor turns into unity. Here, the double sum over n and over m turns into a single sum (n=m). Thus, our vector product, averaged over the unit cell, acquires the form

$$[\overline{\underline{D}\underline{H}^*}]^{cell} = \sum_i \sum_j \exp[2\pi i \Delta k_z^{(ij)} z] \cdot$$

$$\cdot \exp[-\tfrac{1}{2}(\sigma^{(i)} + \sigma^{(j)})z] \sum_n [\underline{D}_n^{(i)} \underline{H}_n^{(j)*}] \quad . \tag{5.9}$$

Now, transform the vector products on the right-hand side of (5.9) to scalar products. To do this, use (2.59) for the magnetic field vector. We take into consideration that moduli of the vectors $\underline{k}_m = \underline{k}_n$ differ from K by values of the order of 10^{-5}. Therefore, if we introduce a unit vector along $\underline{k}_n$

$$\underline{s}_n = \frac{\underline{k}_n}{|\underline{k}_n|}$$

(2.59) can be rewritten

$$\underline{H}_n \approx [\underline{s}_n \underline{D}_n] \quad , \tag{5.10}$$

and further, according to (2.47a),

$$(\underline{D}_n^{(i)} \underline{H}_n^{(j)*}) = \left[\underline{D}_n^{(i)} [\underline{s}_n^{(j)}, \underline{D}_n^{(j)*}] \right] = (\underline{D}_n^{(i)}, \underline{D}_n^{(j)*}) \underline{s}_n^{(j)} - (\underline{s}_n^{(j)}, \underline{D}_n^{(i)}) \underline{D}_n^{(j)*} \quad . \tag{5.11}$$

It is easy to see that the second term on the right-hand side of (5.11) reduces to zero, since the vectors parallel to the wave vectors are in all cases perpendicular to the corresponding induction vectors. Thus, (5.9) will be rewritten

$$\begin{aligned} [\overline{\underline{D}\underline{H}^*}]^{cell} \approx & \sum_i \sum_j \exp(2\pi i \Delta k_z^{(ij)} z) \cdot \\ & \cdot \exp[-\tfrac{1}{2} (\sigma^{(i)} + \sigma^{(j)}) z] \sum_n \left(\underline{D}_n^{(i)} \underline{D}_n^{(j)*} \right) \underline{s}_n \end{aligned} \tag{5.12}$$

and the value of the doubly averaged energy current-density vector (with the use of the real part of (5.12))

$$\begin{aligned} \overline{\overline{S}} \equiv \overline{\underline{S}} = \frac{c}{8\pi} \sum_n \Big\{ & \sum_i |\underline{D}_n^{(i)}|^2 \exp(-\sigma^{(i)} z) + \\ & + \sum_i \sum_j \exp[-\tfrac{1}{2} (\sigma^{(i)} + \sigma^{(j)}) z] \cos[2\pi \Delta k_z^{(ij)} z] (\underline{D}_n^{(i)} \underline{D}_n^{(j)}) s_n \Big\} \quad . \end{aligned} \tag{5.13}$$

Thus, the vector $\overline{\underline{S}}$ is the sum of the vectors directed from the corresponding points of the dispersion surface towards the separate points of the reciprocal lattice. The value of each of these vectors is determined by the expression in the braces. Terms of the type $|\underline{D}_n^{(i)}|^2$ represent the independent contribution of the individual waves from the various sheets of the dispersion surface, and the doubled sum, the corresponding interference terms from each pair of such waves.

The cosine term in (5.13) has an argument with a period corresponding to the extinction distance

$$z = \tau = |\Delta k^{(ij)}|^{-1} \quad . \tag{5.14}$$

The third averaging of the vector $\underline{S}$ over the period τ is interesting. It is convenient to use this value if we take into account that it is difficult to ensure the constancy of the crystal thickness to an accuracy of the order of a micron over any appreciable surface area.

Introduce the vector

$$\overline{\overline{\underline{S}}} = \frac{c}{8\pi} \sum_n \left\{ \sum_j |D_n^{(i)}|^2 \exp(-\sigma^{(i)} z) \underline{s}_n \right\} \tag{5.15}$$

averaged additionally over the extinction distance for a given radiation. Passing on from the multiwave approximation to the two-wave one, we have to rewrite (5.12) and (5.15), putting n = 0, h and i = 1.2, in the form

$$\begin{aligned} \overline{\underline{S}} = \frac{c}{8\pi} \Big\{ & (|D_o^{(1)}|^2 \underline{s}_o + |D_h^{(1)}|^2 \underline{s}_h) \exp(-\sigma^{(1)} z) + \\ & + (|D_o^{(2)}|^2 \underline{s}_o + |D_h^{(2)}|^2 \underline{s}_h) \exp(-\sigma^{(2)} z) + \\ & + [(D_o^{(1)} D_o^{(2)*} + D_o^{(1)*} D_o^{(2)}) \underline{s}_o + \\ & + (D_h^{(1)} D_h^{(2)*} + D_h^{(1)*} D_h^{(2)}) \underline{s}_h] \exp(-\sigma_o z) \cos(2\pi \Delta k_z z) \end{aligned} \tag{5.16}$$

$$\begin{aligned} \overline{\overline{\underline{S}}} = \frac{c}{8\pi} \Big\{ & (|D_o^{(1)}|^2 \underline{s}_o + |D_h^{(1)}|^2 \underline{s}_h) \exp(-\sigma^{(1)} z) + \\ & + (|D_o^{(2)}|^2 \underline{s}_o + |D_h^{(2)}|^2 \underline{s}_h) \exp(-\sigma^{(2)} z) \Big\} \quad , \end{aligned} \tag{5.17}$$

where $\Delta k = |\Delta \underline{k}|$ is defined from (3.35) and (3.37).

Note that (5.16) and (5.17) can conveniently be related to one and the sam state of polarization. Indeed, it should be taken into account that the product $[D_h^{(i)} D_o^{(j)*}]$ of the induction components referring to two different states of polarization reduced to zero. Moreover, the extinction distance τ contains the factor C and will hence be greater for σ polarization.

5.2 The Triply Averaged Poynting's Vector in Transparent Crystal

Vector Components Over the Fields. Ewald-Kato Theorem

We shall introduce into our expressions the average value of the current density vector modulus for the incident vacuum wave:

$$|\bar{\underline{S}}_o| = \frac{c}{8\pi} |D_o^{(a)}|^2 \quad . \tag{5.18}$$

Replacing, in (5.17), $c/8\pi$ by its value from (5.18), using (3.30) and (3.31), and introducing the notation $\ell = \gamma_o/\gamma_h \, |\chi_h/\chi_{\bar{h}}|$, we obtain $(\sigma^{(1)} = \sigma^{(2)} = 0)$

$$\bar{\bar{\underline{S}}} = \frac{|\bar{\underline{S}}_o|}{4\cosh^2\nu} \Big\{ [(y-\sqrt{1+y^2})^2 \underline{s}_o + \ell \underline{s}_h] + [(y+\sqrt{1+y^2})^2 \underline{s}_o + \ell \underline{s}_h] \Big\} \quad , \tag{5.19a}$$

$$\bar{\bar{\underline{S}}} = \frac{|\bar{\underline{S}}_o|}{4\cosh^2\nu} \{[\exp(-2\nu)\underline{s}_o + \ell \underline{s}_h] + [\exp(2\nu)\underline{s}_o + \ell \underline{s}_h]\} \tag{5.19b}$$

$$\bar{\bar{\underline{S}}} = \bar{\bar{\underline{S}}}^{(1)} + \bar{\bar{\underline{S}}}^{(2)} \quad , \tag{5.20a}$$

$$\bar{\bar{\underline{S}}}^{(i)} = \frac{c}{8\pi} [|D_o^{(i)}|^2 \underline{s}_o + |D_h^{(i)}|^2 \underline{s}_h] \quad . \tag{5.20b}$$

It is clear that in (5.19) and (5.20) the full value of the energy current-density vector is represented as the sum of the respective vectors corresponding to the energy current-density of each of the two fields in the crystal.

Another representation of the vector $\bar{\bar{\underline{S}}}$ follows from a different grouping of the terms on the right-hand side of (5.19), and also from (3.30), (3.31), (5.18), and (5.20)

$$\overline{\overline{\underline{S}}} = |\overline{\underline{S}}_o| \left[\sum_{i=1,2} \left| \frac{D_o^{(i)}}{D_o^{(a)}} \right|^2 \underline{s}_o + \sum_{i=1,2} \left| \frac{D_h^{(i)}}{D_o^{(a)}} \right|^2 \underline{s}_h \right] \quad ,$$

$$\overline{\overline{\underline{S}}} = |\overline{\underline{S}}_o| \left(P\underline{s}_o + Q\underline{s}_h \right) \quad . \tag{5.21}$$

Here, the vector $\overline{\overline{\underline{S}}}$ is represented as the sum of the energy current-density vectors of the refracted and diffracted waves in the crystal.

Expressions similar to (5.21) are obtained from (5.20b) if we replace $c/8\pi$ by its value from (5.18) and introduce the variables p_i and q_i:

$$\overline{\overline{\underline{S}}}^{(i)} = |\overline{\underline{S}}_o| \left[\left| \frac{D_o^{(i)}}{D_o^{(a)}} \right|^2 \underline{s}_o + \left| \frac{D_h^{(i)}}{D_o^{(a)}} \right|^2 \underline{s}_h \right] \quad ,$$

$$\overline{\overline{\underline{S}}}^{(i)} = |\overline{\underline{S}}_o| (p_i\underline{s}_o + q_i\underline{s}_n) \quad . \tag{5.22}$$

An essential characteristic of the vectors $\overline{\overline{\underline{S}}}$ and $\overline{\overline{\underline{S}}}^{(i)}$ is their orientation at a given point of the crystal, for instance relative to the vectors $\underline{s}_o$ and $\underline{s}_h$.

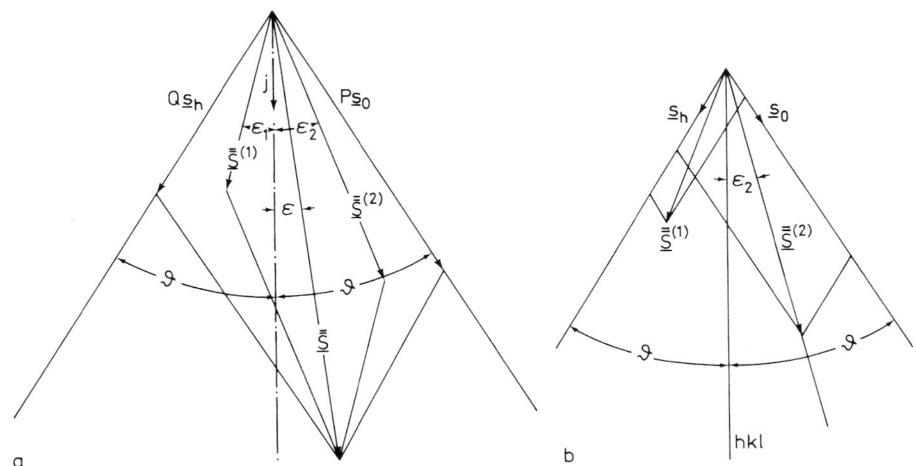

Fig. 5.1 Geometrical construction of vectors $\overline{\overline{\underline{S}}}$ and $\overline{\overline{\underline{S}}}^{(i)}$. a) Geometrical construction of vector $\overline{\overline{\underline{S}}}$; b) Geometrical construction of vector $\overline{\overline{\underline{S}}}^{(i)}$

Figures 5.1a and b show the decomposition of Poynting's vectors along the $\underline{s}_o$- and $\underline{s}_h$-axes. The direction of the vectors $\overline{\overline{\underline{S}}}$ and $\overline{\overline{\underline{S}}}^{(i)}$ is determined by the angles ε and ε_i that these vectors make with the direction $\underline{j}$, and the positive values of ε and ε_i are reckoned in the direction of $\underline{s}_o$. It follows immediately from the figures that, according to (5.21) and (5.22),

$$\frac{\sin(\vartheta+\varepsilon)}{\sin(\vartheta-\varepsilon)} = \frac{P}{Q} = \frac{\sum_i |D_o^{(i)}|^2}{\sum_i |D_h^{(i)}|^2} \quad , \tag{5.23}$$

$$\frac{\sin(\vartheta+\varepsilon_i)}{\sin(\vartheta-\varepsilon_i)} = \frac{|D_o^{(i)}|^2}{|D_h^{(i)}|^2} = \frac{p_i}{q_i} \tag{5.24}$$

hence we determine the angles ε and ε_i:

$$\tan\varepsilon = \frac{P - Q}{P + Q} \tan\vartheta \quad , \tag{5.25a}$$

$$\tan\varepsilon_i = \frac{|D_o^{(i)}|^2 - |D_h^{(i)}|^2}{|D_o^{(i)}|^2 + |D_h^{(i)}|^2} \tan\vartheta \quad . \tag{5.25b}$$

Further, using the relations (3.30) and (3.31), we obtain the following equations:

$$\tan\varepsilon = \frac{2\cosh^2 v - (1+\ell)}{2\cosh^2 v - (1-\ell)} \tan\vartheta = \frac{2y^2 + (1-\ell)}{2y^2 + (1+\ell)} \tan\vartheta \tag{5.26a}$$

for ℓ = I, i.e., for symmetrical reflection and a centro-symmetrical crystal

$$\tan\varepsilon = \tanh^2 v \tan\vartheta = \frac{y^2}{1 + y^2} \tan\vartheta$$

$$\sin\varepsilon = \frac{1}{f}\, y^2 \sin\vartheta \quad , \quad \cos\varepsilon = \frac{1}{f}\,(1+y^2)\cos\vartheta \quad ,$$

$$f = \sqrt{y^4 + (1+2y^2)\cos^2\vartheta} \quad , \tag{5.26b}$$

$$\tan\varepsilon_1(v) = \tan\varepsilon_2(-v) = \frac{\exp(-v) - \ell \exp v}{\exp(-v) + \ell \exp v} \tag{5.27a}$$

and for $\ell = 1$

$$\tan\varepsilon_i = \pm \tanh v \tan\vartheta = \mp \frac{y}{(1+y^2)^{1/2}} \tan\vartheta$$
$$\sin\varepsilon_i = \mp \frac{y \sin\vartheta}{(y^2+\cos^2\vartheta)^{1/2}} \quad . \tag{5.27b}$$

Thus, for symmetrical reflections, in the case of a centro-symmetrical crystal, all three vectors $\overline{\overline{\underline{S}}}$, $\overline{\overline{\underline{S}}}_1$, and $\overline{\overline{\underline{S}}}_2$ lie on the reflecting plane (see further (5.48)). Using (5.23) and (5.24), it is easy to obtain equations for the moduli $\overline{\overline{\underline{S}}}$ and $\overline{\overline{\underline{S}}}^{(i)}$ at each given value of the angles ε and ε_i.

Since the brackets in (5.21) and (5.22), on the right-hand side, contain the sums of the components of the corresponding vectors along the axes $\underline{s}_o$ and $\underline{s}_h$, it is obvious that the moduli of these vectors are equal to the sums of the component projections onto the vector's directions. These projections are obtained by multiplying the components by the cosines of the angles made by the vector with the corresponding axis. Transforming (5.21) and using (5.23), we get

$$|\overline{\overline{\underline{S}}}| = |\overline{\underline{S}}_o| \frac{\ell}{2(1+y^2)} \left[\frac{1 + 2y^2}{\ell} \cos(\vartheta-\varepsilon) + \cos(\vartheta+\varepsilon)\right] \tag{5.28}$$

and, similarly, we have from (5.22) and (5.24)

$$|\overline{\overline{\underline{S}}}^{(i)}| = |\overline{\underline{S}}_o| \frac{\ell \sin 2\vartheta}{4(1+y^2) \sin(\vartheta-\varepsilon_1)} = |\overline{\underline{S}}_o| \frac{\ell}{2(1+y^2)} \left[\frac{(\sqrt{1+y^2} \mp y)^2}{\ell} \cdot\right.$$
$$\left. \cdot \cos(\vartheta-\varepsilon_i) + \cos(\vartheta+\varepsilon_i)\right] \quad . \tag{5.29}$$

Finally, using (5.26a) and (5.27a), we can calculate the values of $\sin(\vartheta-\varepsilon)$ and $\sin(\vartheta-\varepsilon_i)$ appearing in the denominators of (5.28) and (5.29), and derive the following expressions for the moduli of the vectors $\overline{\overline{\underline{S}}}$ and $\overline{\overline{\underline{S}}}^{(i)}$ as a function of only the variable y. In a general case of asymmetrical reflection

$$\bar{\bar{\underline{S}}} = |\bar{\underline{S}}_0| \cdot \frac{\sqrt{(2y^2+1)^2 + 2\ell \cos 2\vartheta(2y^2+1) + \ell^2}}{2(1+y^2)} \tag{5.30a}$$

$$\bar{\bar{\underline{S}}}^{(i)} = |\bar{\bar{\underline{S}}}| \cdot \sqrt{(2y^2+1)\,(1+\ell^2) + 2\ell \cos 2\vartheta \mp (1-\ell^2)\,2y\sqrt{1+y^2}} \cdot \left[4(1+y^2)\,(\sqrt{1+y^2} \pm y)\right]^{-1} \quad . \tag{5.30b}$$

In the case of symmetrical reflection and a centro-symmetrical crystal

$$|\bar{\bar{\underline{S}}}_s| = |\bar{\underline{S}}_0| \frac{[y^4 + (2y^2+1)\cos^2\vartheta]^{1/2}}{1+y^2} \quad , \tag{5.31a}$$

$$|\bar{\bar{\underline{S}}}_s^{(i)}| = |\bar{\underline{S}}_0| \frac{(\sqrt{1+y^2} \mp y)\,(y^2+\cos^2\vartheta)^{1/2}}{2(1+y^2)} \quad . \tag{5.31b}$$

Turning again to (5.20), we note that the expansion of the vector $\bar{\bar{\underline{S}}}$ into components over the separate fields in the crystal is physically substantiated because these components "arise" at the excitation points of the dispersion curve as a result of their intersection by the normal to the entrance surface. It has been shown by EWALD [4.33] and KATO [4.41] that the vectors $\bar{\bar{\underline{S}}}^{(i)}$ are directed along the normals to the branches of the dispersion curve at the corresponding points.

Elementary proof of the above theorem follows.

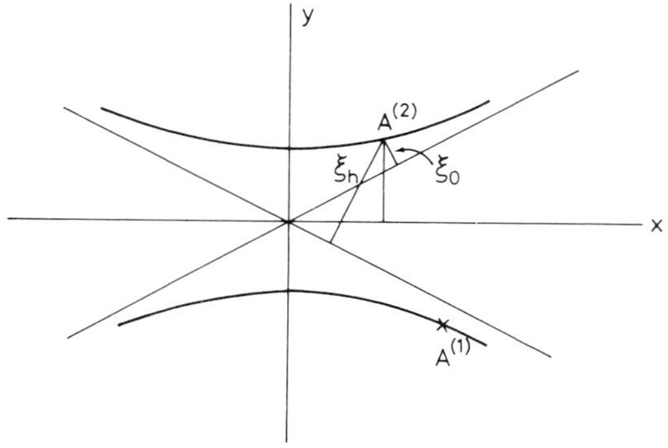

Fig. 5.2 Proof of the Ewald-Kato theorem (after JAMES [1.16])

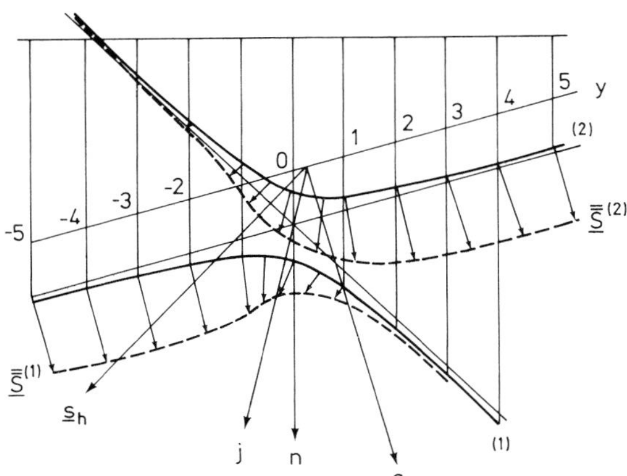

Fig. 5.3 Direction of the vectors $\bar{\bar{S}}^{(i)}$ in the maximum range (after JAMES [1.16])

Figure 5.2 displays the rectangular coordinates x and y in which the dispersion curve is described according to (2.81),

$$x^2 \cos^2\vartheta - y^2 \sin^2\vartheta = \xi_o^{(i)}\xi_h^{(i)} \quad , \tag{5.32}$$

where

$$\xi_o^{(i)} = x \cos\vartheta - y \sin\vartheta \quad , \quad \xi_h^{(i)} = x \cos\vartheta + y \sin\vartheta \quad ,$$

or

$$x = \frac{\xi_o^{(i)} + \xi_h^{(i)}}{2 \cos\vartheta} \quad , \quad y = - \frac{\xi_o^{(i)} - \xi_h^{(i)}}{2 \sin\vartheta} \quad . \tag{5.33}$$

The angle formed by the normal to the dispersion curve (5.32) at the point (x,y) with the x-axis is determined by the condition

$$- \frac{dx}{dy} = \frac{y}{x} \tan\vartheta = \frac{\xi_o^{(i)} - \xi_h^{(i)}}{\xi_o^{(i)} + \xi_h^{(i)}} \tan\vartheta \quad . \tag{5.34}$$

Using (2.82), these equations can be rewritten

$$\frac{dx}{dy} = - \frac{|D_o^{(i)}|^2 - |D_h^{(i)}|^2}{|D_o^{(i)}|^2 + |D_h^{(i)}|^2} \tan\vartheta \quad . \tag{5.35}$$

This equation corresponds precisely to (5.25b) for angles ε_i between the vectors $\bar{\bar{\underline{S}}}^{(i)}$ and the bisector $\underline{j}$, whose direction is opposite to the (positive) direction of the x-axis. Fig.5.3 shows the directions of the vectors $\bar{\bar{\underline{S}}}^{(1)}$ and $\bar{\bar{\underline{S}}}^{(2)}$ at different points throughout the maximum range.

To explain the applicability of the result obtained to a general case of multiwave scattering, we can apply reasoning based on the group velocity concept [see Ref.1.13, p.455].

Consider the general solution of the wave equation in the form of a superposition (train) of Bloch waves (see (2.38) and (2.36)):

$$\underline{D} = \sum_j \underline{D}_j = \sum_j a_j \exp\{2\pi i[v_j t - (\underline{k}_{oj}\underline{r})]\} \sum_m \underline{D}_m \exp[-2\pi i(\underline{h}_m\underline{r})] \quad . \tag{5.36}$$

The spread of the frequencies ν_j and of the wave vectors $\underline{k}_{oj}$ of the separate components in the train is characterized by the conditions

$$|\underline{k}_{oj} - \underline{k}_o| \ll |\underline{k}_o| \quad , \quad (\nu_j-\nu) \ll \nu \quad , \tag{5.37}$$

where $\underline{k}_o$ and ν are certain average values for the train. The expression (5.36) can be rewritten

$$\underline{D} = \exp 2\pi i[\nu t - (\underline{k}_o\underline{r})] \sum_j \{\sum_m a_j\underline{D}_m \exp[-2\pi i(\underline{h}_m,\underline{r})]\} \cdot$$
$$\cdot \exp\{2\pi i[(\nu_j-\nu)t - (\underline{k}_{oj}-\underline{k}_o,\underline{r})]\} \quad . \tag{5.38}$$

In accordance with the conditions (5.37), the rate of motion of the amplitude, more precisely, of the energy associated with the train, is expressed by

$$\underline{u} = -\mathrm{grad}_{-\underline{k}_o},\nu \quad . \tag{5.39}$$

A pictorial representation of the magnitude of the group velocity vector u can be obtained as follows. When constructing the dispersion surface we used the vectors $\underline{k}_o^{(i)}$, which converge on the origin of the coordinates of the reciprocal lattice O. When such a vector rocks within the maximum, at a constant value of the frequency ν, its initial point describes a certain definite sheet of the dispersion surface. If, however, the frequency varies within a certain range $d\nu$, each vector in the given wave field i will be represented by a bundle of wave vectors. The oscillation of these vectors leads to the filling by their initial points of a definite volume inside of which the frequencies ν will be functions of the coordinates of reciprocal space.

The introduced value of group velocity $\underline{u} = \text{grad}_{-k_0}\,\nu$ will be directed along the normal to the surface ν = const, i.e., to the dispersion surface. It can be shown that the magnitude of Poynting's vector is equal to the product of the group velocity of propagation by energy density.

Now consider the qualitative picture of changes in the vectors $\bar{\bar{\underline{S}}}^{(i)}$ in the maximum region, and their contribution to the total value $\bar{\bar{\underline{S}}}$. Here we can compare Figs.5.3 and 3.7 if we consider the changes in the squares of the respective values (Fig.3.7). At large positive values of y or v near the right boundary of the maximum region, the value of $\tan\varepsilon_2$, according to (5.27a) approaches $\tan\vartheta$, i.e., the vector $\bar{\bar{\underline{S}}}^{(2)}$ is close in direction to the vector $\bar{\underline{S}}_0$. The modulus $\bar{\bar{\underline{S}}}^{(2)}$, according to (5.30), acquires the asymptotic value

$$|\bar{\bar{\underline{S}}}^{(2)}|_{y\to\infty} \approx |\bar{\underline{S}}_0| \, \frac{\sqrt{2y^2[(1+\ell) + (1-\ell)]}}{4y^2}\, 2y = |\bar{\underline{S}}_0| \tag{5.40}$$

and, hence, almost completely determines the magnitude of the vector $\bar{\bar{\underline{S}}}$. This is in agreement with Fig.3.7 and (5.22), on right-hand side of which the principal contribution to the value $\bar{\bar{\underline{S}}}^{(2)}$ will be equal to $|D_0^{(2)}|^2/|D_0^{(a)}|^2$, while the contribution of the vector $\bar{\bar{\underline{S}}}^{(1)}$ on the right-hand side of the maximum range is negligible. The ratios described become reciprocal in the extreme left-hand corner of the maximum, where the main contribution is determined by the vector $\bar{\bar{\underline{S}}}^{(1)}$.

Turning again to the vector $\bar{\bar{\underline{S}}}^{(2)}$, we note that with the decrease in y the direction of $\bar{\bar{\underline{S}}}^{(2)}$ gradually changes, because the angle ε_2 decreases. At the same time the modulus of this vector also decreases, and the ratio between its two components in (5.22) changes, since the contribution from the diffracted wave increases. At a certain definite value of y the two components in (5.22) become equal. According to (3.30) and (3.31),

$$\sqrt{1 + y'^2} + y' = |\sqrt{\ell}| \quad , \quad y' = \frac{\ell - 1}{2|\sqrt{\ell}|} \begin{cases} > 0, & \ell > 1 \quad , \\ < 0, & \ell < 1 \quad . \end{cases} \tag{5.41}$$

At y = y' the value of the vector

$$\bar{\bar{\underline{S}}}^{(2)} = |\bar{\underline{S}}_o| \frac{2\ell^2 \cos\vartheta}{(1+\ell)^2} \underline{j} \quad , \tag{5.42}$$

since the equality of the components implies that according to (5.25) $\varepsilon_2 = 0$ and, hence, the vector $\bar{\bar{\underline{S}}}^{(2)}$ is directed along $\underline{j}$. The vector $\bar{\bar{\underline{S}}}^{(1)}$ takes the same value and has equal components at

$$\sqrt{1 + y'^2} - y' = |\sqrt{\ell}| \quad , \quad y' = \frac{1 - \ell}{2|\sqrt{\ell}|} \begin{cases} > 0, & \ell < 1 \quad , \\ < 0, & \ell > 1 \quad . \end{cases} \tag{5.43}$$

A further decrease in y to zero results in coincidence of the directions of all the three vectors $\bar{\bar{\underline{S}}}$, $\bar{\bar{\underline{S}}}^{(1)}$ ans $\bar{\bar{\underline{S}}}^{(2)}$. This follows directly from (5.26a) and (5.27a), which yield, at y = 0, the following general value of the tangent of the angle relative to the vector $\underline{j}$:

$$\tan\varepsilon^{y=o} = \tan\varepsilon_i^{y=o} = \frac{1 - \ell}{1 + \ell} \tan\vartheta \quad . \tag{5.44}$$

Accordingly, (5.30) for the moduli of the energy current-density vector at y = 0 takes the form

$$|\bar{\bar{\underline{S}}}| = |\bar{\underline{S}}_o| \, \tfrac{1}{2}[(1+\ell)^2 \cos^2\vartheta + (1-\ell)^2 \sin^2\vartheta]^{1/2} =$$

$$= 2|\bar{\bar{\underline{S}}}^{(1)}| = 2|\bar{\bar{\underline{S}}}^{(2)}| \quad . \tag{5.45}$$

The general direction of all the three vectors indicated in (5.44), which does not coincide with $\underline{j}$, increasingly deviates from the latter with an increase in the asymmetry factors introduced in Chapter 4 (see (4.111)) (we assume $|\chi_h| \approx |\chi_{\bar{h}}|$)

$$q = \frac{\gamma_h - \gamma_o}{2\sqrt{\gamma_o\gamma_h}} \approx \frac{1 - \ell}{2\sqrt{\ell}} \quad , \quad \tan\varepsilon = \tan\varepsilon_i = -\frac{q\sqrt{\ell}}{(1-q\sqrt{\ell})} \tan\vartheta \quad . \tag{5.46}$$

It is very important to establish the direction of this turn. It is obvious that if

$$\text{for} \quad (hkl)\gamma_o > \gamma_h \quad \text{and} \quad |\psi_o| < |\psi_h| \quad , \quad \tan\varepsilon = \tan\varepsilon_i < 0 \quad , \tag{5.47a}$$

$$\text{for} \quad (\bar{h}\bar{k}\bar{l})\gamma_o < \gamma_h \quad \text{and} \quad |\psi_o| < |\psi_h| \quad , \quad \tan\varepsilon = \tan\varepsilon_i > 0 \quad . \tag{5.47b}$$

The relations (5.47) imply that at $y = 0$ the total wave field energy propagates at $\gamma_o > \gamma_h$ in the direction lying between the reflecting plane and the vector $\underline{s}_h$, and at $\gamma_o < \gamma_h$ between the same plane and the vector $\underline{s}_o$.

It is also clear that in the range of decreasing positive values of y discussed, the vectors $\bar{\bar{\underline{S}}}^{(1)}$ and $\bar{\bar{\underline{S}}}$ have also made a turn: the former from $\underline{s}_h$, where the modulus $\bar{\underline{S}}^{(\bar{1})} \approx 0$, and the latter, from $\underline{s}_o$ to the values $\varepsilon_1^{y=o} = -\varepsilon_2^{y=o} = -\varepsilon^{y=o} \equiv -\varepsilon_o$.

With a further change in y in the direction of increasing negative values within the angular range $(\vartheta + \varepsilon)$ the vectors $\bar{\bar{\underline{S}}}^{(1)}$ and $\bar{\bar{\underline{S}}}$ will rotate up to the direction of $\underline{s}_o$, and the vector $\bar{\bar{\underline{S}}}^{(2)}$, in a smaller angular range $(\vartheta - \varepsilon)$, up to the direction of $\underline{s}_h$, where $|\bar{\underline{S}}^{(\bar{2})}| \approx 0$.

At $\ell = 1$, the value of y at which both components of each of the vectors $\bar{\bar{\underline{S}}}^{(i)}$ are equal vanishes to zero according to (5.41) and (5.43) and thus coincides with the value of y at which all the three vectors of the energy current-density have the same direction. The expression (5.45) acquires a simpler form

$$|\bar{\bar{\underline{S}}}_s| = |\bar{\underline{S}}_o| \cos\vartheta = 2|\bar{\bar{\underline{S}}}^{(1)}| = 2|\bar{\bar{\underline{S}}}^{(2)}| \quad . \tag{5.48}$$

Here, the general direction obviously coincides with $\underline{j}$ and, hence, with the reflecting plane.

Finally, it is important to note that the general value of the turns of the vectors $\bar{\bar{\underline{S}}}^{(i)}$ within the range of variation of y, i.e., within a few seconds of arc, equals 2ϑ, which is equivalent to 20°-90° or more, depending on the indices of the reflecting plane and the radiation used. It can be shown that the value of the derivative $d\varepsilon/d\eta$ contains as a factor the value $|\chi_h|^{-1} = 10^5$-10^6, which fact determines this effect. The interval of turns of $\bar{\bar{\underline{S}}}$ for symmetrical reflection is from $\underline{s}_o$ to $\underline{j}$ and back; for asymmetrical reflection it is from $\underline{s}_o$ to $(\vartheta \pm \varepsilon_o)$ and back. Particularly when using a short wave radiation, for instance that of Mo Kα, the modulus of ε_o is several degrees.

Transform some of the relations obtained, using a system of coordinates naturally connected with the shape of the crystal under investigation in the form of a plane-parallel plate. In this system, the x-axis (Fig.5.4) with a unit vector $\underline{a}$ lies on the plane of the entrance surface, and the z-axis with a unit vector $\underline{n}$ lies along the direction of an inward-drawn normal.

In accordance with the choice of signs for ψ_o and ψ_h made in Chapter 3 (positive when reckoning from n clockwise), we assume the same sign for α. Then the positive direction of the x-axis and the vector $\underline{a}$ is assumed to the left from the origin, as shown in Fig.5.4 for asymmetrical reflection. The relations between the vector $\underline{s}_o$ and $\underline{s}_h$, on the one hand, and $\underline{n}$ and $\underline{a}$, on the other, have the following form

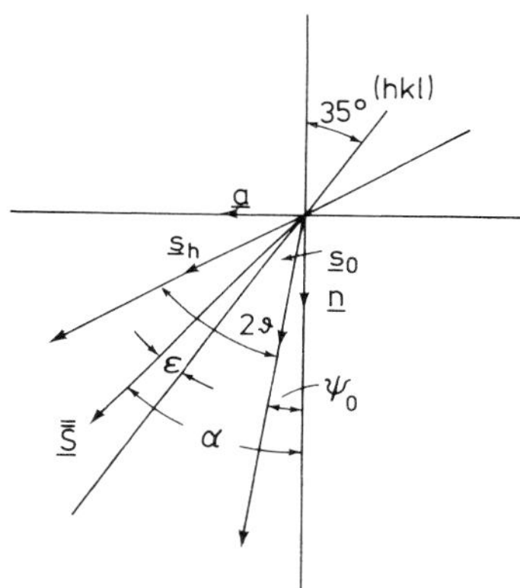

Fig. 5.4 Derivation of (5.50) and (5.51): Si, Mo $K\alpha_1$, reflection 444, face (100) $\varphi = 55^0$

$$\frac{\underline{s}_o}{\gamma_o} = \underline{n} + \underline{a}\ \tan\psi_o \quad , \quad \frac{\underline{s}_h}{\gamma_h} = \underline{n} + \underline{a}\ \tan\psi_h \quad . \tag{5.49}$$

Transform (5.21), factoring out $Q\gamma_o$ and using (5.49)

$$\bar{\bar{\underline{S}}} = |\bar{\underline{S}}_o| \frac{\gamma_o}{2(1+y^2)} [2(1+y^2)\underline{n} + 2y^2\underline{a}\ \tan\psi_o + (\tan\psi_o+\tan\psi_h)\underline{a}] \quad . \tag{5.50}$$

From this expression one can set up the equation of the energy propagation path in a transparent crystal referred to the axes indicated

$$x = z\ \tan\alpha \quad ; \quad \tan\alpha = \frac{y^2}{1+y^2}\tan\psi_o + \frac{\tan\psi_o + \tan\psi_h}{2(1+y^2)} \quad . \tag{5.51}$$

This equation shows that the wave field energy in a transparent crystal propagates along straight lines emanating from the "point" of incidence towards the entrance surface of the incident wave with a limited wave front. The slope of these lines, over the entire range of y, is known from the previous analysis, and the ratio between the angles α and ε corresponds to given reflection.

Examples of such schemes and these relations are given in Figs.5.4 and 5.5. In the case of Fig.5.5

$$-\tan\psi_0 = \tan\psi_h = \pm|\tan\vartheta| \quad . \tag{5.52}$$

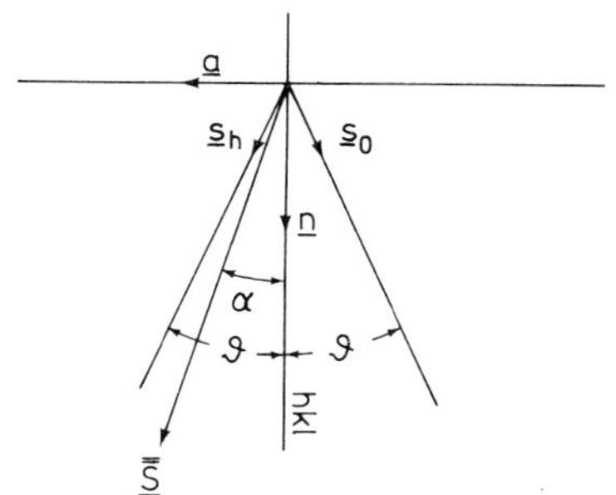

Fig. 5.5 Diagram of symmetrical reflection

As a result, the value of $\tan\alpha$ is defined from a simpler relation

$$\tan\alpha = \frac{y^2}{1+y^2}\tan\psi_0 = \pm\frac{y^2}{y+y^2}\tan\vartheta \tag{5.53}$$

where the plus or minus signs are chosen in accordance with the sign of $\tan\alpha$ or $\tan\psi_0$.

The expressions (5.51) as such do not contribute anything new to the physical picture or geometry of the propagation of the wave field energy in transparent crystals. At the same time these expressions directly indicate the orientation of the vector $\bar{\bar{\underline{S}}}$, whereas (5.26) make it possible to calculate only the position of this vector relative to the reflecting plane or vector $\underline{j}$.

Moreover, the use of the above system of coordinates is absolutely essential when switching to an absorbing crystal in which, in contrast to a transparent crystal, $\tan\alpha$ is a function of the vertical coordinate, or the depth z. As a result, the energy propagation path becomes curvilinear, and the derivation of its equation in explicit form requires integration of a more complex function than (5.51).

The foregoing analysis, which applies to energy propagation in a transparent crystal, is equally applicable both to a centro-symmetrical crystal and to a crystal without a centre of symmetry.

5.3 Triply Averaged Poynting's Vector in Absorbing Centro-symmetrical Crystal

Take down the following expression for the vector $\bar{\bar{\underline{S}}}$ in an absorbing crystal

$$(\sigma_h+\sigma_h') = x \quad , \quad (|\chi_h|/|\chi_{\bar{h}}|=1) \quad :$$

$$\bar{\bar{\underline{S}}} = |\bar{\underline{S}}_o| \frac{\exp(-\sigma_m z)}{2\cosh^2 v_r} \{[\cosh(2v_r+xz)]\underline{s}_o + \frac{\gamma_o}{\gamma_h}(\cosh xz)\underline{s}_h\} \quad . \qquad (5.54)$$

Using the system of coordinates introduced above, with unit vectors $\underline{a}$ and $\underline{n}$, we rewrite (5.54)

$$\bar{\bar{\underline{S}}} = |\bar{\underline{S}}_o| \frac{\gamma_o \exp(-\sigma_m z)}{2\cosh^2 v_r} [\cosh(2v_r+xz)(\underline{n}+\underline{a}\tan\psi_o) + \cosh xz(\underline{n}+\underline{a}\tan\psi_h)] \quad . \qquad (5.55)$$

Using the identities

$$\cosh(2v_r+xz) + \cosh xz = 2\cosh(v_r+xz)\cosh v_r$$

$$\cosh(2v_r+xz) - \cosh xz = 2\sinh(v_r+xz)\sinh v_r$$

it is easy to obtain the following expressions for $\bar{\bar{\underline{S}}}$:

$$\bar{\bar{\underline{S}}} = |\bar{\underline{S}}_o| \frac{\exp(-\sigma_m z)}{2\cosh^2 v_r} \{[2\cosh(v_r+xz)\cosh v_r]\underline{n} + [2\sinh(v_r+xz)\sinh v_r \tan\psi_o]\underline{a} + [\cosh xz(\tan\psi_o+\tan\psi_h)]\underline{a}\} \quad . \qquad (5.56)$$

Hence, the slope of the tangent line to the path at the point with coordinates (x,z) relative to the z-axis

$$\frac{dx}{dz} = \tan\alpha = \tan\psi_o \tanh(v_r+\varkappa z)\tanh v_r + \frac{\tan\psi_o + \tan\psi_h}{2\cosh v_r}\frac{\cosh\varkappa z}{\cosh(v_r+\varkappa z)} \tag{5.57}$$

and the path equation in the chosen axes will be

$$x = \int_0^z [\tanh(v_r+\varkappa z)\tanh v_r \tan\psi_o]dz + \frac{\tan\psi_o + \tan\psi_h}{2\cosh v_r}\int_0^z \frac{\cosh\varkappa z}{\cosh(v_r+\varkappa z)}dz = X_1 + X_2 \quad . \tag{5.58}$$

The first integral is standard

$$X_1 = \frac{\tanh v_r}{\varkappa}\tan\psi_o \ln\frac{\cosh(v_r+\varkappa z)}{\cosh v_r} \tag{5.59}$$

and the second one is easily solved with the aid of substitution $\exp(v_r + \varkappa z) = t$:

$$X_2 = \frac{\tan\psi_o + \tan\psi_h}{2\cosh v_r}\left[\frac{\cosh v_r}{\varkappa}(v_r+\varkappa z)\ln 2\cosh(v_r+\varkappa z)\right]_0^z = \frac{1}{2}(\tan\psi_o+\tan\psi_h)z - \frac{1}{2}(\tan\psi_o+\tan\psi_h)\frac{\tanh v_r}{\varkappa}\ln\frac{\cosh(v_r+\varkappa z)}{\cosh v_r} \quad . \tag{5.60}$$

Finally,

$$x = \frac{y_r}{\varkappa\sqrt{1+y_r^2}}\tan\psi_o \ln\frac{\cosh(v_r+\varkappa z)}{\cosh v_r} + \frac{z}{2}(\tan\psi_o+\tan\psi_h) - \frac{y_r}{2\varkappa\sqrt{1+y_r^2}}(\tan\psi_o+\tan\psi_h)\ln\frac{\cosh(v_r+\varkappa z)}{\cosh v_r}$$

or:

$$x = - \frac{y_r \tan\psi_h}{\varkappa \sqrt{1 + y_r^2}} \ln \frac{\cosh(v_r + \varkappa z)}{\cosh v_r} -$$

$$- \frac{\tan\psi_o + \tan\psi_h}{2} \left[\frac{y_r}{\varkappa \sqrt{1 + y_r^2}} \ln \frac{\cosh(v_r + \varkappa z)}{\cosh v_r} - z \right] \quad . \tag{5.61}$$

In the last expression, the first term corresponds to the energy propagation path in the case of symmetrical reflection, and the second term arises upon transition to the general case of asymmetrical reflection.

The expressions obtained for the wave field energy propagation path (5.61) and for the angle α (5.57) as a function of the coordinate parameter v_r and the depth z are rather complicated.

A qualitative analysis of the shape and position of the paths, and also some basic numerical evaluations, can be performed by using the asymptotic form of (5.57) for a crystal area near the surface ($z \approx 0$) and in the volume for a thick crystal ($z \approx \infty$). In the first case we use the conditions

$$\tanh(v_r + \varkappa z)|_{z \approx o} \approx \tanh v_r \quad , \quad \cos \varkappa z\,|_{z \approx o} \approx 1$$

$$\cosh(v_r + \varkappa z)|_{z \approx o} \approx \cosh v_r \tag{5.62}$$

hence

$$\tan\alpha|_{z \approx o} \approx \tanh^2 v_r \tan\psi_o + \frac{\tan\psi_o + \tan\psi_h}{2 \cosh^2 v_r} =$$

$$= \frac{y_r^2 \tan\psi_o}{1 + y_r^2} + \frac{\tan\psi_o + \tan\psi_h}{2(1+y_r^2)} \quad . \tag{5.63}$$

In the second case we have

$$\tanh(v_r + \varkappa z)|_{z \approx \infty} \approx 1 \quad , \quad \frac{\cosh \varkappa z}{\cosh(v_r + \varkappa z)} \approx \exp(-v_r) \tag{5.64}$$

$$\tan\alpha|_{z\approx\infty} \approx \tanh v_r \tan\psi_o + \frac{\tan\psi_o + \tan\psi_h}{2\cosh v_r \exp v_r} =$$

$$= \frac{y_r \tan\psi_o}{\sqrt{1+y_r^2}} + \frac{\tan\psi_o + \tan\psi_h}{2(y_r+\sqrt{1+y_r^2})\sqrt{1+y_r^2}} \quad . \tag{5.65}$$

Note that (5.63) and (5.65) are similar to those which describe energy propagation in a transparent crystal, in the surface layer with two fields, and inside the crystal bulk with a single field only. Indeed, considering the propagation of the second field in an absorbing crystal, we use (5.22) and (5.49) to obtain

$$\bar{\bar{\underline{S}}}^{(2)} = |\bar{\underline{S}}_o| \frac{\exp[-(\sigma_m - \varkappa)z]}{2(1+y_r^2)} \{[\exp(2v_r) + 1]\underline{n} + [\exp(2v_r)\tan\psi_o + \tan\psi_h]\underline{a}\} \quad . \tag{5.66}$$

From this expression we can directly determine the value of $\tan\alpha$, which coincides with (5.57).

The following analysis is applied to specific schemes of asymmetrical (and symmetrical) recording. The relevant schemes are presented in Figs.5.4-6.

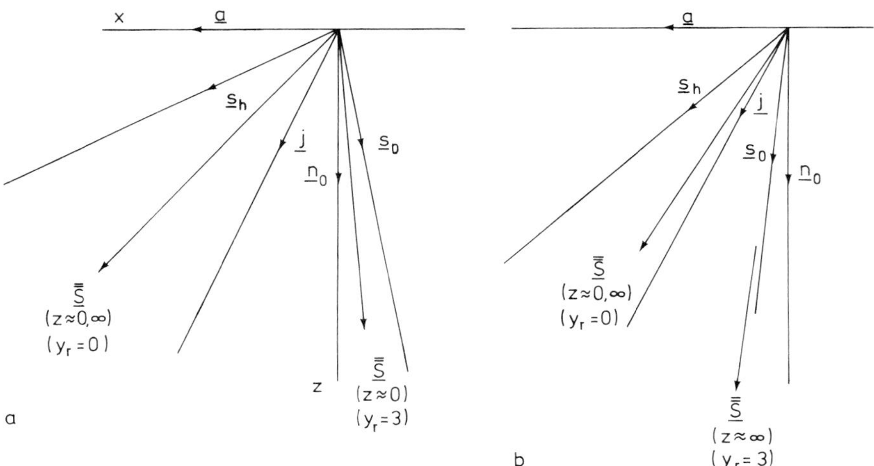

Fig. 5.6a,b Schemes of asymmetrical reflection: a) $\psi_o = 11^\circ$, $\psi_h = 65^\circ$, b) $\psi_o = +6^\circ 30'$, $\psi_h = 50^\circ$

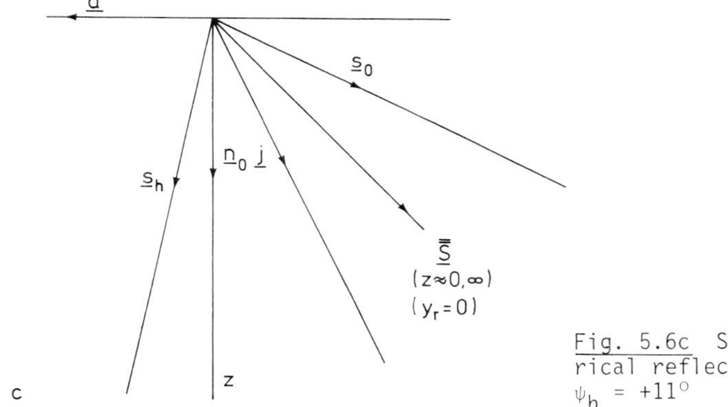

Fig. 5.6c Schemes of asymmetrical reflection: c) $\psi_o = -65^\circ$, $\psi_h = +11^\circ$

In Figs.5.4,5.6b the angles $\measuredangle\, \underline{s}_o \underline{s}_h$ are completely enclosed in one of the quadrants of the coordinates' rectangular system. Hence, at any value of the angular parameter v_r or y_r, and also at any depth, the signs of $\tan\alpha$ and of the angle α will be constant: for the schemes of Figs.5.4 and 5.6b, the signs will be positive. In contrast, the schemes of Fig.5.6a and c, and also the particular case of such schemes of Fig.5.5, correspond to more complicated laws of variation in the shape of the paths of the Poynting's vector with changed above parameters (Table 5.1)

Table 5.1 Values of angles α between the direction of the vector $\overline{\overline{\underline{S}}}$ and the vertical axis $\underline{n}$ in absorbing crystal (Fig.5.6)

Figures	Reflection scheme	Depth	Angular parameter -5	-3	-1	0	1	2	3
5.6a	$\psi_o=-11^\circ$	$z\approx 0$	$-8^\circ30'$	$-4^\circ25'$	$21^\circ20'$	$44^\circ20'$	$21^\circ20'$	$-4^\circ25'$	$-8^\circ30'$
	$\psi_h=65^\circ$	$z\approx\infty$	$64^\circ30'$	$64^\circ40'$	$59^\circ10'$	$44^\circ30'$	$8^\circ30'$	$-7^\circ35'$	$-9^\circ45'$
5.6b	$\psi_o=6^\circ30'$	$z\approx 0$	$7^\circ40'$	$9^\circ30'$	21°	$33^\circ10'$	21°	$9^\circ30'$	$7^\circ40'$
	$\psi_h=50^\circ$	$z\approx\infty$	$49^\circ25'$	$48^\circ50'$	46°	$33^\circ10'$	$15^\circ10'$	8°	7°
5.6c	$\psi_o=-65^\circ$	$z\approx 0$	$-64^\circ30'$	$-63^\circ45'$	$-57^\circ20'$	$-44^\circ20'$	$-57^\circ20'$	$-63^\circ45'$	$-64^\circ30'$
	$\psi_h=11^\circ$	$z\approx\infty$	$10^\circ45'$	$6^\circ10'$	$-8^\circ30'$	$-44^\circ20'$	-61°	$-64^\circ20'$	$-64^\circ45'$

Let us first turn to the schemes of Figs.5.4 and 5.6b, in which $\tan\psi_o$ and $\tan\psi_h$ are positive. The expression (5.63) leads to a decrease in the values of $\tan\alpha$ and α equally at positive and negative increasing values of y_r. From the maximum value corresponding to $y_r = 0$, the angle α, decreasing, approaches the value ψ_o. As for the values of the angle α for $z \approx \infty$, determined from (5.65), here the increase in the positive and negative values of y_r leads to different results. In the first case the values of the angle α approach ψ_o, and in the second case, ψ_h. Here, the range of changes in energy propagation from the surface layer to deeper ones (from $z \approx 0$ to $z \approx \infty$) decreases almost to zero with an increase in the positive values of y_r. With an increase in the negative values of y_r, this range increases to 2ϑ.

In other words, an increase in $|y_r|$ results in the coincidence of the directions of the vectors $\bar{\bar{\underline{S}}}$ and $\underline{s}_o$ in the surface layer, whereas an increase in the positive values of y_r results in the coincidence of the same vectors deep inside the crystal, while an increase in the negative values of y_r results in the coincidence of $\bar{\bar{\underline{S}}}$ and $\underline{s}_h$. Hence, the curvature of the respective paths decreases in the first case ($y_r > 0$) and increases in the second ($y_r < 0$).

Accordingly, the path of the vector $\bar{\bar{\underline{S}}}$ near the right (positive) edge of the maximum (as well as in the centre at $y_r = 0$) is almost rectilinear, while on the left side of the maximum the range of variation $\Delta\alpha$ approaches 2ϑ. This also determines the signs of the angles α, which depend on the signs of ψ_o and ψ_h.

All this also refers to schemes representing the mirror image of Fig.5.6b.

An essential feature of the paths of Poynting's vector when using schemes with an angle $\measuredangle\, \underline{s}_o\underline{s}_h$ overlapping parts of both the lower quadrants of the coordinate field is their curvature on intersecting the vertical axis n. This follows from the changed signs of the angles α inside the crystal bulk as compared with the signs in the surface layer, which is observed (Table 5.1) on the left side of the maximum. The use of the asymptotic expressions (5.63) and (5.65) for numerical estimates given in Table 5.1 does not permit one to trace the shape of such paths and, in particular, determine (in this semi-quantitative analysis) the values of the z-coordinate at which the paths intersect the vertical axis. This is possible only in the case of symmetrical reflection, whose scheme is given in Fig.5.5.

For symmetrical reflection (see (4.101)), C = 1,

$$\psi_o = -\psi_h = \vartheta \quad , \tag{5.67}$$

$$x = \sigma_h = \frac{\mu\varepsilon}{\cos\vartheta\,\sqrt{1 + y_r^2}} \quad . \tag{5.68}$$

The energy propagation path equation (5.61) takes the following form

$$x_s = \frac{y_r \sin\vartheta}{\mu\varepsilon} \ln \frac{\cosh\left(v_r + \dfrac{z\mu\varepsilon}{\cos\vartheta\sqrt{1+y_r^2}}\right)}{\cosh v_r} \quad . \tag{5.69}$$

It is obvious that the sign of x_s will be opposite to that of y_r, because $\sin\vartheta$ in this case ($\psi_0 < 0$) will be a negative value. Finally, (5.57) is also considerably simplified:

$$\tan\alpha_s = \tan v_r \tanh(v_r+\sigma_h z) \tan\vartheta \equiv \frac{y_r \tanh(v_r+\sigma_h z) \tan\vartheta}{\sqrt{1+y_r^2}} \quad . \tag{5.70}$$

The signs of the function $\tan\alpha_s$ obey the same rule as the signs of x_s. The asymptotic forms of (5.70) are as follows:

$$\tan\alpha_s|_{z\approx 0} \approx \frac{y_r^2}{1+y_r^2} \tan\psi_0 = -\frac{y_r^2}{1+y_r^2} |\tan\vartheta| \tag{5.71}$$

$$\tan\alpha_s|_{z\approx \infty} \approx \frac{y_r}{\sqrt{1+y_r^2}} \tan\psi_0 \tag{5.72a}$$

$$\tan\alpha_s|_{z\approx \infty} \approx -\frac{y_r}{\sqrt{1+y_r^2}} |\tan\vartheta| \quad . \tag{5.72b}$$

The expression (5.72a) coincides with (5.53) for symmetrical reflection from a transparent crystal, while (5.72b) corresponds to the case of propagation of the energy of only the second field and is directly obtained from (5.65) by superposing the conditions for symmetrical reflection. Considering the dependence of the signs of the angles α and $\tan\alpha$ on the signs of y_r, we note the following.

At both limiting values of z, positive values of y_r yield negative values of $\tan\alpha$, i.e., at $y_r < 0$ the energy propagation paths fit completely into the right side of the coordinate field at $x < 0$. Here, the greater y_r is, the less is the increase in $|\tan\alpha|$ over the entire range of z. At values of y_r corresponding to the positive boundary of the maximum, the values of $\tan\alpha|_{z\approx 0} \approx \tan\alpha|_{z\approx\infty}$, i.e., the path of the vector $\bar{\bar{\underline{S}}}$ is almost parallel to the direction of $\underline{s}_0$.

At negative values of y_r near the upper boundary of the crystal, $\tan\alpha$ also has negative values, but at large z, $\tan\alpha$ changes its sign, i.e., the path of the vector $\bar{\bar{S}}$ intersects the z-axis. The larger $|y_r|$ is, the greater the change in $|\tan\alpha|$. At large negative values of y_r the change in $|\alpha|$ almost equals 2ϑ.

From the condition $x_s = 0$, according to (5.70), apart from the solution $y_r = 0$, it also follows that at $y_r < 0$ and $v_r < 0$

$$z = 2|v_r|/\sigma_h \quad . \tag{5.73}$$

In other words, with a decrease in y_r or v_r, the point of intersection "rises" to the origin, i.e., to the entrance surface of the crystal. At $y_r = 0$ the energy propagation path coincides with the z-axis, which lies on the reflecting plane.

The indicated features of the geometry of energy propagation in an absorbing crystal are illustrated by some of the curves representing the trajectories in the reflection scheme in Fig.5.7.

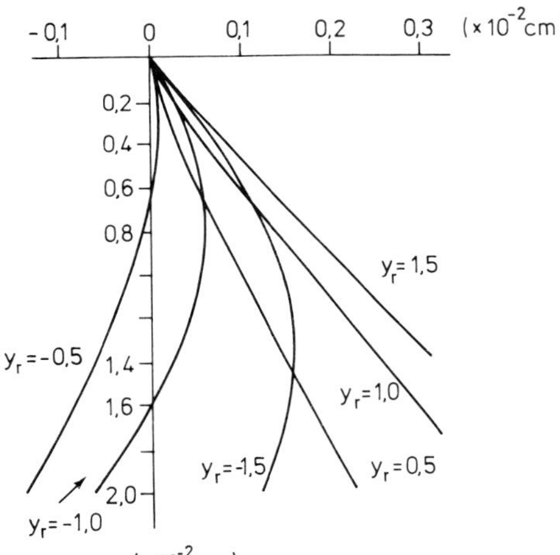

Fig. 5.7 Curvilinear projectory of Poynting's vector in absorbing crystals (after JAMES [1.16])

The above analysis of the geometry of energy propagation in an absorbing crystal deals with relatively small values of the products $\sigma_m t$ for asymmetrical reflection or μt for symmetrical reflection. When dealing with thick crystals we should keep in mind the dependence of absorption on the parameter y_r or introduce into our equations, in place of the absorption coefficient,

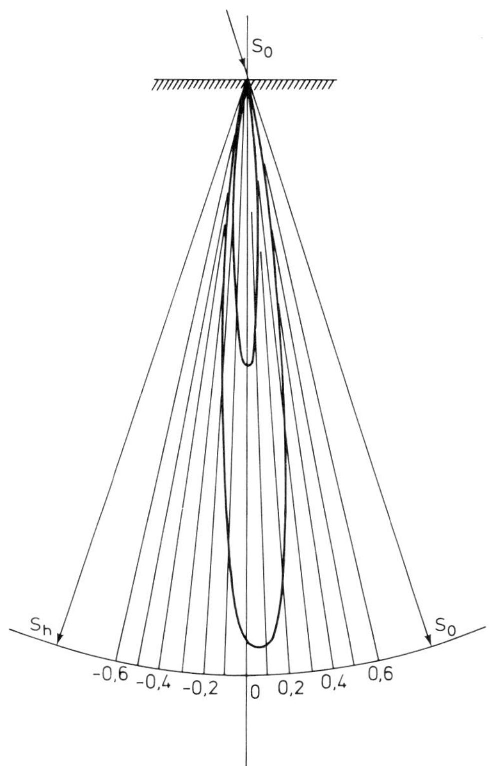

Fig. 5.8 The decrease in the divergence of the X-ray beam in thick crystal

a special function of the linear coefficient of absorption along the direction of the vector $\bar{\bar{\underline{S}}}$, namely $\mu^{\bar{\bar{\underline{S}}}}$. This function may be represented either as a function of y_r (or v_r), or as the angle ε of deviation of $\bar{\bar{\underline{S}}}$ from the reflecting plane.

The simplest way of taking account of the absorption in treatment of energy propagation in a thick absorbing crystal is to replace the expression for $\bar{\bar{\underline{S}}}$ by that for $\bar{\bar{\underline{S}}}^{(2)}$ (5.66).

Thus, the value of the energy current-density vector in a thick absorbing crystal is obtained by using (5.30b) for $\bar{\bar{\underline{S}}}^{(2)}$ and introducing the absorption factor:

$$|\bar{\bar{\underline{S}}}| = |\bar{\underline{S}}_o| \, \frac{\exp[-(\sigma_m - \varkappa)z]}{4(1+y_r^2)} \, N(y_r + \sqrt{1+y_r^2}) \quad , \tag{5.74}$$

where N is the numerator of the fraction (5.30b). In the particular case of symmetrical reflection we use (5.31b):

$$|\bar{\bar{\underline{S}}}| = |\bar{\underline{S}}_o| \frac{\exp[-(\sigma_m - \varkappa)z]}{2(1+y_r^2)} (y_r + \sqrt{1+y_r^2}) \sqrt{y_r^2 + \cos^2\vartheta} \quad . \tag{5.75}$$

Fig.5.8 depicts a polar diagram of the value $|\bar{\bar{\underline{S}}}|/|\bar{\underline{S}}_o|$ as a function of y_r, according to (5.75), for symmetrical 220 reflection of Cu $K\alpha$ radiation from Ge at the values of μt = 17.65 and 35.3. The most important feature of the diagram is the narrowing of the front of a wave field propagating in a sufficiently thick absorbing crystal - to put it more precisely, a decrease in the angular divergence of a beam incident on a crystal. The effect of decreasing the divergence of the beam is due to the dependence of the exponent at the absorption factor in (5.75) on the angular variable y_r or v_r. On the basis of (4.78) and (4.94) the following expression for this can be obtained:

$$\sigma^{(i)} = \sigma_m \pm \sigma_h \pm \sigma_h' =$$

$$= \frac{\mu}{2 \cosh v_r} \left[\frac{1}{\gamma_o} \exp(\mp v_r) + \frac{1}{\gamma_h} \exp(\pm v_r) + \frac{2\varepsilon}{\sqrt{\gamma_o \gamma_h}} \right] \quad . \tag{5.76}$$

It is easy to see that, for instance, in the case of symmetrical reflection $|\bar{\bar{\underline{S}}}^{(2)}|$ decreases from the boundaries of the maximum region towards the midpoint by 4.5-30 percent, depending on the value $\varepsilon = |\chi_{hi}|/|\chi_{oi}|$, which leads to a considerable increase in $\bar{\bar{\underline{S}}}$. As noted in Chapter 9 (Fig.9.4b), this effect is used in diffractometry to obtain beams similar to a plane wave.

Of more general significance is the linear absorption factor, in the form of a function of ε_i, which yields the value of the absorption coefficient along the direction of $\bar{\bar{\underline{S}}}^{(i)}$ for any point of the dispersion curve. This expression can be used in cases where the experimental conditions do not permit each given energy current-density vector to be related to a definite value of the angle η within the maximum upon incidence of a divergent beam or a spherical wave upon the crystal.

Introduce a vector $\underline{L}^{(i)}$ (Fig.5.9), which is defined as follows in accordance with (5.20b).

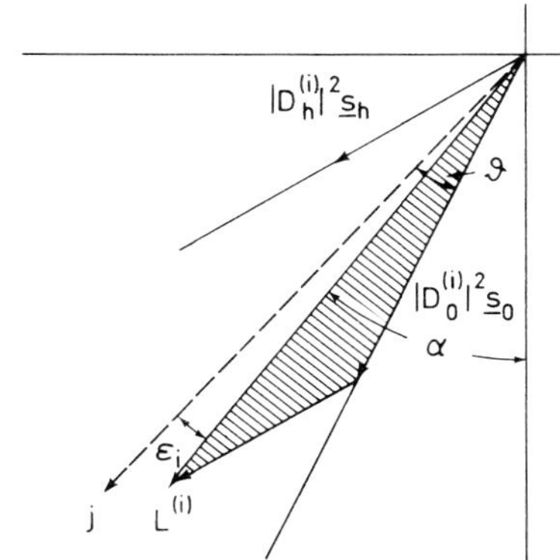

Fig. 5.9 The derivation of (5.84)

$$\underline{L}^{(i)} = \frac{8\pi}{c}\,\overline{\overline{\underline{S}}}^{(i)} = |D_o^{(i)}|^2\,\underline{s}_o + |D_h^{(i)}|^2\,\underline{s}_h \quad , \tag{5.77}$$

hence

$$(\underline{L}^{(i)}\underline{n}) = |\underline{L}^{(i)}|\cos\alpha_i = |D_o^{(i)}|^2\,\gamma_o + |D_h^{(i)}|^2\,\gamma_h \quad . \tag{5.78}$$

On the other hand, from the hatched triangle in Fig.5.9 it follows that

$$|\underline{L}^{(i)}|\sin(\vartheta\pm\varepsilon_i) = |D_{o,h}^{(i)}|^2\,\sin 2\vartheta \quad , \tag{5.79}$$

$$\frac{|D_o^{(i)}|\,|D_h^{(i)}|}{|\underline{L}^{(i)}|} = (\sin 2\vartheta)^{-1}\sqrt{\sin(\vartheta+\varepsilon_i)\,\sin(\vartheta-\varepsilon_i)} = (2\cos\vartheta)^{-1}\sqrt{1-\frac{\sin^2\varepsilon_i}{\sin^2\vartheta}} \quad . \tag{5.80}$$

Projecting the same triangle along the direction $\underline{j}$, we obtain

$$|\underline{L}^{(i)}|\cos\varepsilon_i = [|D_o^{(i)}|^2 + |D_h^{(i)}|^2]\cos\vartheta \quad . \tag{5.81}$$

Further, from (3.18) follows

$$\frac{\exp(\pm v_r)}{\sqrt{\gamma_o}} = \frac{1}{\sqrt{\gamma_h}} \left|\frac{D_o^{(i)}}{D_h^{(i)}}\right| ,$$

$$\frac{\exp(\pm v_r)}{\sqrt{\gamma_h}} = \frac{1}{\sqrt{\gamma_o}} \left|\frac{D_h^{(i)}}{D_o^{(i)}}\right| , \tag{5.82}$$

$$2 \cosh v_r = \frac{|D_o^{(i)}|^2 \gamma_o + |D_h^{(i)}|^2 \gamma_h}{|D_o| \, |D_h| \sqrt{\gamma_o \gamma_h}} .$$

Using the relations derived, as well as the expression (5.76) for σ, we get

$$\mu^{s(i)} = \sigma \cos\alpha_i = (\mu/|\underline{L}|) \, [|D_o^{(i)}|^2 + |D_h^{(i)}|^2 \pm 2\varepsilon |D_o^{(i)}| \, |D_h^{(i)}|] \quad , \tag{5.83a}$$

or, eliminating $\underline{L}$ with the aid of (5.80) and (5.81), we arrive at an expression coinciding with the Laue equation for μ in the direction of the vector $\overline{\overline{\underline{S}}}$ [Ref.1.14, p.417, Eq.(34.39)]

$$\mu^{s(i)} = \frac{\mu}{\cos\vartheta} \left(\cos\varepsilon_i \pm \varepsilon \sqrt{1 - \frac{\sin^2\varepsilon_i}{\sin^2\vartheta}} \right) , \qquad \varepsilon = \frac{|\chi_{hi}|}{|\chi_{oi}|} . \tag{5.83b}$$

Finally, using the equality

$$\frac{1}{\cos\vartheta} \sqrt{1 - \frac{\sin^2\varepsilon_i}{\sin^2\vartheta}} = \frac{\cos\varepsilon_i}{\cos\vartheta} \sqrt{1 - \frac{\tan\varepsilon_i}{\tan\vartheta}}$$

we get

$$\mu^{s(i)} = \left(2\pi K |\chi_{oi}| \pm 2\pi K |\chi_{hi}| \sqrt{1 - \frac{\tan^2\varepsilon_i}{\tan^2\vartheta}} \right) \frac{\cos\varepsilon_i}{\cos\vartheta} . \tag{5.84}$$

If the wave field energy has propagated along a certain direction $\underline{L}$ for a segment ℓ, the loss of this energy by absorption is obviously equal to

$\mu^{s(i)}\ell$. Projecting the path ℓ onto the direction $\underline{j}$ and decomposing the projection obtained into components $\ell_o\underline{s}_o$ and $\ell_h\underline{s}_h$, we have

$$\mu^{s(i)}\ell\cos\varepsilon_i = \mu^{s(i)}(\ell_o+\ell_h)\cos\vartheta = \left(2\pi K|\chi_{oi}| \pm 2\pi K|\chi_{hi}|\sqrt{1-\frac{\tan^2\varepsilon_i}{\tan^2\vartheta}}\right)(\ell_o+\ell_h) \quad . \tag{5.85}$$

From this relation it follows that the expression in the parentheses is the linear coefficient of absorption along a certain optical path in the crystal consisting of infinitesimal segments along $\underline{s}_o$ and $\underline{s}_h$. The expression obtained for the linear coefficient of absorption, as shown by KATO [1.24], possesses a remarkable property: it can be obtained without fulfilling the conditions (4.80), which lie at the basis of the theory of dynamical scattering in an absorbing crystal.

Note the two limiting values on the right-hand side of (5.85) at $\varepsilon_i = 0$ and $\varepsilon_i = \vartheta$. At $\varepsilon_i = 0$ the second term in parentheses acquires the maximum possible value, and hence the linear coefficient of absorption (for the second field) takes on the minimum value. In other words, the minimum of energy absorption occurs when it propagates along the reflecting plane. Conversely, the absorption maximum, corresponding to the value of normal linear absorption $2\pi K|\chi_{oi}| = \mu$ corresponds to the value $\varepsilon_i = \vartheta$; thus definite sharp boundaries of the dynamical wave field inside the crystal correspond to the value $\varepsilon_i = \pm\vartheta$. Therefore, (5.85) describes a wave field strictly localized in a (triangular) area of the absorbing crystal.

At the same time, (5.85), which is of a general nature and refers both to symmetrical and asymmetrical reflections, shows the two component parts of the linear coefficient of absorption of a crystal in the maximum region: the common coefficient μ and the anomalous one - the interference, or Borrmann absorption coefficient

$$\mu_h = \pm 2\pi K\left[\chi_{hi}\chi_{\bar{h}i}\left(1-\frac{\tan^2\varepsilon_i}{\tan^2\vartheta}\right)\right]^{1/2} \quad . \tag{5.86}$$

The interference coefficient of absorption depends solely on the parameter χ_{hi} and is independent of χ_{oi}, which determines the value of μ.

Comparing the expression for the linear coefficient of absorption within the maximum range (5.85) with an equation of the type (4.99)

$$\sigma = \frac{\mu}{2}\left(\frac{1}{\gamma_o}+\frac{1}{\gamma_h}\right) \pm \frac{\mu}{2}\left(\frac{1}{\gamma_o}-\frac{1}{\gamma_h}\right) \tanh v_r \pm \frac{\mu\varepsilon}{\sqrt{\gamma_o\gamma_h}\cosh v_r} \tag{5.87}$$

we note, above all, that if (5.85) is not related to a definite value of the angle of incidence ψ_0 and, hence, to the position of the reflecting plane relative to the entrance face, then (5.87) is not related to a definite point of incidence of the vacuum wave on the crystal. Thus, switching from (5.87) to (5.85), we pass over from the incident-plane-wave to the incident wave with a sufficiently narrow wave front (slit). It is obvious that in the general case the two expressions are not equivalent, since (5.85) yields the coefficient of absorption along any direction in the crystal, whereas (5.87) yields the coefficient of absorption along the normal to the entrance surface. Recall that (5.85) has a general meaning because it is applicable at any ratio of the parameters $|\chi_{hi}|$ and $|\chi_{hr}|$, while (5.87) is applicable only at $|\chi_{hi}| << |\chi_{hr}|$. The wave field in the crystal, which has been induced by an incident wave with a narrow wave front, evidently has a simpler structure in some respects, and the task of determining the linear coefficient of absorption in any direction can be solved fairly simply without using the above approximation. In an important particular case of symmetrical reflection at $v_r = y_r = \varepsilon_i = 0$ the two expressions coincide. More precisely,

$$t\sigma_s^{v_r=0} = (2\pi K|\chi_{oi}| \pm 2\pi K|\chi_{hi}|)\,(\ell_o+\ell_h)t \quad . \tag{5.88}$$

5.4 Energy Propagation in Absorbing Crystal Without a Centre of Symmetry, Taking into Account the Periodic Component of Poynting's Vector. Additional Remarks

As already mentioned above, the theory outlined is applicable to crystals with a centre of symmetry. In the case of an absorbing crystal without a centre of symmetry, an expression corresponding to (5.54) does not permit us to make a detailed qualitative analysis like the one made above.

The expression for the vector of energy flow propagation in an absorbing crystal without a centre of symmetry has the form

$$\overline{\overline{\underline{S}}} = |\overline{\underline{S}}_o| \frac{\exp(-\sigma_m z)}{2\cosh^2 v_r} \left\{ \left[\cosh(2v_r + \varkappa z)\right] \underline{s}_o + \frac{\gamma_o}{\gamma_h} \left|\frac{\chi_h}{\chi_{\bar{h}}}\right| \cosh(\varkappa z)\ \underline{s}_h \right\} \quad . \tag{5.89}$$

Switching to the system of coordinates x,z we obtain

$$\overline{\overline{\underline{S}}} = |\overline{\underline{S}}_o| \frac{\gamma_o \exp(-\sigma_m z)}{2\cosh^2 v_r} \left\{ \left[\cosh(2v_r + \varkappa z) + \left|\frac{\chi_h}{\chi_{\bar{h}}}\right| \cosh(\varkappa z)\right] \underline{n} + \right.$$

$$\left. + \left[\cosh(2v_r + \varkappa z) \tan\psi_o + \left|\frac{\chi_h}{\chi_{\bar{h}}}\right| \cosh(\varkappa z) \tan\psi_h\right] \underline{a} \right\} \quad . \tag{5.90}$$

Hence

$$\left.\frac{dx}{dz}\right|_{z\approx 0} \approx \tan\alpha = \frac{\cosh(2v_r + \varkappa z) \tan\psi_o + \left|\frac{\chi_h}{\chi_{\bar{h}}}\right| \cosh(\varkappa z) \tan\psi_h}{\cosh(2v_r + \varkappa z) + \left|\frac{\chi_h}{\chi_{\bar{h}}}\right| \cosh(\varkappa z)} \tag{5.91a}$$

$$x = \int_0^z \frac{dx}{dz} dz \quad . \tag{5.91b}$$

The asymptotic equations for $\tan\alpha$ at $z \approx 0$ and $z \approx \infty$ have the form

$$\left.\frac{dx}{dz}\right|_{z\approx 0} \approx \left[\cosh(2v_r) \tan\psi_o + \left|\frac{\chi_h}{\chi_{\bar{h}}}\right| \tan\psi_h\right] \cdot$$

$$\cdot \left[\cosh(2v_r) + \left|\frac{\chi_h}{\chi_{\bar{h}}}\right|\right]^{-1} \quad ; \tag{5.92}$$

$$\left.\frac{dx}{dz}\right|_{z\approx \infty} \approx (\exp 2v_r \cdot \tan\psi_o + \left|\frac{\chi_h}{\chi_{\bar{h}}}\right| \tan\psi_h)/(\exp 2v_r + \left|\frac{\chi_h}{\chi_{\bar{h}}}\right|) \quad .$$

In the case of symmetrical reflection

$$\tan\psi_o = -\tan\psi_h = \tan\vartheta \tag{5.93}$$

$$-\tan\alpha = \frac{\cosh(2v_r+\varkappa z) - \left|\frac{\chi_h}{\chi_{\bar h}}\right| \cosh(\varkappa z)}{\cosh(2v_r+\varkappa z) + \left|\frac{\chi_h}{\chi_{\bar h}}\right| \cosh(\varkappa z)} \quad . \tag{5.94}$$

Note that here, in contrast to a centro-symmetrical crystal, even at $y_r = v_r = 0$, $\tan\alpha \neq \tan\vartheta$. Instead of this we obtain from (5.84)

$$\tan\sigma_s^{v_r=0} = \frac{1 - |\chi_h/\chi_{\bar h}|}{1 + |\chi_h/\chi_{\bar h}|} \tan\vartheta \quad . \tag{5.95}$$

Comparing the value of $\tan\alpha$ from (5.95) for two reflections hkl and $\bar h\bar k\bar l$ and using (4.56), we get

$$\tan\alpha_{s,hkl}^{y_r=0} = \frac{-|\chi_{hi}/\chi_{hr}| \sin v_h}{1 + |\chi_{hi}/\chi_{hr}| \sin v_h} |\tan\vartheta| \quad , \tag{5.96a}$$

$$\tan\alpha_{s,\bar h\bar k\bar l}^{y_r=0} = \frac{|\chi_{hi}/\chi_{hr}| \sin v_h}{1 - |\chi_{hi}/\chi_{hr}| \sin v_h} |\tan\vartheta| \quad . \tag{5.96b}$$

Thus, when fixing the point of emergence of the wave field energy on the exit surface of the crystal slab, the difference between reflections hkl and $\bar h\bar k\bar l$ can be established solely *from the geometry of the diffraction effect*, whereas, using the reflection coefficient (see Chap.4), the difference is established by the *intensities*.

In conclusion, we should consider the relations referring to the paths of the twice averaged Poynting's vector $\bar{\underline{S}}$, which contain the periodic components. The need to include these oscillations is particularly obvious in the case of absorbing crystal, since the use of the vector $\bar{\bar{\underline{S}}}$ averaged over the extinction depth τ is limited by the corresponding value of the $\sigma = \exp(-\mu_{int.} \cdot n\,\tau_0)$. Thus, in the case Cu $K\alpha_1$ radiation, 220: n = 100, $\sigma_{Ge} = 0.234$, $\sigma_{Si} = 0.13$; 333: n = 10, $\sigma_{Ge} = 0.23$; n = 20, $\sigma_{Si} = 0.081$. Unfortunately, the calculation of the trajectories of Poynting's vector with an allowance for the periodic components and absorption is rather complicated it can be analyzed qualitatively only in the simplest case of centro-symmetrical structures.

The main specific features of the problem can be demonstrated for the case of scattering from the transparent crystal.

Taking into account the periodic components in the values of the transmission power and reflection power coefficients (see (3.30,32)) we obtain for $\underline{\bar{S}}$, in place of (5.21),

$$\underline{\bar{S}} = |\underline{\bar{S}}_0| \ (1+y^2)^{-1}[(y^2+\cos^2\rho) \ \underline{s}_0 + \frac{\gamma_0}{\gamma_h} \sin^2\rho \ \underline{s}_h] \quad , \tag{5.97}$$

where $\rho = \pi z/\tau$. Switching to the system of coordinates x,z, we rewrite (5.97) as

$$\underline{\bar{S}} = \frac{|\underline{\bar{S}}_0| \ \gamma_0}{1 + y^2} [(y^2+\cos^2\rho)(\underline{n}+\underline{a} \ \tan\psi_0) + \sin^2\rho(\underline{n}+\underline{a}\tan\psi_h)] \tag{5.98}$$

or

$$\underline{\bar{S}} = \frac{|\underline{\bar{S}}_0| \ \gamma_0}{1 + y^2} \{(1+y^2)\underline{n} + [(y^2+\cos^2\rho) \ \tan\psi_0 + \sin^2\rho \ \tan\psi_h] \ \underline{a}\} \quad . \tag{5.99}$$

Hence

$$\tan\alpha = \frac{y^2}{1 + y^2} \tan\psi_0 + \frac{\cos^2 \frac{\pi z}{\tau}}{1 + y^2} \tan\psi_0 + \frac{\sin^2 \frac{\pi z}{\tau}}{1 + y^2} \tan\psi_h \quad . \tag{5.100}$$

Replacing $\sin^2\rho$ and $\cos^2\rho$ by their average values, we get (5.51).

Let us now consider symmetrical reflection in more detail. In this case we obtain, in place of (5.100),

$$\frac{dx}{dz} = \tan\alpha = \frac{y^2}{1 + y^2} \tan\vartheta + \frac{\cos 2 \frac{\pi z}{\tau}}{1 + y^2} \tan\vartheta \quad . \tag{5.101}$$

The first term from the right refers to a nonperiodic component, which coincides with (5.33) for a triply averaged vector and here can be referred to as the energy propagation centre line. The second term describes periodic increases and decreases in angle α compared with the angle for the centre lines. Integrating (5.101) with respect to z, we obtain the trajectory equation:

$$x = \frac{y^2}{1 + y^2} z \tan\vartheta + \frac{\tan\vartheta}{1 + y^2} \frac{\sin 2\rho}{2\rho/z} \quad . \tag{5.102}$$

The analysis of (5.101) and (5.102) is directly connected with the description of the Pendellösung fringe solution for the transport crystal given in Chapter 3 (see (3.44-47)). Indeed, the centre line will intersect with the oscillating line at the depths

$$z = \frac{m}{\Delta k} \quad , \quad z = \frac{2m + 1}{2\Delta k} \quad , \quad \Delta k = \tau^{-1} \quad , \tag{5.103}$$

which correspond to the extinction distance. At such values of z, the slopes of the oscillating curve, according to (5.101) and because $\cos 2\rho = \pm 1$, will be equal to

$$\tan\alpha = \tan\vartheta \quad , \quad \tan\alpha = \frac{y^2 - 1}{y^2 + 1} \tan\vartheta \quad . \tag{5.104}$$

Between the two points of intersection, the curve is spaced from the centre line

$$x_0 = \frac{y^2}{1 + y^2} z \tan\vartheta \tag{5.105}$$

by the distance

$$x - x_0 = \frac{\tan\vartheta}{2\pi\Delta k\ (1+y^2)} \quad . \tag{5.106}$$

Here, $z = (m \pm 1/4)\Delta k$. At these points the energy-current density vector $\underline{\bar{S}}$ is parallel to the z-axis.

It is important to note that the characteristic parameters of the energy flow trajectory, determined by the above equations, are functions of y. In particular, at y = 0 the z-axis becomes the centre line (5.105). On the z-axis, the spacings between the two closest points of intersection $z = m\tau$ and $z = (m + 1/2)\tau$ will be the greatest among the distances corresponding to the various possible values of y. At the points of intersection $z = m\tau$ the vector $\underline{\bar{S}}$ will be directed along $\underline{s}_0$. At the same value of y = 0, the vector $\underline{\bar{S}}$ will be directed along $\underline{s}_h$ at the points $z = (m + 1/2)\tau$. On the centre lines (straight lines) corresponding to $y \neq 0$, the vector $\underline{\bar{S}}$ will also be directed along $\underline{s}_0$ at the points $z = m\tau$, but at the intermediate points of intersection it will deviate from $\underline{s}_h$. This deviation increases with y. A general idea of the oscillations of the direction of the energy flow in the case of symmetrical reflection is given by Fig.5.10, which differs from the corresponding figure for $\overline{\underline{S}}$ (Fig.5.1) by superposition of the oscillating curves.

In the case of the absorbing crystal, the equations for $\tan\alpha$ and x (5.69,70) will include additional components similar to the second terms in the right-hand side of (5.101) and (5.102). These additional terms have the form

$$\tan v_{r,s} = \cos\frac{2\pi z}{\tau}\tan\vartheta/\cosh(v_r+\varkappa z)\cosh v_r \tag{5.107}$$

$$x_{v_r,s} = \frac{\tan\vartheta}{\cosh v_r}\int_0^z \cos\frac{2\pi z}{\tau}\left[\cosh(v_r+\varkappa z)\right]^{-1} dz \quad . \tag{5.108}$$

The integrand in (5.108) oscillates with a depth z. At positive values of $y_r = \sinh v_r$, these oscillations decrease monotonically with depth, and at negative values of y_r they pass a (gently sloping) maximum, and then also fall off. The function $x_{v_r,s}$, which is, according to (5.108), proportional

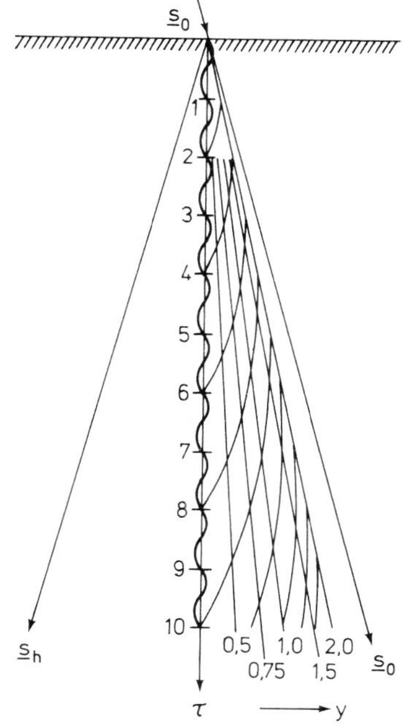

Fig. 5.10 Oscillations of directions of propagation of Poynting's vector (after JAMES [1.16])

to the area under the oscillating curves, behaves similarly. It is obvious that the oscillation period is equal to the ratio z/τ. Some additional data referring to the behaviour of the function $x_{v_r,s}$ can be obtained from JAMES' survey [1.16].

6. Dynamical Theory in Incident-Spherical-Wave Approximation

The dynamical theory of X-ray scattering in perfect crystals outlined in Chapters 2 through 4 has, as one of its starting conditions, the incidence of a *plane monochromatic wave* on a crystal. At the end of Chapter 3 we mentioned the possibility of other experiments, in particular the use of a beam with a limited wave front and a set of directions of incidence overlapping the angular width of the reflection maximum. In Chapter 5 we considered the case of the wave front of an incident wave which was small in comparison with the thickness of the crystal slab.

Experiments in which such a scheme was used (a slit system between the tube and the crystal) were staged in 1959 by KATO and LANG [1.23] for investigation of the Pendellösung fringe solution in transmission of an X-ray beam through a wedge-shaped crystal. These authors obtained X-ray patterns of two types: section (picture taken with a fixed position of the crystal and film relative to the X-ray source) and traverse type diffraction topographs obtained in scanning the crystal and film relative to the source. In section patterns, they obtained the pendulum solution fringes in the form of strongly elongated hyperbolas. With scanning, equal thickness type straight bands were obtained.

In connection with the qualitative, and particularly quantitative, analysis of these experiments, KATO developed a dynamical theory which describes X-ray scattering by a crystal in the experimental conditions illustrated by Fig.3.17b [1.24]. This approach to scattering phenomena is obviously closer to the experimental conditions than is the plane-incident-wave approximation.

It is easy to see that the most essential factor is the ratio between the angular width of the reflection (or transmission) maximum, and the angular width of the X-ray beam emerging from the anode and equivalent (in some respects) to a spherical wave. This width is $\sim 1.5'$, i.e., at least one order larger than the half-width of the maximum. It is interesting to compare these conditions with the electron diffraction patterns from a single-crystal film, obtained in a electron microscope. The divergence of the electron beam in such instruments does not exceed 10^{-3} rad, while the half-width of the first maxima of dynamical scattering is one order higher.

An analysis of this fundamental problem of "dynamical" X-ray experiment is also given in Chapter 9 in connection with the methods for "preparation" of a plane monochromatic wave with the aid of monochromators.

In the title of this chapter and partly in the text we will retain KATO's terminology. Strictly speaking, here we are solving the problem of dynamical scattering by a crystal of a wave packet as was done by KATO in the papers indicated. The fundamental significance of these works is in the successful use of the method of decomposition of an incident wave packet into an (infinite) set of plane waves with subsequent integration with respect to the angular variable.

This method made it possible, in the present set of investigations, to describe with high accuracy the above-mentioned bands of the pendulum solution in the section patterns, which were actually the first experimental realization of the effect predicted by EWALD in his papers of 1916-1917.

An important result of KATO's theory and of the experimental advancement of the section pattern technique is the new method for measuring absolute values $|\chi_{hr}|$ and, hence, structure amplitudes $|F_{hr}|$. Here, the absolute values are opposed to the relative values, which are usually determined from measurements of the intensity of X-ray diffraction. A specific feature of KATO's method is exclusion of the error due to the effect of extinction on the intensity of strong reflections, since only the geometrical parameters of the diffraction pattern are measured on section patterns.

Moreover, in distinction to experiments similar to the conditions of incidence of a plane wave, the difficulty associated with the uncertainty in the value of the angular parameter y_r is almost eliminated.

On the other hand, the field of application of the hyperbola method is restricted to scattering in transparent or weakly absorbing crystals.

The importance of the plane-wave method is by no means limited to the solution of the problem of scattering in a single perfect, plane-parallel or wedge-shaped crystal. The method was used successfully in solving the problem of scattering on a bounded crystal (both Bragg and Laue reflection), as has been briefly described in Section 7.3. It acquired a still greater importance in the analysis of the diffraction pattern of a plane defect in a crystal and particularly in two-crystal interferometric schemes (Chap.10). Its advantages over methods using the influence of the Green's function consist in its pictorialness and, in some cases, greater simplicity.

6.1 Dynamical Theory in a Two-Wave Approximation with Spherical Wave Incident on Crystal. Application to Scattering in Transparent Plane-Parallel and Wedge-Shaped Crystals

A scalar wave in a vacuum propagating in a positive direction along the z-axis is described in the general case by the following integral equation:

$$\Psi_a = \int_{-\infty}^{\infty}\int_{-\infty}^{\infty} F(\hat{\underline{K}}) \exp i(\underline{K}\underline{r}) dK_x dK_y \quad , \tag{6.1}$$

where $F(\underline{K})$ is the weight function referring to the angular spectrum of the wave Ψ_a. In the case of a plane wave the weight function turns into a δ-function:

$$F_p(\hat{\underline{K}}) = \delta(\hat{\underline{K}}-\hat{\underline{K}}_e) \quad . \tag{6.2}$$

In the case of a spherical wave the conventional Fourier representation (with a unit amplitude) can be used:

$$\Psi_s = \frac{\exp i(\underline{K}\underline{r})}{4\pi r} = \frac{1}{8\pi^3}\int_{-\infty}^{\infty}\int_{-\infty}^{\infty} \frac{\exp i(\underline{K}\underline{r})}{K_z} dK_x dK_y \quad . \tag{6.3}$$

Comparing (6.3) and (6.1), we note that the weight function for a spherical wave

$$F_s(\hat{\underline{K}}) = \frac{i}{8\pi^3}\frac{1}{K\cos(\underline{K}\underline{z})} \quad . \tag{6.4}$$

Here, $K\cos(\underline{K}\underline{z}) = K_z = (K^2 - K_x - K_y^2)^{1/2}$ takes positive values either on the real or on the imaginary axis of the complex plane.

In the case of X-rays an expression similar to (6.1) for vector waves should be used. In order to obtain such an expression we consider spontaneous electromagnetic radiation from atomic sources. In a vacuum, away from the source, the electromagnetic field can be fully characterized by assigning any of the vectors E or H related by the equation

$$\underline{E} = iK^{-1} \operatorname{rot} \underline{H} \quad . \tag{6.5}$$

If we introduce the operators

$$\underset{\sim}{M} = [\underline{I}\nabla] \quad , \quad \underset{\sim}{N} = \frac{1}{iK} [[\underline{I}\nabla]\nabla] \tag{6.6}$$

($\underline{I}$ is a constant vector proportional to the matrix element of the current density operator of a spontaneous atomic transition which results in the emission of a gamma-quantum) and retain only the main term (of the order of r^{-1}) in the asymptotic expression of the field strenght, we can obtain the following expressions for the electromagnetic field in the wave zone of radiation:

$$\underline{H}_s \approx \underset{\sim}{M}\Psi_s \quad , \quad \underline{E}_s = \underset{\sim}{N}\Psi_s \quad . \tag{6.7}$$

Note that in the asymptotic region the vectors $\underline{E}$ and $\underline{H}$ are perpendicular to the direction of field propagation. Substituting (6.1) in (6.7), we get

$$\begin{aligned} \underline{H}_s &\approx iK \int_{-\infty}^{\infty}\int_{-\infty}^{\infty} dK_x dK_y F_s(\hat{\underline{K}})[\underline{I}\underline{K}] \exp i(\underline{K}\underline{r}) \quad , \\ \underline{E}_s &\approx iK \int_{-\infty}^{\infty}\int_{-\infty}^{\infty} dK_x dK_y F_s(\hat{\underline{K}})[[\underline{I}\underline{K}]\underline{K}] \exp i(\underline{K}\underline{r}) \quad . \end{aligned} \tag{6.8}$$

Further on, we shall be interested in some definite states of polarization of the electromagnetic field in a crystal with displacement vector oscillations either in the plane of reflection or perpendicular to this plane. Therefore we can restrict ourselves to consideration of the scalar wave field, including the appropriate polarization factors $[\underline{I}\underline{K}]$ and $[[\underline{I}\underline{K}]\underline{K}]$ and the weight function $F(\underline{K})$.

Turning to the expression for the spherical wave in the form of superposition of plane waves, each of which is described by the dynamical theory outlined in the preceding chapters, we introduce into (6.3) the corresponding amplitude coefficients:

$$D_{o,h} = \frac{i}{8\pi^2} \int_{-\infty}^{\infty}\int_{-\infty}^{\infty} K_z^{-1}\, d_{o,h}\, dK_x dK_y \quad . \tag{6.9}$$

The component $d_{o,h}$ may be regarded as the amplitude modulation of the refracted and diffracted waves in the crystal. Recall that a similar interpretation was used by us with respect to (2.36) for the Bloch wave, and also in analyzing the wave field in a crystal (3.49).

The factors $d_{o,h}$ for a transparent crystal (in the two-wave approximation) are treated in detail in Chapter 3. They can also be written as

$$d_o = \sum_{i=1,2} D_o^{(i)} \exp i[(\underline{k}_o^{(i)} - \underline{K})\underline{r}] \tag{6.10}$$

$$d_h = \sum_{i=1,2} D_h^{(i)} \exp i[(\underline{k}_h^{(i)} - \underline{K})\underline{r}] \quad . \tag{6.11}$$

In (6.10) and (6.11), $\underline{r}$ denotes the distance of the observation point from the source, which is conventionally assumed to be a section of an infinitely narrow slit on the entrance surface of the crystal, point 0 in Fig.6.2; the same point is chosen as the origin of the coordinates in a real space.

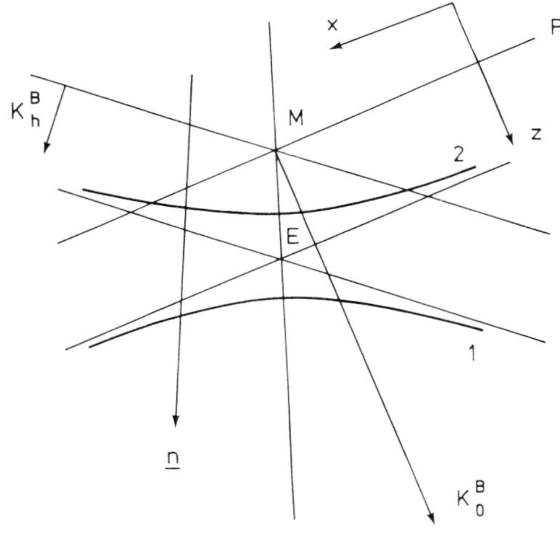

Fig. 6.1 Choice of coordinate axes x and z in reciprocal space (after KATO [1.24])

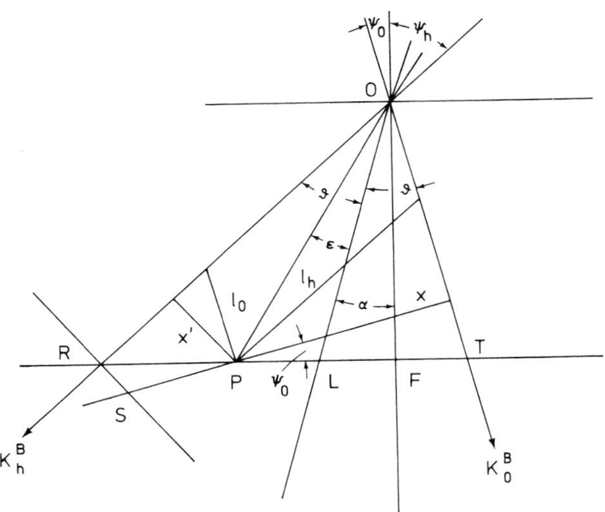

Fig. 6.2 Analysis of the process of wave propagation in plane-parallel crystal. Formation of Δ-Borrmann (after KATO [1.25])

Here, according to (3.1),

$$\underline{k}_o^{(i)} - \underline{K} = -K\delta^{(i)}\underline{n}_o \tag{6.12}$$

is normal to the entrance surface at any point. Thus, in (6.10) and (6.11) the magnitude of the phase in the phase factors is given with relation to an arbitrary point of the entrance surface.

Now we proceed to detailed consideration of the case of the incident spherical wave. For integration in a general expression of the type (6.9) the axes are chosen as follows.

In a reciprocal space, as shown in Fig.6.1 (see also Fig.3.1), the x-axis runs in the direction of reckoning the negative values of the angles η, and the z-axis coincides with the vector $\underline{K}_o^B$, which makes an angle ϑ with the reflecting plane. The origin is chosen at point M.

The analysis of the process of propagation in the plane-parallel crystal is explained in Fig.6.2. The origin is point O. The x-axis is perpendicular to the vector $\underline{K}_o^B$, which has the same meaning as in Fig.6.1 and represents the z-axis. The y-axis is perpendicular to the plane of the drawing, which coincides with the plane of reflection. The vector $\underline{K}_h^B$ is the diffracted wave vector. $\overline{OF}$ is a perpendicular dropped onto the exit surface, $\overline{OL}$ is the trace of the reflecting plane, and $\overline{OP}$ is a variable vector corresponding to the energy current density vector, which will be treated in more detail further on. In order to describe the wave field, as well as its strength, we use the direction of $\underline{\overline{OP}}$ at any point inside the crystal. Any point on this line inside the crystal and the point P on the exit surface are determined by oblique coordinates ℓ_o and ℓ_h, and also by the lengths of the perpendiculars dropped onto the vectors $\underline{K}_o^B$ and $\underline{K}_h^B$ (x and x').

The calculations referring to the theory under review are performed with the aid of the following variables

$$s = -K_x - \frac{K\chi_o\left(\frac{\gamma_h}{\gamma_o} - 1\right)}{2\sin 2\vartheta} \quad , \quad f = \frac{KC}{\sin 2\vartheta}\sqrt{\chi_h\chi_{\bar{h}}}\sqrt{\frac{\gamma_h}{\gamma_o}} \quad ,$$

$$t = [(\underline{r}-\underline{r}_e)\underline{n}] \quad , \quad \alpha = \frac{1}{2\gamma_h}\sin 2\vartheta \quad . \tag{6.13}$$

The relationship between the new and old variables, with an allowance for the direction of the x-axis, is established directly (in this chapter $K_o = 1/\lambda$ and $K = 2\pi/\lambda$; the same refers to the wave vectors inside the crystal).

$$\sqrt{1+y^2} \pm |y| = f^{-1}\left(\sqrt{s^2+f^2} \pm |s|\right) \quad . \tag{6.14}$$

$$K_x = +2\pi K_o\,|\eta| \quad , \quad s = \frac{K\,|\beta|}{2\sin 2\vartheta} \quad , \quad y = \frac{|s|}{f} \quad . \tag{6.15}$$

The complete analytical expression for the wave function of a refracted wave in a transparent crystal in the incident-plane-wave approximation, using the old variables, is

$$d_o^{(i)}(\underline{r}) = \frac{\sqrt{1+y^2} \mp y}{2\sqrt{1+y^2}} \exp - 2\pi i \left\{(\underline{K}_o\underline{r}) + t\left[\frac{K_o\chi_o}{2\gamma_o} \mp \mp \frac{K_oC\sqrt{\chi_h\chi_{\bar{h}}}}{2\sqrt{\gamma_o\gamma_h}}\left(\sqrt{1+y^2} \pm y\right)\right]\right\} \tag{6.16}$$

and the new ones,

$$d_o^{(i)}(\underline{r}) = \frac{1}{2}\,\frac{\mp s + \sqrt{s^2+f^2}}{\sqrt{s^2+f^2}} \exp i\left[(\underline{K}\underline{r}) + \frac{K\chi_o t}{2\gamma_o} + \right.$$

$$\left. + \alpha t(s\pm\sqrt{s^2+f^2})\right] \quad . \tag{6.17}$$

Note also that further on in this chapter the following expression is used to denote a plane wave:

$$\Psi = \Psi_o \exp[-i\omega t+i(\underline{K}\underline{r})] \quad . \tag{6.18}$$

Similar transformations lead to the expression for the wave function of a diffracted wave in a transparent crystal (see (3.32)):

$$d_h^{(i)}(\underline{r}) = \pm\frac{1}{2}\sqrt{\frac{\gamma_o}{\gamma_h}}\;\frac{f\exp i\eta_h}{\sqrt{s^2+f^2}} \exp i\left\{[(\underline{K}+2\pi h)\underline{r}] + \right.$$

$$\left. + \frac{K\chi_o z}{2\gamma_o} + \alpha t(s\pm\sqrt{s^2+f^2})\right\} \quad . \tag{6.19}$$

We shall now proceed in calculating the integral (6.9). We replace, in the amplitude, the value K_z^{-1} by K^{-1}, which is equivalent to replacing the spheres of radius K by tangent planes at their intersection at the Laue point (its trace is the line $\overline{MP}$). The phase factor $\exp i(\underline{K}\underline{r})$ in (6.17) and (6.19) is transformed according to (6.13)

$$\exp i(\underline{K}\underline{r}) = \exp i(K_z z+K_x x) \quad . \tag{6.20}$$

As regards integration with respect to K_y in (6.9), it should be taken into account that the variable K_y appears only in (6.20), since $K_z = (K^2 - K_x^2 - K_y^2)^{1/2}$. Integrating with respect to K_y with the use of the saddle-point method [6.1a], we obtain

$$\int_{-\infty}^{+\infty} \exp iz \sqrt{K^2 - K_x^2 - K_y^2}\, dK_y \approx \int_{-\infty}^{+\infty} \exp i\left(Kz - z\frac{K_y^2}{2K}\right) dK_y =$$

$$= \exp iKz \int_{-\infty}^{+\infty} \exp\left(-iz\frac{K_y^2}{2K}\right) dK_y \quad . \qquad (6.21)$$

Switching from the exponential integrand function to trigonometrical ones, in calculating the integrals obtained, we use the Table of Integral Transformants in SNEDDON's book [6.16]. Reverting to the exponential function, we obtain

$$\exp(iKz)\, D_y = \exp(iKz) \sqrt{\frac{2\pi K}{z}} \exp\left(-i\frac{\pi}{4}\right) \quad . \qquad (6.22)$$

Instead of integrating with respect to K_x we introduce a new variable s related to K_x by (6.13). Taking the expressions containing no s outside the integral, we get the following equations:

$$D_o = \frac{i}{2}(2\pi)^{-3/2}(Kz)^{-1/2} \exp\left(-i\frac{\pi}{4}\right) \exp[i(Kz+P)]U_o \qquad (6.23)$$

$$D_h = \frac{i}{2}(2\pi)^{-3/2}(Kz)^{-1/2} \exp\left(-i\frac{\pi}{4}\right) \cdot$$
$$\cdot \exp\{i[Kz + 2\pi(\underline{h}\underline{r}) + P]\}U_h \qquad (6.24)$$

$$P = \frac{K\chi_o}{2}\left(\frac{t}{\gamma_o} - \frac{x\gamma_h}{\gamma_o \sin 2\vartheta} + \frac{x}{\sin 2\vartheta}\right) = \frac{K\chi_o}{2}(\ell_o + \ell_h) \quad , \qquad (6.25)$$

and the integrals U_o and U_h have the form

$$U_o = \frac{1}{2}\int_{-\infty}^{\infty} [(-s+\sqrt{s^2+f^2}) \exp(i\alpha t\sqrt{s^2+f^2}) + (s+\sqrt{s^2+f^2}) \cdot$$
$$\cdot \exp(-i\alpha t\sqrt{s^2+f^2})]\, (\sqrt{s^2+f^2})^{-1} \exp(-iqst)ds \quad , \qquad (6.26)$$

$$U_h = \frac{1}{2}\int_{-\infty}^{\infty} [\exp(i\alpha t\sqrt{s^2+f^2}) - \exp(-i\alpha t\sqrt{s^2+f^2})] \cdot$$

$$\cdot \frac{KC\sqrt{\chi_h\chi_{\bar{h}}}}{\sin 2\vartheta\sqrt{s^2+f^2}} \exp(-iqst)\, ds \quad , \tag{6.27a}$$

$$U_h = \frac{iKC\sqrt{\chi_h\chi_{\bar{h}}}}{\sin 2\vartheta}\int_{-\infty}^{\infty} \frac{\sin(\alpha t\sqrt{s^2+f^2})}{\sqrt{s^2+f^2}} \exp(-iqst)\, ds \quad . \tag{6.27b}$$

Here,

$$q = \frac{x}{t} - \alpha \quad , \quad x = \ell \sin(\vartheta+\varepsilon) \quad . \tag{6.28}$$

The simpler integral U_h is evaluated directly with the aid of the above-mentioned Table II of Integral Transformants [6.16]. The result is recorded thus:

$$U_h = F(q) = \begin{cases} 0 \quad , & |q| > \alpha \quad , \\ \pi i \dfrac{KC\sqrt{\chi_h\chi_{\bar{h}}}}{\sin 2\vartheta} J_0(ft\sqrt{\alpha^2-q^2}) \quad , & |q| < \alpha \quad . \end{cases} \tag{6.29}$$

The calculation of the integral U_o is facilitated by its obvious relation with U_h (see (6.27a)):

$$U_o = \left(\frac{KC\sqrt{\chi_h\chi_{\bar{h}}}}{\sin 2\vartheta}\right)^{-1} \left[\frac{\partial U_h}{\partial(iqt)} + \frac{\partial U_h}{\partial(i\alpha t)}\right] \quad , \tag{6.30}$$

and also by the recurrence relation

$$J_1(\xi) = -\partial[J_o(\xi)]/\partial\xi \quad . \tag{6.31}$$

As a result we obtain

$$U_o = \begin{cases} 0 \quad , & |q| > \alpha \quad , \\ -\pi f\sqrt{\dfrac{\alpha - q}{\alpha + q}}\; J_1(ft\sqrt{\alpha^2-q^2}) \quad , & |q| < \alpha \quad . \end{cases} \tag{6.32}$$

An analysis of the expressions obtained for the wave field in a crystal can be performed with the aid of Figs.6.2 and 6.3

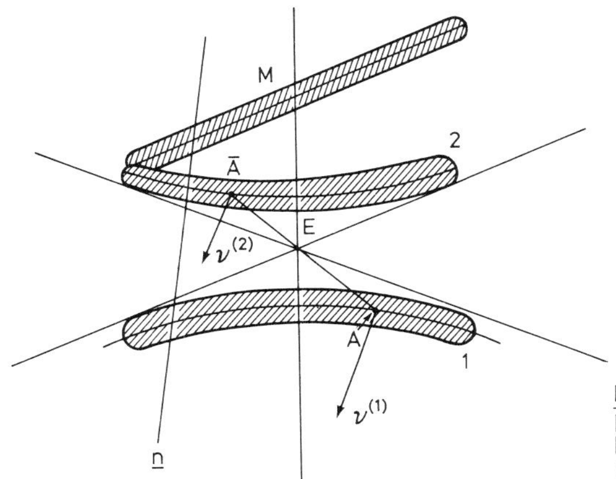

Fig. 6.3 Excitation of all the points of the dispersion hyperbola in spherical wave case (after KATO [1.25b)

First of all, as follows from (6.29) and (6.32), the wave field in a crystal is strictly limited to the triangular area ROT. In the plane-incident-wave approximation, as comprehensively demonstrated in Chapter 5, to each particular value of the angle of incidence $\psi_{ok} \pm \eta$ there corresponds a definite direction of energy propagation, or the energy-current density vector $\bar{\bar{\underline{S}}} = \bar{\bar{\underline{S}}}^{(1)} + \bar{\bar{\underline{S}}}^{(2)}$, the vectors $\bar{\bar{\underline{S}}}^{(i)}$ representing normals to the corresponding branches of the dispersion surface at the excitation points. In the case of a spherical incident wave, when all the points of the dispersion hyperbola are excited, the normals at the corresponding points of some branch or other are the vectors $\underline{\nu}$, which represent the energy-current density of certain ray beams in the crystal. They are no longer associated with any exact value of the angle of incidence, however, and are determined by the angle ε with respect to the reflecting plane or the line $\overline{OL}$ in Fig.6.2. In distinction to the corresponding angles in the case of the plane incident wave, here we have a small angular interval $\pm d\eta/2$ (see Fig.6.4).

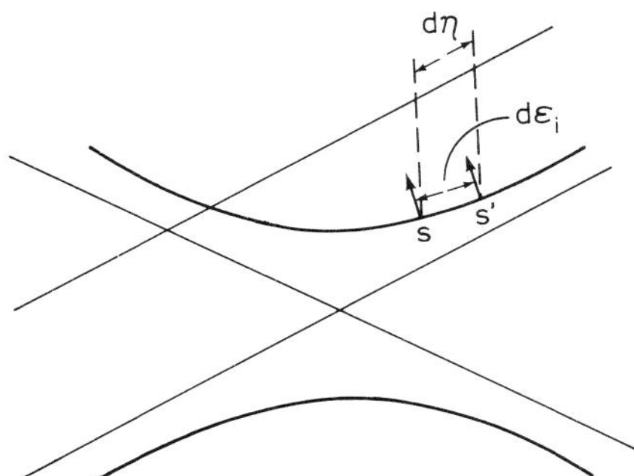

Fig. 6.4 Calculation of magnification M in crystal (after AUTHIER [6.2a])

The common variable vector $\underline{\nu}$ represented in Fig.6.2 by the line $\overline{OP}$ corresponds to the same value of ε. The base P of this variable vector is fixed by oblique coordinates x' and x, whose directions are perpendicular to $\underline{K}_h^B$ and $\underline{K}_o^B$. In their turn, x and x' depend on the common angular parameter q, which is a function of ε. The value of x is represented by (6.28), while the value of x' is determined from the relation following from Fig.6.2:

$$x' = \frac{\gamma_h}{\gamma_o}(\alpha - q)t. = \ell \sin(\vartheta - \varepsilon) \quad . \tag{6.33}$$

Further, using the obvious relations

$$\ell = \overline{OP} = t/\cos(\vartheta - \psi_o + \varepsilon) \quad , \tag{6.34}$$

we obtain for q

$$q = \frac{\sin(\vartheta + \varepsilon)}{\cos(\psi_h - \vartheta + \varepsilon)} - \frac{\sin 2\vartheta}{2\cos\psi_h} \quad . \tag{6.35}$$

Taking into account the above values of the arguments x and x'((6.28) and (6.33)) and the solutions of the integrals ((6.29) and (6.32)), we write out the total values of the wave functions D_o and D_h in a crystal in the following form

$$D_o = \frac{-i}{4}\frac{1}{\sqrt{2\pi}}\frac{1}{\sqrt{Kz}}\exp\left(-\frac{i\pi}{4}\right)\exp i(Kz+P)\bar{f}\sqrt{\frac{x'}{x}}\cdot$$
$$\cdot J_1(\bar{f}\sqrt{xx'}) \quad , \tag{6.36}$$

$$D_h = \frac{-1}{4\sqrt{2\pi}} \frac{1}{\sqrt{Kz}} \exp(-\frac{i\pi}{4}) \exp i[Kz + 2\pi(\underline{hr}) +$$

$$+ \delta + P]\, \bar{f} J_0 (\bar{f}\sqrt{xx'}) \quad , \tag{6.37}$$

$$\bar{f} = f\sqrt{\frac{\gamma_o}{\gamma_h}} \quad . \tag{6.38}$$

The expressions obtained show that the loci of the points corresponding to the constant values of the product

$$xx' = \frac{\gamma_h}{\gamma_o} (\alpha^2 - q^2) t^2 \tag{6.39}$$

form hyperbolas, whose asymptotes are the vectors $\underline{K}_o^B$ and $\underline{K}_h^B$.

Proceeding from (6.36) and (6.37), we pass on to the intensity field inside the crystal. Write out the corresponding expressions for the diffracted wave

$$I_h = T(\bar{f})^2 [J_0(f\sqrt{xx'})]^2 = T(f)^2 \{J_0[t\varphi(\varepsilon)]\}^2 \quad . \tag{6.40}$$

As has already been noted in the introductory part of this chapter, the theory outlined was used for interpreting and calculating the interference effects of the pendulum solution which were observed on section patterns.

It would be natural to connect such oscillations of the above functions with the interference of the fields induced by the excitation of the points on both branches of the dispersion curve. In describing the wave field inside the crystal, use is made of the variable vector $\overline{\underline{OP}}$, which, at each particular value of the angle ε, on the one hand, corresponds to the interference of a definite pair of waves, for points A and A'(Fig.6.3); on the other hand, it serves as a section through the interference pattern of the pendulum solution.

For interference of the waves corresponding to the conjugate points A and A' of the dispersion hyperbola, a dual condition must be fulfilled, i.e., in addition to the coincidence of the directions of propagation of the energy flow ($\underline{\nu}^{(1)}$, $\underline{\nu}^{(2)}$), the given waves must be coherent.

Since the physical model on which the discussion in this chapter is based is associated with the concept of the incidence of a wave train with a narrow wave front (δ-function), a finite spectral interval, and an angular spread exceeding the width of the dynamical maximum, the conditions for coherence of the beams propagating along the directions $\nu^{(1)}$ and $\nu^{(2)}$ need to be clarified. It would be natural to assume that the spectral interval of these beams leads to dispersion, in their propagation in the crystal, no less than the angular divergence of the above vectors. This is only possible due to the amplification effect inherent in crystals (when the waves propagate in the maximum range). The appropriate statement and numerical calculations are due to AUTHIER [6.2].

Indeed (see Fig.6.4), with a certain interval of angles of incidence $d\eta$ available, an interval $d\varepsilon_i$ of angles of vectors with respect to the reflecting plane appears on a small segment ss' of the dispersion hyperbola. The amplification M, which is proportional to the hyperbola curvature on the given segment 1/R, where R is the radius of curvature, equals

$$M = \frac{d\varepsilon_{is}}{d\eta} = \frac{d\varepsilon_i}{dy_s}\frac{dy_s}{d\eta} = \frac{2\sin^2\vartheta\cos^2\varepsilon_{is}K}{\cos\vartheta[1+(s_s^2/f_s^2)]^{3/2}}\tau_o = \frac{K\cos\vartheta}{R\cos\varepsilon_i} \quad . \tag{6.41}$$

The magnitude of the radius of curvature on the hyperbola segment under study

$$R = \frac{[(s^2/f^2) + \cos^2\vartheta]\sqrt{1+(s^2/f^2)}}{2\tau_o\sin^2\vartheta\cos\varepsilon_i} \quad . \tag{6.42}$$

Consider two limiting cases.

1) At the boundary of the maximum $R \to K$, since this boundary corresponds to the transition from the dispersion hyperbola to a circle. Here, $\varepsilon_i \to \vartheta$, and

$$M \approx 1 \quad , \tag{6.43}$$

i.e., there is no amplification.

2) At the exact value of the Bragg angle $\varepsilon_i = y = s_s/f_s = 0$ and

$$R = \frac{1}{2\tau_o\tan^2\vartheta} \tag{6.44}$$

$$M = \frac{2K\tau_0 \sin^2\vartheta}{\cos\vartheta} \quad . \tag{6.45}$$

From the ratio τ_0/λ it can be seen that M will be of the order of 10^4.

This estimate, first of all, yields the necessary relation

$$g_m \gg L2\eta_m \tag{6.46}$$

and, hence, the possibility of wave interference corresponding to point (segment) A of the dispersion hyperbola with, and only with, waves corresponding to the excitation of the conjugate point (segment) $\bar{A}$. Comparing the conditions of formation of the interference patterns of the pendulum solution in the cases of incidence of a spherical and a plane wave, we note a considerable difference. Indeed, in the second case, waves with approximately parallel wave vectors interfere, whereas in the first case interference affects waves with parallel vectors of propagation of energy current-density or rays.

Another conclusion that is to be drawn from the above estimate of the crystal M amplification in the maximum range is of more general significance. Frauenhoffer diffraction, at which the geometrical dimensions of the source are much less than the diffraction pattern, is realized in a crystal with source-diffraction pattern distances about 10^4 times less than in vacuo.

Equation (6.40), which yields the intensity distribution on the plane of the reflection, shows that the quantity I_h actually depends on two independent variables, namely the depth (thickness) t and the angle ε, formed by the variable vector $\overline{OP}$ and the trace of the reflecting plane. The value of $\varphi(\varepsilon)$ in (6.40) is expressed as follows (see (6.14,23,34,38))

$$\varphi(\varepsilon) = \frac{KC|\chi_h|}{\sin 2\vartheta} \sqrt{\sin(\vartheta+\varepsilon)\sin(\vartheta-\varepsilon)} \cos^{-1}(\psi_h-\vartheta+\varepsilon) \quad . \tag{6.47}$$

Experimental investigations of these effects, and particularly their application in determining the absolute values of the structure amplitudes, are based on intensity variations along the line $\overline{OL}$ (trace) of the reflecting plane (Fig.6.2). Denoting by ρ the distance of an arbitrary point on this line from point O on the entrance surface and taking into consideration that along $\overline{OL}$ ε vanishes to zero and $\cos(\psi_h - \vartheta + \varepsilon) = t/\rho$ we get

$$t\varphi(\varepsilon) \to \rho \sin\vartheta \frac{KC|\chi_h|}{\sin 2\vartheta} = \frac{KC|\chi_h|}{2\cos\vartheta}\rho \quad . \tag{6.48}$$

The expression (6.40) is rewritten

$$I_h(\rho) = T \frac{4\pi^2 a^2}{\sin^2\vartheta} [J_o(2\pi a\rho)]^2 \quad , \tag{6.49}$$

where $a = K_o C|\chi_h|/2\cos\vartheta$ is the real semi-axis of the dispersion hyperbola.

As has been shown by experimental investigations of these interference effects, in describing the intensity variation along the line ρ it suffices to (6.49) only on the diffraction pattern area adjoining point 0. Deeper areas correspond to asymptotic expansion of the Bessel function, which takes the following form for the zero-order function [4.39]

$$J_o(\xi) \approx \sqrt{\frac{2}{\pi\xi}} \cos(\xi - \frac{\pi}{4}) \quad . \tag{6.50}$$

In this case, for deeper areas of the interference pattern, the intensity variation law is expressed thus

$$I_h(\rho) \approx A \frac{2a}{\pi\rho \sin^2\vartheta} \cos^2(2\pi a\rho - \frac{\pi}{4}) \quad . \tag{6.51}$$

The distance between the two neighbouring hyperbolas along the line ρ, near point 0 is, in accordance with (6.49),

$$\Lambda_{mh} = \frac{\xi_{m+1} - \xi_m}{2\pi a} \quad , \tag{6.52}$$

where ξ_{m+1} and ξ_m are the values of the argument J_o corresponding to the two adjoining maxima. For the remaining part of the diffraction pattern, at large values of ρ the distance between the adjoining maxima (6.51)

$$\Lambda_{mh} \approx \frac{\pi}{2\pi a} = \frac{\lambda \cos\vartheta}{C|\chi_h|} = \Lambda_{\rho\ell} \quad . \tag{6.53}$$

Despite the coincidence of (6.53) with (3.45) and (3.47) for the extinction distance in the case of an incident plane wave, the differences between the two interference patterns are quite striking when investigating the absolute positions of the maxima, i.e., their location with respect to the entrance surface of the crystal (see Fig.6.5) [1.25a].

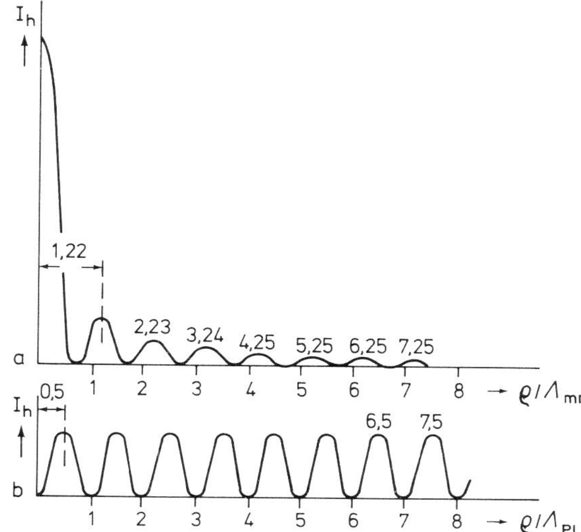

Fig. 6.5a,b Comparing intervals of interference patterns in the Pendellösung fringe solution (after KATO [1.25a]). a) Incident wave packet with sufficient angular spread; b) Plane incident wave. Divisions on the horizontal axes correspond to intervals $\Lambda_{p\ell}$; the figures in a) denote distances from the origin, i.e., from the entrance surface of the wedge

First of all, at $\rho < 0$ ($2\pi a\rho < 0.01$), that is to say, in the immediate vicinity of the entrance surface, according to (6.49), the function J_0, and along with it the value of intensity $I_h(\rho)$, has a maximum, while according to (3.43) the intensity of the diffracted wave in the crystal with an *incident plane wave*

$$I_h \sim |D_0^{(a)}|^2 \sin^2(\pi\Delta kz) \quad , \tag{6.54}$$

at $z \approx 0$, $I_h \approx 0$. The distances between the maxima in the case of an incident plane wave $\Lambda_{\rho\ell}$ can be determined accurately from (6.53) and approximately correspond to the distances Λ_{mh} at large ρ. The distances between the first maxima, however, exceed $\Lambda_{\rho\ell}$. Thus, the distances between the first and second maxima exceed $\Lambda_{\rho\ell}$ by 22 percent. These distances then gradually decrease, approaching $\Lambda_{\rho\ell}$. The maxima are *displaced towards the entrance surface* by $1/4\ \Lambda_{mh}$ as compared with the case of the plane wave, which follows at once from (6.54).

We shall now consider in more detail the physical causes of the differences indicated, in particular the phase shift by $\pi/4$ in (6.54) compared to (6.51) [1.25b]. The complete analytical expression for the phase of each of the two component waves (transmitted and diffracted) has the following form after integration with respect to K_y (see (6.23-26)):

$$T_{o,h}^{(i)} = Kz + P + (\alpha t - x)s \pm \alpha t \sqrt{s^2 + f^2} \quad . \tag{6.55a}$$

On the other hand, the phase for each of the fields varies along the trajectory of the vector $\underline{\nu}^{(i)}$ by the law

$$T^{(i)} = \left(\underline{k}^{(i)}(\underline{\nu}), \underline{\nu}\right) \quad . \tag{6.55b}$$

It is easy to show that the value (6.53) for Λ_{mh} can be obtained from the difference in the vectors

$$\left(\Delta\underline{k}(\underline{\nu}), \underline{\nu}\right) = \left([\underline{k}^{(1)}(\underline{\nu}) - \underline{k}^{(2)}(\underline{\nu})], \underline{\nu}\right) \quad , \quad \Lambda_{mh} = 2\pi/\left(\Delta\underline{k}(\underline{\nu}), \underline{\nu}\right) \quad , \tag{6.56}$$

while the value

$$\Lambda_{p\ell} = 2\pi/(\Delta\underline{k}(\underline{n}), \underline{n}) \quad . \tag{6.57}$$

A more detailed expression for the phase of oscillation along the trajectory of the vector $\underline{\nu}^{(i)}$ can be obtained by the approximate saddle-point method. For the waves transmitted and diffracted separately, we can get

$$T_o^{(i)} = (\tfrac{\pi}{2})_1 + (-\tfrac{\pi}{4})_2 + (\pm\tfrac{\pi}{4})_3 + \left(\underline{k}_o^{(i)}\underline{\nu}\right)_\ell \quad , \tag{6.58}$$

$$T_h^{(i)} = [+\tfrac{\pi}{2} + 2\pi(\underline{h}\underline{r}) + \delta]_1 + (-\tfrac{\pi}{4})_2 + (\pm\tfrac{\pi}{4})_3 + \left(\underline{k}_h^{(i)}\underline{\nu}\right)_\ell \quad . \tag{6.59}$$

Here, the phase denoted by the subscript 1 contains the value $\pi/2$ corresponding to factor i in (6.23) and (6.24). The phase with the subscript 2 has appeared as a result of integration with respect to K_y (see (6.22)). Accordingly, the phase with the subscript 3 appears on integration with respect to K_x. Finally, the term depending on ℓ defines the shape of the hyperbolas and the distances between them.

As regards the phase shift, the presence of two signs in $(\pm\pi/4)_3$ in (6.58) and (6.59) is interesting. If we proceed from the fact that a spherical wave falling on a crystal excites all the points of both branches of the dispersion hyperbola (surface), and vectors $\underline{\nu}^i$ propagate from each point along the normal, then it is obvious that for the first field we must expect the formation of a focus (focal line) of these rays near the entrance surface. As regards the second field, the corresponding branch has the same sign of curvature as the incident wave, and only yields a divergent fan of $\underline{\nu}^{(2)}$ vectors. This can be demonstrated by integrating the expression for a cylindrical wave

$$D(\underline{r}) = \int_{-\infty}^{\infty} \exp i[k_x x + f(k_x)z]dk_x \tag{6.60}$$

using the approximate saddle-point method. The result is

$$\begin{aligned} D(\underline{r}) &= \int_{-\infty}^{\infty} \exp(i/2)[f''(x)]_o\, z[k_x - (k_x)_o]^2 dk_x = \\ &= \left[\frac{2\pi}{(|f''|)_o z_o}\right]^{1/2} \exp[a(i\tfrac{\pi}{4})]\ [\Psi(r)]_o \quad , \end{aligned} \tag{6.61}$$

where the subscript o means the value corresponding to the point of the stationary phase, $a = \pm 1$, depending on the sign of $(f'')_o$. The expression obtained shows that the sign of the phase $i(\pi/4)$ depends on the sign of the second derivative of $f(k_x)$, i.e., on the curvature of our hyperbolas.

At the same time it is known from optics that in passing through a focus the phase changes jumpwise by $\pi/2$. Thus, the phase of the second field remains unchanged and equals $[-i(\pi/4)]_3$, whereas the phase of the first field changes and becomes $[i(\pi/4)]_3$. Detailed discussion of this effect, which follows from the Debye theory, is given in SOMMERFELD's monograph [6.1b].

A significant feature of the interference pattern discussed is the modulation of the intensity of the maxima at intervals which more or less exceed the intervals (6.52) and (6.53). This modulation, as shown by HATTORI, HART et al. [6.3], is due to the beats when summing up the intensities of radiations with mutually perpendicular planes of oscillation of the electrical displacement. Taking into account that the radiation emanating from the X-ray tube is (practically) nonpolarized and considering, for simplicity, Λ_{mh} from (6.53), a distinction must be made between Λ_σ and Λ_π:

$$(\Lambda_\sigma)^{-1} = |\chi_h|/\lambda \cos\vartheta \quad , \quad (\Lambda_\pi)^{-1} = |\chi_h| \cos 2\vartheta/\lambda \cos\vartheta \quad ,$$

$$(\Lambda_\sigma)^{-1} + (\Lambda_\pi)^{-1} = (\Lambda_\sigma)^{-1}(1+\cos 2\vartheta) \quad . \tag{6.62}$$

Eq. (6.51) for the intensity along the reflecting plane is rewritten and further transformed as follows:

$$I_h(\rho) = N_\sigma \rho^{-1} \quad \cos^2\left[\left(\frac{\pi\rho}{\Lambda_\sigma} - \frac{\pi}{4}\right) + \cos 2\vartheta \cos^2\left(\frac{\pi\rho}{\Lambda_\pi} - \frac{\pi}{4}\right)\right] =$$

$$= \frac{N_\sigma}{2\rho}(1+\cos 2\vartheta) + \frac{N_\sigma}{2\rho}(1+\cos 2\vartheta)\cos\{\pi[(\Lambda_\sigma)^{-1} + (\Lambda_\pi)^{-1}]\rho -$$

$$- \pi/2\} \cos\{\pi[(\Lambda_\sigma)^{-1} - (\Lambda_\pi)^{-1}]\rho\} - \frac{N_\sigma}{2\rho}(1-\cos 2\vartheta) \cdot \tag{6.63}$$

$$\cdot \sin\{\pi[(\Lambda_\sigma)^{-1} + (\Lambda_\pi)^{-1}]\rho - \frac{\pi}{2}\} \sin\{\pi[(\Lambda_\sigma)^{-1} - (\Lambda_\pi)^{-1}]\rho\} \quad ,$$

$$N_\sigma = \left(2\frac{Aa}{\pi \sin^2\vartheta}\right)_\sigma \quad .$$

Without discussing here the decrease in amplitude with increasing ρ, which is represented by factor $N/2\rho$, we consider the three terms on the right-hand side of (6.63).

The first term gives us only a general background. The third term yields insignificant modulations, since $\cos 2\vartheta$ is close to unity. Therefore, the interference pattern is essentially determined by the second term, more precisely by the factor $\cos\{\pi[(\Lambda_\sigma)^{-1} + (\Lambda_\pi)^{-1}]\rho - \pi/2\}$. The interval Λ, which is slightly different than Λ_{mh} from (6.53), corresponds to the relation

$$\Lambda^{-1} = (2\Lambda_\sigma)^{-1}(1+\cos 2\vartheta) \quad . \tag{6.64}$$

The amplitudes of the bands N/ρ are modulated by the effect of the beats of two intensity oscillations, which is expressed by the factor

$$\cos[\pi\rho(\Lambda_\sigma)^{-1}(1-\cos 2\vartheta)] \tag{6.65}$$

in the second term. The interference bands are fading and finally disappear completely if the following condition is met:

$$(1-\cos 2\vartheta)\ \rho/\Lambda \approx (2n+1)/2 \quad . \tag{6.66}$$

The number of bands between the two areas of fading is determined by the relation

$$N = (1+\cos 2\vartheta)/2(1-\cos 2\vartheta) \tag{6.67}$$

with the exception for the interval between the origin and the first weakening. Here, the number of bands is N/2. In the region with an even (in the order of magnitude) number, the factor $\cos[\pi\rho(\Lambda_\sigma)^{-1}(1-\cos 2\vartheta)]$ is positive, and with an odd one, negative. The conditions of formation of bands in the intervals between the two weakenings depend on the sign indicated. Where the sign is positive, the band intensity maxima and minima are determined by the conditions

$$\begin{aligned} &\pi\rho\Lambda_\sigma^{-1}(1+\cos 2\vartheta) - \frac{\pi}{2} = 2n\pi - \max \quad , \\ &\pi\rho\Lambda_\sigma^{-1}(1+\cos 2\vartheta) - \frac{\pi}{2} = (2n+1)\pi - \min \quad . \end{aligned} \tag{6.68}$$

Where the sign is negative, (6.68) is reversed and, hence, the positions of the maxima and minima interchange after each weakening interval. The above characteristics are in good quantitative agreement with the experimental data [6.3] (see also [Ref.1.25c,Fig.44]).

Although the above effect of polarization on the interference pattern is a natural and, possibly, elementary manifestation of the transversality of electromagnetic waves of the X-ray range, the pictorial character of the effect and also the possibility of introducing corrections in the observed values of Λ make it important. We associate this with the small shifts of the maxima due both to the smooth background and the weak oscillations (third term from the right in (6.63)). Of importance is the change in the shape of the interference pattern maxima. Indeed, the shape of the maxima is a regular function of the shape of the branches of the dispersion hyperbola and, according to (6.40), should be hyperbolic. The method of experimental study of the shape of the dispersion surface with the aid of the section patterns described here is more direct than the use of the reflection curves. At the same time, the discovery of any deviation from the hyperbolic shape of the

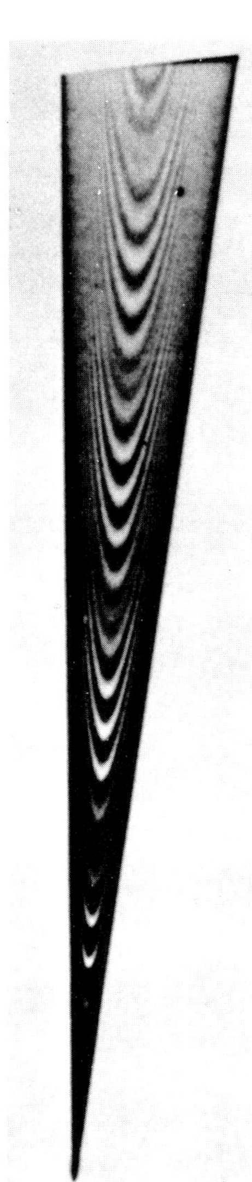

Fig. 6.6 Hyperbola intensity modulation due to interference of polarization waves (after KATO [1.25c])

dispersion curves would be important, since it would point directly to the inapplicability (partial at any rate) of the two-wave approximation.

In accordance with (6.58) or (6.59), the total value of the phase difference between the waves associated with the two different branches of the dispersion hyperbola is

$$\Phi = [\Phi^{(1)} - \Phi^{(2)}] + \left([\underline{k}^{(1)}(\underline{\nu}) - \underline{k}^{(2)}(\underline{\nu})], \underline{\nu} \right) \ell \qquad (6.69)$$

or, taking into account (6.50) and (6.59),

$$\Phi = [\Phi^{(1)} - \Phi^{(2)}] + \frac{2\pi C|\chi_h|}{\lambda \cos\vartheta} \sqrt{xx'} \quad , \qquad (6.70)$$

hence we obtain

$$\sqrt{xx'} = \{\Phi - [\Phi^{(1)} - \Phi^{(2)}]\} \Big/ \frac{2\pi}{\lambda} \frac{C|\chi_h|}{\cos\vartheta} \quad . \qquad (6.71)$$

Thus, the condition for the exact hyperbolic shape of the interference bands is the constancy of the phase difference $[\Phi^{(1)} - \Phi^{(2)}]$. This condition obviously refers to the shape of the dispersion surface or of its section by the reflection plane as well. The experimental checking of the shape of interference bands carried out so far [6.4a] has not revealed any deviation from the hyperbola.

In connection with the use of section patterns in determining the absolute values of the structure amplitudes, in a number of investigations [6.4b,5],

KATO and co-workers developed a method for the recording and accurate measurements of the periods Λ_h given by (6.52) and (6.53). If we turn to the main part of the diffraction pattern, ignoring the parts adjoining the entrance surface, we see that (6.53) must be supplemented by the polarization factor in accordance with (6.64). In this case, the period between the adjoining hyperbola vertices on the *reflecting plane* inside the crystal is equal to

$$\Lambda_h^c = \frac{\lambda}{|\chi_h| \cos\vartheta} \quad . \tag{6.72}$$

Fig.6.7a shows the corresponding diffraction pattern on the lateral side of a plane-parallel slab. The period Λ_h^c is measured along the ρ-axis. On the exit surface of the plane-parallel slab we observe bands corresponding to the hyperbola bases.

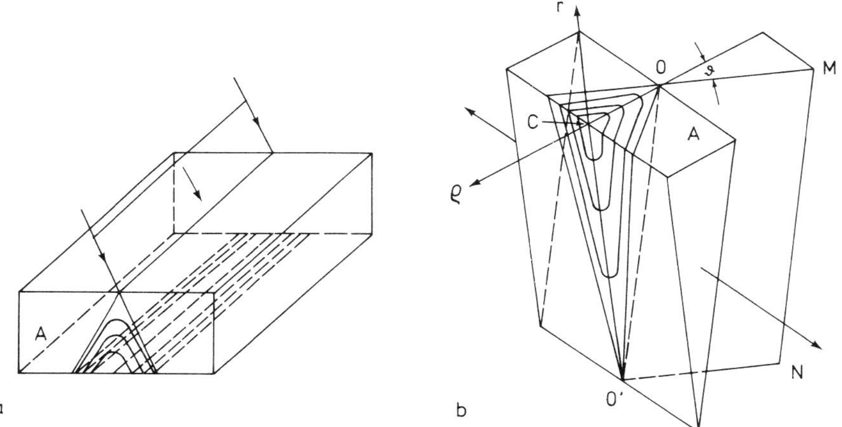

Fig. 6.7a,b Formation of the hyperbola pattern of the pendulum solution during propagation of the diverging beam in crystal: a) on the plane-parallel slice; b) in a wedge-shaped slab

For objective recording of the hyperbolas, a wedge may be used as shown in Fig.6.7b. Here, the diffraction pattern is observed on the exit surface and corresponds to the oblique projection of the image on the face A.

As we noted at the very beginning of this chapter, in investigating dynamical scattering, patterns with scanning, section patterns are also used.

Figure 6.7 shows the direction of displacements of the crystal (and the film) relative to the source in such a technique. It is easy to see that the film records lines which are the loci of the hyperbola vertices forming the section pattern. At the same time these lines are lines of constant, or equal, thickness in a wedge-shaped crystal.

Figure 6.7b exhibits a particular case of recording in symmetrical reflection: the reflecting plane OCO' is perpendicular to the entrance and exit surfaces of the slab. The period Λ_h on the exit surface along the r-axis will be larger than the period Λ_h^c along ρ in accordance with the equation

$$\Lambda_h = \Lambda_h^c \Phi_h \quad , \quad \Phi_h = \frac{\Delta r}{\Delta \rho} \approx \cot\mu \quad , \tag{6.73}$$

where μ is the wedge angle CO'O. The value of Λ_h, with an allowance for (6.72),

$$\Lambda_h \approx \frac{1}{\cos\vartheta} \frac{\lambda}{C|\chi_h|} \cot\mu \tag{6.74}$$

differs from the corresponding value of the period in the incident-plane-wave approximation only by the polarization factor.

Thorough consideration of the images obtained in recording the diffraction pattern on the exit surface of the wedge, however, has made it necessary to introduce certain corrections (see Chap.9).

The value of the integral reflection in the case of the transparent crystal should not differ from the value obtained in the theory for the incident plane wave. According to (3.81b)

$$R_i = \frac{\pi}{2} \int_0^{2A} I_0(x)dx \quad . \tag{6.75}$$

The expression (6.75) represents a damping oscillating function against a background of constant intensity $\pi/2$. The integral is clearly a function of the upper limit

$$2A = \frac{2\pi t C|\chi_h|}{\lambda \sqrt{\gamma_0 \gamma_h}} \quad , \tag{6.76}$$

it has an oscillation period of 2π and yields the interval between adjoining bands of equal thickness t:

$$\Lambda = \frac{\lambda \sqrt{\gamma_o \gamma_h}}{C|\chi_h|} \quad . \tag{6.77}$$

Thus, the use of traverse type patterns, which were first obtained in the investigation of KATO and LANG [Ref.1.23,Chap.1], does not require revision of the theory in the incident-plane-wave approximation. However, as follows from our outline, it is only on the basis of the spherical-wave theory that the mechanism of formation and the quantitative characteristics of the interference bands in patterns can be established.

Here, it would be useful to supplement a comparison of the two types of the pendulum solution, i.e., bands of equal thickness and subsidiary maxima, which was discussed in Section 3.3. The hyperbolic bands on section patterns evidently cannot be applied to the two types indicated. Indeed, according to (6.33-35), the argument of the Bessel function J_o (appearing in (6.40)) for the intensity field of the interference pattern will be

$$\xi = \xi(t,\varepsilon) \quad . \tag{6.78}$$

Hence it follows that the bands of section patterns represent the loci of the points corresponding to the intensity maxima, for a definite pair of values of the angle ε and the thickness (or depth) t. The points of all the hyperbolas on the straight line ρ lying on the reflecting plane correspond to the value $\varepsilon = 0$, i.e., to the direction of the vectors $\underline{\nu}^{(i)}$ or $\underline{\nu} = 2\underline{\nu}^{(i)}$, parallel to the same plane. Scanning of the crystal along the direction indicated in Fig.6.7 naturally results in the formation of lines of equal thickness on the film. (On the exit surface of the crystal, only interference hyperbolas are formed; during scanning the volume of the crystal in which the scattering of the incident spherical wave occurs is changed).

On the other hand, the bands in section patterns, as well as the secondary maxima, are due to interference on an *equal slope*. However, during the formation of subsidiary maxima, interference affects waves with identically directed wave vectors both inside the crystal and in the vacuum. When a spherical wave falls on a crystal, interference occurs between waves which have identically directed vectors $\underline{\nu}$, i.e., rays in the crystal, but differently directed rays, or wave vectors, in the vacuum.

If we take into account (6.56) and (6.57) and study the dispersion hyperbola in the reciprocal space, the following becomes clear. Since the oscillation intervals and the phase difference of the interferring waves in the case of section patterns are defined, according to (6.56), by the projection of the corresponding diameter of the dispersion hyperbola onto the normal ($\underline{\nu}$ not $\underline{n}$!), it is obvious that the phase difference reduces to zero towards the edge of the pattern. This occurs because a normal to the curve becomes perpendicular to the hyperbola diameter near the boundary of the maximum. Thus, in this respect the hyperbolas on section patterns resemble lines of equal thickness rather than lines of equal incidence.

6.2 Application of the Theory Described to Scattering in Absorbing Crystal

The main field in which the theory outlined above is applied to the experimental determination of scattering amplitudes is that of the transparent crystal, but it would also be interesting to generalize the results obtained for the case of the absorbing crystal.

In order to do this, we must assume the parameters χ_o and χ_h, $\chi_{\bar{h}}$ in (6.23-27) to be complex values. Notice that χ_o appears only in the value of the phase P (6.25), while χ_h and $\chi_{\bar{h}}$ enter into the integrand functions in (6.26) and (6.27). Eq. (6.25) can be rewritten

$$P = P_r + iP_i = \frac{K}{2}(\chi_{or}+i\chi_{oi})(\ell_o+\ell_h) \quad , \tag{6.79}$$

and, hence, the absorption factor referring to the pre-integral parts in (6.23) and (6.24) corresponds to the common coefficient of absorption.

Passing on to the integrals U_o and U_h, we notice that in calculating the interference part of the exponent in the absorption factor, as shown in Chapter 4, use is made of an extremely important approximation $|\chi_{hi}| \ll |\chi_{hr}|$ in accordance with the theory of Zachariasen and Laue. While calculating the amplitude factors it is assumed that $\chi_o \approx \chi_{or}$ and $\chi_h\chi_{\bar{h}} \approx (\chi_h\chi_{\bar{h}})_r = \Phi_h$. However, as shown by KATO, when integrating (6.26) and (6.27) it is possible to avoid the limitations indicated and calculate these integrals for a general case of any ratio between the moduli of the real and imaginary parts of the parameters χ_h. For this purpose integration is taken over the complex plane. As before, we begin with the integral U_h. At first this integral is calculated with the condition

$$\alpha > |(x-\alpha \cdot t_o)|/t_o = q \quad . \tag{6.80}$$

The following function of the complex variable z is introduced;

$$I(\pm) = [\pm 1/2(z^2+f^2)^{1/2}]\exp it[-qz \pm \alpha(z^2+f^2)^{1/2}] \quad . \tag{6.81}$$

Here, $(z^2 + f^2)^{1/2}$ will be equal to $(s^2 + f^2)^{1/2}$ on the line $z = s_r + is_i$ for $s_r \geq 0$. The integral contours are shown in Fig.6.8.

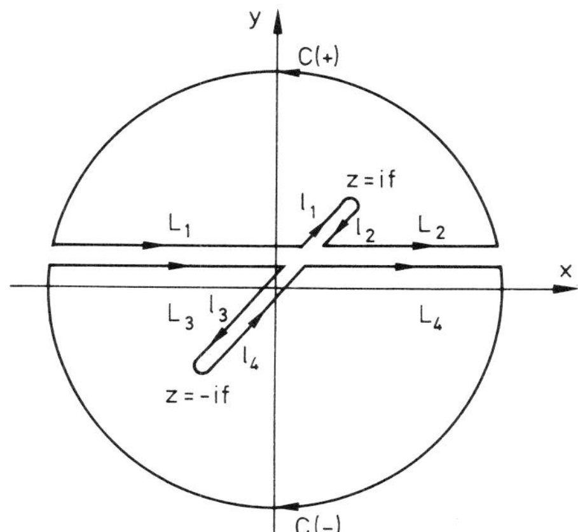

Fig. 6.8 The derivation of (6.89) by differentiating over a complex plane (after KATO [1.24b])

Integration over the lines $L_i(i=1,2,3,4)$ reduces to four integrals

$$\int_{L_1=-\infty}^{0} I(+)dz \quad ; \quad \int_{L_2=0}^{\infty} I(+)dz \quad ;$$
$$\int_{L_3=-\infty}^{0} I(-)dz \quad ; \quad \int_{L_4=0}^{\infty} I(-)dz \quad . \tag{6.82}$$

Then we take integrals over the contours C(+) and C(-). However, since there are no poles inside these latter contours, the linear integrals over infinite semicircles tend to zero under the conditions (6.80) and integral values over infinitesimal circles around the points (poles) $z = \pm if$ are equal to zero. Therefore

$$U_h = \bar{f} \exp i\eta_h \left[\int_{L_1+L_2} I(+)dz + \int_{L_3+L_4} I(-)dz \right] =$$

$$= \bar{f} \exp i\eta_h \left[\int_{\ell_1+\ell_2} I(+)dz + \int_{\ell_3+\ell_4} I(-)dz \right] \quad . \tag{6.83}$$

On these latter lines the following conditions are fulfilled:

$$z = if \sin\varphi \ (\ell_1,\ell_2) \quad ; \quad z = -if \sin\varphi \ (\ell_3,\ell_4) \quad . \tag{6.84}$$

In this case the linear integrals over the paths ℓ_i $(i=1,\ldots,4)$ take the form

$$\int_{\ell_1} I(+)dz = -i/2 \int_{\varphi_0}^{\pi/2} \exp ft(q \sin\varphi - i\alpha \cos\varphi)d\varphi \quad ,$$

$$\int_{\ell_2} I(+)dz = i/2 \int_{\pi/2}^{\varphi_0} \exp ft(q \sin\varphi + i\alpha \cos\varphi)d\varphi \quad ,$$

$$\int_{\ell_3} I(-)dz = -\frac{i}{2} \int_{-\varphi_0}^{\pi/2} \exp ft(-q \sin\varphi + i\alpha \cos\varphi)d\varphi \quad , \tag{6.85}$$

$$\int_{\ell_4} I(-)dz = \frac{i}{2} \int_{\pi/2}^{-\varphi_0} \exp ft(-q \sin\varphi - i\alpha \cos\varphi)d\varphi \quad ,$$

where φ_0 is determined from the relation $f_r \sin\varphi_0 = s_i$. By changing the variable φ so that all the integrals take the form of the first one, for instance by replacing φ with $\pi - \varphi$ in the second integral, we obtain, for the sum of the integrals,

$$U_h = \frac{i}{2}\,\bar{f}\exp i\eta_h \int_{\varphi_o}^{2\pi+\varphi_o} \exp ft(q\sin\varphi - i\alpha\cos\varphi)d\varphi =$$

$$= \frac{i}{2}\,\bar{f}\exp i\eta_h \int_0^{2\pi} \exp[ift(\alpha^2-q^2)^{1/2}\sin\vartheta]d\vartheta \quad . \tag{6.86}$$

$$= \pi i\bar{f}\exp i\eta_h J_o[ft(\alpha^2-q^2)^{1/2}] \quad .$$

It is easy to show that by replacing (6.80) with $\alpha < q$ we obtain $U_h = 0$, and at $\alpha = q$ the integral becomes improper, although it has the main value.

The integral U_o is solved in a similar way.

Thus, we arrive at a surprising and extremely important conclusion. Eqs. (6.36) and (6.37) for the wave functions in a crystal are equally applicable to the transparent and the absorbing crystal if in the second case χ_o and χ_h are assumed to be complex values. No restrictions as to the ratio of the moduli $|\chi_{or}|$ and $|\chi_{oi}|$ or $|\chi_{hr}|$ and $|\chi_{hi}|$ are imposed.

Since at large values of the argument ξ, in particular $\xi(\rho)$, the Bessel functions J_o take on asymptotic values

$$J_o(\xi) = \begin{cases} \sqrt{\dfrac{2}{\pi\xi}}\cos(\xi - \dfrac{\pi}{4}) \quad , & -\dfrac{\pi}{2} < \xi < +\dfrac{\pi}{2} \quad , \\ \sqrt{\dfrac{2}{\pi\xi}}\cos(\xi + \dfrac{\pi}{4}) \quad , & \dfrac{\pi}{2} < \xi < \dfrac{3\pi}{2} \quad , \end{cases} \tag{6.87}$$

(see also (6.50)), we can assume that these expressions describe two plane waves in a crystal. Each wave is associated with a definite branch of the dispersion hyperbola. In an absorbing crystal, these lines have a linear coefficient of absorption in the direction of the normal at a given point of the hyperbola, according to (5.85), on a certain path ℓ:

$$\mu^{\nu}\ell = K\chi_{oi}(\ell_o+\ell_h) \pm 2\bar{f}_i(xx')^{1/2} =$$
$$= \left(\mu \pm 2\pi K_o\sqrt{\chi_{hi}\chi_{\bar{h}i}}\sqrt{1-\frac{\tan^2\varepsilon_i}{\tan^2\vartheta}}\right)(\ell_o+\ell_h) \quad . \tag{6.88}$$

In connection with this expression, Section 5.3 considers the relation between this value and the total value of σ in the theory of scattering in the absorbing crystal on incidence of a plane wave.

Passing over from the expressions for the wave fields to those for the intensities, and tracing the variation in intensities along the reflecting plane (the lengths are reckoned in the direction ρ), we can use (6.36) and (6.37). Thus, the intensities along the direction indicated in a crystal have the following form for a transmitted and a diffracted wave

$$I_o = \frac{1}{32\pi}\frac{1}{Kr} \exp\left(-\frac{\mu\rho}{\cos\vartheta}\right)|\bar{f}|^2 \, |J_1[2a_r\ell + i(\ell_o+\ell_h)\frac{K\mu\varepsilon}{2}\exp i\omega_h]|^2 \quad , \tag{6.89}$$

$$I_h = \frac{1}{32\pi}\frac{1}{Kr} \exp\left(-\frac{\mu\rho}{\cos\vartheta}\right)|\bar{f}|^2 \, |J_o[2a_r\ell + i(\ell_o+\ell_h)\frac{K\mu\varepsilon}{2}\exp i\omega_h]|^2 \quad , \tag{6.90}$$

$$\varepsilon \exp i\omega_h = \frac{\chi_{hi}}{\chi_{oi}} \quad . \tag{6.91}$$

Consider the two most important particular cases (as applied to I_h).

Thin crystal

$$\left|(\ell_o+\ell_h)\frac{K\mu\varepsilon}{2}\right| << 1 \quad . \tag{6.92}$$

Here, we can use the approximate expression

$$J_o(u+iv) \approx J_o(u) - ivJ_1(u) \quad , \quad v << 1 \quad , \tag{6.93}$$

or

$$|J_o(2a\ell)|^2 = |J_o(2a_rb)|^2 + \left|\left(\ell_o+\ell_h\right)\frac{K\mu\varepsilon}{2}\right| |J_1(2a_r\ell)|^2 \quad . \tag{6.94}$$

Substituting this expression in (6.90), we obtain the particular value of I_h needed. We shall now determine the relative shift of the maxima of the function (6.94) and of the total value of I_h relative to the position of the maxima in the case of the transparent crystal according to (6.52). The shift indicated is

$$(\frac{\Delta \ell}{\ell})_m = -x_o/\Lambda_{mh} \quad ; \quad x_o = |\chi_{oi}|/|\chi_{hr}| \quad . \tag{6.95}$$

This value holds good if, and only if, $x_o \Lambda_{mh} << 1$. Note that the maxima shift towards the entrance surface. The numerical estimate of x_o for reflection 220 of Cu Kα radiation from Ge is $\sim$ 5 percent.

Thick crystal

$$|(\ell_o+\ell_h) \frac{K\mu\varepsilon}{2}| >> 1 \quad . \tag{6.96}$$

In accordance with the asymptotic expansion of the zero-order Bessel function J_o, the value of the intensity in the direction of ρ is expressed thus

$$I_h = \frac{1}{32\pi^2} \frac{1}{Kr} \frac{|f|}{\ell \sin\vartheta} \exp(- \frac{\mu\ell}{\cos\vartheta})\{\cos(4a_r - \frac{\pi}{2}) + \\ + \cosh[(\ell_o+\ell_h)\mu\varepsilon \exp i\omega_h]\} \quad . \tag{6.97}$$

An estimate of the relative shift of the maxima in this case shows that in precision measurements this shift need be taken into account only when the values $x\lambda_m < 1$ and $x \approx 0{,}1$. Under these ultimate conditions only the first three maxima can be observed.

The next task in developing the theory in the spherical-incident-wave approximation is the calculation of the integrated intensities, which are here determined by integrating the intensity field on the exit surface. The problem is solved for a wedge-shaped crystal, in particular, in connection with the interpretation of traverse type patterns and their utilization for determining structure amplitudes.

The values of the integrated reflection power and transmission power in the theory outlined are formulated as follows:

$$T_i^\eta = \frac{\sqrt{\gamma_h' / \gamma_o'}}{\sin 2\vartheta} C|\chi_h| \frac{\exp(-\mu t)}{2} \pi W_o \quad ,$$

$$W_o = \left|\frac{KC|\chi_h|t}{2\sqrt{\gamma_o'\gamma_h'}}\right| \int_{-1}^{+1} \frac{1-\sigma}{1+\sigma} \exp\left[\frac{\mu t}{2}\left(\frac{1}{\gamma_o'} - \frac{1}{\gamma_h'}\right)\right] \cdot \tag{6.98}$$

$$\cdot \left|J_1\left(\frac{KC|\chi_h|t}{2\sqrt{\gamma_o'\gamma_h'}} \sqrt{1-\sigma^2}\right)\right|^2 d\sigma$$

$$R_i^\eta = \frac{\sqrt{\gamma_h' / \gamma_o'}}{\sin 2\vartheta} C|\chi_h| \frac{\exp(-\mu t)}{2} \pi W_h \quad ;$$

$$W_h = \left|\frac{KC|\chi_h|t}{2\sqrt{\gamma_o'\gamma_h'}}\right| \int_{-1}^{+1} \exp\left[\frac{\mu t}{2}\left(\frac{1}{\gamma_o'} - \frac{1}{\gamma_h'}\right)\right] \cdot \tag{6.99}$$

$$\cdot \left|J_o\left(\frac{KC|\chi_h|t}{2\sqrt{\gamma_o'\gamma_h'}} \sqrt{1-\sigma^2}\right)\right|^2 d\sigma \quad .$$

The integration variable σ is a normalized variable coordinate of the line RT, which is the trace of the exit surface; it is determined by the expression (see Fig.6.9)

$$\sigma = [\tau - \tfrac{1}{2}(a+b)] \;/\; 1/2(b-a) \quad . \tag{6.100}$$

where $a = TF$, $b = RF$, and τ is the variable distance from the footing F of the perpendicular EF dropped onto the exit surface. Further, $t' = (t/2)\,[(1/\gamma_o') + (1/\gamma_h')]$, γ_o' and γ_h' being the cosines of the angles with the normal to the exit surface.

KATO [1.24b] evaluates the integral in the expression (6.99) and analyzes in detail the expressions obtained for the integral reflections R_i. For R_i^y, he derived the expression

$$R_i^y = \frac{\pi}{2} \exp(-\mu t') \sqrt{\frac{1 + x^2}{1 - g^2}} \left\{ \int_0^{2\bar{A}} J_0(\rho) d\rho + \right.$$

$$\left. + \sum_{r=1}^{\infty} \frac{1}{r!r!} \left(\frac{h}{2}\right)^{2r} g_{2r+1}[2A \sqrt{(1-g^2)}] \right\} \quad , \qquad (6.101)$$

$$2\bar{A} = 2A \sqrt{1 - g^2} \quad .$$

In this expression

$$x = |\chi_{hi}|/|\chi_{hr}| \quad ; \quad g = \frac{\mu t}{2} \left(\frac{1}{\gamma_0'} - \frac{1}{\gamma_h'}\right) (2A)^{-1} \quad , \qquad (6.102)$$

A corresponds to (3.61) with an allowance for the complex value of $|\chi_h|$ and $|\chi_{\bar{h}}|$. Hence, $A = |A| \exp i\alpha$, $\cos\alpha = 1$, $\sin\alpha = x$. Finally, h is given by (4.117).

Comparing (6.104) with the value of the integral reflection in the incident-plane-wave approximation (4.115), (4.121), and (3.81), we notice that the integral in (6.101) differs from the integral W in (4.115) as regards the upper integration limit, and the sum in (6.101) differs from the integral V in (4.115) by the factor g_{2r+1} $(2A \sqrt{1 - g^2})$.

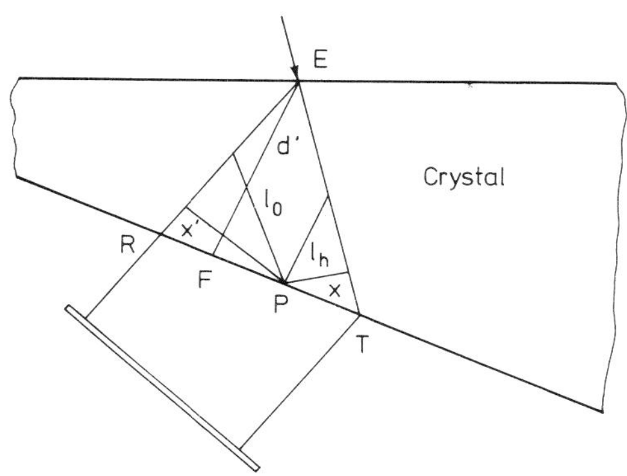

Fig. 6.9 Calculation of the integrated intensities in the case of a wedge-shaped slab (after KATO [1.24b])

This similarity is used in the qualitative analysis of the integration results. In this case the cosines of the angles must be replaced by the normal to the exit surface, γ_0' and γ_h', by the corresponding values referring to the entrance surface.

1) Transparent crystal: $\mu = x = g = 0$. The integrated reflection power describing the intensity distribution already as a function of the angular variable y or the angle η, is the same for the cases of incident plane and spherical waves.

2) Thin absorbing crystal: $h(\approx \mu t') \ll 1$. Assuming the value of $\sum$ in (6.101) to be negligibly small, we conclude that the integrated reflection power in the case of an incident spherical wave differs slightly from the expression for the transparent crystal in factor $[(1 + x^2)/(1 - g^2)]^{1/2}$ and the upper integration limit.

3) Intermediate thickness range: $h(\approx \mu t') \approx 1$. In this case the following value should be taken into account

$$V^* = \sum_{r=1}^{\infty} \frac{1}{r!r!} \left(\frac{h}{2}\right)^{2r} g_{2r+1}(2A\sqrt{1-g^2}) \quad . \tag{6.103}$$

The numerical values of the function $g_m(2\bar{A})$ over a wide range of values of $2\bar{A}$, i.e., over a wide range of products μt, are close to unity for the values of $2\bar{A} < 3$, when x and g are of the order of 0.1 as well. Thus, in this case, too, we can use the theory in the incident-plane-wave approximation.

4) Thick absorbing crystal: $h \gg 1$. Using the asymptotic expansion of the Bessel functions, we can obtain the following expression for the integrated reflection power:

$$R_h^y = \frac{\pi}{2} \frac{1}{\sqrt{2\pi h}} \exp(-\mu t'+h) \left\{1 + \left[\frac{1}{8} + \frac{x^2 + g^2}{2(1+x^2)^2}\right] \frac{1}{h} + \ldots\right\} \quad , \tag{6.104}$$

i.e., a result practically coinciding (at sufficiently large h) with (4.174) for the integrated reflection power in the case of an incident plane wave, provided, naturally, that $\cos\Psi_0$ and $\cos\Psi_h$ in (6.99) are referred to the entrance surface.

In conclusion we will give the linear coefficient of absorption in the direction of the normal to the exit surface for the integrated reflection power:

$$\mu = \frac{1}{2} K|\chi_{oi}| \left(\frac{1}{\gamma_0'^2} + \frac{1}{\gamma_h'^2}\right) + rKC\sqrt{\chi_h\chi_{\bar{h}}}\,[(x^2+g^2)/\gamma_0'\gamma_h']^{1/2} \quad , \tag{6.105}$$

where the parameters χ_h and $\chi_{\bar{h}}$ are assumed to be complex values.

7. Bragg Reflection of X-Rays. I. Basic Definitions. Coefficients of Absorption; Diffraction in Finite Crystal

The theory outlined above refers to X-ray transmission through a crystal with emergence of the reflected wave through the exit, or back, surface of the crystal slice. This scheme of experiment is called the Laue method, or case. Another scheme (Fig.7.1) refers to the Bragg case, in which the reflected wave emerges into the vacuum through the entrance surface.

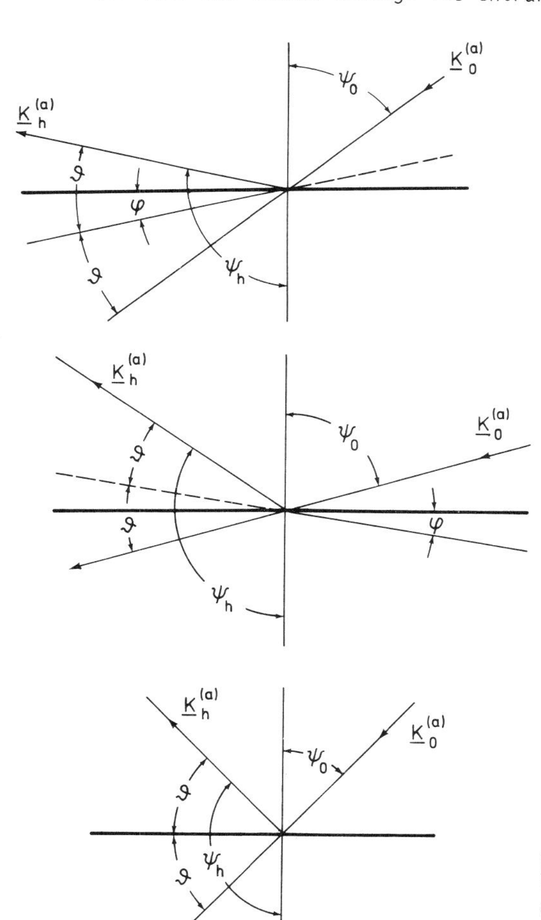

Fig. 7.1a-c Bragg reflection scheme: a) $\gamma_0 > |\gamma_h|$; b) $\gamma_0 < |\gamma_h|$; c) $\gamma_0 = \gamma_h$

The important difference between the physical phenomena occurring here in the crystal from what happens in the Laue case is due to the different conditions for the amplitudes at the crystal-vacuum interface. In the Laue case, the complex nature of polarizability, the angular functions, and the wave vectors inside the crystal express the true absorption of the X-rays. In the Bragg case, along with absorption we encounter extinction, i.e., an interference effect, whose action greatly exceeds that of absorption for certain cases in a definite part of the maximum region. A characteristic feature of Bragg reflection is the difference in the physical mechanism of scattering in different parts of the maximum region.

Historically, the theory of X-ray scattering by perfect crystals for Bragg reflection preceded Ewald's theory. It was first outlined in the works of DARWIN [1.5] and PRINCE [7.1]; the latter included the absorption effect in the consideration. In this chapter we use the same approach as in the preceding chapters, which was suggested by Laue-Zachariasen. WAGNER's paper [7.2], which considered Bragg reflection in a parallel-sided absorbing slice, and also papers by BONSE [1.59a,7.3] are worth mentioning. The last few years have seen the appearance of important theoretical and experimental investigations dedicated to this problem [1.46-48,53-55].

Since the difference between Laue and Bragg reflection is caused by change in the boundary conditions, the text of Chapter 2 referring to the infinite crystal can be used in both cases. A different arrangement of the dispersion surface with respect to the normal to the exit surface of the crystal will be considered separately.

Switching to the semi-infinite crystal (see Chap.3), we notice at once that with the chosen directions of measuring of the angles Ψ_o and Ψ_h the value of $\cos\Psi_h = \gamma_h$ becomes negative in the Bragg case:

$$\Psi_h > \pi/2 \quad , \quad \gamma_h = -|\gamma_h| \quad . \tag{7.1}$$

Accordingly we obtain

$$\beta = 2\alpha - \chi_o\left(1+ \frac{|\gamma_h|}{\gamma_o}\right) \quad , \tag{7.2}$$

$$c^{(i)} = \frac{+\beta \pm(\beta^2-4\chi_h\chi_{\bar{h}}C^2|\gamma_h|/\gamma_o)^{1/2}}{2\chi_{\bar{h}}\frac{|\gamma_h|}{\gamma_o}C} \tag{7.3}$$

It can be seen that the essential difference of this expression from (3.11) lies in the fact that the root in the right-hand side for a definite range of β can assume imaginary values. This range is limited by the conditions

$$-2C|\chi_h|\sqrt{\frac{|\gamma_h|}{\gamma_o}} < \beta < +2C|\chi_h|\sqrt{\frac{|\gamma_h|}{\gamma_o}} \quad . \tag{7.4}$$

As a result, in the indicated range of β the wave vectors of the waves in the lattice take on complex values. By analogy with wave vectors in absorbing crystals this means exponential weakening of the intensities of such waves as they penetrate into the crystal. The corresponding maximum region is called the total-reflection region.

In this chapter we shall also use angular functions β, y, and v of the value η (deviation from the angle Ψ_o). The relationships between the indicated functions, however, will be slightly different in different maximum regions and also in transparent and absorbing crystals.

7.1 Reflection from Transparent Crystal

In order to investigate the different regions of the maximum with Bragg reflection, let us first consider the scheme of experiment where the incident vacuum wave has a wide enough wave front compared with the crystal thickness, while its divergence is negligible and less than the angular width of the maximum.

The entire maximum area can evidently be divided at least into three parts: the total-reflection region, and the two adjacent regions - towards the smaller and larger angles $\Psi_{ok} \pm \eta$. To analyze the wave fields inside the crystal, we use (7.3) and (3.15)

$$\sqrt{\frac{|\gamma_h|}{\gamma_o}}\,\frac{D_h^{(i)}}{D_o^{(i)}} = \sqrt{\frac{\chi_h}{\chi_{\bar{h}}}}\,(y_n \pm \sqrt{y_n^2-1}) = \mp\sqrt{\frac{\chi_h}{\chi_{\bar{h}}}}\,\exp(\pm v) \quad , \tag{7.5}$$

where $i = 1, 2$ is the usual designation of numbers of the wave fields in the crystal, $n = \mathrm{I}, \mathrm{III}$ refer to the maximum regions for $y < -1$ or to region I for $y > +1$, or to region III, respectively. For Bragg reflection, the value of v is determined thus:

$$|y| = |\cosh v| > 0 \quad . \tag{7.6}$$

For the transparent crystal, in both indicated maximum regions the values of $c^{(i)}$ will be real; the same refers to the values of y and v. A characteristic feature of Bragg reflection is a different contribution of the two fields to the total wave field in the two indicated regions.

Indeed, in region I, at negative values of y, an increase in its modulus, $|y_I|$ results in the vanishing of the coefficient modulus of (7.5) only for the first field, in accordance with experiment. The second modulus of (7.5) for the second field increases infinitely, and thus we conclude that in region I only the first field is excited in the crystal.

The reverse is true for region III at positive values of y_{III}. In this range of angles of incidence, only the second field is induced in the crystal.

In the maximum region or in the range of angles of incidence from y = -1 to y = +1, the values of (7.5) become complex. Assuming for this region

$$y_{II} = \cos v \quad , \tag{7.7}$$

we can write

$$\sqrt{\frac{|\gamma_h|}{\gamma_o}}\frac{D_h^{(i)}}{D_o^{(i)}} = \sqrt{\frac{\chi_h}{\chi_{\bar{h}}}}\,(y_{II} \pm i\sqrt{1-y_{II}^2}) = \sqrt{\frac{\chi_h}{\chi_{\bar{h}}}}\exp \pm iv \quad . \tag{7.8}$$

It is obvious that in absorption, the variables y and v become complex values. Also, the angular width of the total-reflection region, which will be determined later, corresponds to the half-width of the maximum in the case of Laue reflection and equals, for the transparent crystal, from ~ 10-13" for strong reflections to ~ 2-3" for weak ones.

For the purpose of further analysis we have to explain the three possible Bragg arrangements differing in the ratio between γ_o and the modulus $|\gamma_h|$. Fig.7.1a depicts the case $\gamma_o > |\gamma_h|$; $[\Psi_h - (\pi/2)] < [(\pi/2) - \Psi_o]$, $[(\pi/2) - \Psi_o] > \vartheta$. Fig.7.1b shows the arrangement for $\gamma_o < |\gamma_h|$, $[\Psi_h - (\pi/2)] > [(\pi/2) - \Psi_o]$, $[(\pi/2) - \Psi_o] < \vartheta$. Fig.7.1c illustrates symmetrical Bragg reflection where the reflecting plane is parallel to the entrance surface of the crystal slice and $\gamma_o = |\gamma_h|$, $\Psi_o = \pi - \Psi_h$.

We will now compare (7.5-8) in the three maximum regions with an analysis of the mutual positions of the dispersion surface and the normals $\underline{n}$ to the entrance surface of the crystal in the Bragg case (see Fig.7.2). The reflecting plane is represented here by the straight line passing through points E and M, being the real axis of the dispersion hyperbola. The most important difference of this scheme from Fig.3.1 or the other schemes referring to the Laue case is the position of the normal $\underline{n}_0$ which either

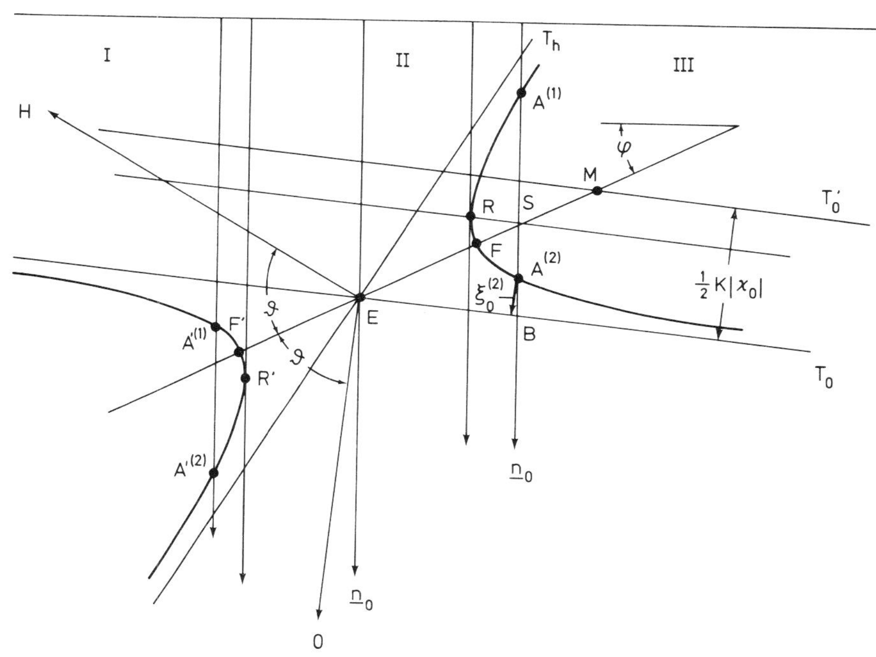

Fig. 7.2 Dispersion surface in reciprocal space for Bragg reflection

intersects one of the hyperbola branches or passes between them, depending on the value of angle η. It is easy to see that the above-mentioned three maximum regions are represented here by the following parts of the scheme: region I - to the left of point R', region II - between points R' and R, and region III - to the right of point R. Indeed, in region II the value of β depends on (7.4), in accordance with which y takes on values from -1 to +1. The boundaries of the region inside which $(y^2 - 1)^{1/2} = i(1 - y^2)^{1/2}$ will be (with the given direction of $\underline{n}$) points R' and R, where $(y^2 - 1)^{1/2} = 0$.

In regions I and III the normals n intersect the corresponding hyperbola branch at two points, $A^{(1)}$ and $A^{(2)}$, $[A'^{,(1)}, A'^{,(2)}]$, which, as η reduces, get closer together and then merge into a single point R' or R.

Using (2.82) for the amplitude coefficient inside the crystal, we find in Fig.7.2 the confirmation of our conclusion about the existence of only one field inside the crystal in regions I and III; since

$$c^{(i)} = \sqrt{\frac{|\xi_o^{(i)}|}{|\xi_h^{(i)}|}} \quad , \tag{7.9}$$

then in region III, as the normal moves to the right, i.e., away from the centre of the maximum (point E), and the distance of point $A_{III}^{(1)}$ from the straight line T_o increases, the value of $\xi_o^{(1)}$ increases infinitely, while $\xi_h^{(1)}$ asymptotically tends to zero. The opposite is the case with point $A_{III}^{(2)}$ and also point $A_I^{(1)}$.

It can also be shown that if we use for $c^{(2)}$ an expression equivalent to (7.9),

$$c^{(i)} = \frac{\overline{A^{(2)}B}}{\overline{SB}} \quad ; \quad \overline{A^{(2)}B} = \frac{\xi_o^{(2)}}{\gamma_o} \quad ; \quad \overline{SB} = \frac{a\cos\vartheta}{\gamma_o} = \frac{\sqrt{\xi_o^{(2)}\xi_h^{(2)}}}{\gamma_o} \quad . \tag{7.10}$$

Thus, for the *semi-infinite or infinitely thick crystal* under consideration, the boundary conditions for Bragg reflection can be written:

$$\begin{aligned} &\text{in region III:} \quad D_o^{(a)} = D_o^{(2)} \quad ; \quad D_h^{(a)} = c^{(2)}D_o^{(2)} \\ &\text{in region I:} \quad D_o^{(a)} = D_o^{(1)} \quad ; \quad D_h^{(a)} = c^{(1)}D_o^{(1)} \quad . \end{aligned} \tag{7.11}$$

So, in each of these regions the values of the corresponding amplitude of the reflected waves, $D_h^{(i)}$, increase with decreasing modulus $|\eta|$ or $|y|$, reaching a maximum at $y = \pm 1$, i.e., at the boundaries of the total-reflection region II. For $\chi_h = \chi_{\bar{h}}$, these maximum values, according to (7.5), at $c^{(i)} = 1$ lead to the following conditions in regions I and III:

$$\exp(-v) = -\sqrt{\frac{|\gamma_h|}{\gamma_o}} \quad ,$$

$$\exp(-v) = \sqrt{\frac{|\gamma_h|}{\gamma_o}} \quad . \tag{7.12}$$

These conditions (7.12), however, can be fulfilled only when $|\gamma_h| < \gamma_o$, since v is always positive - in other words, when the scheme of experiment corresponds to Fig.7.1a. In the case of the scheme of Fig.7.1b, the dispersion surface in the reciprocal space will be positioned somewhat differently with respect to the normals than in Fig.7.2. With a decrease of $|\eta|$ in a similar scheme, points $A^{(2)}$ in region III, and $A'^{(1)}$ in region I, moving along the hyperbola, will reach the points of contact R and R', respectively, before they attain the apex of hyperbola F (or F'), where, and only where, the coefficient $c^{(i)}$ can turn into unity. As a result, (7.12) will not be satisfied.

Passing over to the reflection coefficient, we obtain the identical value for regions I and III (see (7.5)):

$$R_{I,III} = \frac{|\gamma_h|}{\gamma_o} \left| \frac{D_h^{(a)}}{D_o^{(a)}} \right|^2 = \exp(-2v) \quad . \tag{7.13}$$

As regards region II, it follows from (7.14) and (7.15) that for both fields

$$R_{II} = |y \pm i\sqrt{1 - y^2}|^2 \tag{7.14}$$

Thus, for the transparent crystal, the reflection coefficient is equal to unity over the entire range of total reflection from y = -1 to y = +1, and falls off rapidly in regions I and III from unity to zero by the law (7.13). The corresponding reflection curve (see Fig.7.3) was first calculated by DARWIN [1.5], who used the well-known first version of the dynamical theory.

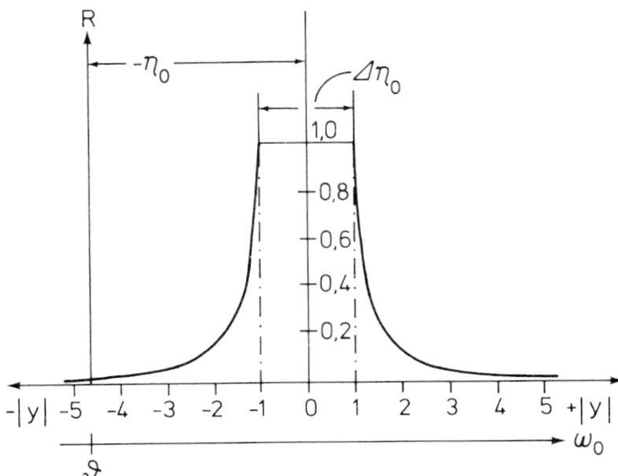

Fig. 7.3 Profile of maximum of Bragg reflection from transparent, semi-infinite crystal with asymmetric arrangement

It is obvious that the effect of total reflection of the incident wave in region II is associated with exponential decrease of the wave field in the crystal with depth. This decrease, as well as absorption, can be described by introducing a complex wave vector.

Since in region II we cannot construct the tangential projection of the wave vector in the crystal, we will make use of the formal (3.1), replacing $\xi_o^{(2)}$ by the coefficient $c^{(2)}$ according to (2.82). Thus, we obtain the boundary condition for the wave vectors in the form

$$\underline{k}_o^{(2)} = \underline{K}_o^{(a)} - K\delta^{(2)}\underline{n}_o \quad , \tag{7.15}$$

$$|K\delta^{(2)}\underline{n}_o| = \frac{1}{\gamma_o}[-\xi_o^{(2)} - \tfrac{1}{2}K\chi_o] = \frac{1}{\gamma_o}[-\tfrac{1}{2}K\chi_{\bar{h}}Cc^{(2)} - \tfrac{1}{2}K\chi_o] \quad , \tag{7.16}$$

or, substituting the value of $c^{(2)}$ from (7.8), we obtain, for the imaginary part of the wave vector due to the total reflection from the transparent crystal,

$$|K\delta_i^{(2)}\underline{n}_o| = -\frac{K\chi_{\bar{h}}C}{2\gamma_o}\sqrt{\frac{\gamma_o}{|\gamma_h|}}\sqrt{\frac{\chi_h}{\chi_{\bar{h}}}}\sqrt{1-y^2} \quad . \tag{7.17}$$

By analogy with the absorption coefficient (2.58) we introduce the extinction coefficient

$$\sigma_e = 4\pi k_{oi}^{(2)} = 2\pi KC|\chi_h| \sqrt{1 - y^2} / \sqrt{\gamma_o|\gamma_h|} \quad . \qquad (7.18)$$

As can be seen from (7.18), the extinction has a maximum at $y = 0$ and drops to zero at the edges of the total-reflection region at $y = \pm 1$. Substituting the value of χ_h in (7.18), we get (see (2.34))

$$\sigma_e^{max} = 2\lambda Ce^2 F(h) / \sqrt{\gamma_o|\gamma_h|} mc^2 \Omega \quad . \qquad (7.19)$$

By substituting $c^{(1)}$ for $c^{(2)}$ in (7.16) we obtain, for the extinction coefficient, according to (7.8),

$$\sigma_e' = -2\pi KC|\chi_h| \sqrt{1 - y^2} / \sqrt{\gamma_o|\gamma_h|} \quad . \qquad (7.20)$$

If the positive value of σ_e, according to (7.19), means exponential weakening of the wave field penetrating into the crystal, the negative value of σ_e' corresponds to exponential strengthening of this field with depth and is therefore physically inacceptable, because in region II the entire energy of the incident wave is reflected back into the vacuum. This result suggests that region II contains only the second field.

In Chapter 9 we will use, in additions to the angles Ψ_o and Ψ_h with respect to the normal, angles ω_o and ω_h with respect to the entrance surface, or the glancing angles. An increase in angle Ψ_o, or the value η, laid out from the normal in the clockwise sense, means a decrease in glancing angle ω_o.

With symmetrical reflection (Fig.7.1c), the maximum value of extinction is

$$\sigma_{e,s}^{max} = 2 \frac{e^2}{mc^2} \frac{\lambda C}{\sin\vartheta} \frac{|F_h|}{\Omega} \qquad (7.21)$$

and, hence, it depends exclusively on the crystal structure, the nature of the atoms-scatterers, the reflection indices, and the wavelength λ.

Let us compare the values of extinction and common absorption. The greatest excess of σ_e over μ takes place for the very first reflections and in the cases where common absorption is at a minimum, i.e., for penetrating radiation and for atoms of elements with a low atomic number. These requirements are satisfied, for example, by reflection of Mo $K\alpha_1$ radiation from the plane (200) Si. In this case, assuming $C = 1$, we get

$$\sigma_{e,s}^{max} = 9200 \text{ cm}^{-1} \quad , \quad \mu \approx 15 \text{ cm}^{-1} \quad ,$$
$$(\sigma_{e,s}^{max}/\mu) \approx 613 \quad . \tag{7.22}$$

It is more correct to compare σ_e with the value of μ referred to the normal to the entrance surface; then the extinction-absorption ratio will be as low as $\sim$ 112.

To make the characteristic of the intensity maximum for Bragg reflection more complete, we will give the expressions for the angular width and the midpoints of the total-reflection region. According to (7.4), the angular width of this region is

$$\Delta\beta = 4C|\chi_h|\sqrt{\frac{|\gamma_h|}{\gamma_o}} \tag{7.23}$$

or, switching to the angular argument η,

$$\Delta\eta_o = (\sin 2\vartheta)^{-1} 2C\sqrt{\frac{|\gamma_h|}{\gamma_o}}\,|\chi_h| \quad . \tag{7.24}$$

The presence of the factor C in some cases substantially reduces the angular width $\Delta\eta_o$ for π-oscillations of vector $\underline{D}$. The subscript o indicates that the given angular range refers to the incident wave. The midpoint of the maximum, η_o, corresponds to the value $\beta = 0$, i.e., is displaced with respect to Ψ_{ok}

$$\beta = 2\eta_o \sin 2\vartheta - \chi_o\left(1 + \frac{|\gamma_h|}{\gamma_o}\right) = 0 \quad ,$$
$$|\eta_o| = |\chi_o|\left(1 + \frac{|\gamma_h|}{\gamma_o}\right)(2 \sin 2\vartheta)^{-1} \quad . \tag{7.25}$$

The value η_o, by (7.25), coincides with the angular displacement of the reflection maximum in the Laue case, according to (3.83), with replacement of $-\gamma_h$ by $|\gamma_h|$. The remarks referring to the expression (3.33) are partly applicable to the Bragg case. This is true of the statement concerning the dependence of the displacement effect exclusively on χ_o, but not on χ_h. Note that because $\chi_o = -|\chi_o|$, (7.25) indicates a decrease in angle Ψ_o, i.e., an increase in angle ω_o. Finally, (7.25) can be transformed as was done in the Laue case, yielding an expression similar to (3.85)

$$\eta_o = -\frac{\delta_D}{\sin 2\vartheta}\left(1+\frac{|\gamma_h|}{\gamma_o}\right) \tag{7.26}$$

and in the case of symmetrical reflection

$$\eta_{o,s} = -\frac{2\delta_D}{\sin 2\vartheta} = \frac{\chi_o}{\sin 2\vartheta} \quad , \tag{7.27}$$

the expression obtained by DARWIN.

It is obvious that the displacement of the angle of incidence (7.25) corresponding to the midpoint of the maximum leads to a displacement of the reflection corresponding to the same point. From the main condition relating the variable vectors $\underline{K}_h^{(i)}$ and $\underline{K}_o^{(i)}$, $\underline{K}_o^{(i)} - \underline{K}_h^{i} = h$ it follows that

$$K(\sin\Psi_o' - \sin\Psi_h') = |\underline{h}| \quad , \tag{7.28}$$

where Ψ_o' and Ψ_h' are variable angles

$$\Psi_o' = \Psi_o + \eta_o \quad , \quad \Psi_h' = \Psi_h + \eta_h \quad . \tag{7.29}$$

Thus,

$$K[(\cos\Psi_o)\eta_o - (\cos\Psi_h)\eta_h] = 0 \quad , \tag{7.30}$$

since we assume $\cos\Psi_o' \approx \cos\Psi_o$, $\cos\Psi_h' \approx \cos\Psi_h$ and $\sin\eta_o \approx \eta_o$, $\sin\eta_h \approx \eta_h$. From (7.38) we get

$$\eta_h = \frac{\gamma_o}{|\gamma_h|}\eta_o = -\frac{\delta_D}{\sin 2\vartheta}\left(1+\frac{\gamma_o}{|\gamma_h|}\right) \quad . \tag{7.31}$$

The change in the deviation of the incident ray in Bragg reflection is

$$\sum \eta = \eta_0 + \eta_h = \frac{\chi_0(\gamma_0+|\gamma_h|)^2}{2(\gamma_0|\gamma_h|)\sin 2\vartheta} \quad . \tag{7.32}$$

This expression shows that $\sum \eta$ can assume particularly large values and, hence can be measured with a high accuracy when either the incident or the reflected ray is nearly parallel to the entrance surface. In order to ensure such conditions, the crystal surface is ground down until the appropriate angle φ with the reflecting plane is obtained. Fig.7.3 shows the Bragg maximum profile according to (7.24) and (7.26). The figure also shows the glancing angle scale for the incident beam. This shape of the maximum, which was first obtained by DARWIN, is sometimes called the Darwin reflection curve. Its characteristic features are: a flat top corresponding to the reflection coefficient R = 1 and a symmetry about the middle point ordinate. These features are typical of reflection from the transparent crystal and disappear on transition to absorbing crystals.

The remarkable ability of the transparent crystal to completely reflect X-rays impinging on it in region II of the maximum is very valuable in handling the fundamental experimental task, i.e., obtaining sufficiently intensive radiation as close as possible in its characteristics to the ideal model of the plane monochromatic wave. An essential feature here is a reduction in angular width of the reflected beam, which may slightly exceed 10" for strong reflections, as has been indicated above. In the case of symmetrical reflection, the angular widths of the incident and reflected (in region II of the maximum) beams are identical. The angular width in this case is determined by an expression which can be obtained from (7.32) and from the condition $\gamma_0 = \gamma_h$

$$\Delta\eta_s = \frac{2C|\chi_h|}{\sin 2\vartheta} \quad . \tag{7.33}$$

Hence, reverting to asymmetrical reflection, we again obtain, for the angular width of an incident beam reflected within region II of the maximum,

$$\Delta\eta_0 = \Delta\eta_s \sqrt{\frac{|\gamma_h|}{\gamma_0}} = \Delta\eta_s \sqrt{b^{-1}} \quad . \tag{7.34}$$

It is obvious that the angular width of a beam reflected in the same region of the maximum will be related with $\Delta\eta_s$ by an expression in which γ_0 and γ_h change places. Using (7.28), we have for the scheme of Fig.7.1b:

$$\Delta\eta_h = \Delta\eta_s \sqrt{b} = b\Delta\eta_o \quad ; \quad b = \frac{\sin(\vartheta-\varphi)}{\sin(\vartheta+\varphi)} \quad . \tag{7.35}$$

Since we can change the angles φ between the entrance surface of the crystal and the reflecting plane by grinding, and also use repeat reflection from different crystals, the divergence $\Delta\eta_h$ of the reflected beam can be greatly reduced. This method of asymmetrical diffraction has actually proved one of the most effective ways for obtaining beams with the necessary properties (see Chap.9).

Variation in the angular width of beams as a result of the change in the angles of incidence and reflection is vividly illustrated by the rocking curves (Fig.7.4a) and also by the scheme of the dispersion surface in the reciprocal space (Fig.7.4b).

Note that in the case shown in Fig.7.1a, $b > 1$ and the divergence $\Delta\eta_h > \Delta\eta_s$. Obviously the value of the displacement η_o according to (7.26) decreases on switching from the scheme of Fig.7.1b to that of Fig.7.1a.

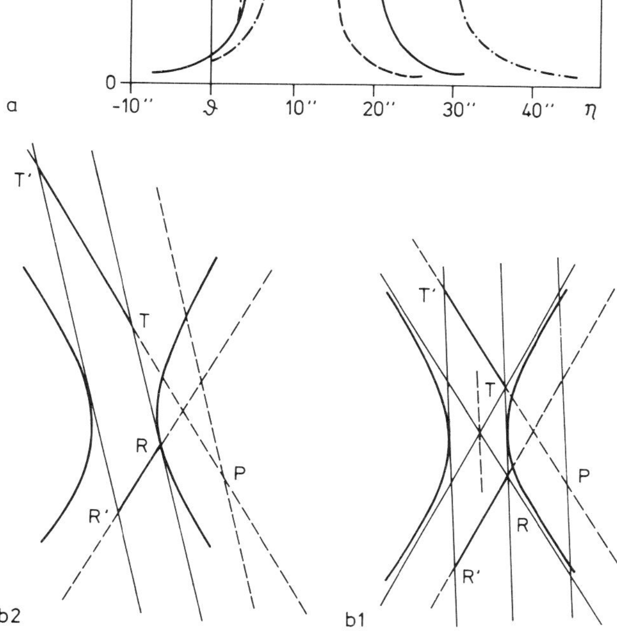

Fig. 7.4a,b Change in angular width of reflected beams in total-reflection region as a result of changes in angles of incidence: a) rocking curves (III Cu $K\alpha_1$ from Ge), obtained at different values of the angle φ: 1: $\varphi = 0$, symmetric reflection; 2: $\varphi = \vartheta - \omega_o = +4.5^\circ$, correspond to Fig. 7.1b; 3: $\varphi = \vartheta - \omega_o = -4.5^\circ$, correspond to Fig. 7.1a (after BUBAKOVA [7.4]). b) schemes in reciprocal space; 1: symmetric reflection; 2: asymmetric reflection (after RENNINGER [7.6])

These correlations are discussed in detail in Section 9.4 in connection with the design of spectrometers in asymmetrical arrangement.

We will now make a few remarks referring to the summary wave field in the crystal (for σ-polarization) under the indicated reflection conditions, i.e., on incidence of a plane monochromatic wave with a wide wave front. This field is described by (3.49)

$$D^{(i)} = \{D_o^{(i)} + D_h^{(i)} \exp[-2\pi i(\underline{hr})]\} \exp 2\pi i\{\nu t - [\underline{k}_o^{(i)}\underline{r}]\} \quad , \tag{7.36}$$

and its total value is determined by the ratio between the phases of the transmitted and diffracted waves in the crystal. In distinction to the Laue case, in which both fields of the two-wave approximation are present, in the case under discussion only one of the fields is effective in each of the regions: the first field in region I, and the second in regions II and III. In order to determine the phase relationships it is necessary to adjust the phase η_h of the ratio $(\chi_h/\chi_{\bar{h}})^{1/2} = \exp i\eta_h$ in (7.5) and (7.8). The values $\eta_h = \pi$ adopted in (3.17) correspond to the simplest case of the primitive translation lattice with a single atom in the unit cell; the conditions of anomalous dispersion are excluded. Thus, the ratio D_h/D_o takes on the following values (with an accuracy to the factor $(\gamma_o/\gamma_h)^{1/2}$):

$$\begin{aligned} &\text{I} \quad - \exp(-v) \quad , \\ &\text{II} \quad - \exp[i(\pi - v)] \quad , \\ &\text{III} - \exp(i\pi - v) \quad . \end{aligned} \tag{7.37}$$

In other words, in region I the two waves are in phase, in region III in counterphase, and in region II the phase shift increases monotonically from 0 at $y = -1$ to π at $y = +1$.

As follows further from (7.36), the maxima and minima of the total wave field in the crystal are located on planes parallel to the atomic planes. The values of the field intensity extrema are $[1 \pm (\gamma_o/\gamma_h) \exp(-v)]^2$ for regions I and III and $[1 \pm (\gamma_o/\gamma_h)]^2$ for region II. It is also obvious that if, under the adopted conditions, there are maxima on the atomic planes in region I, in region III there will be minima on the same planes. In region II, the positions of the maxima and minima change monotonically from one extreme point to the other. These data clearly indicate the direction of energy propagation in each of the regions. While in the lateral regions of the maximum the summary vector $\overline{\overline{S}}$ is directed along $\underline{K}$, in the total-reflection region II the wave field energy propagates parallel to the entrance surface of the crystal. In accordance with the shifts of phases of the electric induction D_0 and D_h of the oscillations for the three regions, it is possible to plot a graph of the total field intensity in the crystal, which is presented in Fig.7.5. The dotted line in region II shows the value of intensity with an allowance for some adopted extinction value.

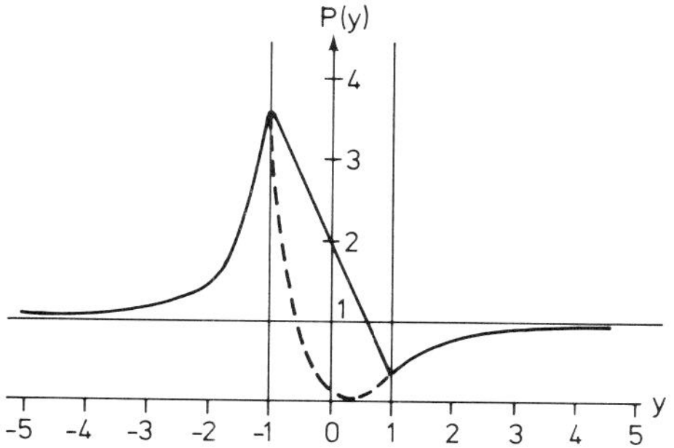

Fig. 7.5 Variation in wave field intensity in crystal as a function of angular variable y in three maximum regions (after JAMES [1.16])

Attention is drawn to the sharp asymmetry of the maximum with respect to the midpoint of this intensity distribution inside the crystal, whereas the (Darwin) reflection curve depicted in Fig.7.3 is symmetrical.

Some deviations from the above-described pattern take place on transition from σ- to π-polarization; in particular, the field intensity in region III increases.

We now wish to make a general remark of great importance. Although we have assumed reflection from a transparent crystal with a wide front of the incident plane wave, actually the given interpretation of the semi-infinite crystal implied that the reflecting crystal is sufficiently thick. In this case, even at infinitely small absorption, the wave fields in regions I and III completely damp out in propagation inside the crystal.

The situation is different when reflection occurs from a crystal slice without absorption or with vanishingly small absorption. The wave fields induced in the crystal at angles of incidence corresponding to regions I or III reach the opposite exit surface, where different boundary conditions obtain.

Thus, since the reflected wave does not emerge into the vacuum on the exit surface, the boundary condition for diffracted waves can be rewritten:

$$D_h^{(d)} \exp[-2\pi i(\underline{K}_h \underline{r})] = \exp[-2\pi i(\underline{K}+\underline{h}),\underline{r}] \cdot$$
$$\cdot \sum_{i=1,2} D_h^{(i)} \exp[2\pi K\delta^{(i)} t] = 0 \quad . \tag{7.38}$$

So, the second field is induced additionally in region I and the first field, in region III. To establish the direction of propagation of the energy of the newly formed fields, we write down the expression for the scalar product of the vectors $\bar{\bar{\underline{S}}}^{(i)}$ by the normal $\underline{n}_o$. Using (5.20b) and (7.5), we readily obtain the following relation:

$$(\bar{\bar{\underline{S}}}^{(i)} \underline{n}_o) = \frac{c}{8\pi} |D_o^{(i)}|^2 \gamma_o \left[1 - \left|\frac{\chi_h}{\chi_{\bar{h}}}\right| \exp(\pm 2v)\right] \quad . \tag{7.39}$$

Applying this relation, for instance, to region III and taking into condiseration that $v > 0$, we get a positive value of the scalar product for the second field for $-2v$ and a negative value for $2v$, i.e., for the first field. Since the inward-drawn normal $\underline{n}_o$ is directed from the entrance surface to the exit one, it follows from (7.39) that the second field propagates in the same direction and the first one, in the opposite direction, i.e., towards the entrance surface of the crystal.

It is obvious further that with a sufficiently wide front of the incident plane wave (at an angle of incidence corresponding to one of the lateral regions of the maximum), interference effects appear inside the crystal, as well as on the entrance surface. The constant value of the relative phase shift of the interferring waves of the two fields, which determines the distance between the maxima or minima of the interference pattern

$$\Delta \underline{k} = \underline{k}_h^{(2)} - \underline{k}_h^{(1)} \quad , \tag{7.40}$$

corresponds to the *chord* joining two points $A^{(i)}$ or $A'^{(i)}$ on the common branch of the dispersion hyperbola. These interference effects are represented by subsidiary maxima of the pendulum solution in the Bragg case. They correspond to the fine structure of the lateral regions of the maximum. This fine structure shape gradually disappears with increasing thickness of the reflecting crystal plate and is replaced by a monotonic decrease in reflection coefficient from R = 1 to R = 0 on the Darwin curve (Fig.7.3).

7.2 True Absorption in Bragg Reflection. Investigation of Coefficient of Absorption σ from Plane-Parallel Plate

In distinction to the Laue case, the value of the absorption coefficient here is substantially different in the three maximum regions. We will begin our discussion with the absorption coefficient referred to the normal $\underline{n}_o$ to the entrance surface of the crystal plate

$$\sigma = \mu/\gamma_o \quad , \tag{7.41}$$

where μ is the linear coefficient of absorption in an arbitrary direction on incidence of a vacuum wave at an angle of $\Psi_{ok} \pm \eta$ in the maximum region. Here, σ is the total value of the absorption coefficient, which includes both the normal and anomalous components. Further, σ^n at n = I,II,III are the values referring to the respective maximum regions. Both the precise and approximate expressions for σ are calculated similarly to Chapter 4 with corrections necessitated by the specific features of the Bragg case.

Since σ is directly related with the imaginary part of the wave vector arising in connection with $\delta_i^{(i)}$, we obtain, in accordance with (2.58,3.1, 4.67,68),

$$\sigma = -4\pi|\underline{k}_{oi}| = 4\pi k\delta^{(i)} = 4\pi\left[-\frac{K\chi_{oi}}{2\gamma_o} - \frac{K}{4|\gamma_h|}(\beta_i \pm W_i)\right] \quad , \tag{7.42}$$

where

$$\beta_i = |\chi_{oi}|\left(1+\frac{|\gamma_h|}{\gamma_o}\right) \quad , \quad W = \sqrt{\beta^2 - 4C^2\chi_h\chi_{\bar{h}}\frac{|\gamma_h|}{\gamma_o}} \quad . \tag{7.43}$$

Below, we give the accurate and approximate values of the real and imaginary parts of the root W similarly to (4.65,66,86,87). The accurate values of W_r and W_i for any ratio of $|\chi_{hi}|$, $|\chi_{hr}|$, $|\chi_{oi}|$ and $|\chi_{or}|$ are determined according to (4.65), but with slightly different values of a and b. It is obvious that in the Bragg case

$$a = \beta_r^2 - \beta_i^2 - 4C^2\Phi_h\frac{|\gamma_h|}{\gamma_o} \quad , \quad b = 2\beta_r\beta_i - 4C^2\Psi_h\frac{|\gamma_h|}{\gamma_o} \tag{7.44}$$

or, as a function of y,

$$W_r = \ell W_r(y) \quad , \quad W_i = \ell W_i(y) \quad , \quad \ell = 2C\sqrt{\Phi_h}\sqrt{\frac{|\gamma_h|}{\gamma_o}} = \frac{\beta_r}{y_r} \quad . \tag{7.45}$$

In this case

$$a(y) = y_r^2 - g^2 - 1 \quad , \quad b(y) = -2y_r g - \frac{\Psi_h}{\Phi_h} \quad , \quad g = -\frac{\beta_i}{\ell} \quad . \tag{7.46}$$

In calculating the *approximate* values of W_r and W_i we use analogy with the expressions (4.86) and (4.87)

$$W_r \approx \ell\sqrt{y_r^2 - 1} \quad ,$$

$$W_i \approx \frac{\beta_r\beta_i - 2C^2(|\gamma_h|/\gamma_o)\Psi_h}{[\beta_r^2 - 4C^2(|\gamma_h|/\gamma_o)\Phi_h]^{1/2}} = \frac{\mu|\gamma_h|}{2\pi K\sqrt{y_r^2 - 1}} \cdot \tag{7.47}$$

$$\cdot \left[y_r\left(\frac{1}{|\gamma_h|} + \frac{1}{\gamma_o}\right) - \frac{2\varepsilon\cos\nu_h}{\sqrt{\gamma_o|\gamma_h|}}\right] .$$

Thus, for the *approximate* value of σ in regions I and III we obtain

$$\sigma = \frac{\mu}{2}\left\{\left(\frac{1}{\gamma_o} - \frac{1}{|\gamma_h|}\right) \mp \frac{1}{\sqrt{y_r^2 - 1}}\left[y_r\left(\frac{1}{\gamma_o} + \frac{1}{|\gamma_h|}\right) - \frac{2\varepsilon\cos\nu_h}{\sqrt{\gamma_o|\gamma_h|}}\right]\right\} , \tag{7.48a}$$

or

$$\sigma = \frac{\mu}{2} \frac{1}{\sqrt{y_r^2 - 1}} \left[\frac{1}{\gamma_o} (\sqrt{y_r^2-1} \mp y_r) - \frac{1}{|\gamma_h|} (\sqrt{y_r^2-1} \pm y_r) \pm \frac{2\varepsilon \cos\nu_h}{\sqrt{\gamma_o |\gamma_h|}}\right] =$$

$$= \frac{\mu}{2 \sinh v_r} \left(\frac{\mp\exp \mp v_r}{\gamma_o} - \frac{\pm\exp \pm v_r}{\gamma_h} \pm \frac{2\varepsilon \cos\nu_h}{\sqrt{\gamma_o |\gamma_h|}} \right) . \tag{7.48b}$$

Keeping in mind that in region I, $y_r = -|y_r|$ and in region III, $y_r = |y_r|$, and introducing the notation

$$\alpha = \frac{\mu}{2\sqrt{y_r^2 - 1}} \quad , \quad \beta = |y_r| + \sqrt{y_r^2 - 1} = \exp v_r$$

$$\Delta = |y_r| - \sqrt{y_r^2 - 1} = \exp(-v_r) \qquad \delta = \frac{2\varepsilon \cos v_h}{\sqrt{\gamma_o |\gamma_h|}} \quad , \tag{7.49}$$

we write out separately the values of σ for the two fields. Region I - the first field

$$\sigma^{I} = \alpha \left(\frac{\beta}{\gamma_o} + \frac{\Delta}{|\gamma_h|} + \delta \right) \quad , \tag{7.50a}$$

the second field

$$\sigma^{I} = -\alpha \left(\frac{\Delta}{\gamma_o} + \frac{\beta}{|\gamma_h|} + \delta \right) \quad . \tag{7.50b}$$

Region III - the first field

$$\sigma^{III} = -\alpha \left(\frac{\Delta}{\gamma_o} + \frac{\beta}{|\gamma_h|} - \delta \right) \quad , \tag{7.50c}$$

the second field

$$\sigma^{III} = \alpha \left(\frac{\beta}{\gamma_o} + \frac{\Delta}{|\gamma_h|} - \delta \right) \quad . \tag{7.50d}$$

Comparing the last ratios with the similar expression (5.76) for the Laue case, we notice a radical difference in absorption effects.

In the Laue case, the value of σ is always positive, although it may differ quite considerably for the two fields. In the Bragg case, σ may take on either positive or negative values, and the values in parentheses in (7.50) are always positive. The absolute value σ^i (i = I,III) differs little for the two fields and is identical in the case of symmetrical reflection.

Equation (7.50a-d) are in line with the analysis of the wave fields in a parallel-sided plate carried out at the end of Section 7.1 (7.38,39).

It is obvious that (7.50a) and (7.50d) refer to fields arising on the entrance (upper) surface and propagating towards the lower one. Eqs. (7.50b) and (7.50c) yield the values of the absorption coefficients σ for fields arising on the lower surface, with the opposite direction of propagation.

For the time being we will discuss only the positive values of σ in regions I and III. As would be expected from the analysis carried out in Section 7.1, the plus signs refer to the first field in region I and to the second field in region III. If we assume $2\varepsilon \cos\nu_h > 0$, then $\sigma^I > \sigma^{III}$. Thus, a maximum profile asymmetry arises in reflection from the absorbing crystal. Near the boundary of the total-reflection region, at values of the angular function $y_r > 1$, the value of the absorption coefficient σ^{III} is less than σ^I, which corresponds to the other lateral side of the maximum. As the angle of deviation increases, i.e., with an increase in $|y_r|$, the relative weight of the first term in parentheses in (7.50) increases, and the entire expression asymptotically approaches the value μ/γ_o.

If $2\varepsilon \cos\nu_h < 0$, regions I and III have the opposite ratios of the coefficients σ. For a more detailed investigation of region III with lower absorption, we determine, by differentiating the expression (7.50) with respect to y_r, the minimum of the function σ^{III} and the corresponding arguments:

$$\sigma^{III}_{min} = \frac{\mu}{2}\left[\frac{1}{\gamma_o} - \frac{1}{|\gamma_h|} \mp \sqrt{\left(\frac{1}{\gamma_o} + \frac{1}{|\gamma_h|}\right)^2 - \frac{4\varepsilon^2 \cos^2 v_h}{\gamma_o|\gamma_h|}}\right] ,$$

$$y_{r,min} = \frac{1}{2\varepsilon \cos v_h}\sqrt{\frac{\gamma_o}{|\gamma_h|}}\left(1 + \frac{|\gamma_h|}{\gamma_o}\right) \qquad (7.51)$$

and in the case of symmetrical reflection

$$\sigma^{III}_{min,s} = \frac{\mu}{\gamma_o}\sqrt{1 - \varepsilon^2 \cos^2 v_h} \quad , \quad y_{r,min,s} = (\varepsilon \cos v_h)^{-1} \quad . \tag{7.52}$$

From (7.52) it follows that for σ-polarizations of the displacement vector the absorption coefficient will be slightly less than for π-polarizations. For numerical estimation of $\sigma^{III}_{min,s}$, we will give its value for symmetrical reflection 220 of Cu Kα radiation from Ge. Under these conditions $\varepsilon^2 \cos^2 \nu_h \approx 0.925$ (at room temperature), and σ will equal $\sim 0.275\ \mu/\gamma_o$.

We shall now determine the value of σ in the total-reflection region. Since in this maximum region

$$W = i\sqrt{4C^2 \frac{|\gamma_h|}{\gamma_o} \chi_{hr}\chi_{\bar{h}r} - \beta^2} = iW_r - W_i \quad , \tag{7.53}$$

we can write, on the basis of (7.42) and (7.44)

$$\sigma^{II} = 4\pi\left(-\frac{K\chi_{oi}}{2\gamma_o} - \frac{K\beta_i}{4|\gamma_h|} \pm \frac{K}{4\gamma_h} W_r\right) = \frac{\mu}{2}\left(\frac{1}{\gamma_o} - \frac{1}{|\gamma_h|}\right) \pm$$

$$\pm \frac{\pi K \ell}{|\gamma_h|}\left\{\frac{1}{2}[\sqrt{a^2(y) + b^2(y)} + a(y)]\right\}^{1/2}$$

or

$$\sigma^{II} = \frac{\mu}{2}\left(\frac{1}{\gamma_o} - \frac{1}{|\gamma_h|} \pm \frac{2C|\chi_{hr}|}{|\chi_{oi}|\sqrt{\gamma_o|\gamma_h|}}\left\{\frac{1}{2}[\sqrt{a^2(y)+b^2(y)}+a(y)]\right\}^{1/2}\right) \quad . \tag{7.54}$$

In cases where $b \ll a$ we can approximate the radicand in (7.54) by a simpler expression and obtain, for the absorption coefficient,

$$\sigma^{II} \approx \frac{\mu}{2}\left\{\frac{1}{\gamma_o} - \frac{1}{|\gamma_h|} \pm \frac{2C|\chi_{hr}|}{|\chi_{oi}|\sqrt{\gamma_o|\gamma_h|}}\sqrt{1 - y_r^2}\ \cdot \right.$$

$$\left. \cdot \left[1 + \frac{b_1^2}{8(1-y_r^2)}\right]\right\} \quad . \tag{7.55}$$

Finally, directly on the boundaries of the total-reflection region, at $y_r = \pm 1$, we use, for calculating σ, the exact equations (4.65), and (7.45) for W_i. Eq. (4.65) is simplified if we assume the value $g^2 \approx \beta_i^2$ in (7.46) to be small, and hence a vanishes to zero for the y values under discussion. For W_i we get

$$\pm W_i \approx \ell \sqrt{\pm i |b(y)|} = \pm \ell \sqrt{\frac{1}{2}|b(y)|} \quad , \tag{7.56}$$

whence, for $y_r = \pm 1$

$$\sigma_{-1} = \frac{\mu}{2}\left[\frac{1}{\gamma_o} - \frac{1}{|\gamma_h|} \pm \frac{2C|\chi_{hr}|}{|\chi_{oi}| \sqrt{\gamma_o |\gamma_h|}} \sqrt{\frac{1}{2}|b(y)|}\right] \quad , \tag{7.57}$$

$$\sigma_{+1} = \frac{\mu}{2}\left[\frac{1}{\gamma_o} - \frac{1}{|\gamma_h|} \mp \frac{2C|\chi_{hr}|}{|\chi_{oi}| \sqrt{\gamma_o |\gamma_h|}} \sqrt{\frac{1}{2}|b(y)|}\right] \quad . \tag{7.58}$$

Here,

$$|b(y)| = \frac{1}{C}\left|\frac{\chi_{oi}}{\chi_{hr}}\right|\left(\sqrt{\frac{\gamma_o}{|\gamma_h|}} + \sqrt{\frac{|\gamma_h|}{\gamma_o}}\right) \pm 2x \cos\nu_h \quad ,$$

$$x = \left|\frac{\chi_{hi}}{\chi_{hr}}\right| \quad . \tag{7.59}$$

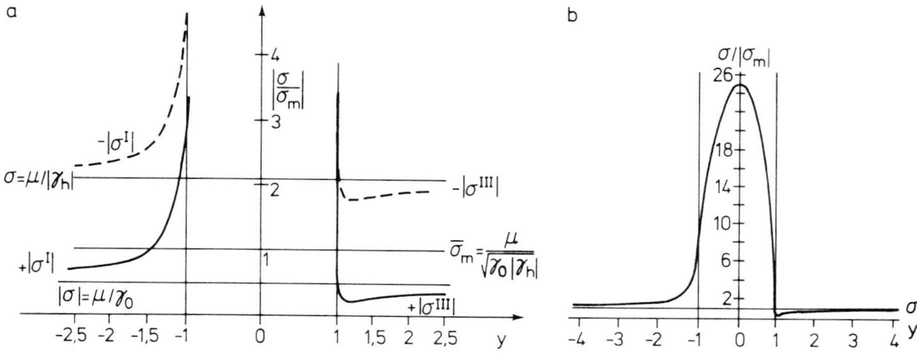

Fig. 7.6a,b Variation of relative coefficient of absorption σ/σ_m in maximum region. a) asymmetric reflection (after WAGNER [7.2]); b) symmetric reflection including extinction in the range of total reflection (after JAMES [1.16])

The further investigation of (7.57-59) is given in [1.16,7.2]. The general picture of variation of σ in all the maximum regions is illustrated by the two graphs in Fig.7.6a describing the variation of σ/σ_m in the lateral regions of the maximum and Fig.7.6b, where σ/σ_m is given also inside the perfect-reflection region. It is easy to see that in the above equations for σ (7.48a,54,55,57,58), $(\gamma_o^{-1} - \gamma_h^{-1})$ vanishes in symmetrical reflection. As regards asymmetrical reflection, this difference is considerable if the incident or reflected wave forms small glancing angles. In all cases, however, this difference is less than the second term in the braces, and consequently the signs of the absorption coefficient σ for the first and second fields, obtained from all the indicated equations, will be opposite to each other both in symmetrical and asymmetrical reflection.

Figure 7.6a refers to reflection 333 of Cu $K\alpha_1$ radiation from Ge, and the reflecting plane (111) is inclined at 35° to the entrance surface of the crystal. Here, $\vartheta = 45^o$, $\gamma_o^{-1} = 1.015$, $\gamma_h^{-1} = 5.75$. When calculating the variation of σ with the angular variable y_r in regions I and III, it is convenient to use (7.48a), which yields sufficiently accurate results from large y_r to y_r of about ± 1, we assume C = 1. The curve in Fig.7.6a gives the relative values of σ/σ_m, where

$$\sigma_m = \mu / \sqrt{\gamma_o |\gamma_h|} \tag{7.60}$$

corresponds to the geometrical mean of the values referring to the propagation of radiation in the crystal in the directions of $\underline{k}_o$ and $\underline{k}_h$: μ/γ_o and μ/γ_h. Inspecting Fig.7.6a, we notice that the solid curves calculated with the aid of (7.48a), with a minus in front of the second term in the braces for region I and a plus for region III, refer to absorption for reflection from a thick crystal. Indeed, as we have shown previously, in a thick crystal only the first field is excited in region I and only the second in region III. In region I we observe interference absorption exceeding normal absorption μ/γ_o beyond the maximum. The absorption value of the first field in region I rises abruptly towards the boundary of the total-reflection region. In region II, which will be discussed again in connection with Fig.7.6b, we see the superposition of two effects: photoelectric (true) absorption and extinction. On transition to region III absorption drops drastically, and abnormal transmission takes place whose maximum value, corresponding to the minimum of σ, can be calculated with the aid of (7.51). Naturally, as $|y_r|$ increases on both boundaries of the maximum, the values of σ asymptotically approach μ/γ_o - in region I, from values exceeding this value, and in region III, from values less than this limiting value.

The solid curves in Fig.7.6a correspond to the positive values of σ/σ_m for the first field in region I (7.50a) and for the second in region III (7.50d). The absorption coefficients for the additional fields represented by the dotted curves in Fig.7.6a are negative and refer, respectively, to the second field in region I (7.50b) and to the first in region III (7.50c). On the relative values scale of Fig.7.6a, they will be much higher than the values $+|\sigma^{I}|$ and $+|\sigma^{III}|$ considered previously.

The curve of variation in relative absorption value σ/σ_m for symmetrical reflection (Fig.7.6b) was calculated as follows. For regions I and III we used the equations

$$\frac{\sigma^{I,III}}{\sigma_{m,s}} = C \left|\frac{\chi_{hr}}{\chi_{oi}}\right| \sqrt{\tfrac{1}{2}[\sqrt{a^2(y_r) + b^2(y_r)} - a^2(y_r)]} \quad , \tag{7.61}$$

$$a(y_r) \approx y_r^2 - 1 \quad , \tag{7.62}$$

$$b(y_r) = 2 \left|\frac{\chi_{oi}}{\chi_{hr}}\right| [(y_r/C) - \varepsilon \cos\nu_h] \quad . \tag{7.63}$$

The approximate equation was obtained from (7.48a); beyond the areas directly adjoining the boundaries of region II,

$$\left(\frac{\sigma^{I,III}}{\sigma_{m,s}}\right)_s = \frac{|y_r - \varepsilon \cos\nu_h|}{\sqrt{y_r^2 - 1}} \quad . \tag{7.64}$$

For values of $y_r = \pm 1$, (7.57-59) lead, in the case of symmetrical reflection, to the form

$$\frac{\sigma^{\pm 1}}{\sigma_m} = \sqrt{C} \sqrt{\left|\frac{\chi_{hr}}{\chi_{oi}}\right| (1 \mp \varepsilon \cos\nu_h)} \quad , \tag{7.65}$$

where, as usual, for the positive value of $\varepsilon \cos\nu_h$ the upper sign refers to $y_r = 1$, i.e., to the first field.

Finally, in the total-reflection region one can use (7.54) and (7.55) with an allowance for the conditions of symmetrical reflection, and for $\sigma_m = \mu/\sin\vartheta$. The more precise equation, (7.54), as in the case of regions I and III, must be applied to areas located near the region boundaries.

We will now dwell on the characteristic features of the curve of variation of absorption within the maximum presented in Fig.7.6b. As regards regions I and III, we can note a similarity with the curve of Fig.7.6a. It is also significant that both in these regions and on the whole, despite the symmetrical reflection, the maximum retains its asymmetrical shape. The anomalous absorption in region I and the anomalous penetration of the wave field in region III can be better understood by analysing the phase relationships between the transmitted and diffracted waves in the crystal (see Fig.7.5). It was indicated there that, due to the coincidence of the phases, in region I the maxima of the total wave field lie on the atomic planes, and due to the phase shift by π, in region III the maxima lie between the atomic planes.

Reverting to the analysis of the absorption curve in Fig.7.6b, we notice a drastic difference in the absorption values on the boundaries of the total-reflection region. When considering this region, one should keep in mind that in the centre of the region and of the entire maximum as a whole, at the maximum value of primary extinction, the penetration of the wave field inside the crystal is negligibly small, and hence the contribution of true absorption to the effect of attenuation of the wave field in the crystal is insignificant. On transition to the lateral parts of region II, extinction falls off abruptly, and absorption becomes predominant.

It could be noted that in the case of asymmetrical reflection, region II must contain two absorption curves corresponding to the solid and dotted curves in regions I and III shown in Fig.7.5.

7.3 Diffraction in Finite Crystal in Incident-Spherical Wave or Incident-Wave Packet Approximation

As has already been noted, we have previously considered Bragg reflection from a crystal bounded either by the entrance surface alone (infinitely thick crystal) or by two parallel surfaces (parallel-sided slice). An important starting condition of the foregoing analysis was the approximation of an incident plane wave with a front width exceeding the slice thickness. Recall that the theory for the case of Laue reflection outlined in Chapters 2, 3, and 4 also refers to a parallel-sided or wedge-shaped crystal of some thickness or other.

The fact that in the Ewald-Laue-Zachariasen dynamical theory consideration is restricted to plane crystals, neglecting the boundary conditions on the lateral surfaces, was thought by some authors to be a specific limitation of this theory. This approach was opposed by the kinematical, or geometrical, theory, which describes X-ray scattering on finite crystals.

At the same time, the methods of the dynamical theory, especially in its contemporary forms (Chaps.6,10, and 11), can be used successfully for solving the corresponding problems, and the condition for comprehensive consideration is in this case a departure from the classical scheme, i.e., the incident-plane-wave approximation.

Multiple reflection from the boundaries, which is typical of scattering in a finite crystal, was considered by WAGNER [7.2] using a parallel-sided slice unbounded from the sides and taking account of absorption. The scheme of this reflection is shown in Fig.7.7.

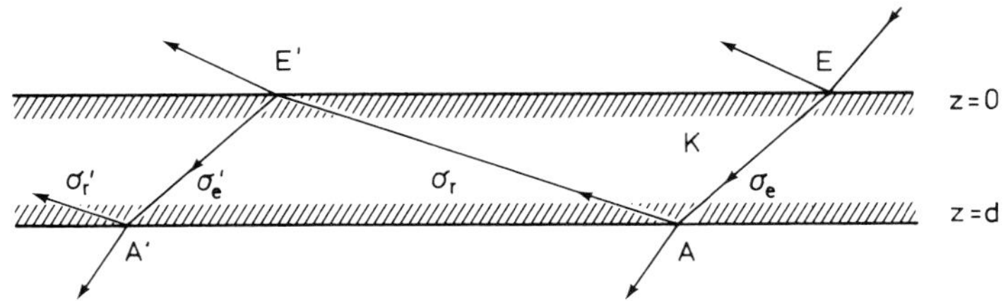

Fig. 7.7 Scheme of multiple Bragg reflections and transmissions in parallel-sided slice (after WAGNER [7.2])

Although this scheme is based on the model of a wave with a narrow wave front incident at point E, WAGNER used the theory in the incident-plane-wave approximation for considering the wave field inside the crystal and the waves emerging from it (for instance, at points A and E'). Using successively the standard boundary conditions for Bragg reflection of the type (7.11), this author obtained expressions for Poynting's vectors and the intensities of reflected and transmitted waves at points E, A, E', ... It is obvious that these expressions, in the more comprehensive theory in the incident-spherical-wave approximation, are integrated over the angles [1.48].

It should be noted that WAGNER's paper [7.2] gives equations for the absorption coefficient as a function of the angles ε_i (i = 1,2) formed by Poynting's vectors with the reflecting plane. Similarly (5.83-85), as applied to the Laue case, were considered in Chapter 5 and are of interest because they are readily applicable to the spherical wave theory.

For the Bragg case, one can obtain the following equations for the absorption coefficient along the directions of Poynting's vectors:

$$\mu^{-\bar{\bar{S}}(i)} = \frac{\mu}{\cos\vartheta}\left(\cos\varepsilon_i - \varepsilon\cos\nu_h\sqrt{1-\frac{\sin^2\varepsilon_i}{\sin^2\vartheta}}\right). \tag{7.66}$$

Comparing (7.66) with (5.83), we notice that the plus sign in front of the second term in the braces of (5.83), when applied to (7.66), would mean transition from region III to region I of the maximum. One must also take into account the signs of ε_i. They are considered positive if the vector products $[\underline{k}_o\underline{k}_h]$ and $[\underline{j}\bar{\bar{S}}^{(i)}]$ are parallel. The minimum of the function $\mu^{\bar{\bar{S}}}$ is determined in relation to the ratio between the $\varepsilon\cos\nu_h$ and $\sin^2\vartheta$.

1) $\varepsilon|\cos\nu_h| > \sin^2\vartheta$, minimum at $\varepsilon_i = 0$.

$$\mu^S_{min} = \frac{\mu}{\cos\vartheta}(1-\varepsilon\cos\nu_h) \quad . \tag{7.67}$$

2) $\varepsilon|\cos\nu_h| < \sin^2\vartheta$, minima at two values of ε_i determined from the equation

$$\sin^2\varepsilon_i = \frac{\sin^4\vartheta - \varepsilon^2\cos^2\nu_h}{\sin^2\vartheta - \varepsilon^2\cos^2\nu_h} \quad ,$$

$$\mu^{\bar{\bar{S}}}_{min} = \mu\sqrt{1-\frac{\varepsilon^2\cos^2\nu_h}{\sin^2\vartheta}} \quad . \tag{7.68}$$

Using (7.67) and (7.68), it is possible to plot a polar diagram of minimum absorption or maximum penetration $(\mu^{\bar{\bar{S}}(2)})^{-1}$ of rays penetrating into the crystal on incidence in the Bragg case and reflected back at point A (in Fig.7.7). Such a diagram is shown in Fig.7.8.

In the figure NN represents the reflecting plane. The minimum attenuation corresponds to ray ET, where T is the point of tangency of the straight line parallel to the lower surface. Part of the scheme above the line EL relates to the wave field reflected from point A. AR||ET' corresponds to the back-reflected ray with penetration $(\mu^{\bar{\bar{S}}(1)})^{-1}$.

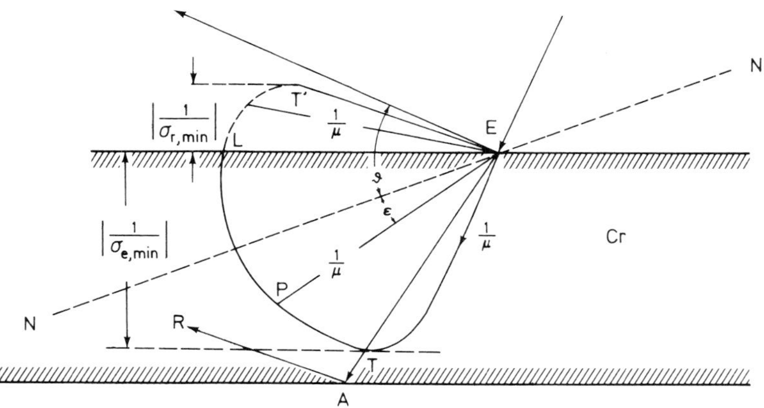

Fig. 7.8 Polar diagram of maximum penetration of rays penetrating into crystal and reflected into crystal from exit surface (see text)(after WAGNER [7.2])

WAGNER [7.2], as well as URAGAMI [1.53,54], considers the conditions for transition from the Bragg case to the Laue case and vice versa.

When using an incident beam with a sufficient angular width in reflection from a system of planes with angle $\varphi \approx \vartheta$, Bragg reflections ($\varphi < \vartheta$) and Laue reflections ($\varphi > \vartheta$) may take place simultaneously. In such an experiment the reflection planes of the two types do not coincide.

Another scheme of such transitions, which is discussed in [1.55], is based on the use of a spherical incident wave with an angular width of the order of the width of the dynamical maximum. Here, both Laue- and Bragg-type reflections occur either due to the deviation of the entrance surface from a plane shape or as a result of subsequent reflection from other boundaries of the crystal.

In his papers [1.53,54], URAGAMI considers transition from Bragg to Laue reflection for WAGNER's scheme with the aid of the indicated method.

X-ray scattering in a bounded crystal was studied by BORRMANN and LEHMANN [1.46,47]. An incident X-ray beam with a narrow wave front was directed onto the smaller surface of a crystal of rectangular cross section. The wave field formed inside the crystal in the angular range of 2ϑ between the directions of s_0 and s_h propagated partly without reflection up to the opposite exit surface, and in the lateral portions was reflected on the crystal boundaries. Thus, in this experiment, in the first place, the conditions of Laue and Bragg reflections were satisfied simultaneously (the larger surfaces corresponded to planes with small indices) and, secondly, there was interference between the different parts of the wave field. In the case described

(Cu $K\alpha_1$ radiation, Si crystal) the scattering corresponded to a thick absorbing crystal and the effects of the pendulum solution were largely suppressed. More general cases of scattering were considered theoretically in subsequent papers by URAGAMI [1.53-55] and KATO and co-workers [1.48] and partly, experimentally, in the indicated papers of URAGAMI. In the schemes investigated by these authors, the "points" of incidence of the vacuum wave lay both between the edges of the smaller surface and at the vertices of this cross section. Besides, the trapezoidal cross section of the crystal was considered. In distinction to LEHMANN and BORRMANN [1.46,47], who analyzed the wave field in the crystal in the incident-plane-wave approximation, the authors of [1.48,53-55] used the above-mentioned contemporary versions of the dynamical theory.

It was shown in Chapter 6 that wave fields excited in a crystal by an incident X-ray beam (transmitted through a slit system) can be considered in the incident-spherical-wave approximation. In KATO's theory outlined above, the final expressions for the wave functions are obtained by integrating with respect to the functions of the angles or by Fourier transformation of the expressions for plane waves. This method was used by KATO with reference to Laue reflection.

When considering wave fields in a bounded crystal KATO and co-workers [1.48] successfully applied the same method of expansion in plane waves to the Bragg case.

The schemes considered in the first of these papers is presented in Fig. 7.9.

On the entrance surface we have Laue reflection at point E. Accordingly, the expressions for plane waves in the crystal coincide with (6.16). The diffracted wave falls on the exit surface at point $\bar{A}$, and here it is subjected to Bragg reflection and transmission (further, at points $\bar{A}_{2n+1}$ and B_{2n}, multiple reflections and transmission).

From the boundary conditions on this surface

$$E_{h,t} \exp[i(\underline{K}_{h,t}\underline{r}_a)] = d_h \exp[i(\underline{k}_h\underline{r}_a)] + d_{h,r} \exp[i(\underline{k}_{h,r}\underline{r}_a)] \quad , \qquad (7.69)$$

$$0 = d_o \exp[i(\underline{k}_o\underline{r}_a)] + d_{o,r} \exp[i(\underline{k}_{o,r}\underline{r}_a)] \qquad (7.70)$$

(the subscripts t and r stand for transmitted and reflected waves) and from the corresponding conditions for the wave vectors from which the values of accomodations $\delta^{(i)}$ are determined, we can calculate the final values of the

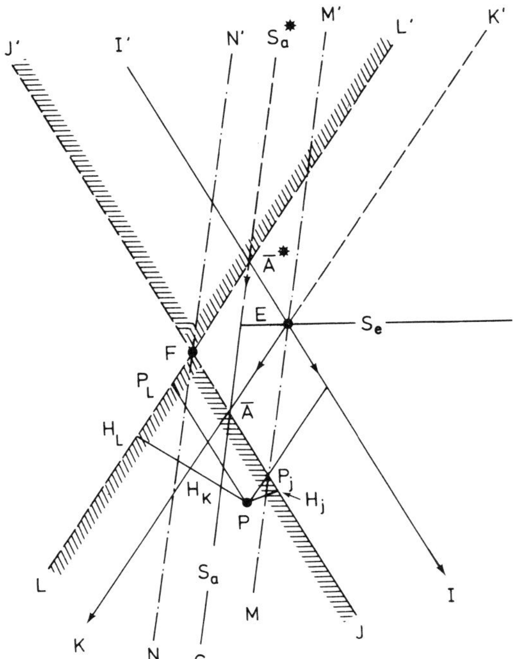

Fig. 7.9 Formation of wave fields in bounded crystal. (Laue-Bragg reflections) (after SAKA et al. [1.48])

wave functions inside the crystal upon reflection from the exit surface. When switching to the incident-plane-wave approximation, one must take into consideration the difference between Figs.7.9 and 6.2. The paper under review considers the incidence of a vacuum wave on the surface bounded to the left of point E. Therefore the use of expression of the type of (6.16) (as in the investigations described in Chap.6) must be regarded just as a mathematical device which permits solving the problem by subsequent integration. The values of $\Phi_{o,r}$ and $\Phi_{h,r}$, already in the incident-spherical-wave approximation, are obtained by integrating the values $d_{o,r}$ and $d_{h,r}$ in the complex plane. The expressions for $\Phi_{o,r}$ and $\Phi_{h,r}$ are closely similar to (6.36) and (6.37) for a single Laue reflection. In the notation adopted in Chapter 6, the values of the wave functions for the Laue-Bragg case are equal to (the first line in (7.71) and (7.72) refers to the condition $x_{o,r}x_{h,r} > 0$, and the second, to the condition $x_{o,r}x_{h,r} < 0$)

$$\Phi_{o,r} = \begin{cases} \pi\bar{f}\,\dfrac{\gamma_h'}{|\gamma_o'|}\sqrt{\dfrac{x_{o,r}}{x_{h,r}}}\,J_1(\bar{f}\sqrt{x_{o,r}x_{h,r}})B_oE_e \quad , \\[2ex] 0 \quad , \end{cases} \tag{7.71}$$

$$\Phi_{h,r} = \begin{cases} i\pi\bar{f}\sqrt{\dfrac{\chi_h}{\chi_{\bar{h}}}}\ \dfrac{\gamma_h'}{|\gamma_o'|}\ b\ \dfrac{x_{o,r}}{x_{h,r}}\ J_2(\bar{f}\sqrt{x_{o,r}x_{h,r}})B_hE_e \quad , \\ \\ 0 \quad , \end{cases} \tag{7.72}$$

b = ±1, depending on the sign of $x_{o,r}$,

$$B_o = \frac{i}{8\pi^2}\sqrt{\frac{2\pi}{Kz}}\exp[i(-\tfrac{\pi}{4}+P+Kz)] \quad ,$$

(7.73)

$$B_h = B_o\exp[2\pi i(hr)] \quad .$$

The values γ_o' and γ_h' are the cosines of the angles formed by the wave vectors $\underline{K}_{o,t}$ and $\underline{K}_{h,t}$ of the emerging vacuum waves with the normal $\underline{n}_a$ to the exit surface.

The rectangular coordinates of the arbitrary observation point P, $x_{o,r}$ and $x_{h,r}$ represent the distances of this point from the straight lines LL' and JJ' in Fig.7.9. These lines are constructed as follows. The wave vectors $\underline{K}_o$ and $\underline{K}_h$ of the incident wave and the wave reflected at point E intersect the real boundary of the crystal at point $\bar{A}$ and its (imaginary) continuation at point $\bar{A}$*. The straight lines LL' and JJ' are parallel to $\underline{K}_h$ and $\underline{K}_o$, respectively.

For the wave $\Phi_{h,t}$ emerging into the vacuum we obtain the expression

$$\Phi_{h,t} = \begin{cases} i\pi\bar{f}\sqrt{\chi_h/\chi_{\bar{h}}}\ c\left\{\dfrac{x_h}{x_{h,r}}\ J_2[\bar{f}\sqrt{(|\gamma_o'|/\gamma_h')x_hx_{h,r}}] + \right. \\ \\ \left. + J_o[\bar{f}\sqrt{(|\gamma_o'|/\gamma_h')x_hx_{h,r}}]\right.\ B_{h,t}E_e \quad , \quad x_hx_{h,r} > 0 \quad , \\ \\ 0 \quad , \quad x_hx_{h,r} < 0 \end{cases} \tag{7.74}$$

c = ±1, depending on the sign of x_h; E_e is the amplitude of the incident wave.

Without dwelling on detailed analysis of the results obtained, we will make the following two remarks.

Actually, the waves (rays) excited in the *crystal* are reflected, as a result of Laue reflection at point E, and then subjected to Bragg reflection on the exit surface of the crystal S_a exclusively within the angle KEM. In contrast to this, the waves propagating within the angle IEM correspond to the imaginary reflection from S_a^*. The same difference is observed between the areas inside JFN (real reflection) and LFN (imaginary reflection) for the *vacuum*. Although (7.71) and (7.72) formally completely take into account the areas inside the angles IEK and JFL, they yield the correct solution, since the waves reflected imaginarily from S_a^* actually do not interfere with the wave fields in the area inside the angle C$\bar{A}$J to which the indicated expressions refer.

A similar problem is solved in the same paper for the right-hand (with respect to point E) boundary of the crystal. Here, the ray path in the crystal and the vacuum is mirror-symmetrical to the one presented in Fig.7.9; the symmetry plane passes through some point lying on the entrance surface S_e. In reciprocal space the scheme considered above is shown in Fig.7.10.

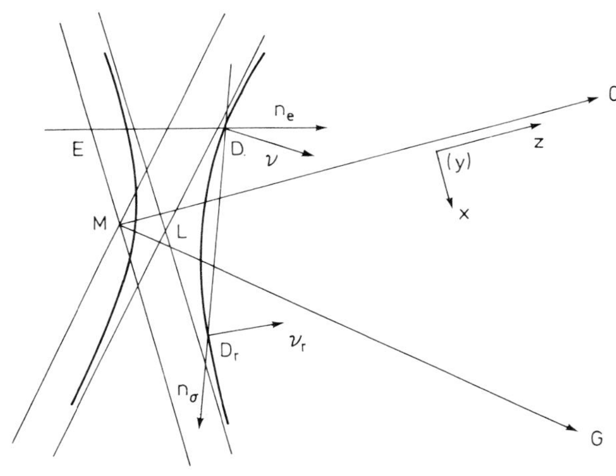

Fig. 4.10 Formation of wave fields in bounded crystal. Ray path in reciprocal space (after SAKA et al. [1.48])

It is also interesting to consider the correctness of the plane-wave metho because the integration of expressions of the type of (6.16) and (6.19) is carried out from $-\infty$ to $+\infty$. The same question arises in solving the Laue reflection problem outlined in Chapter 6. The physical aspect has already been mentioned above when indicating the areas corresponding to the real and imaginary reflection.

The area corresponding to Laue-Bragg reflection inside and outside the crystal is evidently actually bounded by the angles between the dotted lines with hatching where the wave fields are superimposed, according to (7.71), (7.72), and (7.74).

We will now mention the remarks made in this work concerning LEHMANN and BORRMANN's [1.46,47] experiment and the interference bands observed by them. Deeming it necessary to consider the Lehmann-Borrmann scheme in the spherical-incident-wave approximation, the authors of [1.48] give the corresponding expressions for the waves transmitted through the crystal without reflection on the inner side of the lateral surfaces (Laue-Laue) and for the waves which have experienced such reflection (Laue-Bragg-Laue). In the spherical-incident-wave approximation, the asymptotic forms of the Bessel functions appearing in the indicated expressions have the following form in the Laue-Laue case for the transmitted wave

$$J_1(\rho_1) \simeq \sqrt{1/2\pi\rho_1}\sqrt{\frac{x_h}{x_o}}\,\{\exp[i(\rho_1+\tfrac{\pi}{4})] + \exp[-i(\rho_1+\tfrac{\pi}{4})]\} \quad ; \tag{7.75}$$

for the diffracted wave

$$iJ_o(\rho_1) \simeq \sqrt{1/2\pi\rho_1}\,\{\exp[i(\rho_1+\tfrac{\pi}{4})] + \exp[-i(\rho_1+\tfrac{\pi}{4})]\} \quad . \tag{7.76}$$

In the Laue-Bragg-Laue case for the transmitted wave

$$J_1(\rho_2) \simeq \sqrt{1/2\pi\rho_2}\sqrt{\frac{x_{o,r}}{x_{h,r}}}\,\{\exp[i(\rho_2-\tfrac{3\pi}{4})] + \exp[-i(\rho_2-\tfrac{3\pi}{4})]\} \quad ; \tag{7.77}$$

for the diffracted wave

$$iJ_2(\rho_2) \simeq \sqrt{\frac{1}{2\pi\rho_2}}\ \frac{x_{o,r}}{x_{h,r}}\ \{\exp[i(\rho_2 - \frac{3\pi}{4})] + \\ + \exp[-i(\rho_2 - \frac{3\pi}{4})]\} \quad , \tag{7.78}$$

$$\rho_1 = \bar{f}\sqrt{x_o x_h} \quad , \quad \rho_2 = \bar{f}\sqrt{x_{o,r} x_{h,r}} \quad . \tag{7.79}$$

Comparing (7.75-78), it is easy to obtain conclusions which agree with LEHMANN and BORRMANN's experimental observations.

With mutual interference of the transmitted as well as the diffracted wave the phase shift is expressed by an identical value ($\rho_1 - \rho_2 + \pi$), since the second terms in the braces are negligibly small in the case of thick absorbing crystals. Hence it follows that both kinds of waves have minima and maxima at the same points and, in particular, on the common exit surface, where $\rho_1 = \rho_2$ have a common minimum. Here, the minimum corresponds to the zero amplitude only for the transmitted waves.

The foregoing discussion of the interference effect in [1.48] illustrates the possibilities of the theory developed by these authors. As has already been noted, LEHMANN and BORRMANN's experiments show practically no pendulum solution effects, and this considerably simplifies the calculation. The authors of [1.48] emphasize that the general case of interference effects in multiple reflection is extremely complicated and therefore is not considered.

By contrast, URAGAMI dedicated his investigations [1.53-55] to interference effects in conditions of multiple reflection of an arbitrary-type incident wave and the presence of both Laue and Bragg reflection on the entrance (narrow) surface.

We have already mentioned in our historical survey (Chap.1) that URAGAMI, along with the Soviet authors AFANASIEV and KOHN [1.62], applied, in solving the Bragg reflection problem, the generalized dynamical theory based on Tagaki-type equations. Chapter 11 briefly outlines the solution of this problem by the generalized theory method, and the integral equation for a Bragg-reflected wave is solved with the aid of the Fourier transformation of this function (D_h).

In [1.54] URAGAMI considers a more complex case, i.e., an incident wave packet which simultaneously experiences Bragg and Laue reflections. Here, the interface need not be plane, and this is an important generalization as compared with the classical dynamical theory. Indeed, Laue [1.13] stresses that the standard conditions on the boundary for the values of D and H refers precisely to a plane interface.

In the generalized dynamical theory, as shown in Chapter 11, in order to determine the wave field at any point of the crystal it is necessary to calculate the integral over some contour C of the Green function (the influence function) G (11.43). The necessary condition of solution is the assignment of the values of D_o, D_h, and their normal derivatives on the perimeter.

URAGAMI calculates the corresponding integral for the perimeter TPQRS (Fig.7.11) on the surface RT of the crystal; the segment AB represents a

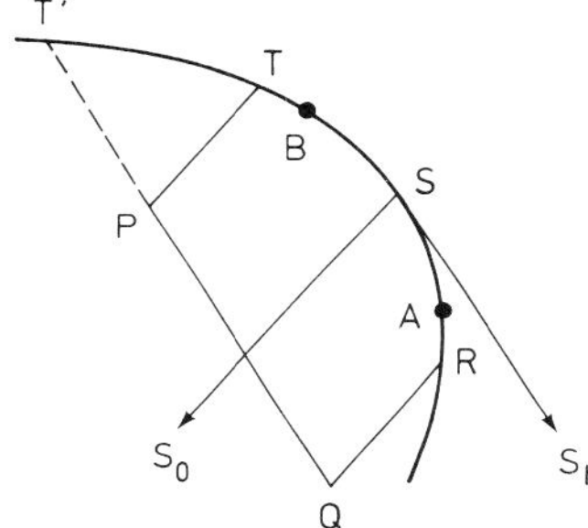

Fig. 7.11 Laue-Bragg reflection with curvilinear crystal surface (after URAGAMI [1.53])

limited front of the incident wave packet, i.e., a group of waves with spread-out directions of propagation. The segment AB is divided into two parts by the point S: on AS the incidence conditions correspond to the Bragg case, and on SB, to the Laue case. The investigation carried out in this work shows that on incidence of a narrow beam on an edge of a wedge-shaped crystal, a wave field arises in the crystal which represents Laue and Bragg scattering and is accompanied by the formation of pendulum solution bands. Using the relations of the generalized theory, URAGAMI obtained expressions for the transmitted and diffracted waves in the crystal, which include all the indicated effects.

Of fundamental importance is the result obtained in both [1.62] and [1.54] and referring to the so-called fringe effect. It has been established experimentally that when using wide (as regards the angle widths) beams, the intensity of a Bragg-reflected beam reaches a maximum on the side facing the incident beam. In accordance with this experimental result the authors of [1.62] and [1.54] obtained, although large, finite values for the amplitude of the reflected wave at the point of incidence.

In [1.54], URAGAMI gives an X-ray pattern obtained by him, which corresponds qualitatively to the theory.

In another paper [1.53], he considered multiple reflection similar to WAGNER's scheme, but for the general case of the wedge-shaped plate. Here, results similar to those obtained by WAGNER are derived in a more rigorous way.

Finally, in his third paper [1.55], URAGAMI considers a multiple reflection scheme similar to the one which was later investigated in the paper of KATO and co-workers [1.48]. In this work, URAGAMI calculates, mainly for the transparent crystal, both the pendulum solution for multiple reflections and the intensity distribution of the waves Bragg-reflected on one of the lateral walls.

The most typical case is the pattern of V-shaped pendulum solution bands formed by interference of waves singly reflected from both lateral boundaries of the plate, which was calculated theoretically and observed experimentally (in the same investigation).

8. Bragg Reflection of X-Rays. II. Reflection and Transmission Coefficients and Their Integrated Values

In this chapter we derive expressions for calculating the reflection and transmission coefficients and their integrated values in the general form: both in the sense of applicability to any one of the three maximum regions, and with an allowance for the two fields in the crystal. The general expressions obtained are then applied to particular problems.

In deriving the general relations, we compare equations obtained by different authors, in particular those given in ZACHARIASEN's book [Ref. 1.14, Chap. 1], which are widely used in the literature (see also [8.1, 2]).

The expressions given here are applicable to the case of incidence of a plane vacuum wave.

8.1 Deriving General Expressions for Reflection and Transmission Coefficients

Let us derive expressions for the coefficients R and T in Bragg reflection with an allowance for both fields in the crystal without trying to predict the real or complex nature of the parameters of the dynamical problem. Consider reflection and transmission in the case of a parallel-sided crystal plate in which we distinguish the entrance and exit surfaces. The derivation that follows is close to the one given in the paper of RAMACHANDRAN and KARTHA [8.3].

The boundary conditions on the entrance surface

$$D_o^{(1)} + D_o^{(2)} = D_o^{(a)} \quad , \quad D_h^{(1)} + D_h^{(2)} = D_h^{(a)} \quad . \tag{8.1}$$

The conditions on the exit surface

$$D_o^{(1)} \exp\left[-2\pi i k_{oz}^{(1)} t\right] + D_o^{(2)} \exp\left[-2\pi i k_{oz}^{(2)} t\right] = D_o^{(d)} \exp\left[-2\pi i k_{oz}^{(d)} t\right] \quad , \tag{8.2}$$

$$D_h^{(1)} \exp\left[-2\pi i k_{hz}^{(1)} t\right] + D_h^{(2)} \exp\left[-2\pi i k_{hz}^{(2)} t\right] = 0 \quad . \tag{8.3}$$

Recall that $D_h^{(a)}$ is the amplitude of the wave reflected into the vacuum on the entrance surface, and $D_o^{(d)}$ is the amplitude of the wave emerging into the vacuum through the exit surface.

According to (3.1), the wave vectors of the waves in the crystal and the wave vector of the vacuum wave $\underline{K}$ are related by the following conditions

$$\underline{k}_o^{(i)} = \underline{K}_o^{(a)} - K\delta^{(i)}\underline{n}_o \quad , \quad \underline{k}_h^{(i)} = \underline{K}(1 + \alpha) - K\delta^{(i)}\underline{n}_o \quad , \tag{8.4}$$

where

$$\delta^{(i)} = -\frac{\chi_o}{2\gamma_o} + \frac{\beta}{4|\gamma_h|} \pm \sqrt{\frac{\beta^2}{16|\gamma_h|^2} - \frac{\chi_h\chi_{\bar{h}}C^2}{4\gamma_o|\gamma_h|}} \quad , \tag{8.5}$$

$$\delta^{(1)} - \delta^{(2)} = -2|\gamma_h|^{-1}\sqrt{\beta^2 - 4C^2\chi_h\chi_{\bar{h}}\frac{|\gamma_h|}{\gamma_o}} \quad . \tag{8.6}$$

Introduce, for all three maximum regions, the common notation (7.6)

$$y = \cosh v \tag{8.7}$$

and also the value

$$f = \sqrt{\chi_h/\chi_{\bar{h}}}\sqrt{\gamma_o/|\gamma_h|} \quad . \tag{8.8}$$

For the amplitudes $D_h^{(i)}$ we write down the relations (see (7.5) (without indicating the signs of the arguments))

$$D_h^{(1)} = fD_o^{(1)}\exp(v) \quad , \quad D_h^{(2)} = fD_o^{(2)}\exp(-v) \quad . \tag{8.9}$$

Eliminating $D_o^{(2)}$ from (8.1) and substituting (8.9) and (8.4) in (8.3), we get

$$D_o^{(1)} = -\frac{\exp[-v + 2\pi iKt\delta^{(2)}]D_o^{(a)}}{\exp[v + 2\pi iKt\delta^{(1)}] - \exp[-v + 2\pi iKt\delta^{(2)}]} \quad . \tag{8.10}$$

Introduce the variable

$$x = i\pi Kt[\delta^{(1)} - \delta^{(2)}] \quad . \tag{8.11}$$

We cancel the right-hand side of (8.10) by

$$\exp\left\{2\pi iKt\left[\delta^{(1)} - \frac{(\delta^{(1)} - \delta^{(2)})}{2}\right]\right\} \quad ;$$

the expression for $D_o^{(1)}$ takes the form

$$D_o^{(1)} = - \frac{\exp[-(v + x)]D_o^{(a)}}{\exp(v + x) - \exp[-(v + x)]} = - \frac{\exp[-(v + x)]D_o^{(a)}}{2 \sinh(v + x)} \quad . \tag{8.12}$$

Further, using (8.1), we obtain

$$D_o^{(2)} = \frac{\exp(v + x)D_o^{(a)}}{2 \sinh(v + x)} \quad . \tag{8.13}$$

Substituting the values of $D_o^{(i)}$ in (8.9) and introducing, with the aid of (8.1), the value $D_h^{(a)}$, we get the following expression for the reflection coefficient

$$R = \left| f \frac{D_h^{(a)}}{D_o^{(a)}} \right|^2 = |f|^2 \left| \frac{D_h^{(1)} + D_h^{(2)}}{D_o^{(a)}} \right|^2 = |f|^2 \frac{|\sinh x|^2}{|\sinh(v + x)|^2} \quad . \tag{8.14}$$

In order to obtain the transmission coefficient T we substitute in (8.2) the values of $D_o^{(i)}$ from (8.12) and (8.13) and the values of $k_{oz}^{(i)}$ from (8.4):

$$T = \left| \frac{D_o^{(d)}}{D_o^{(a)}} \right|^2 = \left| \frac{\exp(v + x)}{2 \sinh(v + x)} \exp\left[-2\pi iK_{oz}^{(a)}t + 2\pi iKt\delta^{(2)}\right] - \right.$$
$$\left. - \frac{\exp[-(v + x)]}{2 \sinh(v + x)} \exp\left[-2\pi iK_{oz}^{(a)}t + 2\pi iKt\delta^{(1)}\right] \right| \quad . \tag{8.15}$$

This expression, with an accuracy to the factor

$$\exp\left[-2\pi it\left(K_{oz}^{(a)} - K \frac{\delta^{(1)} + \delta^{(2)}}{2}\right)\right] \quad ,$$

which yields unity on multiplication by the conjugate complex value, takes the following simple form:

$$T = \frac{|\sinh v|^2}{|\sinh(v + x)|^2} \quad . \tag{8.16}$$

Before analyzing (8.14) and (8.16) as applied to more particular problems, we will give a similar derivation of expressions for R and T, using ZACHARIASEN's notation [1.14].

Introduce the variables q, z, W, $c^{(i)}$, and $n^{(i)}$:

$$q = -C^2 \chi_h \chi_{\bar{h}} \frac{\gamma_o}{\gamma_h} \ , \qquad z = -\frac{\beta}{2} \frac{\gamma_o}{|\gamma\ |} \ , \qquad \sqrt{z^2 + q} = \frac{W}{2} \frac{\gamma_o}{|\gamma_h|} \ ,$$

$$c^{(i)} = \frac{-z \pm \sqrt{z^2 + q}}{\chi_{\bar{h}}} \tag{8.17}$$

$$n^{(i)} = \exp\left[-i2\pi \frac{K}{\gamma_o} \frac{t}{2} (\chi_o - z \pm \sqrt{q + z^2})\right] = \exp[2\pi i K t \delta^{(i)}] \ . \tag{8.18}$$

The boundary conditions on the exit surface (8.2) can be rewritten as follows with an accuracy to the phase factor $\exp\left[-2\pi i\left(K_{oz}^{(a)} t\right)\right]$:

$$n^{(1)} D_o^{(1)} + n^{(2)} D_o^{(2)} = D_o^{(d)} \tag{8.19}$$

and (8.3) with an accuracy to the factor $\exp\left[-2\pi i\left(1 + K_{oz}^{(a)} t\right)\right]$

$$c^{(1)} n^{(1)} D_o^{(1)} + c^{(2)} n^{(2)} D_o^{(2)} = 0 \ . \tag{8.20}$$

Solving (8.1), (8.12), and (8.20) simultaneously, we obtain

$$R = \frac{|\gamma_h|}{\gamma_o} \left| \frac{c^{(1)} c^{(2)} [n^{(1)} - n^{(2)}]}{c^{(2)} n^{(2)} - c^{(1)} n^{(1)}} \right|^2 \ ,$$

$$T = \left| \frac{n^{(1)} n^{(2)} \left[c^{(2)} - c^{(1)}\right]}{c^{(2)} n^{(2)} - c^{(1)} n^{(1)}} \right|^2 \ . \tag{8.21}$$

The expression for R in (8.21) was transformed by ZACHARIASEN for the general case of the thin and thick crystal, as well as for the transparent and absorbing one.

The division into the real and imaginary parts of the corresponding parameters yields the following:

$$\sqrt{q + z^2} = \ell + iw \ . \tag{8.22a}$$

We also introduce the notation

$$a = 2\pi Kt \frac{|\gamma_h|}{(\gamma_o + |\gamma_h|)} \quad . \tag{8.22b}$$

In this case the value $n^{(i)}$ from (8.18) is written:

$$n^{(i)} = \exp[-ia(\chi_o - z \pm \ell)] \exp(\pm aw) \quad . \tag{8.23}$$

After introducing in (8.21) the values $c^{(i)}$ from (8.18) and $n^{(i)}$ from (8.23), we write the expression for the reflection coefficient

$$R = \frac{|\gamma_h|}{\gamma_o} \left| \frac{\chi_{\bar{h}}^{-2}(-z + P)\ (-z - P)\left\{\exp[2\pi iKt\delta^{(1)}] - \exp[2\pi iKt\delta^{(2)}]\right\}}{\chi_{\bar{h}}^{-1}\left\{(-z - P)\ \exp[2\pi iKt\delta^{(2)}] - (-z + P)\ \exp[2\pi iKt\delta^{(1)}]\right\}} \right|^2 \quad . \tag{8.24}$$

Here, $P = \sqrt{q + z^2}$.

The numerator of this expression is transformed as follows:

$$(-z + \sqrt{q + z^2}) - (z - \sqrt{q + z^2}) = -q \quad , \quad |q\chi_{\bar{h}}^{-2}|^2 = \frac{\gamma_o}{|\gamma_h|^2}$$

$$|n^{(1)} - n^{(2)}|^2 = 4(\sin^2 a\ell + \sinh^2 aw) \quad . \tag{8.25}$$

In the denominator of (8.24) we calculate the squares of the moduli

$$|-z \pm \sqrt{q + z^2}|^2 \tag{8.26}$$

and use the identities

$$z(\sqrt{q + z^2})^* \pm z^* \sqrt{q + z^2} = \sqrt{|(|q + z^2| \pm |z|^2)^2 - |q|^2|} \quad . \tag{8.27}$$

As a result, performing obvious transformations, we derive the following general Zachariasen equation for the reflection coefficient:

$$R = \frac{\gamma_o}{|\gamma_h|}|\chi_h|^2 C^2(\sin^2 a\ell + \sinh^2 aw)/[D + (D + z^2)\sinh^2 aw - $$
$$- (D - z^2) \sin^2 a\ell + \frac{1}{2}M \sinh 2 aw + \frac{1}{2}N \sin 2 a\ell] \quad , \tag{8.28}$$

$$D = |q + z^2| \quad , \quad M = |(D + |z|^2)^2 - |q|^2|^{1/2} \quad ,$$

$$N = |(D - |z|^2)^2 - |q|^2|^{1/2} \quad .$$

The text of the subsequent sections of this chapter consists of transformation, qualitative analysis and, partly, numerical calculations of the obtained general equations (8.21) and (8.28) as applied to the important particular cases.

8.2 Bragg Reflection from Transparent Crystal

In this case the variable x in (8.11) can be represented as (see (8.6))

$$x = i\pi Kt[\delta^{(1)} - \delta^{(2)}] = -iA\sqrt{y^2 - 1} \quad , \quad A = \frac{\pi CKt\sqrt{\chi_h\chi_{\bar{h}}}}{\sqrt{\gamma_o|\gamma_h|}} \quad . \tag{8.29}$$

We will now consider the various regions of the maximum separately.

Region II

$$i\sqrt{y^2 - 1} = -\sqrt{1 - y^2} \quad , \quad x = A\sqrt{1 - y^2} \quad . \tag{8.30}$$

From (8.14) it follows that for the relative intensity of the reflected wave

$$\frac{I_h^{(a)}}{I_o^{(a)}} = R\,\frac{\gamma_o}{|\gamma_h|} = \frac{|\sinh x|^2}{|\sinh(v + x)|^2}$$

$$= \frac{|\sinh(A\sqrt{1 - y^2})|^2}{|\sinh(A\sqrt{1 - y^2} + v)|^2} \quad . \tag{8.31}$$

Using further the identities

$$|\sinh(A\sqrt{1 - y^2} + v)|^2 = \sinh^2 A\sqrt{1 - y^2} + \sin^2 v$$

$$(\sinh\alpha)^{-2} = \coth^2\alpha - 1 \tag{8.32}$$

and taking into account (7.7) and (7.8), we obtain

$$\frac{I_h^{(a)}}{I_o^{(a)}} = \frac{\sinh^2(A\sqrt{1-y^2})}{1-y^2+\sinh^2(A\sqrt{1-y^2})}$$

$$= \frac{1}{y^2+(1-y^2)\coth^2(A\sqrt{1-y^2})} \quad . \tag{8.33}$$

Note that in our expressions, in addition to those coinciding with ZACHARIASEN's equations, the variable y is given by (7.5):

$$y = \frac{\beta}{2C|\chi_h|\sqrt{|\gamma_h|/\gamma_o}} \quad , \tag{8.34}$$

i.e., it actually does not differ from the function y, which we used in the chapters dealing with Laue reflection, because $\gamma_{h,L} = |\gamma_{hB}|$. As for ZACHARIASEN's equations, one must keep in mind that

$$y_z = -y \tag{8.35}$$

and in the transparent crystal, $y^2 = z^2/q$.

Regions I and III

In this case $x = -iA(y^2-1)^{\frac{1}{2}}$, $\sinh x = -\sinh[iA(y^2-1)^{\frac{1}{2}}]$, $|\sinh x|^2 = \sin^2[A(y^2+1)^{\frac{1}{2}}]$.

Using the identities (8.32), we obtain

$$\frac{I_h^{(a)}}{I_o^{(a)}} = \frac{\sin^2[A(y^2-1)^{\frac{1}{2}}]}{y^2-1+\sin^2[A(y^2-1)^{\frac{1}{2}}]} =$$

$$= \frac{1}{y^2+(y^2-1)\cot^2[A(y^2-1)^{\frac{1}{2}}]} \quad . \tag{8.36}$$

The expressions (8.33) and (8.36) given in ZACHARIASEN's form and notations are also readily obtained from the general equation (8.28). Indeed, region II corresponds to the condition $(q+z^2) < 0$ and regions I and III, to $(q+z^2) > 0$.

1) $(q+z^2) < 0$. Since $\ell = 0$, then $\sin^2 a\ell$ and $\sin 2a\ell$ vanish to zero as well as $|[(q+z^2)+(z)^2]^2-(q)^2|^{\frac{1}{2}}$. Further, $|q+z^2| = q(1-y^2)$ and, finally,

$$aw = aq\sqrt{1 - y^2} = A\sqrt{1 - y^2} \quad . \tag{8.37}$$

As a result, from (8.28) we obtain $R\,\gamma_o/|\gamma_h|$ in accordance with (8.33).

2) $(q + z^2) > 0$, then $w = 0$, $a\ell = A\sqrt{y^2 - 1}$, $\sin^2 aw = \sin 2aw = 0$, $(|q + z^2| - |z^2|)^2 - |q|^2 = 0$, $q + z^2 = q(y^2 - 1)$, and (8.28) transforms to $I_h^{(a)}[|\gamma_h|/\gamma_o]$ in accordance with (8.36).

Note that, taking into account the identities

$$\sinh^2\alpha = -\sin^2 i\alpha \quad , \quad \coth^2\alpha = -\cot^2 i\alpha \quad , \tag{8.38}$$

(8.33) and (8.36) must be regarded as two forms of the same general expression.

Inspecting (8.36), we notice that it describes the subsidiary maxima of the pendulum solution for the Bragg case which form in the lateral regions of the intensity maximum. We have already mentioned the possibility of appearance of such maxima when we were considering the phenomena in a crystal slice. The likely cause of it is the interference between the second wave field penetrating inside the crystal and the first field, which arises on reflection from the exit surface and propagates back into the crystal.

Transform (8.36), partially replacing the variable y by v in the denominator

$$R\,\frac{\gamma_o}{|\gamma_h|} = \frac{I_h^{(a)}}{I_o^{(a)}} = \frac{1 - \cos 2A\sqrt{y^2 - 1}}{\cosh 2v - \cos 2A\sqrt{y^2 - 1}} \quad . \tag{8.39}$$

Fig. 8.1 illustrates the shape of the maximum, (R values) in accordance with (8.39), for several values of the parameter $A = C\pi|\chi_h|t/(\lambda\sqrt{\gamma_o\gamma_h})$. The expressions (8.36) and (8.39) vanish to zero provided that

$$2A\sqrt{y^2 - 1} = 2\pi m \quad . \tag{8.40}$$

For large m (practically already at $m \geq 2$), the angular interval between the two minima corresponds to

$$\Delta y \approx \frac{\pi}{A} = \frac{\lambda\sqrt{\gamma_o|\gamma_h|}}{Ct|\chi_h|} \quad , \tag{8.41}$$

which formally coincides with (3.70); the latter yields the angular intervals of the subsidiary maxima in the Laue case.

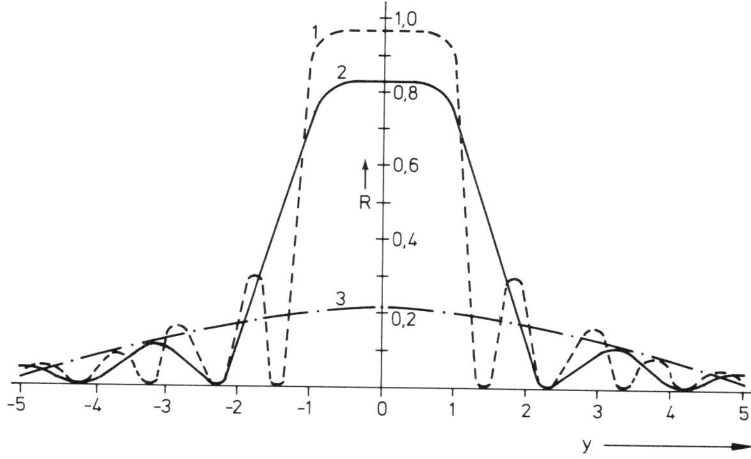

Fig.8.1 Curves of Bragg reflection from transparent crystal of finite thickness t with different parameter values (after ZACHARIASEN [1.14]). $A = f(t)$: I: $A = \pi$; 2: $A = \pi/2$; 3: $A = 0.5$

It is easy to see that in the lateral regions of the maximum in Bragg reflection we must also have another type of pendulum solution, namely bands of equal thickness.

As has already been noted, the difference in the mechanism of formation of pendulum solution effects in both cases of reflection is due to the fact that the points of excitation of the interferring waves in the Laue case lie on different branches, while in the Bragg case they lie on the common branch of the dispersion hyperbola.

The modulus of the vector $|\Delta k|$, which represents, in reciprocal space, the extinction distance τ^{-1}, is given in the Laue case by (3.37):

$$|\Delta k| = \frac{C|\chi_h|}{\lambda\sqrt{\gamma_o\gamma_h}} \sqrt{1+y^2} = \tau^{-1} \quad . \tag{8.42}$$

In Bragg reflection, the normal to the entrance surface in the maximum regions beyond the total-reflection region intersects some one branch of the dispersion hyperbola at two points.

The length of the chord $\overline{A^{(1)}A^{(2)}}$ in Fig. 7.2 is determined directly from the expression

$$\overline{A^{(1)}A^{(2)}} = |\Delta k|_B = \left|\left(\xi_o^{(1)} - \xi_o^{(2)}\right)\right|\gamma_o^{-1} \quad , \tag{8.43}$$

where the values of $\xi_o^{(i)}$ differ slightly from (3.14) if we take into consideration the negative sign of γ_h:

$$\xi_o^{(i)} = \mp \frac{KC|\chi_h|}{2} \sqrt{\frac{\gamma_o}{|\gamma_h|}} \left(\sqrt{y^2 - 1} \pm y\right) \quad . \tag{8.44}$$

Thus, in the Bragg case

$$|\Delta k|_B = \frac{C|\chi_h|}{\lambda\sqrt{\gamma_o|\gamma_h|}} \sqrt{y^2 - 1} \approx \frac{\sqrt{y^2 - 1}}{\Delta y t} \quad . \tag{8.45}$$

For symmetrical reflection we obtain, in the Laue case (see (3.73)),

$$\Delta y_s \approx \frac{\lambda \cos\vartheta}{Ct|\chi_h|} \quad , \quad \Delta\eta_s = \frac{d_{hk\ell}}{t} \quad . \tag{8.46}$$

In the Bragg case

$$\Delta y_s \approx \frac{\lambda \sin\vartheta}{Ct|\chi_h|} \quad , \quad \Delta\eta_s = \frac{d_{hk\ell}}{t} \tan\vartheta \quad . \tag{8.47}$$

The foregoing consideration of the maxima of the Pendellösung fringe solution in the Bragg case showed their close analogy with the same phenomenon in Laue reflection.

The distribution of the corresponding maxima, as in the Laue case, can be established, in principle, by using, in place of a parallel-sided crystal plate, a wedge-shaped plate with a fixed value of the angle of incidence, i.e., the value η.

We will finally note that the numerical estimate of the values Δy and $|\Delta k| = \tau^{-1}$ given in Chapter 3 for the Laue case is, as regards its order of magnitude, applicable to the Bragg case as well. This follows at once from a comparison of the expressions (8.42) and (8.45 - 8.47).

Transition to Thick Crystal. DARWIN and EWALD Equations

It is interesting to consider the case of a thick crystal but with such a low absorption that it can be assumed transparent.

As follows from (8.29), (8.33) and (8.36), which contain the quantity A, depend on the crystal thickness t. In region II of the maximum, according to (8.33), the value of the relative intensity tends to unity together with $\coth^2[A(1 - y^2)^{\frac{1}{2}}]$ with an increase in A; if $|y| < 1$ and $A >> 1$, then

$$\frac{I_h^{(a)}}{I_0^{(a)}} = 1 \quad . \tag{8.48}$$

As far as the maximum regions I and III are concerned, the task of calculating the relative intensities for the infinite thickness is more difficult because the presence of the trigonometric functions in (8.36) makes the value of intensity an oscillating function of the thickness. To calculate the desired value it is necessary to average out $\cot[A(y^2 - 1)^{\frac{1}{2}}]$ over a range of A greater than $\pi/2$. Integration with respect to A leads to the following expression for $|y| > 1$ and $A >> 1$:

$$R = \frac{|\gamma_h|}{\gamma_0}\frac{I_h^{(a)}}{I_0^{(a)}} = (1 - \sqrt{1 - y^{-2}}) \quad . \tag{8.49}$$

Now compare the expressions for the coefficients for reflection from the transparent crystal obtained in this chapter (8.36, 49) with (7.14).

For regions I and III the transition from (8.36) to (8.49) applicable to a thick crystal did not eliminate the difference from (7.20), which can be written in the following form if we switch from the variable v to y:

$$R_{I,III} = \left(\mp |y| - \sqrt{y^2 - 1}\right)^2 \quad , \tag{8.50}$$

where the upper sign refers to region I.

The cause of this difference is quite evident. Equation (7.14) (and (8.50), equivalent to it) was derived in Section 7.1, allowing only for one first field in region I of the maximum and only one second field in region III. As has already been noted, this derivation was based on the concept of a thick, slightly absorbing crystal. In such a crystal, the wave field excited on the entrance surface propagates into the crystal and damps out before reaching the exit surface, where an additional field directed backwards may arise.

In contrast to this, (8.49) was obtained on the basis of a concept according to which, even in a sufficiently thick crystal, the wave field directed inside the crystal reaches the exit surface and induce an additional field. Indeed, the initial equations (8.14) and (8.24) were derived for a wide front of the incident wave and superposition of the two fields throughout the maximum region.

ZACHARIASEN named equation (8.50) the solution of DARWIN, who derived it originally, and (8.49), EWALD's solution. It is obvious that EWALD's solution is applicable to thinner crystals, in which the absorption is negligibly small, and DARWIN's formula must be the limiting expression for the reflection coefficient. Hence it follows that a rigorous consideration requires a more accurate allowance for the complex nature of the parameters appearing in (8.14) and (8.24) (see the analysis given in Section 8.4 and (8.113 - 116).

Here, we will give the expressions for integrated reflection in DARWIN's and EWALD's approximations. It is obvious that in both cases integrated reflection for region II of the maximum

$$R^y_{i,II} = 2 \quad . \tag{8.51}$$

For the DARWIN solution

$$\begin{aligned} R^y_{i,I} + R^y_{i,III} &= 2\int_0^\infty \exp(-2v)dy = \\ &= 2\int_0^\infty \exp(-2v)\sinh v dv = \\ &= \int_0^\infty [\exp(-v) - \exp(-3v)]dv = \frac{2}{3} \quad , \end{aligned} \tag{8.52}$$

whence the total value of integrated reflection from a thick crystal

$$R^y_i = \frac{8}{3} \quad , \qquad R^\eta_i = \frac{8C|\chi_h|}{3\sin 2\vartheta}\sqrt{\frac{|\gamma_h|}{\gamma_0}} \quad . \tag{8.53a}$$

It is obvious that both in (8.53a) and in the subsequent expression for integrated reflection (8.55) with the incidence of nonpolarized radiation on the crystal, C is replaced by the known value of the polarization factor

$$R_i^\eta = \frac{8}{3} \frac{|\chi_h|}{\sin 2\vartheta} \frac{1 + |\cos 2\vartheta|}{2} \sqrt{\frac{|\gamma_h|}{\gamma_o}} \tag{8.53b}$$

The expressions (8.53) were obtained by DARWIN. Depending on the condition $2\vartheta > \pi/2$ or $2\vartheta < \pi/2$ or $\cos^2\vartheta \gtrless \sin^2\vartheta$, integrated reflection acquires one of the following two particular values

$$R_i^\eta = \begin{cases} 4/3|\chi_h| \sqrt{\dfrac{|\gamma_h|}{\gamma_o}} \cot\vartheta \quad , \quad 2\vartheta < \dfrac{\pi}{2} \quad , \\ \\ 4/3|\chi_h| \sqrt{\dfrac{|\gamma_h|}{\gamma_o}} \tan\vartheta \quad , \quad 2\vartheta > \dfrac{\pi}{2} \quad . \end{cases} \tag{8.53c}$$

For EWALD's solution, we can either integrate (8.49), obtained by averaging (8.36) over a sufficiently wide range of thicknesses (and A values), or integrate (8.33) and (8.36) with an allowance for the conditions (8.38), and then pass over to the thick crystal. This second method was used by LAUE by integrating over the complex plane of variation in y. Without dwelling on the process of integration, which is described in detail in LAUE's book [1.13], we will give the final result:

$$\int_{-\infty}^{\infty} \frac{dy}{y^2+(1 - y^2)\coth^2\left(A\sqrt{1 - y^2}\right)} = \pi\tanh A \quad . \tag{8.54}$$

With large values of A and, hence, of the thickness of the reflecting crystal t, $\tanh A \approx 1$. For this limiting value, it will be sufficient to have $A \geq 3$. In this case

$$R_i^y = \pi \quad , \quad R_i^\eta = \frac{\pi C|\chi_h|}{\sin\vartheta} \sqrt{\frac{|\gamma_h|}{\gamma_o}} \quad . \tag{8.55}$$

Comparing this value of R_i with integrated reflection in the Laue case according to (3.82), we notice that in the Bragg case integrated reflection is twice as large.

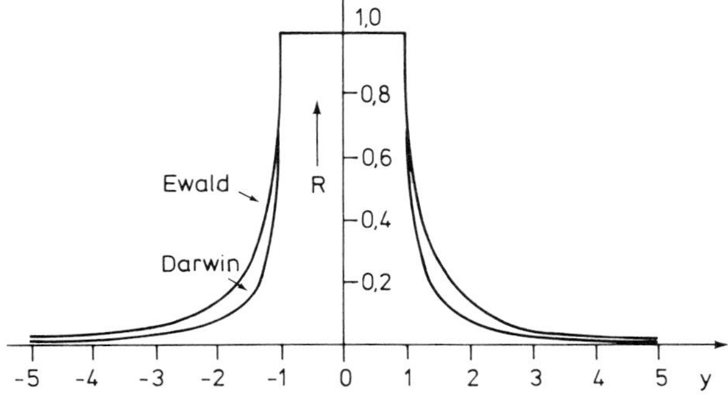

Fig. 8.2 Maximum of Bragg reflection curves as calculated by DARWIN's and EWALD's formulae (after ZACHARIASEN [1.14])

Figure 8.2 depicts the shapes of the maxima corresponding to DARWIN's and EWALD's formulae. The difference in these shapes finds its expression in the difference of the half-widths. Using the condition $R_{I,III} = 1/2$ we find, from (8.50) and (8.49),

$$w_D = \Delta y_{\frac{1}{2}} = \frac{3\sqrt{2}}{2} \quad , \tag{8.56a}$$

$$w_E = \frac{4\sqrt{3}}{3} \quad . \tag{8.56b}$$

8.3 Bragg Reflection from Thick Absorbing Crystal

The expressions for R and T in this case can be obtained either from the general equations (8.14) and (8.16) or from ZACHARIASEN's equation (8.28).

We will first give the extended version of the expression for the complex variables x and v in (8.14) and (8.16).

According to (8.11) and (8.6)

$$x = -i\pi Kt \frac{W}{2|\gamma_h|} \quad , \tag{8.57}$$

where the precise and approximate values of W_r and W_i for the Bragg case are given by (7.44 - 47). Note that from (8.57)

$$x_r = \frac{\pi Kt}{2|\gamma_h|} W_i \quad , \quad x_i = -\frac{\pi Kt}{2|\gamma_h|} W_r \quad . \tag{8.58}$$

We will write out the approximate values of x_r and x_i, using (7.47)

$$x_r = \frac{\mu t}{4\sqrt{y_r^2 - 1}}\left[y_r\left(\frac{1}{\gamma_o} + \frac{1}{|\gamma_h|}\right) - \frac{2\varepsilon \cos v_h}{\sqrt{\gamma_o|\gamma_h|}}\right], \tag{8.59}$$

$$x_i = -A_r\sqrt{y_r^2 - 1} \quad , \quad A_r = \frac{\pi K C t|\chi_{hr}|}{\sqrt{\gamma_o|\gamma_h|}} \quad . \tag{8.60}$$

Let us now transform the general equations (8.14) and (8.15) using the identity

$$|\sinh\alpha|^2 = \frac{1}{2}\left[\cosh(2\alpha_r) - \cos(2\alpha_i)\right] \quad . \tag{8.61}$$

The expressions for R and T take the form

$$R = \left|\frac{\chi_h}{\chi_{\bar{h}}}\right| \frac{\cosh(2x_r) - \cos(2x_i)}{\cosh(2x_r + 2v_r) - \cos(2x_i + 2v_i)} \tag{8.62}$$

$$T = \frac{\cosh(2v_r) - \cos(2v_i)}{\cosh(2x_r + 2v_r) - \cos(2x_i + 2v_i)} \quad . \tag{8.63}$$

Equations (8.62) and (8.63) are applicable throughout the maximum region and for crystals of arbitrary thickness as well. It is interesting to consider the limiting case of the infinitely thick absorbing crystal, because in this case (8.62) is considerably simplified. Indeed, with an increase in crystal thickness we can replace the hyperbolic cosines in (8.62) by exponents according to the equations

$$\cosh(2x_r) \underset{x_r \to \infty}{\approx} \frac{1}{2}\exp 2x_r \quad ; \quad \cosh(2x_r + 2v_r) \underset{x_r \to \infty}{\approx} \frac{1}{2}\exp 2x_r \exp 2v_r \tag{8.64}$$

and neglect the trigonometric cosines. As a result, after cancelling by $\exp(2x_r)$ we get

$$R_\infty = \left|\frac{\chi_h}{\chi_{\bar{h}}}\right| \exp(-2v_r) = \left|\frac{\chi_h}{\chi_{\bar{h}}}\right| \left(L - \sqrt{L^2 - 1}\right) \quad . \tag{8.65}$$

We have introduced the symbol $L = \cosh(2v_r)$ here. The transmission coefficient in this limiting case tends to zero, because it contains a large value, exp(2x), only in the denominator. Eq. (8.65) was obtained by a number of authors in different ways. Before we discuss the various investigations, we will express the value $L = \cosh(2v_r)$ in terms of y. We proceed from the identity

$$L = \cosh(v + v^*) = \cosh v \cosh v^* + \sinh v \sinh v^* = |\cosh v|^2 + |\sinh v|^2 \qquad (8.66)$$

whence, with an allowance for the definition of y in (8.7), we obtain

$$L = |y|^2 + |y^2 - 1| = \sqrt{1+(y_r^2 + y_i^2)^2 - 2(y_r^2 - y_i^2)} + y_r^2 + y_i^2 \quad . \qquad (8.67)$$

We will now write down the expressions for y_r and y_i in the most general case of a crystal without a centre symmetry with an arbitrary ratio $x = |\chi_{hi}|/\chi_{hr}|$ [8.4]. In distinction to (4.42) and (4.43), we will write down the complex values χ_{hr} and χ_{hi} as follows

$$\chi_{hr} = \chi'_{hr} + i\chi''_{hr} \quad , \quad \chi_{hi} = \chi'_{hi} + i\chi''_{hi} \qquad (8.68)$$

where, according to (4.37) and (4.36),

$$\chi'_{hi} = -r \sum_j \mu_{aj}^{(h)} \cos\varphi \quad ; \quad \chi''_{hi} = -r \sum_j \mu_{aj}^{(h)} \sin\varphi$$

$$\chi'_{hr} = -\Gamma \sum_j f_{rj} \cos\varphi \quad , \quad \chi''_{hr} = -\Gamma \sum_j f_{rj} \sin\varphi \quad . \qquad (8.69)$$

In this case

$$\chi_h = \chi'_{hr} - \chi''_{hi} + i(\chi''_{hr} + \chi'_{hi}) \quad ,$$

$$|\chi_h|^2 = |\chi_{hr}|^2(1 + x^2 + 2s) \qquad (8.70)$$

where

$$s = \frac{\chi''_{hr}\chi'_{hi} - \chi'_{hr}\chi''_{hi}}{|\chi_{hr}|^2} \quad . \tag{8.71}$$

Similarly,

$$\chi_{\bar{h}} = \chi'_{hr} + \chi''_{hi} - i(\chi''_{hr} - \chi'_{hi}) \quad ; \quad |\chi_{\bar{h}}|^2 = |\chi_{hr}|^2(1 + x^2 - 2s) \quad . \tag{8.72}$$

According to (8.70) and (8.72), we obtain

$$\chi_h\chi_{\bar{h}} = |\chi_{hr}|^2(1 - x^2 + i2p) \quad ;$$

$$|\chi_h\chi_{\bar{h}}| = |\chi_{hr}|^2\sqrt{(1 + x)^2 + 4p^2} \tag{8.73}$$

where (see (8.68)):

$$p = \frac{\chi'_{hr}\chi'_{hi} + \chi''_{hr}\chi''_{hi}}{|\chi_{hr}|^2} = x \cos\nu_h \quad . \tag{8.74}$$

Note that the variables Φ_h and Ψ_h used in (4.52) are equal to

$$\Phi_h = |\chi_{hr}|^2(1 - x^2) \quad ; \quad \Psi_h = 2p|\chi_{hr}|^2 \quad . \tag{8.75}$$

The values of y_r and y_i are calculated thus

$$y = \frac{\beta}{2C\sqrt{\chi_h\chi_{\bar{h}}}\ \sqrt{|\gamma_h|/\gamma_o}} = y_r + iy_i \quad . \tag{8.76}$$

Introduce the variables

$$y_z = -\frac{\beta_r}{2C|\chi_{hr}|\sqrt{|\gamma_h|/\gamma_o}} \quad ; \quad g = -\frac{\beta_i}{2C|\chi_{hr}|\sqrt{|\gamma_h|/\gamma_o}} \quad . \tag{8.77}$$

Then (8.76) takes the form

$$y = -(y_z + ig)[(1 - x^2) + 2ip]^{-\frac{1}{2}} \tag{8.78}$$

from which follows

$$|y|^2 = (y_z^2 + g^2)/\sqrt{(1 - x^2)^2 + 4p^2} \ . \tag{8.79}$$

Calculating further $|y^2 - 1|$, we obtain the general expression for L

$$L_c = \frac{\sqrt{(y_z^2 - g^2 - 1 + x^2)^2 + 4(y_z g - p^2)} + y_z^2 + g^2}{\sqrt{(1 - x^2)^2 + 4p^2}} \ . \tag{8.80}$$

The ratio $|\chi_h|/|\chi_{\bar{h}}|$ is readily calculated with the aid of (8.70) and (8.73), and the final expression for R_∞ has the form

$$R_{\infty c} = \frac{1 + x^2 + 2s}{\sqrt{(1 - x^2)^2 + 4p^2}} \left(L_c - \sqrt{L_c^2 - 1}\right) \tag{8.81}$$

Eqs. (8.80) and (8.81) solve the problem. The expressions (8.80) and (8.81) were derived in the paper of COLE and STEMPLE [8.4].

It is obvious that for a crystal with a symmetry center s = 0 and $p = x$. In this case (8.80) and (8.81) are somewhat simplified. We have

$$L_H = \frac{\sqrt{(y_z^2 - g^2 - 1 + x^2)^2 + 4(y_z g - x)^2} + y_z^2 + g^2}{1 + x^2} \tag{8.82}$$

$$R_{\infty H} = \left(L_H - \sqrt{L_H^2 - 1}\right) \ . \tag{8.83}$$

Eqs. (8.82) and (8.83) were given in the paper of HIRSCH and RAMACHANDRAN [8.5]. In the case of a weakly absorbing crystal, when $x \ll 1$, we can neglect the terms of the order of x^2. Here, for a crystal with a symmetry center, (8.82) is simplified still further:

$$L_Z = \sqrt{(y_z^2 - g^2 - 1)^2 + 4(y_z g - x)^2} + y_z^2 + g^2 \tag{8.84}$$

$$R_{\infty Z} = \left(L_Z - \sqrt{L_Z^2 - 1}\right) \ . \tag{8.85}$$

Eqs. (8.84) and (8.85) are given by ZACHARIASEN [1.14].

It should be noted that (8.80) and (8.85) can also be obtained from ZACHARIASEN's general equation (8.28). This equation should be regarded as a general expression applicable to crystals of any thickness. In the case of thick crystals, aw >> 1 (see (8.23)). Hence, we can discard all the terms which do not contain aw and replace $\sinh^2$ aw by 1/2 exp(2aw), and sinh (2aw) by 1/2 exp(2aw). After cancelling by exp(2aw) we get

$$R_{\infty} = \frac{(\gamma_o/|\gamma_h|)C^2|\chi_h|^2}{|q + z^2|+|z|^2+\sqrt{(|q + z^2|+|z|^2)^2 - q^2}} \quad . \tag{8.86}$$

From this equation it is already easy to obtain (8.80), (8.81), and the subsequent ones.

BONSE [7.3] uses for calculating the reflection coefficient the values of the amplitude ratio inside the crystal (see (3.11, 15)) for the Bragg case

$$c^{(i)} = \frac{D_h^{(i)}}{D_o^{(i)}} = \sqrt{\frac{\gamma_o}{|\gamma_h|}}\sqrt{\frac{\chi_h}{\chi_{\bar{h}}}}(y \pm \sqrt{y^2 - 1}) \quad . \tag{8.87}$$

In this equation, as usual, the upper sign in the parentheses refers to the first field and the lower one to the second. The use of the values $c^{(i)}$ for the reflection coefficients of vacuum waves is justified in the Bragg case by the fact that (as shown in Chapter 7) in an infinitely thick absorbing crystal, there is only one field in each of the three maximum regions.

Thus, similarly to (7.11), the following equalities hold good:

$$\frac{D_h^{(a)}}{D_o^{(a)}} = \frac{D_h^{(i)}}{D_o^{(i)}} \quad , \quad i = 1 \text{ or } 2 \quad , \tag{8.88}$$

and the reflection coefficient is expressed by the simple equation

$$R_B = \frac{|\gamma_h|}{\gamma_o}\left|\frac{D_h^{(i)}}{D_o^{(i)}}\right|^2 = \left|\frac{\chi_h}{\chi_{\bar{h}}}\right|\left|y \pm \sqrt{y^2 - 1}\right|^2 \quad . \tag{8.89}$$

Equation (8.89) has a slightly different form than (8.65); it is easy to see that the two equations yield identical results. Indeed, putting $y = \cosh v$, we have

$$R_{\infty} = \left|\frac{\chi_h}{\chi_{\bar{h}}}\right| |\cosh v - \sinh v|^2 = \left|\frac{\chi_h}{\chi_{\bar{h}}}\right| \exp(-2v_r) \quad . \tag{8.90}$$

The sign in front of sinh v is easy to determine, keeping in mind that $v_r > 0$; also, the inequality $R < 1$ is physically obvious.

While (8.80 - 85) refer to the case of the infinitely thick crystal, (8.62) is more general and valid for crystals of arbitrary thickness. For calculations by (8.62), along with x_r and x_i, which are determined by (8.57 - 60), it is necessary to calculate the values of v_r and v_i. When deriving expressions for the complex value v we use the identity

$$y = \cosh v = \cosh(v_r + iv_i) = \cosh v_r \cos v_i + \sinh v_r \sin v_i \tag{8.91}$$

$$y_r = \cosh v_r \cos v_i \quad , \quad y_i = \sinh v_r \sin v_i \quad . \tag{8.92}$$

As in the case of the transparent crystal, we assume that $v_r > 0$ throughout the maximum region. Then, in region III, $y_r > 1$, $\cos v_i > 0$, and, hence, $-\pi/2 < v_i < \pi/2$, and in region I, $y_r < -1$; therefore $\pi/2 < v_i < 3\pi/2$. For a weakly absorbing crystal, in regions far removed from the centre of the maximum, v_r is large, and $v_i \approx 0$ in region III and $v_i \approx \pi$ in region I. Therefore, for approximate values of v_r and v_i we have the formulas

$$\cosh v_r \approx \pm y_r \quad , \quad \sin v_i \approx y_i/\sqrt{y_r^2 - 1} \quad . \tag{8.93}$$

In region II, on the contrary, $v_r \approx 0$, and v_i changes abruptly from 0 to π on transition from region III to region I. In this case the approximate equations have the form

$$\cos v_i \approx y_r \quad , \quad \sinh v_r \approx y_i/\sqrt{1 - y_r^2} \quad . \tag{8.94}$$

Note that the role of v in (7.12) is played here by v_i, not by v_r as in regions I and III.

In the case of thin absorbing crystals, an important feature of (8.62) is the presence of subsidiary maxima, as has already been noted in Section 8.2. As a result of absorption, at the minima corresponding to (8.40) the function R does not vanish to zero. Another important property of (8.62) is asymmetry about the midpoint ($y_r = 0$). This can be demonstrated as follows.

For simplicity's sake, we will consider symmetrical reflection. Transform (8.59), putting $v_r \gg 1$:

$$x \approx \frac{\mu t}{2 \sin\vartheta} - \frac{\mu t \varepsilon \cos\nu_n}{2y_r \sin\vartheta} \quad . \tag{8.95}$$

When (8.40) is fulfilled,

$$2x_i = -2A\sqrt{y_r^2 - 1} = 2\pi m \tag{8.96}$$

and we obtain in the numerator of (8.62) $\cosh(\alpha - \beta)-1$ for $y_r > 0$ and $\cosh(\alpha + \beta)-1$ for $y_r < 0$, where

$$\alpha = \mu t/\sin\vartheta \quad , \quad \beta = \mu t\varepsilon \cos\nu_n/2y_r \sin\vartheta \quad . \tag{8.97}$$

This asymmetry results from the asymmetry of the absorption coefficient.

As the crystal thickness increases, as a result of absorption the wave field penetrating into the crystal in regions I and III of the maximum damps out, and the reflection maximum, acquires the characteristic Darwin shape with a flat top and smoothly falling away lateral parts (see Fig. 7.3).

The consideration of the shape of the Bragg maximum in the absorbing crystal amounts to an analysis of transformation of the Darwin curve in relation to the parameters $x = |\chi_{hi}|/|\chi_{hr}|$ and

$$g = -\frac{\chi_{oi}\left(1 + \frac{|\gamma_h|}{\gamma_o}\right)}{2C|\chi_{hr}|\sqrt{\frac{|\gamma_h|}{\gamma_o}}}$$

for a crystal with a centre symmetry; for a crystal without a centre of symmetry in relation to x, g, p, s (8.83, 85, 81); (see [8.5 - 7]).

Figure 8.3 shows some typical shapes of maxima at various values of x and g plotted in accordance with calculations using (8.83). Note that the shapes of the maximum at $x > 0$ and $x < 0$ for a given value of x are mirror-symmetrical about the ordinate axis. It can be seen from the curves that an increase in x leads to a stronger asymmetry of the maximum and an increase in g leads to a decrease in the area of the maximum or of integrated reflection. At values $x = 0.1 - 0.2$ and $g < -0.3$, one (right-side) edge of the asymmetrical maximum remains almost unchanged with respect to the Darwin curve. This feature of the reflection curves in the case of the absorbing crystal can be explained on the basis of an analysis of the wave field in the crystal. It has been shown previously that the relative positions of the planes of nodes and antinodes of the wave field, on the one hand, and of the atomic planes, on the other, are different in regions I and III, changing smoothly from one edge to the other in the total-reflection

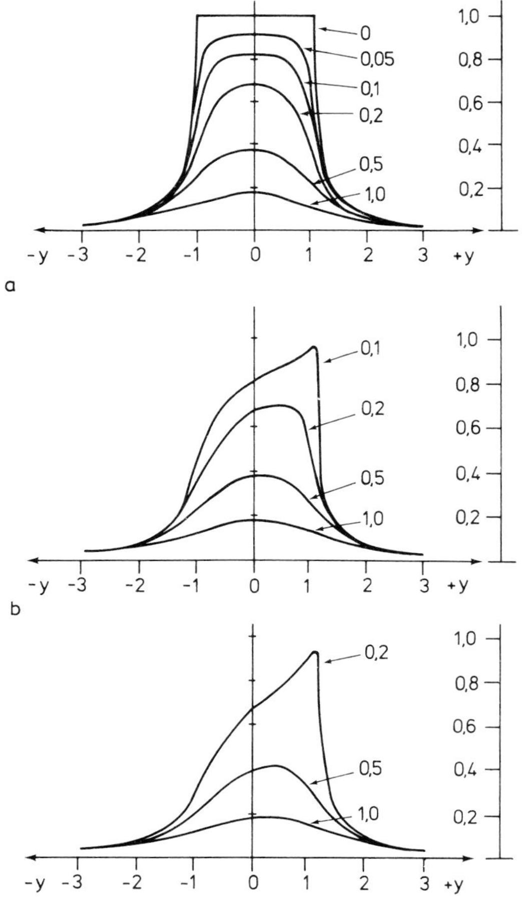

Fig.8.3a-c Curves of Bragg reflection from absorbing crystal with centre of symmetry corresponding to a set of x and g values (after HIRSCH and RAMACHANDRAN [8.5]) a) $x = 0$; $-g = 0$; 0.05; 0.1; 0.2; 0.5; 1.0. b) $x = 0.1$; $-g = 0.1$; 0.2; 0.5; 1.0. c) $x = 0.2$; $-g = 0.2$; 0.5; 1.0

region. If we assume that in region I at $y_r < -1$ the atomic planes coincide with the node planes of the wave field, then on this "left-hand" side the maximum drops drastically and the "left-hand" edge of region II will correspond to $R < 1$. In accordance with the displacement of the node planes, an increase in y diminishes the absorption effect and the remaining part of the maximum shows a significant increase in reflection power coefficients.

The wide range of the parameters g and x presented in Fig. 8.3 corresponds to poorly investigated scattering conditions, for instance in the immediate vicinity of the absorption edge or to extremal wavelengths of the X-ray range.

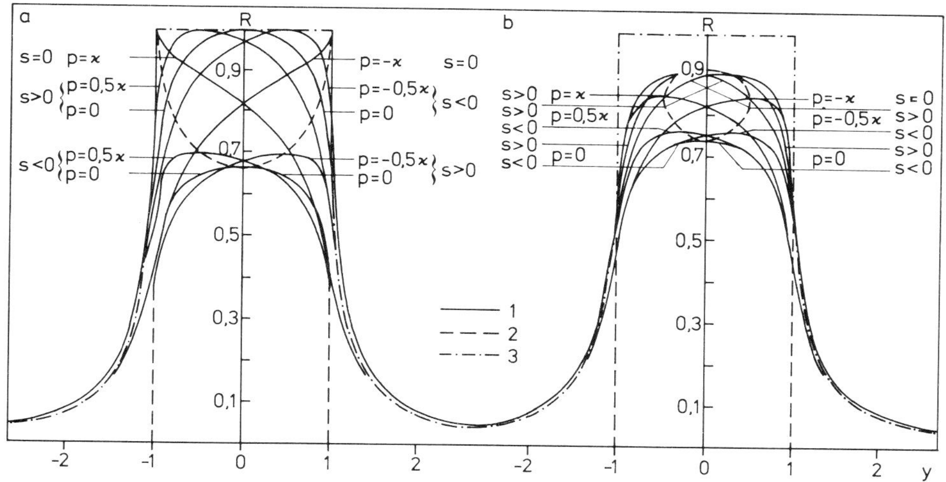

Fig. 8.4 a and b Curves of Bragg reflection from absorbing crystal without centre of symmetry (after BUCKSCH et al. [8.7]). a) $x = -g = 0.1$; b) $x = -g/2 = 0.05$. I: curves R; 2: limiting values of R_{max}; 3: Darwin curve for semi-infinite crystal

Figure 8.4 illustrates changes in the shape of the maximum for the reflection from a crystal without a centre of symmetry (the parameters p and s are given by (8.74) and (8.71)).

Many details of such variations in reflection curves can be interpreted with the aid of the model of a crystal in which the effects of scattering and, hence, of extinction, on the one hand, and the effects of absorption, on the other, are associated with two different scattering centres distributed in a definite way in the structure [8.7].

8.4 Integrated Reflection from Absorbing Crystal in Bragg Case

In spite of the fact that the dynamical theory, as applied to Bragg reflection, has been developing for about 60 years, the analytical expression for integrated reflection in the most complex case of a thick absorbing crystal was derived quite recently in the paper of AFANASIEV and PERSTNEV [8.8].

We will proceed from Zachariasen's equation for the reflection power coefficient (8.104), which, with the aid of the new variable L_A (see also (8.17))

$$L_A = \frac{|q + z^2| + |z|^2}{|q|^2} =$$

$$= \frac{\left|\gamma_o\alpha - \frac{1}{2}\chi_o(\gamma_o + |\gamma_h|)\right|^2 + \left|\left[\gamma_{o\alpha} - \frac{1}{2}\chi_o(\gamma_o + |\gamma_h|)\right]^2 - C\gamma_o|\gamma_h|\chi_h\chi_{\bar{h}}\right|}{C^2\gamma_o|\gamma_h||\chi_h\chi_{\bar{h}}|} \quad (8.98)$$

$$2\alpha = 2\eta \sin 2\vartheta \quad ,$$

is reduced to the form

$$R_i^\eta = \frac{1}{\sin 2\vartheta}\left|\frac{\chi_h}{\chi_{\bar{h}}}\right| \int_{-\infty}^{\infty}\left(L_A - \sqrt{L_A^2 - 1}\right)d\alpha \quad , \quad (8.99)$$

or

$$R_i = \frac{C|\chi_h|}{\sin 2\vartheta}\sqrt{2\,\frac{|\gamma_h|}{\gamma_o}\left|\frac{\chi_h}{\chi_{\bar{h}}}\right|}\int_{L_o}^{\infty}\frac{(L_a - \sqrt{L_A^2 - 1})\sqrt{L_A - L_o}}{L_A - L_o + 2s^2}\,dL_A \quad (8.100)$$

and then, introducing the variables

$$L_A - L_o = 2s^2(1 - q^2)\,\frac{z^2}{1 - z^2} \quad ,$$

$$s = \frac{\gamma_o + |\gamma_h|}{2\sqrt{\gamma_o|\gamma_h|}}\;\frac{|\chi_{oi}|}{|C|\sqrt{|\chi_h\chi_{\bar{h}}|}} \quad , \quad q = \frac{2\sqrt{\gamma_o|\gamma_h|}}{\gamma_o + |\gamma_h|}\;\frac{|C|\,\mathrm{Im}\sqrt{\chi_h\chi_{\bar{h}}}}{|\chi_{oi}|}$$

$$(8.101a)$$

$$k = \frac{1}{1 + s^2(1 - q^2)} \quad , \quad L_o = 1 + 2s^2(1 - q^2) \quad .$$

We obtain

$$R_i = \frac{8}{3\sin 2\vartheta}\,C|\chi_{hr}|\sqrt{\frac{|\gamma_h|}{\gamma_o}}\sqrt{\left|\frac{\chi_h}{\chi_{\bar{h}}}\right|}\,P(s,q) \quad , \quad (8.101b)$$

where the integrals to be evaluated have the form

$$P(s,q) = \frac{3\sqrt{1-k^2}}{2k^3(1-q^2)}(I_1 + I_2) \quad ;$$

$$I_1 = \int_0^1 [2 - k - k^2z^2 - 2\sqrt{(1-k^2)(1-k^2z^2)}] \frac{q^2dz}{(1-q^2z^2)\sqrt{1-z^2}} , \quad (8.102)$$

$$I_2 = \int_0^1 [2 - k^2 - k^2z^2 - 2\sqrt{(1-k^2)(1-k^2z^2)}] \left[\frac{1-q^2}{(1-z^2)^{5/2}} - \right.$$

$$\left. - \frac{1}{(1-z^2)^{3/2}}\right]dz \quad . \quad (8.103)$$

I_1 can be expressed directly through complete elliptical integrals of the first type K and of the third type Π [8.9]:

$$I_1 = \frac{\pi}{2}k^2 + \frac{\pi}{2\sqrt{1-q^2}}[2q^2 - k^2(1+q^2)] + 2\sqrt{1-k^2} \cdot$$

$$\cdot [(k^2 - q^2)\Pi(-q^2k) - k^2K(k)] \quad . \quad (8.104)$$

I_2 is reduced to standard elliptical integrals after integration by parts:

$$I_2 = \int_0^1 [2 - k^2 - k^2z^2 + 2\sqrt{(1-k^2)(1-k^2z^2)}] \cdot$$

$$\cdot d\left\{\frac{z}{3}\left[\frac{1-q^2}{(1-z^2)^{3/2}} - \frac{1+2q^2}{\sqrt{1-z^2}}\right]\right\} = \frac{2}{3}k^2[(1-q^2)I_3 + I_4] \quad , \quad (8.105)$$

where integrals I_3 and I_4 are reduced to the elliptical integrals

$$I_3 = \frac{E(k)}{\sqrt{1-k^2}} - \sqrt{1-k^2}K(k) \quad ,$$

$$I_4 = (1-q^2)\left[\sqrt{1-k^2}K(k) - \frac{\pi}{2}\right] + \quad (8.106)$$

$$+ (1+2q^2) + \left\{\frac{\sqrt{1-k^2}}{k^2}[K(k) - E(k)] - \frac{\pi}{4}\right\} \quad .$$

In (8.106), E is the complete elliptical integral of the second type.

The final function P(s,q) takes the following value:

$$P(s,q) = (1 - s^2 - 2q^2s^2)\frac{E(k)}{k} - \frac{3\pi}{4}s(1 - 2q^2s^2) + \\ + ks^2(1 - q^2)[3(1 - q^2s^2)\Pi(-q^2k) - \\ - (2 - s^2 - 2q^2s^2)K(k)] \quad . \tag{8.107}$$

An important feature of the calculation performed in this work is the general nature of the complex parameters χ_o and χ_h, $\chi_{\bar{h}}$, which appear in the expression for the integration variable z (8.101) via q, s, and L. No restrictions associated with the ratio of the values $|\chi_{hi}|$ and $|\chi_{hr}|$, as well as $|\chi_{oi}|$ and $|\chi_{or}|$, or with the presence or absence of a centre of symmetry in the crystal-scatterer are imposed here. For the product $\chi_h\chi_{\bar{h}}$ we can use either the general expressions (8.73) or a more special one, for instance, the expression corresponding to crystals with a symmetry center. In this simpler case, the main parameters s and q take the following values:

$$s_s = -\frac{|\chi_{oi}|[1+(|\gamma_h|/\gamma_o)]}{2C(|\gamma_h|/\gamma_o)^{\frac{1}{2}}|\chi_h|} = g\sqrt{1 + x^2} \quad ,$$

$$q_s \approx \frac{2C(|\gamma_h|/\gamma_o)^{\frac{1}{2}}}{[1+(|\gamma_h|/\gamma_o)]} \quad \varepsilon = \frac{x}{g} \quad , \quad x \approx s_sq_s \quad , \tag{8.108}$$

where g is determined from (8.77), $x = |\chi_{hi}|/|\chi_{hr}|$ and $\varepsilon = |\chi_{hi}|/|\chi_{oi}|$.

Integrated reflection in the general case can be tabulated with the aid of (8.101) and (8.107) and the elliptical-integral tables. However, since the attention is now focused on cases of reflection at (relatively) small χ_{oi} and χ_{hi} and, hence, $s \approx g \ll 1$, $k \approx 1$ with $q \leq 1$, the function P(s,q) can be simplified. Using the expression for complete elliptical integrals of the first and second type and the expansion

$$(1 - q^2)\Pi(-q^2,k) \approx \frac{1}{2}\ln\left(\frac{8}{1 + q}\,\frac{1 - q}{1 + k}\right) + \frac{1 - q}{4}\ln\frac{1 + q}{1 - q} \quad ,$$

$$(1 - k) \ll 1 \quad , \tag{8.109}$$

we can obtain

$$P(s,q) \approx 1 - \frac{3\pi}{4} s(1 - 2q^2s^2)+3s^2 \left[\frac{1 + q^2}{2} \ln \frac{4}{s(1 + q)} - \right.$$

$$\left. - \frac{1 + 3q^2}{4} + \frac{(1 - q)^2}{4} \ln \frac{1 + q}{1 - q}\right] \quad . \tag{8.110}$$

This expression yields, for $s \leq 0.2$, a result differing by no more than ~ 1 percent from the accurate one. Finally, for $s \lesssim 0.05$, the main term in (8.110) is equal to

$$P(s,q) \approx 1 - \frac{3\pi}{4} s \tag{8.111}$$

and integrated reflection is found to be

$$R_i^\eta = \frac{8}{3} \frac{C|\chi_{hr}|}{\sin 2\vartheta} \sqrt{\frac{|\gamma_h|}{\gamma_o}} \left|\frac{\chi_h}{\chi_{\bar{h}}}\right| (1 - 2{,}355|g|) \quad , \tag{8.112}$$

which coincides almost precisely with the empirical equation of HIRSCH and RAMACHANDRAN proposed by them for the case $g < 0.1$, and is also similar to ZACHARIASEN's formula $R_i \approx 8/3(1 - 2|g|)$. Finally, if we assume $\chi_{oi} = 0$ in the expression for s (8.100), then (8.112) yields DARWIN's equation (8.53).

The paper reviewed [8.8] lists the values of integrated reflection over wide ranges of parameters q from 0.1 to 1 and s from 0.2 to 5 (Table 8.1). DE MARCO and WEISS [3.4] calculated the values of R_i^y by computerized integration of (8.83) for a wide range of parameters g from 0 to 3 and x from 0 to 1. The great number of the calculated values (1291) and the small ranges of these parameters reduce the error to ~ 0.1 percent when applying interpolation (see Appendix B).

Using this table, we can proceed from values of g and in any approximation corresponding to the particular example and the required calculation accuracy. At the same time, in considering reflections from a crystal without a centre of symmetry, in accordance with (8.80) and (8.81), the use of the two indicated parameters alone is, generally speaking, insufficient for determining the value of R_i^y, although in some cases this can be done with satisfactory accuracy. Appendix B reproduces the result of DE MARCO and WEISS' calculation. The small table corresponds to a definite range of $g = x$.

Table 8.1 Function P(s,q) for various s and q

q	s							
	0.20	0.25	0.30	0.35	0.40	0.45	0.50	0.60
1.0	0.716	0.672	0.633	0.599	0.568	0.541	0.516	0.473
0.9	0.708	0.661	0.620	0.584	0.552	0.523	0.496	0.451
0.7	0.696	0.646	0.602	0.564	0.529	0.499	0.471	0.423
0.5	0.687	0.636	0.591	0.551	0.515	0.484	0.455	0.406
0.3	0.682	0.630	0.583	0.543	0.507	0.475	0.446	0.397
0.1	0.680	0.627	0.580	0.539	0.503	0.470	0.442	0.392

q	s							
	0.70	0.80	0.90	1.00	1.50	2.00	3.00	5.00
1.0								
0.9	0.412							
0.7	0.383	0.349						
0.5	0.366	0.332	0.304	0.280				
0.3	0.356	0.323	0.295	0.271	0.190			
0.1	0.352	0.318	0.290	0.266	0.187	0.143	0.097	0.059

As distinct from the data of Appendix B, the comparatively simple expression (8.110) permits calculating the value of integrated reflection for any general case of a crystal without a centre of symmetry over a wide range of parameters $|\chi_{oi}|$, $|\chi_{hi}|$, and $|\chi_{hr}|$, and also carrying out a qualitative analysis of the expected behaviour of R_i.

A particular problem which is associated with calculation of integrated reflection and is of fundamental interest was solved in WEISS' paper [8.10], namely a comparison of (8.50) (DARWIN's solution) and (8.49) (EWALD's solution) for the reflection coefficient in the case of the thick "nonabsorbing" crystal.

We proceed from the most general expression of ZACHARIASEN (8.28). We will give the values of the arguments $a\ell$ and aw in explicit form. On the basis of (8.17), (8.77), and (8.78), we can calculate them most accurately by using the rule of taking the root of a complex number:

$$a\ell = A(q + z^2)^{\frac{1}{2}}_r \quad , \quad aw = A(q + z^2)^{\frac{1}{2}}_i \quad , \tag{8.113}$$

$$(q + z^2)^{\frac{1}{2}} = C|\chi_r| \sqrt{\frac{\gamma_o}{|\gamma_h|}} \; [(y_r^2 - g^2 - 1 - x^2)+i2(y_z g - x)]^{\frac{1}{2}} \quad , \qquad (8.114)$$

$$a\ell = A\{[(y_r^2 - g^2 - 1 + x^2)^2+4(y_z g - x)^2]^{\frac{1}{2}} + +(y_r^2 - g^2 - 1 + \;^2)\}^{\frac{1}{2}} \quad ,$$

$$aw = A\{[(y_r^2 - g^2 - 1 + x^2)^2+4(y_z g - x)^2]^{\frac{1}{2}} - -(y_r^2 - g^2 - 1 + x^2)\}^{\frac{1}{2}} \quad . \qquad (8.115)$$

$$A = \frac{KC|\chi_{hr}|\pi t}{(\gamma_o + |\gamma_h|)} \sqrt{\gamma_o|\gamma_h|} \quad . \qquad (8.116)$$

EWALD's approximation can be formulated as the condition g = 0, and DARWIN's approximation, as aw >> 1. The computer was used to calculate the value $R_i^y = \int_{-\infty}^{\infty} R dy$; R was taken according to (8.28). R_i^y was calculated as a function of the thickness $\mu t = 2gA$.
For $g(x) = x$ WEISS assumed the value 0.01 (Fig. 8.5).

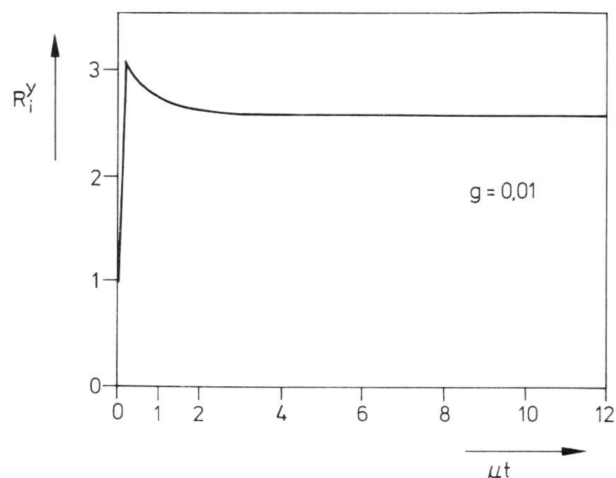

Fig.8.5 Change of integrated reflection from crystal with negligibly small absorption vs thickness (kinematical scattering range, EWALD and DARWIN dynamical solution (after WEISS [3.4])

The curve of R_i^y shows that with very small thicknesses ($\mu t < 0.001$) we have a kinematic region, at $\mu t \approx 0.03$ we approach EWALD's solution. At the maximum, $R_i^y = \pi$, which precisely corresponds to EWALD's approximation. As the crystal thickness increases, the curve asymptotically approaches the value $R_i^y = 8/3$, i.e., DARWIN's solution. This value of integrated reflection is reached at $\mu t > 8$; i.e., it actually corresponds to a thick crystal with vanishingly small absorption.

Numerical estimation of the obtained result, as applied to a particular experiment, can be illustrated for the case of symmetrical reflection 220 of Ag $K\alpha_1$ radiation from a Si crystal. Here,

$$\mu = 11.2\ \mathrm{cm}^{-1} \quad , \qquad |\chi_{hr}| = 1.266\cdot 10^{-6} \quad ,$$

$$|\chi_{oi}| = 1\cdot 10^{-8} \quad , \qquad |\chi|_{hi} = 0.6\cdot 10^{-8} \quad ,$$

$$g \approx 0.008 \quad , \qquad x \approx 0.005 \quad .$$

The maximum of the curve of Fig. 8.5 will be close to the silicon plate thickness t = 0.03 mm, and the value $\mu t = 8$ corresponds to a thickness of t = 7.5 mm.

We will now make a few general remarks concerning integrated reflection in the Bragg case. As follows from the foregoing, the value of integrated reflection is determined, above all, by the parameters g and x, as well as by the ratio $|\gamma_h|/\gamma_o$. We will first give the numerical estimates of the values of g_s for symmetrical reflection, and also of x, using Tables 4.4 and 4.5a.

The values of g_s in most cases increase for weak reflections and equal from ~ 0.05 for 111 to ~ 0.09 for 333 for Ge and Cu $K\alpha_1$ radiation, and from ~ 0.09 for 111 to ~ 0.16 for 333 and Mo $K\alpha_1$ radiation.

For Si and Cu $K\alpha_1$ radiation in the same range, they equal from ~ 0.04 to ~ 0.08, and for Mo $K\alpha_1$ radiation, from ~ 0.01 to ~ 0.02. The values of x for Ge and Cu $K\alpha_1$ radiation equal 0.03-0.06, for Mo $K\alpha_1$ $\sim$ 0.08-0.1. For Si, respectively, $\sim$ 0.04-0.05, and 0.01. The data of Table 8.1 and Appendix B indicate low sensitivity of R_i to variations in $|g|$ and x. An error of $\sim$ 1 percent in the measurement of R_i results in an error of $\sim$ 10-12 percent in the values of g_s. At the same time, the values of the parameters χ_{oi} and χ_{hr} given in Tables 4.4 and 4.5a are accurate to $\sim$ 0.2-0.3 percent. We must add that, in a certain range of disturbances in the ideal structure of the crystal-scatterer, the value of R_i may remain constant within the indicated limits, at about $\pm$ 0.5-1 percent. In any case, an increase in x and $|\chi_{hi}|$ implies increased interference "contribution" to absorption or extinction enhancement

$$\sigma_{e,s} = \frac{2\pi KC|\chi_h|}{\cos\vartheta} = \frac{2\pi KC|\chi_{hr}|(1 + x^2)^{\frac{1}{2}}}{\cos\vartheta} . \tag{8.117}$$

At the same time we must note incomparably lower sensitivity of integrated reflection to variation in parameters x or ε in the Bragg case as compared with the Laue case. In the corresponding equations for the Laue case, for instance (4.165), ε appears in the exponent of the absorption factor, and for strong reflections an increase or decrease in $|\chi_{hi}|$ changes the value of R_i by several orders.

An incomparably stronger effect on R_i in the Bragg case is exerted by the parameter g. One can show that the value of g, in accordance with (8.89), corresponds to the absorption/extinction ratio. Indeed, if we consider the path traversed by the incident and diffracted waves in the crystal, we obtain, for a certain effective depth t_a,

$$\sigma t_a = \mu(\gamma_o^{-1} + |\gamma_h|^{-1})t_a \quad . \tag{8.118}$$

On the other hand, extinction "absorption" at a certain depth of penetration t_e and $y_r \approx x \approx 0$ will be

$$\sigma_e t_e = \frac{2\pi KCt_e|\chi_{hr}|}{(\gamma_o|\gamma_h|)^{\frac{1}{2}}} \quad , \tag{8.119}$$

whence

$$\frac{\sigma}{\sigma_e} = \frac{t_e}{t_a} = \frac{2\pi K|\chi_{oi}|[1+(|\gamma_h|/\gamma_o)]}{2\pi KC|\chi_{hr}|(|\gamma_h|/\gamma_o)^{\frac{1}{2}}} = -2g > 0 \quad . \tag{8.120}$$

Thus, an increase in true absorption or a decrease in depth of penetration due to absorption leads to an increase in parameter g. Large extinction, i.e. a small depth of penetration of the wave field into the crystal due to interference effects, at low absorption, corresponds to a small g. In this connection we can note that in Fig. 8.3a a change in the shape of the curve at small values of $|g|$ affects only the perfect-reflection region, which illustrates the connection between this parameter and the extinction effect. We will also emphasize that the indicated interpretation of the parameter g makes it possible to formulate the physical conditions for transition from dynamical scattering to scattering on a mosaic crystal [8.5].

An interesting property of the function R_i $(|\gamma_h|/\gamma_o)$ (at $g \leq 0.2$) is the fact that it changes its value appreciably on transition from symmetrical to asymmetrical reflection. Indeed, comparing the respective values of g

$$g_s = \left|\frac{\chi_{oi}}{\chi_{hr}}\right| \quad , \quad g_{as} = \left|\frac{\chi_{oi}}{\chi_{hr}}\right| \left[\frac{1}{2} \left(\frac{\gamma_o}{|\gamma_h|}\right)^{\frac{1}{2}} + \left(\frac{|\gamma_h|}{\gamma_o}\right)^{\frac{1}{2}}\right] \quad , \tag{8.121}$$

we note that with a deviation from the value $\gamma_h/\gamma_o = 1$ to 4-5 or to 0.25-0.20 the value of g increases about 2-2.2 times (equally in both cases). This entails some decrease in R_i^y. But since

$$R_i^\eta = \frac{C|\chi_{hr}|}{\sin 2\vartheta} \left|\frac{\chi_h}{\chi_{\bar{h}}}\right| \left(\frac{|\gamma_h|}{\gamma_o}\right)^{\frac{1}{2}} R_i^y \quad ,$$

then, due to the effect of the factor $(\gamma_h/\gamma_o)^{\frac{1}{2}}$, the integrated reflection R_i ultimately either decreases still further (if $|\gamma_h| < \gamma_o$) or increases (if $|\gamma_h| > \gamma_o$). The increase may reach 70-80 percent, while the decrease is always greater.

With such an increase in g, to the above-indicated change in R_i by ~ 1 percent there will already correspond a change of only ~ 5 percent in $|g|$.

Comparing the values of integrated Bragg and Laue reflection, one can draw the following conclusions.

Integrated Bragg reflection power coefficient in similar cases exceeds Laue reflection ones more or less, sometimes by several orders, which is very important to designers of crystal monochromators.

On the other hand, the complicated nature of the dependence of R_i^y on such parameters as $|\chi_{hr}|$ and $|\chi_{hi}|$ minimizes the efficiency, not only of their experimental determination by measuring integrated reflections, but also of accurate and unambiguous verification of the theory.

9. X-Ray Spectrometers Used in Dynamical Scattering Investigations. Some Results of Experimental Verification of the Theory

The dynamical theory of X-ray scattering in perfect crystals outlined in the preceding chapters is based on two possible physical models of an incident vacuum wave, which excites a wave field in the crystal; most of the text is based on the incident-plane-wave approximation, and the remaining part, on the incident-spherical-wave approximation, or, more precisely, on an approximation assuming restriction of the front of the incident wave by a narrow slit (wave packet).

Turning to experimental investigations of dynamical scattering dealing with quantitative verification and utilization of the theory, we shall, first of all, consider the difficulties involved in attempting to realize the incident-plane-wave approximation. These difficulties are mainly due to the use of the characteristic radiation of the X-ray tube. Indeed, it is well known that the wavelength spread of this radiation, as well as its angular spread, considerably exceeds the corresponding parameters of the dynamical maxima, and, of course, of their fine structure (subsidiary maxima).

The devices which make it possible to transform the radiation of sources in the range of X-rays will be called monochro-collimators (MCC) because of their dual function, namely obtaining a beam which is as close as possible to a *plane monochromatic* wave. In the diffractometer[1], this beam or any other one is used to obtain the curves of reflection or transmission from single-crystal specimens of some shape or other. At present, as well as the above-mentioned conventional source, i.e., the characteristic radiation of the X-ray tube with a fixed anode, tubes with a rotating anode and, finally, synchrotron radiation, are being used in experiments. In their power these sources exceed the former by about three orders, which is in some respects of fundamental importance.

---

[1]In the subsequent text we shall also retain the traditional term spectrometer.

Indeed, the considerable decrease in the spectral and angular range attained by MCC when using sources of characteristic radiation of conventional X-ray tubes inevitably leads to a sharp decrease in initial intensity.

Note that the use of powerful radiation sources is above all important in topography; it makes this method effective for controlling various processes affecting the real structure of crystals in growing or annealing and in various conditions of use.

At the same time, MCC systems and various types of diffractometers can now measure, with a previously unattainable degree of sensitivity and accuracy, the quantitative parameters of the structure of highly perfect crystals and crystals with slight distortions.

Such information can be obtained by measuring the characteristics of experimental curves of reflection or transmission, which are also called rocking curves. Since a beam falling on a crystal differs more or less from a plane monochromatic wave, the rocking curves differ from theoretical ones calculated to this approximation. They can be represented as a convolution of the theoretical function $R(y)$ or $T(y)$ and some superfunction $C(\lambda,\alpha)$ referring to the beam which emerges from the MCC. As well as the application indicated and in connection with the procedure described in Chapter 6, the convolution of the theoretical functions with the functions corresponding to an incident wave packet or spherical wave is also of interest.

Much attention is given in this chapter to MCC which decrease the angular and spectral ranges of the beam falling on the crystal under investigation. This problem can also be formulated as the development of methods for decreasing the convolution effect and obtaining rocking curves with characteristics as close as possible to those of the theoretical curves.

We shall now give some estimates of the interferring effects due to the finite wavelength spread of the characteristic line and to its angular divergence.

9.1 Estimating Wavelength Spread and Angular Divergence of X-Ray Tube Radiation

When using one line or another of characteristic radiation as a source of a monochromatic wave, the natural wavelength spread of these lines $\Delta\lambda$ should be taken into account. Within this interval there is a certain distribution of intensity which can be approximated, for instance, by the Gauss distribution. The classical dispersion theory [1.6, 9.1] attributes the finite wavelength spread of the line to the attenuation of the electron oscillator. Detailed consideration of this problem leads to the following equation for the spectral width of the line at half-height:

$$w_\lambda \equiv \Delta\lambda = \frac{4\pi e^2}{3mc^2} = 0.118\ \mathrm{X} \qquad (\mathrm{X} = 10^{-11}\ \mathrm{cm}) \quad . \tag{9.1}$$

This result means that the wavelength spread is a universal constant and is, in particular, independent of the wavelength λ_0 of the middle of the range $\Delta\lambda$. Passing on from the wavelength spread to the effective oscillator life time τ (during which the line intensity reduces by a factor of e), we obtain

$$\tau = \frac{3mc}{8\pi^2 e^2}\lambda_0^2 \approx 4.5\ \lambda_0 \tag{9.2}$$

which yields, for instance, λ_0 = 1537 X, $\tau = 1.064\cdot 10^{-15}$ s for Cu $K\alpha_1$, and λ_0 = 708 X, $\tau = 2.26\cdot 10^{-16}$ s for Mo $K\alpha_1$.

It is well known, however, that the classical X-ray dispersion theory diverges considerably from the experiment in estimating the natural width of the spectral line; the universal character of (9.1) has not yet been confirmed, and the absolute value of $\Delta\lambda$ is, generally speaking, underrated. Quantum-mechanical consideration establishes a relationship between the width of spectral lines and that of the levels determining the corresponding transition:

$$(\Delta E)_{if} = h(\Delta\nu)_{if} = (\Delta E)_i + (\Delta E)_f \quad . \tag{9.3}$$

The numerical estimate of the values $(\Delta E)_{if}$ is actually made from experimental data. The values given in the monographs (for instance [9.1a] should be considered tentative. Since the measurements on a two-crystal spectrometer (see below) directly yield the half-width of the spectral line, we prefer to use these data. Thus, according to [9.1a], the $\Delta\lambda$ values for the $K\alpha_1$ line monotonically decrease from 1.60 X for Ca to 0.152 X for W; $\Delta\lambda$ for $L\alpha_1$, from 4.5 X for Ag to 0.88 X for U. New and more accurate data, although for a limited number of lines, are listed in Table 9.1 in connection with the analysis of the results obtained on a two-crystal spectrometer [9.2].

Reverting to the theoretical estimates, we note that the width of the spectral line

$$(\Delta E)_{if} \equiv \Gamma \approx \frac{h}{2\pi\tau} \quad , \tag{9.4}$$

Table 9.1 Experimental X-ray data on the natural half-width of the spectral lines w (in X units) obtained by measurements on two-crystal spectrometer [9.2]

Line	Reflecting crystal	w_+^a	w_-^b	w	Author
Cu $K\alpha_1$	Quartz $11\bar{2}0$	0.475	0.038	0.44	PARRATT
	Quartz $22\bar{4}0$	0.44	0.010	0.43	PARRATT
	Calcite 211	0.56	0.142	0.42	PARRATT
	Calcite 422	0.455	0.012	0.44	PARRATT
Ti $K\alpha_1$	Quartz $11\bar{2}0$	0.91	0.075	0.84	PARRATT
	Calcite 211	1.09	0.22	0.87	PARRATT
Fe $K\alpha_1$	Quartz $11\bar{2}0$	0.82	0.04	0.78	BROGREN
	Calcite 211	1.00	0.19	0.81	ALLISON
Mo $K\alpha_1$	Quartz $11\bar{2}0$	0.27	0.029	0.24	PARRATT
	Quartz $11\bar{2}0$	0.263	0.027	0.24	BROGREN
	Calcite 211	0.306	0.076	0.23	PARRATT

[a]Half-width according to scheme (1.1). [b]Half-width according to scheme (1,-1)

where τ is the lifetime of the state, is inversely proportional to the decay probability per unit of time.

For the shape of the spectral line, the classical theory supplies the following expression:

$$J_\omega = \frac{J_{\omega_0}}{1+\left[\frac{2(\omega - \omega_0)}{W_\omega}\right]^2} \quad . \tag{9.5}$$

Quantum-mechanical calculation yields the same equation to a first approximation. Here, however, the asymmetry of the effect of the chemical bond, and of other finer effects, is neglected. Eq. (9.5) is of special interest because it is similar to the equation describing the reflection curve obtained on a two-crystal spectrometer when the crystals are in the so-called antiparallel position (see below).

It is further obvious that a certain value of dispersion in the reflection from the given crystal corresponds to each given half-width $\Delta\lambda$ of the characteristic radiation line. This value is determined by differentiating the Bragg equation

$$\Delta\vartheta = \frac{\Delta\lambda}{2d\cos\vartheta} \quad . \tag{9.6}$$

For crystals of cubic symmetry and the simplest structural types (for instance, Si and Ge crystals) we can indicate the characteristic ratio between the values of dispersion $\Delta\vartheta$, according to (9.6), and the half-width of the reflection maxima

$$w_\vartheta = 2\Delta\eta_{\frac{1}{2}} = 2|\chi_{hr}|\sqrt{\gamma_h/\gamma_o}(\sin 2\vartheta)^{-1} \tag{9.7}$$

for increasing orders of reflection from a net of planes.

Indeed, in this case the dispersion $\Delta\vartheta$ increases, whereas the half-width w_ϑ reduces with the condition $(\gamma_h/\gamma_o) \ll 1$, i.e., with symmetrical reflection, or with asymmetrical reflection from planes with positive indices.

A different situation is observed with asymmetrical reflection from planes with negative indices.

Now we shall compare the numerical estimates of the values of dispersion and the half-widths of the maxima in dynamical reflection. Neglecting the absorption, in the symmetrical arrangement we obtain for Si, Cu $K\alpha_1$, radiation, and reflection 220: $\Delta\vartheta \approx 25''$, $w_\vartheta \approx 5''$; the same reflection with Mo $K\alpha_1$ radiation: $\Delta\vartheta \approx 13''$, $w_\vartheta \approx 2.2''$. For Ge, Cu $K\alpha_1$ radiation, reflection 220: $\Delta\vartheta \approx 24''$, $w_\vartheta \approx 11.8''$; reflection 333: $\Delta\vartheta \approx 57''$, $w_\vartheta \approx 3.9''$. With reflection 333, taking into account absorption and $\mu t = 0.35$, $w_\vartheta \approx 3.8''$.

At the same time it is significant that the dispersion of the rocking curves due to the indicated values of the spectral intervals of the characteristic radiation is far less important in diffractometry than is the angular divergence of the X-ray beams. This is due to the remarkable property of the so-called parallel arrangement of the crystals, which will be discussed in detail in the subsequent sections of this chapter.

We shall now discuss the angular divergence of X-ray tube radiation (see also [1.24]).

Let us formulate the conditions needed to apply the plane-incident-wave approximation, assuming that the tube radiation possesses sufficient monochromaticity. If we denote by Ω a (plane) angle inside which the radiation intensity is still appreciable, the first condition will be

$$w_\vartheta = 2\eta_{\frac{1}{2}} \gg \Omega \quad . \tag{9.8}$$

As for the width of the coherent front of the incident wave, the ratio between this width and the thickness of the crystal slice is, generally speaking, essential. But besides, when using the boundary conditions on the entrance surface for the tangential components of the wave vectors, it is necessary to prevent any possibility of diffraction events at the edges of the coherent region of the wave front or the slit. This condition is formulated thus:

$$w_{\vartheta} >> \lambda S^{-1} \quad , \tag{9.9}$$

where S is the effective transverse dimension of the entrance surface. The desired condition, can also be written

$$f >> S \quad , \tag{9.10}$$

where f is the linear width of the coherent front.

In an experiment where the crystal under investigation is hit by a beam of rays from the anode of an X-ray tube which has passed through a system of slits, (9.8) and (9.10) cannot be fulfilled. The value of Ω usually exceeds the angular half-width of the dynamical maximum by two orders.

9.2 Two-Crystal Spectrometer, Using Bragg Reflections in Both Crystals (Bragg-Bragg Scheme)

As is well known, ever since 1917 [1.6] practical researchers have been using two-crystal spectrometers, in which a beam of rays from an X-ray tube is consecutively reflected from two crystals, thereby closely approximating the conditions of dynamical scattering. Although such devices have found wide application in solving a number of other physical and technological problems, we shall consider the two-crystal spectrometers only from the point of view of experimental verification of the dynamical theory and the real structure of crystal-scatterers as regards the parameters of the dynamical maximum and interference effects.

The diffractometric equipment was later improved along the following lines:

1) the use of the Bragg-Laue and Laue-Laue scheme in BROGREN's investigations [9.2, 3];

2) construction and use of the three-crystal spectrometer, which was conceived and first used in practice by RENNINGER [8.6], while the relevant theory and extremely valuable results are due to BUBAKOVA et al. [9.4];

3) asymmetrical arrangement in different schemes realized by RENNINGER [9.5, 6], BUBAKOVA [7.4], and KOHRA et al. [9.7-9];

4) the two-crystal Laue-Laue diffractometer with the use of the Borrmann effect in the first crystal in the investigations of AUTHIER [1.15] and LEFFELD-SOSNOVSKA and MALGRANGE [9.10].

A detailed description of the two-crystal spectrometer theory is presented in the books of COMPTON and ALLISON [1.6] and JAMES [1.9]; certain important conclusions of this theory are discussed in LAUE's book [1.13]. The application of the theory to three-crystal spectrometer designs is developed in [9.4].

The designs of the two-crystal spectrometer for which the theory was developed (Fig. 9.1) are based on combining two crystals with Bragg reflection in a symmetrical position, i.e., from a system of planes parallel to the entrance surface. The X-ray beam impinging on the first crystal or the crystalmonochomator A, having emerged from the tube, passes through two parallel

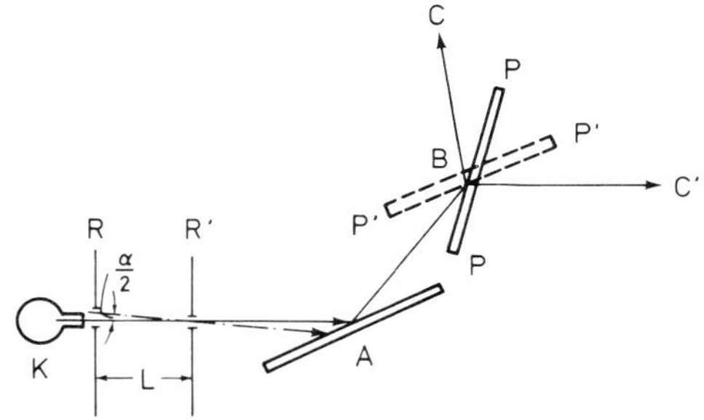

Fig.9.1 Scheme of double-crystal spectrometer

slits, as a result of which it is characterized by two values of divergence or plane angles: α - horizontal divergence, which corresponds to the slit width (in the plane of the drawing), and φ - vertical divergence corresponding to the slit height. Thus, the maximum values of these angles are

$$a_m = c/z \quad , \quad \varphi_m = h/z \quad , \tag{9.11}$$

where c is the width and h, the height of the slits, z being the distance between them. The spectrometer diagram shown in Fig. 9.2 refers to two different settings of the second crystal B. The normals to the reflecting planes of the crystals - the fixed one A and the rotating one B in the positions PP or P'P' - lie on a common horizontal plane.

The fixed crystal A is so set up that a certain central ray in the incident beam (i.e., the ray passing through the loci of both slits) forms with

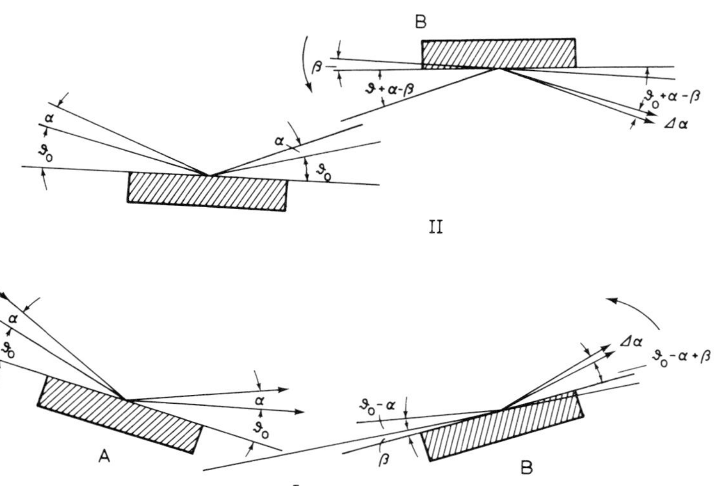

Fig. 9.2 Rotation of crystal B in double crystal spectrometer with anti-parallel (I) and parallel (II) arrangement

the plane face of the crystal an angle $\vartheta_0 = \vartheta + \eta_0$ corresponding to the middle of the total reflection region (see Fig. 7.3).

The crystal B is set up in the initial position. Rocking the crystal B about the vertical axis, which lies on the reflecting plane, we record the beams reflected from this crystal and thus obtain the rocking curve. An important task of the theory is to establish the relationship between this curve and the true shape of the dynamical maximum from crystal B alone. We shall first consider the values of the angles formed by various rays in the incident beam with the crystals A and B at different values of λ.

If the central ray impinging on the crystal A is characterized by the parameters ϑ_0, λ_0, $\alpha = 0$, $\varphi = 0$, n_A, where n_A is the coefficient of the reflection from crystal A, the angular deviation from the central ray in reflection of an arbitrary ray in the beam with parameters $(\vartheta, \lambda, \alpha, \varphi, n_A)$ will be (glancing angles ω):

$$\alpha - \frac{1}{2}\varphi^2 \tan\omega(\lambda_0,n_A)-(\lambda - \lambda_0)\left(\frac{\partial\omega}{\partial\lambda}\right)_0^1 \quad . \tag{9.12}$$

The second term of this expression corresponds to the deviation owing to the vertical divergence, the third one, to the deviation due to the nonmonochromaticity. The last term takes this form due to the natural assumption that within the spectral width of the incident beam the reflection angles change only slightly. The suffixes 1 and 0 mean that the value of this derivative is taken near the middle angle ϑ_o (for the first crystal) and the average wavelength λ_o.

Turning now to the crystal B, we notice that if in the zero position the angle of incidence of the ray on the second crystal is also ϑ_o, then in the case of crystal rotation by an angle β within the maximum region, for instance, counterclockwise, the angle of reflection from B will change by $\vartheta + \beta$ (position I-PP) or by $\vartheta_o - \beta$ (position II-P'P') (Figs. 9.1, 2). For the angle α, the opposite change of signs in the positions I and II will take place. Accordingly, the angle of reflection of the ray from the second crystal (λ, α, φ) will differ from that of the standard ray by

$$\pm\beta\mp\alpha - \frac{1}{2}\varphi^2 \tan\omega(\lambda_o,n_B) - (\lambda - \lambda_o)\left(\frac{\partial\omega}{\partial\lambda}\right)_o^2 \quad . \tag{9.13}$$

Here, the upper signs in front of β and α refer to position I, and the lower ones, to position II, n_B denotes the coefficients of reflection from the second crystal, and the suffixes 2 and 0 refer to the value of the derivative near the angle of reflection of the central ray from the second crystal near λ_o.

Now we must take into account the intensity distribution functions for the original beam in relation to the divergence of the beams α and φ and the wavelength λ. For the time being, we assign some functions

$$G(\alpha, \varphi) \quad , \quad J(\lambda - \lambda_o) \quad , \tag{9.14}$$

which are so normalized that the intensities within a definite interval of the argument ($d\alpha$, $d\varphi$, and $d\lambda$, respectively) are determined by multiplying by these intervals.

As a result, we can write the following expression for the total power of the radiation reflected from the second crystal, as a function of the angle of rotation β of crystal B from the initial position, if the horizontal width of the incident beam lies between the limits $-\alpha_m$ and α_m:

$$P(\beta) = \int_{-\varphi_m}^{\varphi_m}\int_{\lambda_{min}}^{\lambda_{max}}\int_{-\alpha_m}^{\alpha_m} G(\alpha,\varphi)J(\lambda - \lambda_o)C_A\left[\alpha - \frac{1}{2}\varphi^2 \tan\omega_1 -\right.$$

$$\left. - (\lambda - \lambda_o)\left(\frac{\partial\omega}{\partial\lambda}\right)_o^1\right] C_B\left[\pm\beta\mp\alpha - \frac{1}{2}\varphi^2 \tan\omega_2 -\right.$$

$$- (\lambda - \lambda_o) \left(\frac{\partial\omega}{\partial\lambda}\right)_o^2 \Big] d\alpha d\varphi d\lambda \quad . \tag{9.15}$$

In this expression, C_A and C_B are functions corresponding to the indices of the reflections n_A and n_B from the crystals A and B.

The analysis and some simplification of the rather complex expression (9.15) can be started by considering one of the most important characteristics of the spectrometer, namely the dispersion, which can be expected because of the finite spectral range in the radiation used.

Passing on to the limiting case of the infinitely narrow maxima of functions C_A and C_B, which will, consequently, differ from zero only if their arguments vanish, it is easy to obtain the dispersion value

$$\frac{d\beta}{d\lambda} \equiv D = \frac{n_A}{2d\,\cos\vartheta_1} \pm \frac{n_B}{2d\,\cos\vartheta_2} \quad , \tag{9.16}$$

where the upper sign refers to position I and the lower one, to position II of crystal B, as before. This relation leads directly to the following important results. In position I the dispersion in the double-crystal spectrometer consists of the dispersion values corresponding to the two crystals taken separately. Conversely, in position II the total dispersion equals the difference between the same values.

In a certain important case where the two crystals are identical and the reflection indices $n_A = n_B$, position I yields double dispersion, whereas II yields zero dispersion. This property of position II explains its predominant use in many investigations.

The further discussion of the characteristics of the double-crystal spectrometer is given separately for the two designs.

Position II, or the parallel arrangement of the crystals - scheme (n, -n) (Fig. 9.2, II). One important case is the arrangement with two identical crystals, assuming that both crystals are sufficiently (and to the same extent) perfect and the reflection coefficients

$$n_A = n_B \quad . \tag{9.17}$$

Besides, it is obvious that here

$$C_A = C_B = C_1 \quad , \qquad \omega_1 = \omega_2 = \omega \quad . \tag{9.18}$$

In this case (9.15) takes the form

$$P(\beta) = \iiint\limits_{\varphi\lambda\alpha} G(\alpha,\varphi)J(\lambda - \lambda_o)C\left[\alpha - \frac{1}{2}\varphi^2 \tan\omega - (\lambda - \lambda_o)\left(\frac{\partial\omega}{\partial\lambda}\right)_o\right]C\left[\alpha - \beta - \frac{1}{2}\varphi^2 \tan\omega - (\lambda - \lambda_o)\left(\frac{\partial\omega}{\partial\lambda}\right)_o\right]d\alpha d\lambda d\varphi \quad . \tag{9.19}$$

A qualitative analysis of this expression reveals additional characteristics of the parallel arrangement.

The starting point of this analysis is a very small area of argument values measured by just a few seconds of arc or by the absolute value of the angles of the order of 10^{-5} at which the integrad function C differs from zero. In this case the following can be demonstrated.

1) The function $G(\alpha, \varphi)$ can be represented as a product of two functions

$$G(\alpha, \varphi) = G_1(\alpha)G_2(\varphi) \quad . \tag{9.20}$$

Although the values of the functions G_i are different from zero in angle ranges of the order of one minute of arc ($\sim 10^{-3}$ in absolute units), in accordance with the magnitude of the term $1/2\ \varphi^2 \tan\omega$, the effective region of variation of the function G_2 is of about the same order as that of the function C. Moreover, it can be shown that with the parallel arrangement of the crystals, the shape of the resultant reflection curve is completely independent of the vertical divergence of the beam impinging on the first crystal.

2) For each of the monochromatic components of the incident beam the effective range of the argument is very small and equals approximately

$$\alpha_e \approx (\lambda - \lambda_o)(\partial\omega/\partial\lambda)_o \quad . \tag{9.21}$$

This estimate corresponds to the above statement that the crystal A transforms the incident beam into a set of parallel beams for the monochromatic components.

3) The effective range of the wavelengths taking part in the formation of the maximum inside the reflection curve is estimated at

$$\lambda_o \pm (\partial\lambda/\partial\omega)^o \alpha_m \quad . \tag{9.22}$$

In typical cases this amounts to $\sim \lambda \pm (2\text{-}3)$X, i.e., the wavelength range used more or less considerably exceeds the half-width of many spectral lines (see Table 9.1). It can be seen from that expression that the estimate (9.22) is independent on β.

4) The function $P(\beta)$ is different from zero in a very narrow range of the argument. In other words, in the scheme (n,-n) the half-width of the reflection curve is comparable to the half-width of the dynamical maximum from one crystal. For some typical conditions it exceeds the latter by about 1.3 times.

5) Comparing the narrow range of the functions C, which are actually taken into account in integrating, and the relatively wide ranges of the arguments φ, α, λ (the functions G_1, G_2, and J change gradually within seconds of arc), we arrive at the following expression for the reflection curve:

$$P(\beta) = K \int_{-\infty}^{\infty} C(\alpha)C(\alpha - \beta)d\alpha \quad , \tag{9.23}$$

where

$$K = \int_{p_1}^{p_2} \int_{-\varphi_m}^{\varphi_m} G_1\left[\frac{\partial\omega}{\partial\lambda_o} (\lambda - \lambda_o)\right] G_2(\varphi) J(\lambda - \lambda_o) d\varphi d\lambda \quad , \tag{9.24}$$

$$p_1 = \lambda_o - \left(\frac{\partial\lambda}{\partial\omega}\right)^o \alpha_m \quad , \quad p_2 = \lambda_o + \left(\frac{\partial\lambda}{\partial\omega}\right)^o \alpha_m \quad .$$

The question posed above with regard to the relationship between the experimental reflection curve $P(\beta)$ and the true profile of the dynamical maximum of reflection from the crystal B was specially considered by LAUE [9.11]. It was shown that no direct transition is possible.

Moreover, if, with the appropriate choice of the function (or functions) C, the integral (9.23) calculated by one method or another is in good agreement with the experimental reflection curve, this result will not be unambiguous.

It should further be noted that simple inspection of (9.23) immediately reveals that the reflection curve with the use of the scheme (n,-n) is symmetrical about the point $\beta = 0$, even if the true curves of dynamical reflection from the crystals A and B are nonsymmetrical, for instance as in the case of Bragg reflection from the absorbing crystal. In particular, for the symmetry indicated, the dynamical maxima A, B identity is not a necessary condition.

It is further obvious that (9.23) refers to the case where the radiation falling on the crystal A is polarized. With nonpolarized radiation, the function P is written as follows:

$$P(\beta) = K \int_{-\infty}^{\infty} [C_\sigma(\alpha)C_\sigma(\alpha - \beta)+C_\pi(\alpha)C_\pi(\alpha - \beta)]d\alpha \quad . \tag{9.25}$$

This expression yields the total power reflected by the crystal B in position β. The reflection coefficient in this case will be expressed by

$$R(\beta) = \frac{\int_{-\infty}^{\infty} [C_\sigma(\alpha)C_\sigma(\alpha - \beta)+C_\pi(\alpha)C_\pi(\alpha - \beta)]d\alpha}{\int_{-\infty}^{+\infty} [C_\sigma(\alpha)+C_\pi(\alpha)]d\alpha} \quad . \tag{9.26}$$

Incidentally, by the reflection coefficient the magnitude $P(\beta)$ (9.25) is often meant. It is obvious that since the integrand function in (9.25) is, for any given value of α, the product of the ordinates of the two curves $C(\alpha)$ and $C(\alpha - \beta)$, the maximum value of this product at $\beta = 0$ and with parallelism of the two crystals will correspond to the value (of the percent reflection)

$$R(0) = \frac{\int_{-\infty}^{\infty} [C_\sigma(\alpha)]^2 d\alpha + \int_{-\infty}^{\infty} [C_\pi(\alpha)]^2 d\alpha}{\int_{-\infty}^{\infty} [C_\sigma(\alpha)+C_\pi(\alpha)]d\alpha} \quad , \tag{9.27}$$

which is the maximum (relative) value of the reflection coefficient (in the two-crystal spectrometer).

Finally, in measurements on a two-crystal spectrometer the value of the integrated reflection is calculated from the following expression:

$$R_i = \int_{-\infty}^{\infty} R(\beta)d\beta = \frac{\left[\int_{-\infty}^{\infty} C_\sigma(\alpha)d\alpha\right]^2 + \left[\int_{-\infty}^{\infty} C_\pi(\alpha)d\alpha\right]^2}{\int_{-\infty}^{\infty} C_\sigma(\alpha)d\alpha + \int_{-\infty}^{\infty} C_\pi(\alpha)d\alpha} \quad . \tag{9.28}$$

Position I, or the antiparallel arrangement of crystals - scheme (n, n) (Fig. 9.2, I). In this case we can write, similar to (9.19),

$$P(\beta) = \iiint\limits_{\varphi\lambda\alpha} G(\alpha,\varphi)J(\lambda - \lambda_0)C\left[\alpha - \frac{\varphi 2}{2}\tan\omega - \right.$$

$$- (\lambda - \lambda_0)\left(\frac{\partial\omega}{\partial\lambda}\right)_0 C\left[\beta - \alpha - \frac{\varphi 2}{2}\tan\omega - \right. \tag{9.29}$$

$$\left. - (\lambda - \lambda_0)\left(\frac{\partial\omega}{\partial\lambda}\right)_0\right] d\alpha d\lambda d\varphi \quad .$$

A qualitative analysis of this equation, similar to the one referring to the case of the parallel arrangement, provides the following results.

1) In this case, too, the effective values of α are estimated at

$$\alpha \approx (\lambda - \lambda_0)\left(\frac{\partial\omega}{\partial\lambda}\right)_0 \quad . \tag{9.30}$$

2) Estimating the range of β for which the function $P(\beta)$ still has an appreciable value, we arrive at the conclusion that in this case the range considerably exceeds the angular width of the dynamical maximum for one crystal. Quantitative estimates mean establishing the relationships between the quantity β and the spectral range which makes a contribution to the formation of the maximum.

3) The argument of the second function C in the integral (9.29) can be rewritten as follows

$$\left[\beta - 2\left(\frac{\partial\omega}{\partial\lambda}\right)_0(\lambda - \lambda_0)\right] - \left[\alpha \quad \left(\frac{\partial\omega}{\partial\lambda}\right)_0(\lambda - \lambda_0)\right] - \frac{\varphi^2}{2}\tan\omega \quad . \tag{9.31}$$

It is easy to see that since the last term in this expression, as well as the second term in the brackets (according to (9.30)) is small, the first term in the brackets must also be small so that the value of C for the second crystal differs appreciably from zero. Thus, in distinction to the parallel arrangement of the crystals (see (9.22)), if the crystal B deviates from the middle position by an angle β, both crystals select a narrow spectral band

$$\Delta\lambda \approx \lambda_0 + \frac{1}{2}\left(\frac{\partial\lambda}{\partial\omega}\right)^0 \beta \tag{9.32}$$

out of the entire wave spectrum in the incident beam.
This result is extremely important, since it enabled RENNINGER to use precisely the scheme (n,n), as a monochromator in a three-crystal spectrometer. Indeed, as already mentioned, during the formation of the maximum of $P(\beta)$

with parallel arrangement of crystals, a contribution to the value of the function $P(\beta)$ in each given position β is made by the monochromatic components of the beam over a comparatively wide range of wavelengths: in a typical case, 2-3 X, according to (9.22). In contrast to this, in each given position of crystal B, a spectral range which is β/α_m times less (i.e., up to 0.01-02 X wide in the indicated typical case) takes part in the scheme (n,n).

As for dispersion in the double crystal spectrometer, according to (9.16), when the crystals are in the (n,n) position, it is equal to the sum of the dispersions from both crystals and, in this particular case where $n_A = n_B$ and the two crystals are identical, the arrangement indicated yields double dispersion as compared with that from the one-crystal spectrometer.

4) Finally, the result of verification of the symmetry of the function $P(\beta)$ in this scheme is important.

Considering (9.29), we notice that its transformation, similar to the one applied to (9.19), cannot be carried out here. As regards the variable α, according to (9.30), its effective range is small in comparison with α_m, and therefore, here too, we can extend to $\pm\infty$ the limits of integration with respect to α. The effect of vertical divergence cannot be eliminated, nor can the function $J(\lambda - \lambda_0)$. But assuming that our spectrometer operates during incidence on the crystal A of a monochromatic beam with the divergence of (α,φ) indicated and that we can neglect the dependence on φ, we obtain

$$P'(\beta) \approx \int_{-\infty}^{\infty} C(\alpha)C(\beta - \alpha)d\alpha \quad , \tag{9.33}$$

where the proportinality factor is omitted. Careful inspection of this formula shows that it differs from (9.23) in that we do not obtain an equivalent result by replacing (β) with $(-\beta)$. Hence, a reflection curve taken during the rotation of the crystal B in the scheme (n,n) does not show parasitic or artifact symmetry about the point $\beta = 0$ (or any other value of β). This property of the reflection curves taken in the scheme (n,n) is also of fundamental importance for subsequent discussion, although it cannot be used in the present double-crystal spectrometer design because of the strong dispersion effect.

9.3 Three-Crystal Spectrometer

The demerits of both double-crystal diffractometer designs outlined in Section 9.2, such as a more or less considerable broadening of the rocking curves and decrease in R(0) or in percent reflection compared with the theoretical values for one crystal, as well as artifact symmetry of the curves, considerably limit the importance and field of application of such devices. This refers not only to precision verification of the theory and determination of the scattering parameters of perfect crystals, but also to the study and control of defects in specimens with small defects, which are widely used in contemporary technology.

Over the last few years, various authors have increased the number of reflecting crystals, and thereby considerably improved the angular and spectral characteristics of the beam incident on the crystal under investigation as well as a number of parameters of the rocking curve.

The idea of three- and multi-crystal diffractometers was first advanced by DU MOND [9.12], who proposed a convenient and very expressive graphical method for analyzing multi-crystal schemes. RENNINGER [8.6] used this method for qualitative analysis of the problem of designing more precise diffractometers and developed a three-crystal design, using both the parallel and antiparallel arrangement of adjacent crystals.

In the RENNINGER design of the three-crystal spectrometer with symmetrical reflection, which is shown in Fig. 9.3, the first two crystals should be regarded as MCC.

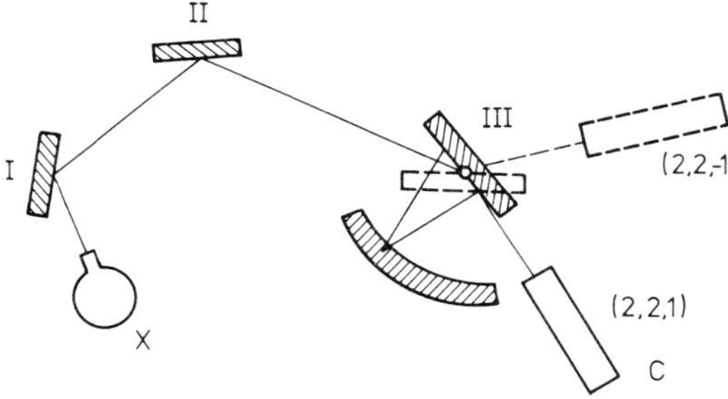

Fig. 9.3 Scheme of three-crystal spectrometer (after BUBAKOVA et al. [9.4])

An essential stage in the further development of diffractometry is the theory of the three-crystal spectrometer (diffractometer) developed by BUBAKOVA et al., 1961 [9.4]. The theory was developed along lines similar to the above-described theory of the two-crystal spectrometer.

The following conditions must be fulfilled:

1) the crystals are to be set up so that the normals to their entrance surfaces lie on a common plane;

2) the crystals must be perfect in structure, so that their reflection functions correspond to the dynamical theory;

3) symmetrical reflection from all the three crystals must be considered. In this case, using the arrangement (n,n) in the MCC (Fig. 9.3) and proceeding from the reasoning given when deriving (9.15), we obtain

$$\begin{aligned} P(\pm\gamma) = \iiint_{\varphi\lambda\alpha} & G(\alpha,\varphi)J(\lambda - \lambda_o)C_1\left[\alpha - \frac{1}{2}\varphi^2 \tan\vartheta_1 - \right. \\ & \left. - (\lambda - \lambda_o)\left(\frac{\partial\omega}{\partial\lambda}\right)_o^{I}\right]C_2\left[-\alpha - \frac{1}{2}\varphi^2 \tan\vartheta_2 - \right. \\ & \left. - (\lambda - \lambda_o)\left(\frac{\partial\omega}{\partial\lambda}\right)_o^{II}\right]C_o\left[\pm(\gamma - \alpha) - \frac{1}{2}\varphi^2 \tan\vartheta_3 - \right. \\ & \left. - (\lambda - \lambda_o)\left(\frac{\partial\omega}{\partial\lambda}\right)_o^{III}\right]d\alpha d\lambda d\varphi \quad . \end{aligned} \tag{9.34}$$

Here, the angle γ represents the deviation of crystal 3 under investigation from the initial position, and the signs "$\pm$" correspond to the antiparallel and parallel positions of crystal 3 relative to 2, respectively. The symbols of these arrangements, adopted on RENNINGER's suggestion, are (n_1,n_2,n_3) and $(n_1,n_2,\bar{n}_3)$.

To simplify the general expression (9.34) we introduce the assumptions

$$\begin{aligned} C_1 &= C_2 \equiv C_o \quad , \\ \omega_1 &= \omega_2 \equiv \omega_o \end{aligned} \tag{9.35}$$

$$C_3 \equiv C_1 \quad , \qquad \omega_3 \equiv \omega_1 \tag{9.36}$$

$$G(\alpha,\varphi) = G_1(\alpha)G_2(\varphi) \quad . \tag{9.37}$$

The functions $J(\lambda - \lambda_o)$ and $G_1(\alpha)$ vary only slightly within the angular coordinates corresponding to the dynamical maxima C_1 and C_o. Therefore they can be taken out of the integral, and the limits of integration with respect to the variables λ and α can be extended to $\mp\infty$. Introduce the new functions

$$\frac{1}{2}\varphi^2 \tan\omega_1 = \psi^2 k \quad , \qquad k = \tan\omega_1 / \tan\omega_o ,$$
$$\psi_m = \varphi_m \sqrt{\frac{\tan\omega_o}{2}} \quad . \tag{9.38}$$

If we denote by D_o the values of dispersion for crystals 1 and 2, and by D_1, those for crystal 3 and use the Bragg equation, we can write

$$\frac{D_1}{D_o} = \left(\frac{\partial\omega}{\partial\lambda}\right)_o^1 \left(\frac{\partial\omega}{\partial\lambda}\right)_o^o = \frac{d_o \cos\omega_o}{d_1 \cos\omega_1} = \frac{\tan\omega_1}{\tan\omega_o} = k \quad . \tag{9.39}$$

Further,

$$(\lambda - \lambda_o)\left(\frac{\partial\omega}{\partial\lambda}\right)_o^o = \beta \quad , \qquad (\lambda - \lambda_o)\left(\frac{\partial\omega}{\partial\lambda}\right)_o^1 = k\beta \quad . \tag{9.40}$$

Assuming $\omega_o \geq \omega_1$, we obtain $0 < k \leq 1$. Finally, we introduce

$$x = -\psi^2 - \beta \quad . \tag{9.41}$$

Eq. (9.34) takes the form

$$P(\pm\gamma) = A \int_{-\psi_m}^{\psi} \int_{-\infty}^{\infty} \int_{-\infty}^{\infty} G_2(\psi) C_o(\alpha + x) C_o(-\alpha + x) \cdot$$
$$\cdot\ C_1[\pm(\gamma + \alpha) + kx]\, dx\, d\alpha\, d\psi \quad . \tag{9.42}$$

Since $G_2(\psi)$ is independent of α and x, (9.42) can be represented thus

$$P(\pm\gamma) = ABP'(\pm\gamma) \quad , \tag{9.43}$$

$$B = \int_{-\psi_m}^{\psi} G_2(\psi)\, d\psi \quad , \tag{9.44}$$

$$P'(\pm\gamma) = \iint_{-\infty}^{\infty} C_o(\alpha + x) C_o(-\alpha + x) C_1(\pm\gamma \pm \alpha + kx)\, d\alpha\, dx \quad . \tag{9.45}$$

An important result of (9.43), (9.44), and (9.45) is that the shape of the reflection curve taken on the three-crystal spectrometer, with the above assumptions, is independent of the vertical divergence of the incident beam (function $G_2(\psi)$). For subsequent analysis of the relations obtained, we introduce a new variable

$$y = \pm \alpha + kx \quad , \tag{9.46}$$

whence

$$\alpha + x = \pm y + x(1 \mp k) \quad , \quad -\alpha + x = \mp y + x(1 \pm k) \quad , \tag{9.47}$$

and (9.45) transforms as follows

$$P'(\pm\gamma) = \iint_{-\infty}^{\infty} C_o[y + x(1 - k)]C_o[- y + x(1 + k)] \cdot \cdot C_1(\pm \gamma + y)dxdy \quad . \tag{9.48}$$

Integration can be performed separately, and we obtain two functions

$$P'(\pm\gamma) = \int_{-\infty}^{\infty} C_1(\pm \gamma + y)[V(y)]dy \quad , \tag{9.49}$$

$$V(y) = \int_{-\infty}^{\infty} C_o[y + x(1 - k)]C_o [- y + x(1 + k)]dx \quad . \tag{9.50}$$

Here, (9.50) depends on the properties of the two-crystal monochromator. Hence, $P'(\pm\gamma)$ represents the convolution of two functions, as in the case of the two-crystal spectrometer, C_1, the function of dynamical scattering from crystal 3, and $V(y)$, the "spread-out" function. A noteworthy property of function $P'(\gamma)$ is that a change of sign in the scheme (n_1,n_2, n_3) yields two *reflection curves*, which are *mirror-symmetrical*. Of still greater importance is the substantiation of the property of the three-crystal design indicated by RENNINGER, namely, the absence of the parasitic symmetry of the rocking curves. Since mirror symmetry is independent of $G_2(\psi)$, it may serve as a method of verifying the setting of a three-crystal spectrometer. Hence it follows that a deviation from mirror symmetry should indicate a difference (for instance, in the degree of perfection) between the crystals making up the monochromator, of course provided the setup meets all the other requirements. The subsequent analysis refers to a particular case: $n_1 = n_2 = n$; $n_3 = m$.

Another important property of the device is the ratio between the half-width of the reflection curve obtained on this setup and the half-width of the dynamical maximum of crystal 3. Let us investigate the function V(y) represented in (9.50), at the various possible values of k.

For k = 1 we have

$$V_1(y) = \int_{-\infty}^{\infty} C_0(y)C_0(-y + 2x)dy = pC_0(y) \quad ,$$
$$p = \int_{-\infty}^{\infty} C_0(-y + 2x)dx = \int_{-\infty}^{\infty} C_0(2x)dx \quad . \tag{9.51}$$

It follows that the half-width of the function $V_1(y)$ is equal to the half-width of $C_0(y)$:

$$w_{V1} = w_0 \quad . \tag{9.52}$$

On the other hand, at k = 0

$$V_0(y) = \int_{-\infty}^{\infty} C_0(y + x)C_0(-y + x)dx$$
$$= \int_{-\infty}^{\infty} C_0(z)C_0(z - 2y)dz \quad . \tag{9.53}$$

The form of the function $V_0(y)$ on the extreme right-hand side in (9.53) corresponds to the parallel arrangement (n, -n), but on a "half-scale". If the half-width in the parallel arrangement is w_π, than

$$w_{V0} = 1/2\ w_\pi \quad . \tag{9.54}$$

Approximating the function C by Gauss' error curve, we get $w_0 = 2^{-\frac{1}{2}}w_\pi \approx \approx 0.71\ w_\pi$. In this case the variation limits of the half-width of the function w_V within the range of k will be

$$0.50\ w_\pi < w_V < 0.71\ w_\pi \quad . \tag{9.55}$$

The quantity w_V gives an estimate of the *angular* divergence of the ray emerging from a double-crystal monochromator and used as a probe for plotting the reflection curve from crystal 3.

RENNINGER [8.6] estimated w_V from DU MOND's graph at 2/3 w_π, which is in agreement with (9.55).

In order to calculate the half-width of the reflection curve obtained on the three-crystal spectrometer, we can use the same approximation of the Gauss curve for the function C_1. In this case, the following relation can be obtained:

$$w^2 = w_1^2 + \frac{w_o^2}{2}(1 + k^2) \quad , \tag{9.56}$$

where w, w_1, and w_o denote the values of the half-width of the reflection curves and of the functions of C_1 and C_o, respectively.

It is easy to see that, according to (9.56), the best approximation to the dynamical curve from the crystal 3 under investigation is achieved when using a high-order reflection yielding a sharp maximum in the monochromator, and a first-order reflection from the crystal under study. Thus, in [9.4] the authors used the scheme (3,3,1), and the monochromator consisted of two single crystals of Si. A Ge crystal was investigated with reflections 111 (Cu $K\alpha_1$ radiation). In this experiment the values of w_1 and k were 8" and 0.24, respectively. Using w_o 1.89", we obtain

$$w = \left[w_1^2 + \frac{w_o^2}{2}(1 + 0.06)\right]^{\frac{1}{2}} \approx 8.06'' \approx 8.1'' \quad . \tag{9.57}$$

Thus, with the appropriate choice of reflections, the three-crystal spectrometer enables one to obtain a reflection curve which almost coincides in shape with the true profile of the dynamical maximum, particularly if we consider that this device does not distort the true curve because of "parasitic" symmetry.

Further considering the various particular relationships between the three crystals forming the spectrometer, it is possible to obtain the main characteristics of the reflection curves expected - first and foremost, the half-width values (see [9.4]).

Finally, similar to the two-crystal spectrometer, in the case of the three-crystal device, the value of R(0) in the particular case of $(n_1, n_2, \pm n_3)$ is determined from the equation

$$R^{III}(0) = \frac{\iint\limits_{-\infty}^{\infty\,\infty} C_o(\alpha + x)C_o(-\alpha + x)C_1(\pm\gamma \pm \alpha + kx)d\alpha dx}{\iint\limits_{-\infty}^{\infty} C_o(\alpha + x)C_o(-\alpha + x)d\alpha dx} \tag{9.58}$$

In the first investigation by RENNINGER [8.6] using a three-crystal spectrometer, the rocking curves $2\bar{1}\bar{3}\bar{1}$, Cu $K\alpha_1$, from calcite were obtained. The author used reflection in MCC (the first two crystals) from the same system of planes as for the third crystal but in the second order. In accordance with (7.32) (see also Fig. 7.3), the decrease in χ_{hr} resulted in a reduced angular width of the beam reflected from the second crystal. Obtained rocking curves confirmed the expected effect of elimination of parasitic symmetry; they showed asymmetry typical of the curves of the Bragg reflection from absorbing crystals. When calculating the theoretical curves it was assumed that the beam incident on crystal 3 has a σ-component of intensity $\sim$ 85 percent. In spite of the considerable difference in the profiles of the theoretical and experimental curves, the values of the integrated reflections proportional to the area bounded by the curve are similar:

$$R_i^T = 37.7 \cdot 10^{-6} \quad , \quad R_i^E = 36.5 \cdot 10^{-6} \quad .$$

The similarity of these values indicates that in X-ray reflection from near-perfect crystals the value of integrated intensity depends only slightly on the degree of deviation from the ideal structure.

9.4 Other Types of Diffractometers

9.4.1 Double-Crystal Spectrometers of the Bragg-Laue and Laue-Laue Type

In their investigations into the reflection curves of calcite, quartz, and then Ge and Si, BROGREN and ADELL [9.3] used double crystal spectrometers with Bragg-Laue and Laue-Laue schemes (Fig. 9.4a).

The Bragg-Laue spectrometer (Fig. 9.4a) uses symmetrical Bragg reflection in crystal 1 and symmetrical Laue reflection with the same indices in the crystal 2 under investigation. To make a comparison with the experimental reflection curves, these authors used (9.25 - 28), in which the values of the functions $C_\sigma(\alpha)$ and $C_\pi(\alpha)$ from crystal 1, and $C_\sigma(\alpha - \beta)$ from crystal 2 corresponded to the coefficients of reflection and transmission, depending on the design selected. Thus, in the case of the Bragg-Laue scheme, the authors used Bragg reflection coefficients for crystal 1 and Laue reflection or transmission coefficients for crystal 2. Although the use of the equations indicated in the Bragg-Laue and Laue-Laue arrangements was not

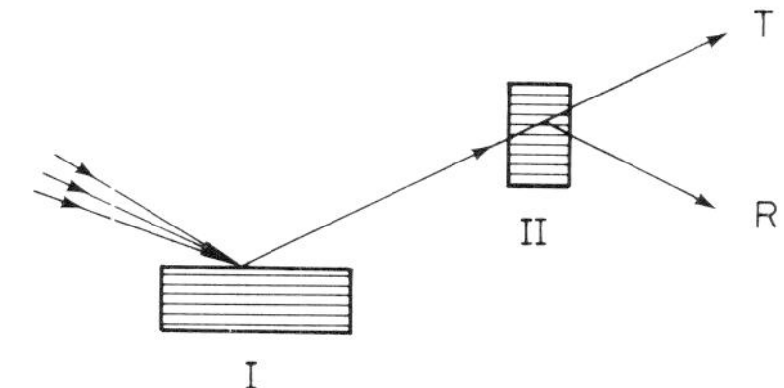

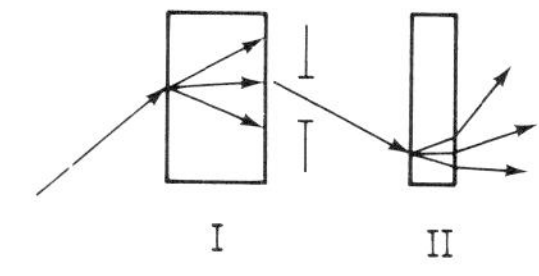

b

Fig. 9.4 a and b Double crystal scheme of diffractometers and MCC: a) Bragg-Laue (after BROGREN and ADELL [9.3]. b) Thick-crystal scheme (after MALGRANGE and AUTHIER [9.38]).

substantiated by a special analysis, the legitimacy of this calculation was confirmed by the good agreement between the theoretical (i.e., calculated as above) and the experimental reflection curves over a wide spectral range. (At the same time it should be noted that the theoretical estimate of such basic characteristics of the reflection curves obtained in the Bragg-Laue scheme as the presence or absence of "parasitic" symmetry and of the ratio between the half-width of the dynamical function and the reflection curve still remains unclear). The same scheme was used by ELISTRATORY, EFIMOV and co-workers to measure the integrated intensity [4.15] (ibid., reference to the investigations of these authors).

The MCC design with Laue reflection for studying the reflection curves, separation of the wave fields in the crystal, and interference effects in dynamical scattering has been used by AUTHIER and co-workers ever since 1961 [1.15, 9.10] (Fig. 9.4b). As noted in Chapter 5, when radiation passes through a thick crystal, the wave front narrows down drastically and, hence, the divergence of the emerging reflected and transmitted beams decreases. The simplicity of AUTHIER's MCC is noteworthy. Theoretical consideration of this design is contained in the paper of LEFELD-SOSNOVSKA and MALGRANGE [9.10]. The efficiency of such an MCC was shown when measuring the fine structure of the Laue transmission and reflection curves.

9.4.2 Multi-Crystal Diffractometers with MCC, with Symmetrical and Asymmetrical Bragg Reflections

During development of the best possible diffractometric schemes, the attention of the investigators was focussed on the use of Bragg reflection, which ensures high intensity of the beam at the MCC output. Further success in reducing the angular divergence of the beams and their spectral range as compared with the three-crystal devices described in Section 9.3 was achieved by RENNINGER [9.5] in 1961 and, independently, by KOHRA [9.13] in 1962. These authors used the effects of variation in the angular width of the total-reflection range in asymmetric arrangements, which were mentioned in Section 7.1, and also the reduction of the spectral range in the antiparallel arrangement of the crystals forming the MCC.

Obviously the development of high-precision multi-crystal designs means that account must be taken not only of the variation in angular width, but also of the displacement of the maxima and variations in the width of the wave from accompanying the asymmetrical reflection from each of the crystals.

It is also important to estimate the spectral and angular parameters of the beam emerging from the MCC. This can be done either on the basis of calculating the main parameters given in (7.1) or by using DU MOND's pictorial method [9.12] and schemes of the type shown in Fig. 9.5. It should be noted that such analysis is rather of a qualitative or semi-quantitative nature.

Let us first consider the effects accompanying asymmetrical Bragg reflection fron one crystal. It is convenient to use variable angles or angular coordinates in the region of the maximum of ω_o and ω_h whose values can be represented with the aid of the asymmetry parameter b.

From (7.35) and Fig. 7.1b it follows that

$$b = \frac{\sin(\vartheta - \varphi)}{\sin(\vartheta + \varphi)} = \frac{\gamma_o}{|\gamma_h|} < 1 \quad . \tag{9.59}$$

In this case, using Fig. 7.3, we can write

$$\omega_o = \vartheta + A(1 + b^{-1}) + Byb^{-\frac{1}{2}} \quad . \tag{9.60a}$$

In accordance with (7.39), (7.41), and (7.43) we get

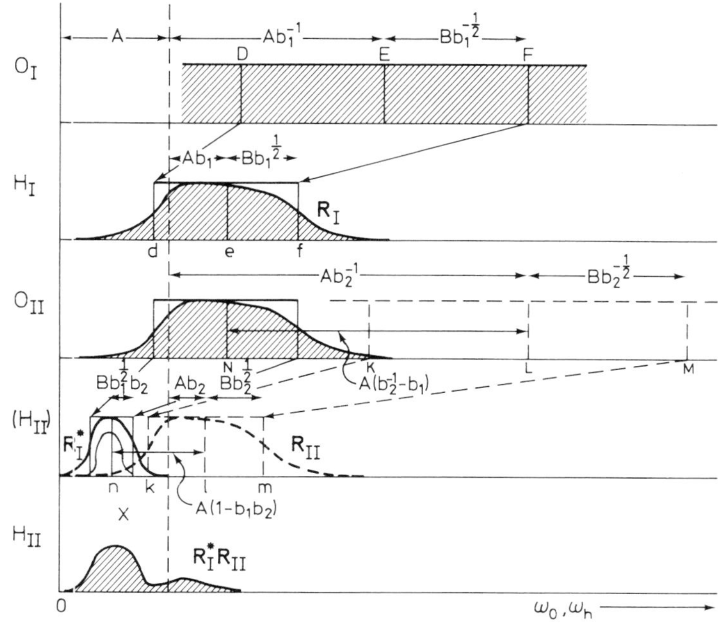

Fig. 9.5 Illustrating effect of the angular decrease of X-ray beam by double asymmetric successive diffraction in antiparallel arrangement of reflecting crystals (after KIKUTA and KOHRA [9.14b]). O_I = angular range of beams incident on crystals (I = I,II). H_I: angular range of beams reflected by these crystals

$$\omega_h = \vartheta + A(1 + b) + Byb^{\frac{1}{2}} \tag{9.60b}$$

$$A = |\chi_{or}|/2 \sin 2\vartheta \quad , \quad B = C|\chi_{hr}|/ \sin 2\vartheta \quad . \tag{9.60c}$$

It is obvious that the second terms from the right in these equations determine the angular displacements of the midpoint of the total-reflection range with respect to the angle ϑ while the third terms determine the angular positions of any point of the maximum relative to the above-mentioned midpoint. The remarkable property of asymmetrical reflection noted in Section 7.1, namely the decrease in the angular width of the total-reflection range, means that the wider angular range in the incident beam as compared with symmetrical reflection contributes to the formation of the maximum of the beam reflected. If we take into account the fact that the corresponding angular range in the beam reflected decreases simultaneously, the total value of "angular compression" in a single asymmetrical reflection will be equal to b in accordance with (7.43). In this case the efficiency of the MCC with respect to the beam used also increases by a factor of $b^{-\frac{1}{2}}$.

However, the decrease in angular width in asymmetrical reflection, a decrease in b or in the glancing angle ω_0, is accompanied by an increase in the width of the projection of the front of the incident beam onto the reflecting plane and, hence, of the excitation region of the beam reflected. As a result the linear width of the wave reflection front increases. With the linear width of the front of the incident beam equal to $L|\Delta\eta_0|$, where L is the distance from the source to the crystal, the width of the reflected beam front will be:

$$L|\Delta\eta_0|b^{-1} + L'|\Delta\eta_h| \approx L|\Delta\eta_0|b^{-1} \quad , \tag{9.61}$$

i.e., it will be increased by a factor of b^{-1}.

The broadening of the reflected beam front may cause difficulties in recording the integrated intensity or power, in accordance with equations of the type (9.58), i.e., it requires an excessively wide window for the recording counter. On the basis of a thorough analysis of the two-crystal MCC operation [9.14-16] it was possible to partly eliminate the second weakening of intensity due to the reflection from the second crystal, at the same time retaining the effect of the decrease in the beam's angular divergence. The intensity increase at the MCC output was achieved due to the fact that the author succeeded in using the reflection from the second crystal of beams whose direction is beyond the maximum reflected by the first crystal. The foregoing can be explained with the use of the scheme of Fig. 9.5.

In Fig. 9.5 the angular range of the beam O_I incident on the first crystal with a coefficient of asymmetry b_1, which makes a contribution to the formation of the total-reflection range, is denoted by DEF. The angles represented by the segments $(A+Ab^{-1})$ and $Bb^{-\frac{1}{2}}$ correspond to (9.60a) at $y = 1$.

The reflected beam H_1 is represented by the maximum R_1; the distance from the centre of the total-reflection range def from the angle ϑ (or from the ordinate axis) is $A(1 + b_1^{-1}) . b_1 = A(b_1 + 1)$, and the half-width of the maximum is $Bb_1^{\frac{1}{2}}(y = 1)$, according to (9.60b). The distortions in the shape of the maxima in Fig. 9.5 must be explained by the absorption.

Passing over to the reflection from the second crystal, one should take into account that within a sufficiently wide range of angles KLM, the beam, reflected by the first crystal, in accordance with the coefficient of asymmetry of the second crystal b_2 is reflected from it, forming a less intensive maximum klm denoted by R_{II}. At the same time, the radiation in the region of the maximum of the beam 0_{II}, being reflected from the second crystal, yields a maximum R_I^*. For R_{II} the angular position of the midpoints of these maxima correspond to a single reflection, and for R_I^*, to a double reflection, and are shifted with respect to each other by the value

$$b_2\left[A_1(1 + b_2^{-1}) - A(1 + b_1)\right] = A(1 - b_1b_2) \tag{9.62}$$

The reflection curve thus obtained is the convolution of the maxima R_{II} and R_I^*; its characteristics shape, which has been observed experimentally, is characterized by a main and a subsidiary maximum.

If one rotates the second crystal additionally, turning it through the appropriate angle in one way or another, the maxima R_{II} and R_I^* can be brought together; this was actually done by the authors or [9.14b].

In the investigation cited, the MCC consisted of a single-crystal block, and the reflecting crystals were joined by a thin strip, which allowed for slight elastic deformation and, hence, for additional rotation of crystal 2 through an angle corresponding to (9.61). Furthermore, since according to (9.60c) the value of B will be much less for the π-component because of the presence of factor C, incomplete rotation by the angle indicated will result in the attenuation of the π-component.

Similar results have been obtained by KOVIEV and BATURIN [9.17] with the aid of a spectrometer in which the MCC consisted of two separate crystals. The MCC produced successive symmetrical-asymmetrical diffraction, which would eliminate difficulties when using a monolithic-block MCC. These authors obtained a curve of symmetrical reflection 333 from a silicon crystal. In the first crystal of the MCC they used an identical reflection (this leads to the complete absorption of the π-polarized beam), and in the second crystal, asymmetrical reflection 511 with b = 0.0262 and a parallel arrangement of the two crystals. In both the above-mentioned investigations, the half-widths of the reflection curves for the crystals under review were 1.998 ± 0.003 percent and 2.04 ± 0.03 percent and maximum reflections, 87 and 88 percent, respectively.

The scheme of the double or multiple X-ray beam reflections from the lateral walls of a rectangular-section channel in a single-crystal block suggested by BONSE and HART [9.18] is worth mentioning. As a result of such successive reflections, the maximum loses its tails, which extend far beyond the total-reflection region. This shape of the maximum is especially important in the small-angle scattering investigations (Fig. 9.6).

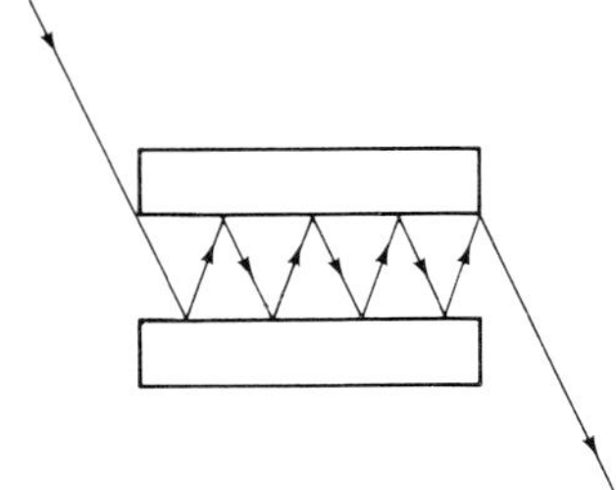

Fig. 9.6 Scheme of multiple X-ray beam Bragg reflection in monolithic block (after BONSE and HART [9.18])

An analysis of the above scheme is conveniently illustrated by DU MOND's graphs [9.12] (Fig. 9.7). The relationship between the quantities λ and ϑ is presented as a rectangular coordinate system. Because of the negligible values of the intervals $\Delta\lambda$ and $\Delta\vartheta$ in the equipment described, the corresponding curves in the superposition region are depicted as straight lines, or more precisely as strips. The cross sections of these strips, which are strongly magnified for clarity's sake and parallel to the axes, correspond to the angle and wavelength intervals. In the case of *symmetrical reflection*, the width of the strips, parallel to the ϑ-axis, will be identical.

To the parallel setting of the crystals in the two-crystal spectrometer there corresponds the superposition of the strips of the two crystals with identical displacement directions along the ϑ-axis.

This arrangement of the crystals implies the opposite direction of the λ-axis, which results in zero dispersion.

With the antiparallel arrangement and symmetrical reflection, the displacements (and slopes) of the strips will be symmetrical about some vertical axis, and the angular, as well as spectral, width of the beam reflected from the second crystal will correspond to the respective dimensions of the rhombus shaded in Fig. 9.7.

With *asymmetrical reflection*, the ϑ-intervals of the strips corresponding to the incident and reflected beams will be different (see Fig. 9.7c). By way of illustration, in Fig. 9.8 we present a graphical analysis of the design of a diffractometer with a three-crystal MCC borrowed from the paper [1.19] . The beam H_1, reflected from the first crystal, is superimposed on a wider strip of the beam H_2 reflected from the second crystal; this broadening is due to the above mentioned effect of reflection by the second crystal of the radiation which did not penetrate into the first region and because $b_2 > 1$. Finally, the beam H_3, reflected by the third crystal, has a very small divergence, but a considerable spectral interval, which is represented by the shaded region *abcd*.

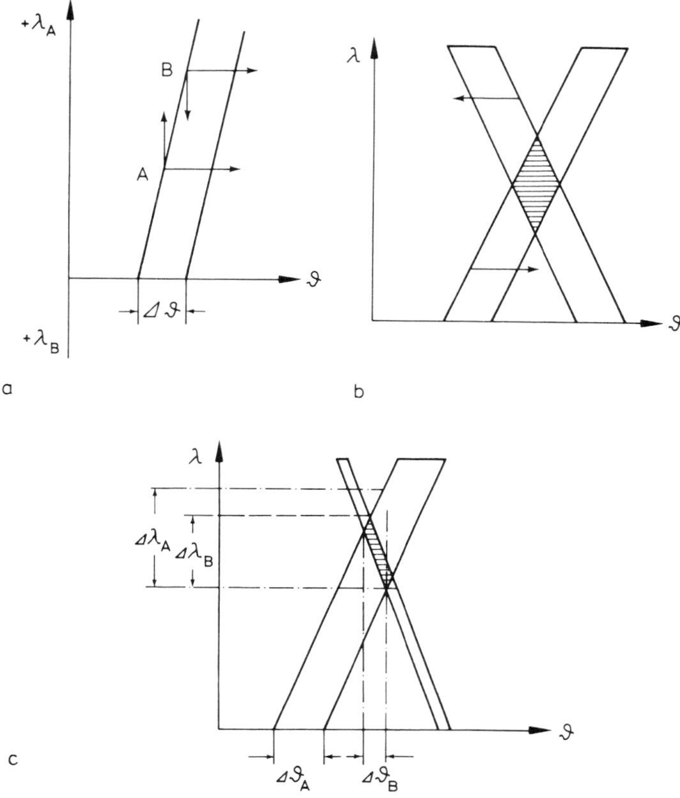

Fig. 9.7a-c DU MOND's scheme for two-crystal MCC: a) parallel arrangement, symmetric reflection; b) antiparallel arrangement, symmetric reflection; c) antiparallel arrangement, asymmetric reflection

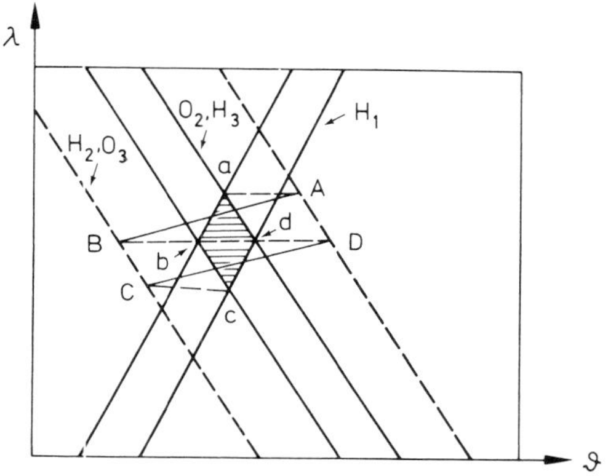

Fig. 9.8 DU MOND's scheme illustrating changes of angular and spectral divergence of beams in three-crystal MCC (see text) (after NAKAYOMA et al. [1.19])

9.4.3 Rigorous Theory of X-Ray Diffractometers

As can be seen from considerations above, the theory of diffractometers with two- and three-crystal Bragg reflection is of a phenomenological nature. The standard checkup equations of the type of (9.15) or (9.14) are not accurate enough, since the integrand functions are represented by reflection coefficients derived in the incident-plane-wave approximation.

PINSKER and CHUKHOVSKY [1.56] have briefly formulated a more rigorous theory of such diffractometers. In their paper, the radiation falling on the first crystal is not superimposed on the restrictions with regard to the angular divergence or the width of the wave front. Here, for the space distribution of a singly and a doubly reflected beam, use is made of expressions of the type (II.65) and (II.69) derived in Chapter 11 with the aid of the generalized theory on the basis of the Takagi equations. The wave function distributions indicated take the form

$$D_h^{(I)}(z) = \exp(iz\chi_o/2) \int_{-\infty}^{+\infty} dz' G^{(I)}(z - z';\alpha) D_o^{(i)}(z') \exp(-iz'\chi_o/2) \quad (9.63)$$

$$\begin{aligned} D_h^{(II)}(z) &= \exp(iz\chi_o/2) \int_{-\infty}^{+\infty} dz' G^{(II)}(z - z';\, \alpha + \beta) D_h^{(I)}(z') \exp(-iz'\chi_o/2) \\ &= \exp(iz\chi_o/2) \cdot \int_{-\infty}^{+\infty} dz' \int_{-\infty}^{+\infty} dz'' G^{(II)}(z - z';\, \alpha + \beta) G^{(I)}(z' - z'';\alpha) \\ &\qquad D_o^{(i)}(z'') \exp(-iz''\chi_o/2) \quad . \end{aligned} \quad (9.64)$$

Here, α and β are the angular deviations of crystal I from ϑ and of crystal II from parallelism with I, respectively.

Similarly to (9.64) one can write the expression for a diffractometer consisting of N crystals, which contains a N-tuple interval and Green or influence functions from $G^{(I)}$ to $G^{(N)}$.

To describe the incidence of a plane wave on the first crystal, (9.64) is reduced to the convolution equation (9.15). Of practical interest is the transmission of a narrow beam (the width of the wave front of the incident beam is $\delta(z)$, δ, delta-function) through a double-crystal Bragg-Bragg diffractometer. In this case, using the asymptotic representation of a general-type integral, the reflection coefficient can be analyzed $R_h^{(II)}(z;\beta) = |D_h^{(II)}(z;\beta)|^2$ in relation to the disorientation of β in various ranges of the variable z. For the wave function we get:

$$D_h^{(II)}(z) = \exp\left(\frac{iz\chi_o}{2} - \frac{iz\alpha}{4} - \frac{iz\beta}{8}\right)$$

$$\begin{cases} -\left(\frac{\chi^2}{2\beta}\right)\sin\frac{z\beta}{8} \quad , \quad |z\chi/2| \ll 1 \\ \frac{4\chi^{1/2}}{\pi^{1/2}z^{3/2}}\left[\frac{\sin\left(\frac{z\chi}{2}+\frac{z\beta}{8}+\frac{\pi}{4}\right)}{\sqrt{\beta(\beta+4\chi)}+2\chi+\beta} - \frac{\sin\left(\frac{z\chi}{2}-\frac{z\beta}{8}+\frac{\pi}{4}\right)}{\sqrt{\beta(\beta-4\chi)}-2\chi+\beta}\right] , \quad |z\chi/2| \gg 1 \quad . \end{cases} \tag{9.65}$$

This result corresponds to the oscillation of the reflected wave, with a reduction in the amplitude on increase in the spatial z and angular β variables. The spatial oscillation period is defined by the value $(\chi_{hr}\chi_{\bar{h}r})^{\frac{1}{2}}$ which makes it possible to use the oscillation period to determine the structure amplitude.

It should also be noted that, in principle, one can obtain from (9.65) exact equations for the reflection curve with the given experimental parameters, the angular and spectral distribution of the amplitude. This, in turn, opens up prospects for precision experimental determination of these parameters.

9.4.4 Investigation and Utilization of Bragg-Reflection Curves

The results of investigations into the reflection curves from more or less perfect single crystals with the use of spectrometers for up to the 1940's and 1950's are described in the monographs [1.6,9,13]. Of the investigations carried out in the 1950's and early 1960's we shall mainly dwell on the works of BROGREN and co-workers [9.2,3,19,20] and BONSE [7.3]. BROGREN and co-workers obtained rocking curves of transmission and reflection from calcite and quartz (1954) and also from quartz, germanium, and silicon (1962-1963) over a wider spectral range than ever before. In plotting the reflection curves, use was made of spectra obtained mainly by the Bragg-Laue scheme, and, to a lesser extent, by the Bragg-Bragg and Laue-Laue schemes. Characteristic radiation, as well as white radiation from a tungsten anode with a crystal monochromator, was used. An important result of these investigations was the plotting of curves for transmission and reflection from crystals over a wide range of thicknesses and wavelengths. These curves apply to both K- and L-absorption, and their shape depends exclusively on the product μt in accordance with the theory outlined in Chapter 4. This can be seen from Fig. 9.9, which shows similar reflection curves from calcite crystals of different thicknesses obtained in Mo $K\alpha_1$ and Au $L\beta_1$ rays with μt = 2.6 and 2.1. We also note the characteristic shape of the transmission curves, which reveal the presence of the Borrmann effect (Fig. 9.9,I). When discussing the results, the authors compared the experimental and calculated (for the convolution curve) parameters of the curves obtained: half-width, R^{max}, T^{max}, and their ratio, the integrated values R_i, T_i (and the displacements of the maxima of R from the maxima of T for thick, and from the minima of T for thin specimens) (Fig. 9.10). In some cases no theoretical data could be obtained.

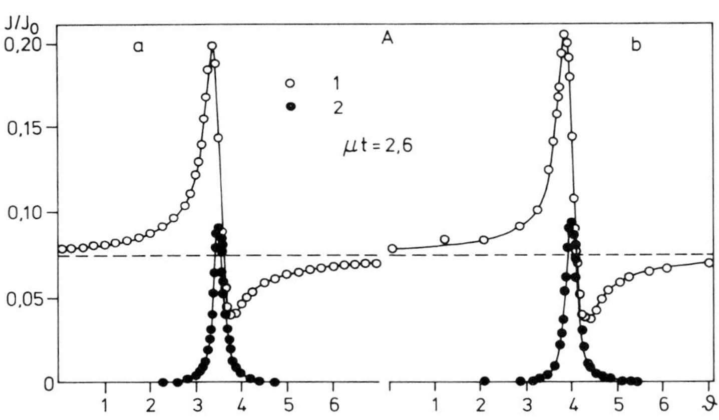

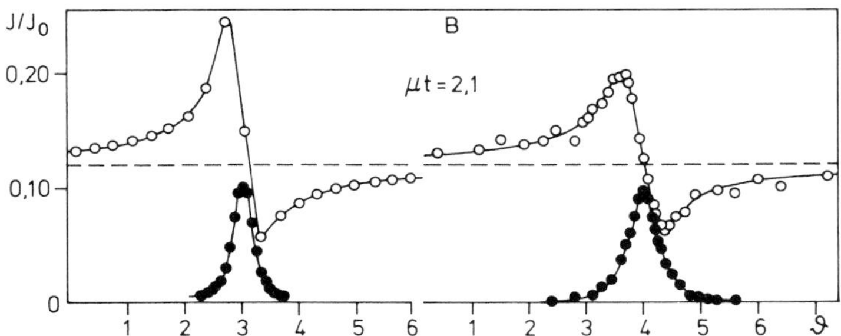

Fig. 9.9 a and b Transmission (I) and reflection (2) measurements of 2$\bar{1}$3$\bar{1}$ from calcite. A) Scheme Laue-Laue, radiation Au$L\beta_1$; B) Bragg-Laue, Mo $K\alpha_1$. a: Theoretical curve; b: Rocking curve (after BROGREN and ADELL [9.20])

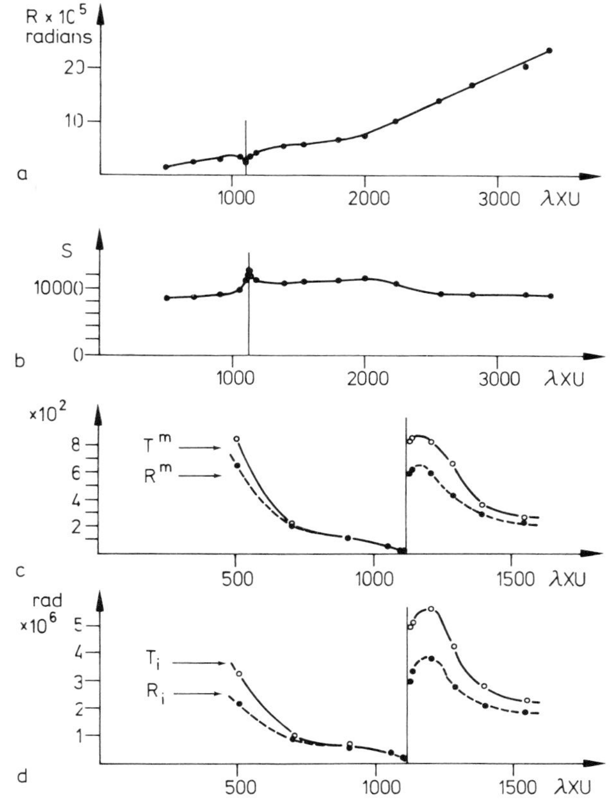

Fig. 9.10a-d Change in reflection and transmission parameters for 220 from Ge vsλ. a) Reflection coefficients and b) resolving power in Bragg case; c) reflection and transmission coefficients and d) integrated values in Laue case (after BROGREN and HÖRNSTRÖM [9.20])

The discontinuities in the values of the half-width, the coefficient, and integrated value of reflection on the absorption edges are of interest. These discontinuities are more pronounced in Laue reflection than in Bragg reflection. As shown by COLE and STEMPLE [8.4], Bragg reflection from a mosaic crystal is accompanied by a large discontinuity on the absorption edge than in dynamical scattering. This difference can be used in checking the degree of perfection of crystal specimens (see also [9.21]).

Systematic experimental verification of the dynamical theory in the case of Bragg reflection and transmission from Ge has been carried out by BONSE [7.3]. He used a classical two-crystal spectrometer according to the scheme (n,-n). The results of his measurements obtained for the values of transmission T_{Br} are of interest. Using a computer, he selected the values of χ_{hi} for reflections 444 and 333 at which the theoretical curve, according to the convolution formula (9.26) for $T(\beta)$, coincides best of all with the experimental one (Fig. 9.11).

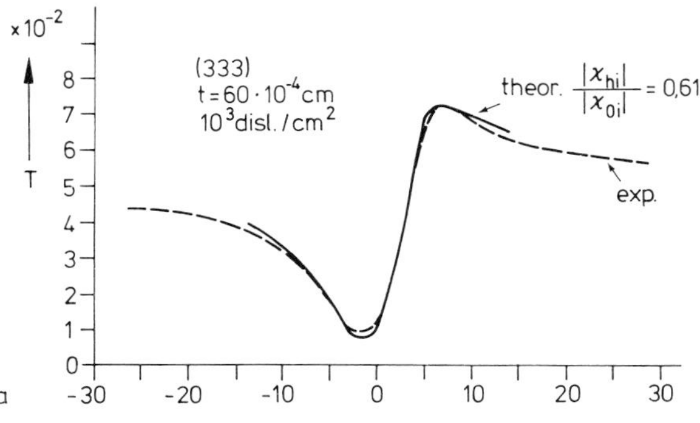

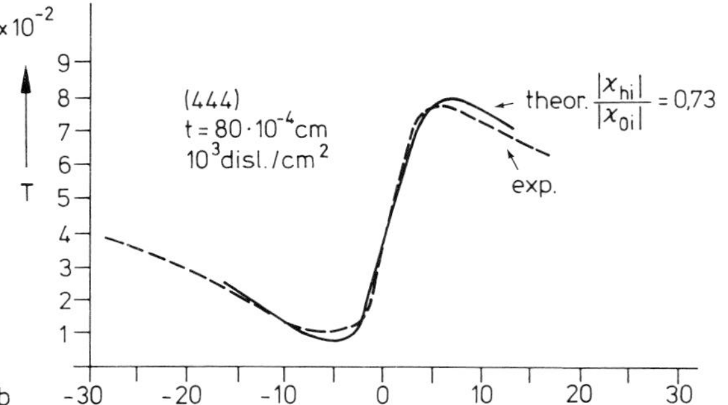

Fig. 9.11 Comparison of experimental and theoretical transmission curves in Bragg case. Ge, $CuK\alpha_1$, radiation (after BONSE [7.3])

BONSE obtained $|\chi_{hi}|/|\chi_{oi}|$ values of 0.61 for 333 and 0.73 for 444.

It should be noted that the convergence of the experimental and theoretical values of such parameters as the half-width and the height of the maximum of the reflection curves in the investigations of BROGREN and BONSE is in many cases of the order of 10 to 15 percent. In some cases the experimental values of R_{max} exceeded the theoretical ones.

We have previously mentioned the results of investigations aimed at improving the diffractometers designed to obtain rocking curves as close as possible to the dynamical maximum from one crystal, and also to the utilization of the parameters of similar curves.

It would seem that these curves could be used directly when checking slight defects in crystals of a sufficiently high degree of perfection. But there are at least two limitations. First, such defects as point defects of an extremely diverse nature and their aggregations make a considerable contribution to diffuse scattering, which often manifests itself only by an increase in intensity or in the coefficient of reflection on maxima tails.

Second, with a sufficiently perfect structure, which can, for instance, be simultaneously checked by the cruder method of X-ray topography, the quantitative treatment of the maxima shapes is difficult because of the insufficiently accurate data on such parameters as χ_{hr}, χ_{hi} and, particularly, on the dispersion correction Δf and the temperature factor exp(-M).

At the same time, the researchers, who had at their disposal perfect specimens of Ge and Si crystals, used the reflection curves to determine precisely the absolute values of the structure (and atomic) scattering amplitudes in these crystals.

Such determinations were carried out with the aid of the Bragg reflection curves of Cu $K\alpha_1$, Mo $K\alpha_1$, and Ag $K\alpha_1$ radiations. The principal experimental data in these investigations were the angular half-width of the curve corrected for the deviation of the crystal from the exact position for symmetrical arrangement, and the percent reflection or $R_{max} \equiv R_o$. This latter value served as the criterion of whether the test specimen was sufficiently perfect.

When determining the structure amplitude, some authors make use of the equation relating to the half-width of the maximum $2\eta_{\frac{1}{2}}$ and $\Delta y_{\frac{1}{2}}$:

$$w \equiv 2\eta_{\frac{1}{2}} = \Delta y_{\frac{1}{2}} |\chi_{hr}| / \sin 2\vartheta \quad .$$

At the same time, the shape of the maximum and hence the half-width in Bragg-reflection from the absorbing crystal depends on x and g as well. Although the numerical values of x and g given in Section 8.3 for ordinary Ge and Si experiments are rather low, they may prove to be essential in precision measurements.

The most objective method is evidently the one used in the above-mentioned paper of BONSE [7.3], i.e., selection of scattering parameters corresponding to the best agreement between the theoretical and experimental curves.

NAKAYAMA et al. [4.23] used this method to determine the atomic scattering amplitudes for Si, reflections 111, 422, and 333, and Ge, reflections 111 and 220, in Cu $K\alpha_1$ radiation. These determinations should be considered quite precise. The values of f obtained differ by $\sim$ 0.1-0.4 percent from those given in the best investigations using the Pendellösung fringes, the interference pattern from a crystal with a small gap.

Similar determinations were performed by KIKUTA [9.16], MATSUSHITA et al. [9.14] (see Fig. 9.12) and also MATSUSHITA and KOHRA [4.22]; the latter paper was used in Section 4.2. These authors also attained values of the maximum (percent) reflection differing by $\sim$ 1 percent from the theoretical ones.

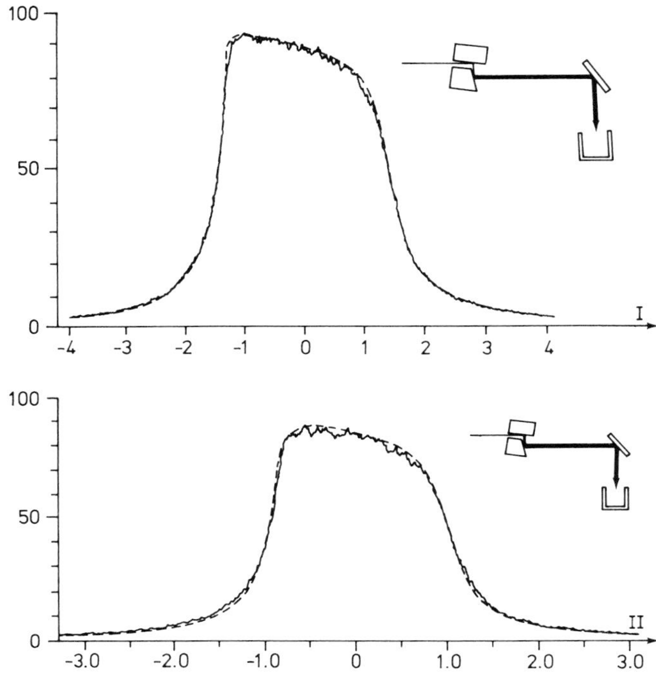

Fig. 9.12 Rocking curve with minimum convolution effect obtained with three-crystal diffractometer. $W_{exp} - W_{th} = 0.014 - 0.020''$, $R^{max}_{th} - R^{max}_{exp} = 1.5$ %. I) Symmetric reflection $CuK\alpha_1$, 422 Si. II) Symmetric reflection $CuK\alpha_1$, 333 Si. Dotted line - theoretical maximum (after MATSUSHITA et al. [9.14a])

As well as determination based on the use of the parameters of the shape of the reflection maximum, some authors determined the atomic scattering amplitudes on the basis of the measured values of integrated Bragg reflection. Among them are the papers by DE MARCO and WEISS [4.24], JENNINGS [4.25], and also a series of papers by ELISTRATORY and EFIMOV, whose main objective was verification of the degree of perfection of the crystal by the "integrated characteristics" in Laue reflection [4.15]. The difficulties involved in such determinations are due both to the need for parallel and independent verification of the degree of perfection of the test specimen by another method and to the low sensitivity of the integrated reflection to variations in $|\chi_{hr}|$ (in Bragg reflection). Nevertheless, owing to its relative simplicity, this method may be regarded as the most universal one, applicable to a wide range of materials. Note the paper by KOVIEV et al. [9.22]. The authors make use of the measured values of maximum and integrated half-width of the rocking curve. KOVIEV and BATURIN [9.23] proposed a method for reconstruction of the reflection curve on the basis of experimental values of R (X) referring to the curve tail.

In such a reconstruction, the value of R(y) (8.50) is used, and the equation is solved for y (the adopted symbol is y_r). Moreover, the authors showed that the convolution effect need not be taken into account.

In this case, by using (3.26) at b = 0 we obtain the *linear relationship* between the angular half-width of the maximum w and the function X(R) for large y_r:

$$w \equiv 2\eta_{\frac{1}{2}} = \frac{2|\chi_{hr}|}{\sin 2\vartheta} X \quad ; \quad X = \frac{n+1}{2\sqrt{n}} \quad ; \quad n = \left|\frac{\chi_h}{\chi_{\bar{h}}}\right| R\ \mathrm{exp} \quad . \tag{9.66}$$

The value of w is determined for X = 1. The results of measurements of f for reflections 422 and 333 of Cu $K\alpha_1$ radiation from Si obtained by the authors are in agreement with the best measurements in other investigations; an appreciable divergence is observed for 220.

When comparing methods of determining the scattering parameters based on the experimental curve of Bragg reflection with other methods, (it should be noted that measurements of the periods of the pendulum solution or interference measurements) a number of advantage indicate the reflection curves, particularly of those obtained in diffractometers with a high-precision MCC. These advantages include the possibility of studying crystals with a known range of degree of perfection (the reflection curve is simultaneously the "quality certificate" of the crystal), and also over a wide range of absorption values. At the same time, the method under review controls a small surface area of $\sim 5\ \mathrm{mm}^2$, which is also an advantage.

Finally, despite the complexity of the equipment, this method is undoubtedly more efficient; it makes it possible to obtain the principal experimental material, i.e., the *reflection* curve, more rapidly (as compared with photographic recording) and easily. Later on (in Chapter 10, in connection with interference effects) we shall revert to determination of the absolute values of the atomic amplitudes.

At this point we must give a general evaluation of such investigations. Obviously they can be carried out only with single-crystal specimens of the appropriate compounds which have a sufficiently perfect structure. The growing of such crystals has been continuously expanding in connection with the utilization of an increasing range of substances, which are of interest from the point of view of physical investigations and technological applications.

In all cases, accurate measurements of the scattering parameters are important for the study of dynamical scattering in the presence of some types of defects in the ideal structure, and hence for quality control of the scattering crystals. Of great interest also is the reliable verification of the typical deviations in the values of the atomic amplitudes (and, hence, in the electron density distribution) of the atoms in crystals of different structure as compared with the latest theoretical values for free atoms. An extremely important parameter in controlling the degree of perfection of a crystal specimen is the lattice spacing. The procedure for absolute determination of lattice spacing was developed by BOND [9.24]. This author made his measurements with an accuracy of the order of 10^{-6}-10^{-7}, which corresponds to the accuracy with which the wavelengths of scattered X-radiation have been determined.

Dislattes and Hennings used X-ray and optical interferometry (see Section 10.4) to measure the absolute value of the silicon lattice spacing with an accuracy of $\sim 1.10^{-8}$ of the value of the spacing, i.e., about one order more precisely than in BOND's determination.

However, reliable and very accurate relative determinations, using reference specimens, are more essential for investigating the relationship between the slightest change in the lattice spacing of the crystal and the changes in its composition, structure, and some of its properties.

Without dwelling on other methods of making such relative measurements, we shall mention the so-called two-beam diffractometry proposed by HART in 1969 [9.25]. In this work he considered measurement schemes, using Bragg and Laue reflections.

The principal idea and scheme of two-beam diffractometry is vividly illustrated by the measurements made in the investigation of KOVAL'CHUK et al. [9.26]. These authors used an ingenious scheme of Bragg measurements, which is in some respects more convenient in practical work (see Fig. 9.13).

An X-ray beam (1) passes twice through the crystal under investigation (B) and is reflected from the reference crystal (A) in the maximum region. The reflected beam (II) will be recorded by the counter (5). Its intensity will be reduced by I_0 exp-$2\mu\ell$, where ℓ is the length of the respective oblique path segments in the crystal B. If (as might be expected) the lattice spacing a and the spacing d_{hkl} of crystal B differ from these values for crystal A, then, rotating crystal B, first to the reflecting position for beam 1, then for beam 2, each time we shall record minima against a background corresponding to reflection from the fixed crystal A with an allowance for absorption losses in B. The angular distance between these mini-

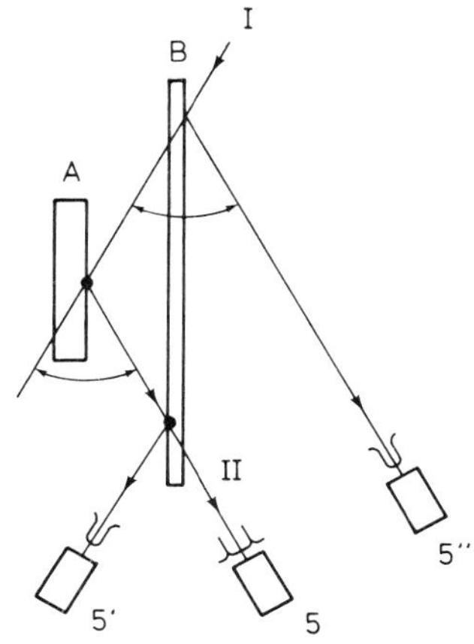

Fig. 9.13 Scheme illustrating the application of two-beam diffractometry to high-precision relative measurements of lattice spacing [9.26]

ma (Fig. 9.14) enables us to determine the ratio $\Delta d_{hkl}/d_{hkl}$. To do this, we can use the following relation with sufficient accuracy:

$$\frac{\Delta d_{hkl}}{d_{hkl}} \approx - \Delta\vartheta \cot\vartheta \left(1 + \frac{2|\chi_{or}|d_{hkl}}{\lambda^2}\right) \quad . \tag{9.67}$$

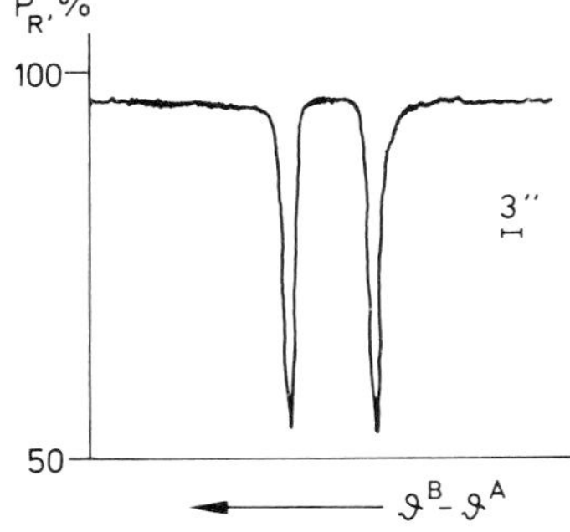

Fig. 9.14 "Extinction" minima obtained by means of scheme represented in Fig. 9.13. Reflection 333 $MoK\alpha_1$, from Si. Displacement of left-hand minimum relative to the right-hand one makes 1.20 ± 0.5 and correspond to the value of d/d = $(1.20 \pm 0.05)\cdot 10^{-4}$. This amount of lattice spacing change results in boron doping

In the angular range $\vartheta \approx 90^0$, $\cot\vartheta \approx 0.01''$ with the counting accuracy $\Delta\vartheta \approx 0.01''$, the range of the ratios Δ d/d measured reaches $\sim 10^{-9}$. As an example of HART's measurements we give Fig. 9.16, illustrating a change in lattice spacing for Si of $\Delta d/d \sim 1.10^{-4}$ which results from phosphorous doping $\sim 10^{-5}$.

Along with relative measurements of the lattice spacings, BAKER and HART [9.27] used one of the schemes described in [9.25] for absolute and high-precision measurement of the lattice spacing, using the above-mentioned determination by Dislattes and Hennings. With an appropriate Si crystal as a reference specimen, in a scheme with one X-ray source (Fig. 9.15) and a single crystal of Ge as a test specimen, the authors selected reflection of Mo $K\alpha_1$ radiation form the two crystals with similar angles ϑ, to reduce the dispersion effect. Reflections 355 Si and 800 Ge were used. The result obtained was recorded as

$$\frac{d_{800}\ \mathrm{Ge}}{d_{355}\ \mathrm{Si}} = 1.0002458\ (\pm 0.0000016)\ \text{at}\ 25^{\circ}\mathrm{C} \quad .$$

Using the above-mentioned measurement of the silicon lattice spacing, the authors of [9.27] obtained for the spacing:

$$\alpha_{Ge} = (5.65790 \pm 0.0000092)\ \mathrm{\AA} \quad .$$

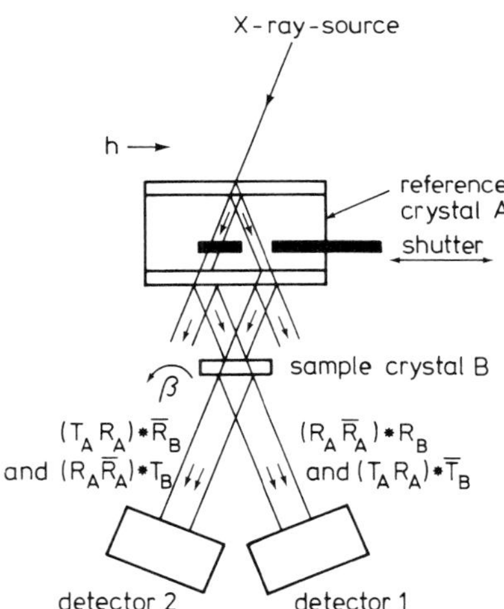

Fig. 9.15 HART's scheme for absolute and high precision measurements of lattice spacing (after HART [9.25])

Incidentally, it should be noted that as regards our knowledge of the mechanism of the effect of the minutest impurities and various experimental conditions on the values of the crystal lattice periods, as well as the possibility of reproducing such measurements on another crystal specimen, we are still very far from interpretation of such unique measurements.

It can be seen from the foregoing that with the present level of accuracy in measuring the reflection curves from single crystal specimens, as

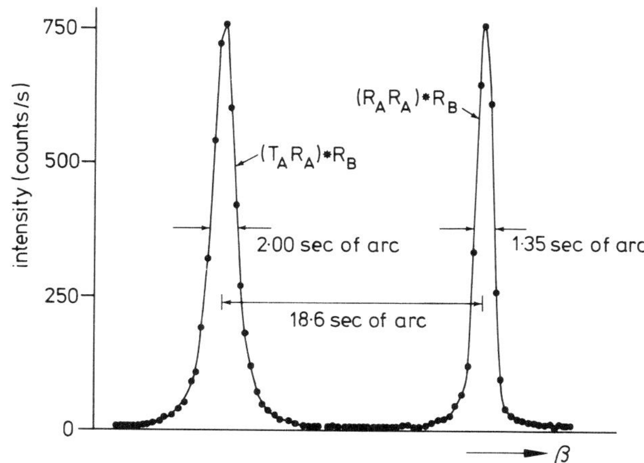

Fig. 9.16 Rocking curves obtained with the arrangement illustrated in Fig. 9.15. ± 440 from Si, MoKα_1. Reference crystal (right curve) undoped and sample doped to P concentration of $7.5 \cdot 10^{19}$ atoms cm^{-3}; d/d = $1.10 \cdot 10^{-4}$ ± $0.02 \cdot 10^{-9}$ (after HART [9.25])

well as with the use of two-beam diffractometry and other methods of high-precision measurements of crystal lattice spacings, the diffractometers have to meet very high requirements. As regards monochro-collimating, even in the two-crystal version of the diffractometer, after a single symmetrical reflection it is necessary to obtain a beam with an angular width in a total-reflection range of the order of 0.1". This width produces a negligibly small convolution effect with a theoretical half-width of the curve of reflection from the crystal under investigation equal to 3-4" or more.

However, in recording the hyperfine structure, for instance, of the subsidiary maxima of the pendulum solution and in recording weak reflections, particularly from an absorbing crystal with an angular width of ~ 1" or less, one must use at least a two-crystal MCC to obtain an exploring beam with an angular width of the order of 0.01". The practice of making such measurements has evidently shown that single-block MCC are less convenient than designs with separate crystals.

This imposes some other requirements on the diffractometer, namely the possibility of using it for measuring curves over a wide range of reflection angles up to angles close to 90^0.

Finally, such methods as two-beam diffractometry, recording of intensity curves on moiré patterns, and some others require adjustment of the two crystals with an infinitely small spacing between them.

In this connection it should be notet that the diffractometer (spectrometer) designs described in the literature, which amount to two basic types, those of ALLISON and WILLIAMS [9.28], and ROSS [9.29] and BROGREN [9.30], do not meet the requirements indicated and, moreover, require much adjustment time in each new experiment. One can, for example, refer to the procedure for the adjustment of the three-crystal spectrometer outlined by BUBAKOVA [9.31]. The authors of [9.26] and [9.32] describe the designs of two- and three-crystal diffractometers which use the principle of coincidence of the rotation axes of the two crystals (Fig. 9.17).

In the two-crystal spectrometer [9.32], the rotation axis of the crystal-analyzer B is combined to the rotation axis of the monochromator crystal A (0^{AB}-axis). In the initial setup, in order to find the zero position of the two crystals, the X-ray tube anode (1) and the collimator (2) are alligned on the optic axis of the device. The zero positions of the crystals (first B, then A) are found by linear displacement of each of their mechanisms $L_{\perp}$ and $L_{\parallel}$ of them, first to the instrument axis and then by turning in the goniometric head and slightly rotating near the reflecting position. Recording is done by counter 5.

After this, the source with the collimator is moved away by the experiment-assigned segment L, and crystal A is set in the reflecting position by two mutally perpendicular displacements so that the ray reflected from it passes through the 0^{AB}-axis. The setting is terminated by reverse displacement of crystal B to the 0^{AB}-axis and additional adjustment for reflection by a slight turn. Then the curve of reflection from B is recorded. In the antiparallel arrangement, only the angular positions of crystal B and counter 5 are changed.

Switching to other diffraction angles of the characteristic or continuous spectrum is effected simply by consecutive rotation of the crystals and changing the components of the displacement of crystal A along two directions ($L_{\perp}$ and $L_{\parallel}$ in Fig. 9.17), while retaining the value of the total displacement L. Because of the coincidence of the rotation axes of A and B, the axes nonparallelism error on switching to other diffraction angles is of the order of 1". It is also easy to see that the design described above readily permits asymmetric arrangement, recording of reflections up to $2\vartheta \approx 170^{o}$, and infinitely close mutual positions of the two crystals. In the three-crystal spectrometer, to begin with crystal C is set in the reflecting position. To do this, source 1 is turned by an angle of $2\vartheta_C$ and the first crystal, by ϑ_C.

We also note that the above-described scheme of two-beam diffraction used for relative determination of lattice-spacings is realized very conveniently in these diffractometers.

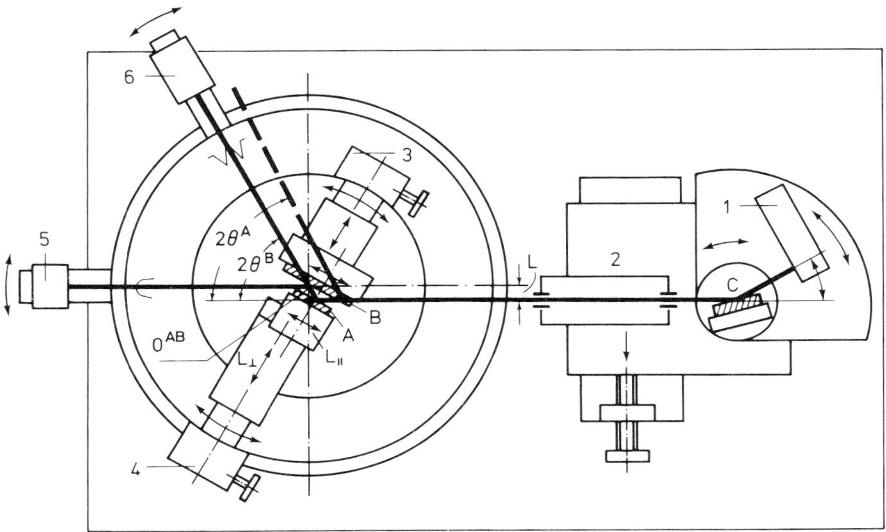

Fig. 9.17 Scheme of three-crystal diffractometer with coinciding rotation axes for the second and third crystals (A and B) [9.26, 32]

Keen interest has recently been shown in the utilization of X-ray synchrotrone radiation. Some of its properties have been outlined [9.33, 34], namely:

1) A continuous spectral range, which can extend from 0.1 Å to the visible range of the electromagnetic scale, depending on the maximum energy of the electrons accelerated.

2) High intensity, both for the given spectral range $\Delta\lambda$ and for the given space angle $\Delta\Omega$.

3) The comparatively small angular spread of the radiation.

4) The radiation is almost completely polarized; only a component with an electric vector parallel to the plane of the electron orbit is left.

The main problem, whose solution is essential for the successful utilization of the synchrotrone radiation in the dynamical scattering of X-rays, is monochromatization with the purpose of obtaining a beam with a given wavelength λ and a (small) spectral range $\Delta\lambda$. This X-ray source must, according to all the available data, exceed in intensity the conventional tubes in use by several orders.

The papers so far published discuss or apply various principles for monochromatization of this radiation. We mention the following:

1) Successive diffractions used in the parallel groove in a highly perfect single crystal of Si (or Ge). As already mentioned above, this method results in almost complete absorption of the tails of the reflection curves.

2) A combination of a single crystal with a filed-through groove, and an absorption filter, or the use of a detector with sufficient spectral dispersion.

3) The use of special synchrotrone operating conditions in order to obtain a more convenient spectral characteristic of the X-radiation.

4) The use of total external reflection from a plate at very small glancing angles of incidence.

The principal difficulty which could not be overcome by using the methods indicated (in any case the first three) is the presence, along with the main wavelength (preferably $\lambda \approx 1.54$ Å), of a number of harmonics: $\lambda/2$, $\lambda/3$, $\lambda/4$, etc., which pass through the diffraction devices.

It should be emphasized, nevertheless, that there are some examples of successful utilization of the radiation monochromatized by the methods indicated for X-ray topography, with exposure reduced from tens of hours (under ordinary conditions) to ~ 1 s. See HART's paper [9.35].

A new method, outlined and checked experimentally in the work of BONSE et al. [9.33], may prove highly promising. Its essence lies in the utilization of the scheme of the double-crystal spectrometer with an almost parallel arrangement of (both fixed) crystals, for which the reflection angles ϑ are slightly different. In this case, an increase in the difference between the displacement of the maximum midpoints from the Bragg angle η_0 (see (7.25)), from a small one for the principal wavelength to an appreciable one for harmonics, will result in the fading of the latter. Experimental investigation, which has been carried out, for instance, with reflection 220 from a Si-Ge pair, showed that the first harmonic yields a maximum of the reflection coefficient, which constitutes $3 \cdot 10^{-3}$ of the main one, while the higher harmonics yield $\sim 10^{-4}$. The investigations were conducted on the DESY accelerator in Hamburg.

9.4.5 Investigations into the Interference Effects of the Pendulum Solution

Following the paper of KATO and LANG [7.23], many investigations have been carried out in which the authors succeeded in realizing and studying, more or less throughly, both lines of equal thickness and subsidiary maxima of the pendulum solution. Owing to the work of KATO and his colleagues, topograms with hyperbolas associated with the interference effect during the incidence of a sperical vacuum wave have been studied very carefully. In Chapter 6, such section patterns are described in detail in connection with the theory developed by KATO.

Determining the Absolute Values of Atomic Amplitudes

A large series of investigations carried out over the last seven years is dedicated to absolute determinations of the atomic scattering amplitudes with the aid of topograms with hyperbolic bands of the pendulum solution. These investigations have been made predominantly by KATO, and also by HART, in cooperation with their colleagues.

Determination of atomic amplitudes from the pendulum solution was carried out, using a wedge as described in Chapter 6. However, direct application of equations of the type of (6.76) and (6.77) is difficult, since the geometrical factor Φ_h used (when it is necessary to deviate from the simplest geometry of the experiment) requires special investigation. In this connection, as far back as 1967, KATO et al. [6.4,5] developed an experimental procedure under which one and the same specimen of a crystal wedge is used successively, first to obtain a section pattern with hyperbolic maxima, and then to determine χ_{or} and F_{or} with the aid of interferograms. The latter are obtained by interposing the wedge in the path of one of the two coherent beams in the interferometer described in Chapter 10. The transmission of the X-ray beam through the wedge when it falls beyond the maximum entails a periodically varying path difference between the coherent waves and the formation of an interference pattern, whose period Λ_o is determined (similarly to Λ_h in (6.76)) by the equation

$$\Lambda_o = (\lambda/|\chi_{or}|)\Phi_o \tag{9.68}$$

where Φ_o is a geometrical factor similar to Φ_h.

Switching from χ_{hr} and χ_{or} to F_{hr} and F_{or}, we note that if the experiment setup is such that transition from the section pattern to the interference pattern is performed with a minimal and definite number of operations, for the ratio of the two structure amplitudes we can write

$$\frac{F_h}{F_o} = \frac{1}{2\cos\vartheta}\frac{\Lambda_o}{\Lambda_h} B \quad , \quad B = \frac{\Phi_h}{\Phi_o} \quad . \tag{9.69}$$

In this way we ensure more accurate correction of sources of error in determining the geometrical factor.

The most accurate absolute measurements of the atomic amplitudes with the aid of the pendulum solution hyperbolas, however, were made in the investigation by ALDRED and HART [4.21] already cited in Section 4.2. These authors conducted their experiment in conditions of symmetrical reflection - more precisely, with the reflecting planes perpendicular to the entrance surface of the wedge. A thorough study of strains in silicon specimens was carried out. As had previously been shown by KATO [9.36] and HART [9.37], for small deformations and certain limitations for stress gradient, the order of the band (hyperbola) and the corresponding range depend on some parameter p:

$$p = \frac{S_n \; t \; \tan\vartheta}{C\sqrt{\chi_{hr}\chi_{\bar{h}r}}} \tag{9.70}$$

where S_n is the average value of the stress normal to the reflecting plane. A topographic investigation of the specimens made it possible to select the most favourable conditions for obtaining patterns corresponding to certain reflections. The use of a stepped wedge, freely resting on a support, provided both a constant value of the gradient in a small volume and the absence of stresses which might have arisen during the fastening of the crystal. The measurements were made at two temperatures, i.e., room temperature (293 K) and 90 K. Much attention was given to controlling the specimen temperature during exposures. Small temperature fluctuations were taken into account when reducing the data obtained to each of the indicated values. As has already been noted in Section 4.2, when calculating the temperature factor the authors proceeded from the Debye temperature, Θ_D = 543 K.

Table 9.2 compares ALDRED and HART's values with the theoretical data of DOYLE and TURNER [4.6]. A considerable divergence can be noted between the theoretical and experimental values for the first reflections. This divergence undoubtedly indicates distortions of the electron density distribution in the valency shell of the atoms in crystals. The error in determining the experimental f values in [4.21] with an allowance for the inaccuracies in

Δf and exp(-M) is probably no more than 0.2-03 percent. The considerable deviation for reflection 555 may be due to some feature of the temperature factor for this reflection, which is noted by the authors of [4.21].

We believe that the paper of ALDRED and HART [4.21] is one of the most reliable experimental proofs of the distortion of the valency shells of the free atoms under the effect of the chemical bond in the crystal. The authors analyze these distortions in connection with DOWSON's papers [4.3,4,7].

Table 9.2 Amplitudes of X-ray scattering by Si atoms at 0 K (in electronic units)

(1)	(2)	(3)	(4)	(5)	(6)	(7)	(8)
$\frac{\sin\vartheta}{\lambda}$	hkl	Absolute experimental determinations[a]	Theoretical data HFCl[b]	Theoretical data RHF[c]	in % (4)-(5)	in % (3)-(4)	in % (3)-(5)
0.159	111	10.739	10.549	10.572	-0.22	+1.3	+1.6
0.26	220	8.651	8.712	8.733	-0.24	-0.70	-0.95
0.305	311	8.024	8.166	8.178	-0.14	-1.75	-1.9
0.368	400	7.444	7.511	7.520	-0.12	-0.90	-1.0
0.401	331	7.247	7.187	7.192	-0.07	+0.83	+0.8
0.451	422	6.711	6.707	6.709	-0.03	+0.06	+0.03
0.478	333	6.439	6.444	6.451	-0.01	-0.25	-0.3
0.521	440	6.041	6.040	6.045	-0.08	0.02	-0.07
0.638	444	4.982	4.976	4.992	-0.3	+0.12	-0.2
0.797	555	3.754[d]	3.762	3.772	-0.3	-0.11	-0.5

[a]Ref. [4.21]; [b]Ref. [4.2,3]; [c]P.A. Doyle, P.S. Turner. Acta Cryst., 1968, A24, 390; [d]see in [4.21] the author's remarks concerning the uncertainty in recalculation of the experimental data to zero Kelvin.

Some Other Pendulum Solution Investigations

In concluding this chapter, we shall dwell briefly on some other works dealing with the pendulum solution effects.

MALGRANGE and AUTHIER [9.38] obtained, in a single experiment, two scattering ranges differing in the ratio between the front width of the incident plane wave and the thickness of the crystal slice. The authors used a monochromator based on the Borrmann effect (see Fig. 9.4b).

An X-ray beam in the shape of a band fell on a Si crystal wedge so that the beam covered both the thick and the thin parts of the wedge (Fig. 9.18).

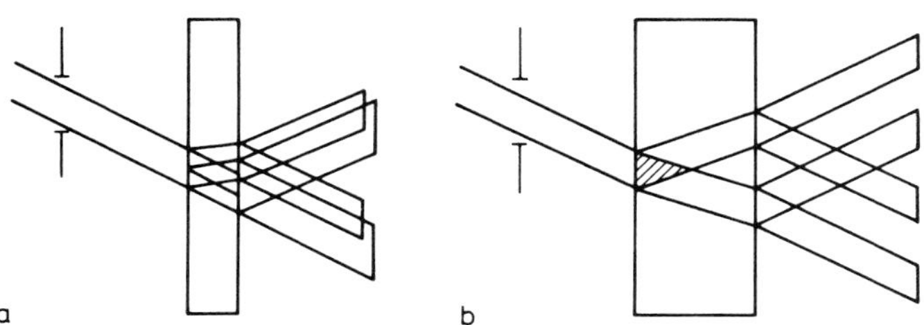

Fig. 9.18 a and b Scheme of MALGRANGE and AUTHIER experiment [9.38]; a) transmission of X-ray beam through thin part of crystal wedge leading to formation of pendulum solution bands. b) transmission through thick part of wedge and division of two wave fields forming in a crystal

In the thin part (thickness ~μ50 m) the wave fields overlapped, and an interference pattern was formed in agreement with the theory given in Section 3.3. Reflection 220 in Mo $K\alpha_1$ rays was studied. The calculation of the fringe spacing was based on (3.108) for $\gamma'_h \approx \cos\mu$. In this case

$$\Lambda = \lambda(\gamma_0\gamma_h)^{\frac{1}{2}}(C|\chi_{hr}| \tan\mu\sqrt{1 + y_r^2})^{-1} \quad . \tag{9.71}$$

Substituting $(\gamma_0\gamma_h)^{\frac{1}{2}} = 0.983$, $y_r \approx 0.64$, $\tan\mu = 0.21$, and $|\chi_{hr}| = 1.90\cdot10^{-6}$, we obtain $\Lambda = 147\mu m$ as against the experimental value of 132μ.

In the thick part of the wedge, the two fields separated. In order to determine the distance between the areas of the two fields on the exit surface we can use (5.27b) for the angles ε_1 and ε_2 formed by the vectors $\bar{\bar{\underline{S}}}^{(1)}$ and $\bar{\bar{\underline{S}}}^{(2)}$ with the bisector j of the angle 2ϑ between $\underline{s}_0$ and $\underline{s}_h$. The use of (5.27b), which was derived for symmetrical reflection, is justified by the closeness of the ratio $|\gamma_h|/|\gamma_0|_0$ to unity in this experiment. Thus

$$\tan\varepsilon_i = \mp \frac{y_r}{\sqrt{y_r^2 + 1}} \tan\vartheta \approx \mp 0.098 \quad .$$

This divergence of the vectors $\bar{\bar{\underline{S}}}^{(i)}$ on the exit surface of the slice 0.6 mm thick is equal to 120μm (in Fig. 9.19 with magnification 38X, it amounts to ≈ 5 mm).

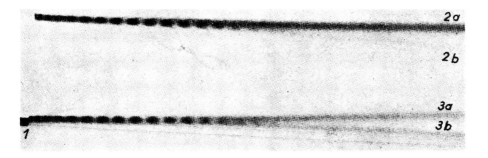

Fig. 9.19 Patterns obtained by MALGRANGE and AUTHIER [9.38]. Left-hand view of bands shows (constant) thickness maxima, right-hand view corresponding to transmission through thick region of wedge represents the beams of two wave fields (see text)

While the authors of this investigation observed lines of equal thickness, in another work LEFELD-SOSNOVSKA and MALGRANGE [9.10] and in the short note RENNINGER [3.3b] obtained subsidiary maxima of the pendulum solution from a near-plane-parallel silicon slice in Cu $K\alpha_1$ and Ag $K\alpha_1$ rays. Reflection 220 was investigated. The authors of [9.10] used a monochromator of the same type as in AUTHIER and MALGRANGE's work. Laue reflection and transmission curves with subsidiary maxima were obtained from slices of thickness from 13 to 47μm. These slices were in fact the bottoms of pits etched in a single crystal, and therefore they exhibited sufficient rigidity during the experiment. It can be seen from Fig. 9.20 that the reflection and transmission curves obtained showed sufficient resolving power; the angular divergences of the maxima were in agreement with the theory. At the same time, the insufficient contrast of the diffraction pattern deserves attention. The strong diffusion background may be due either to the inhomogeneity of the slice thickness or to the effect of the slit diaphragm setup between the crystal-monochromator and the test specimen. The effect of the diaphragm is analyzed in the paper.

Similar results were obtained for Bragg reflection in 9.39 . Using the single-crystal monochromator with asymmetrical Bragg reflection, the authors obtained a curve of reflection from a plane-parallel slice. It is evident from Fig. 9.21 (see (7.38-40)) that interference in the slice arises between the second field, which moves from above, and the first field reflected from the lower face. For these fields to overlap, the front width of the inci-

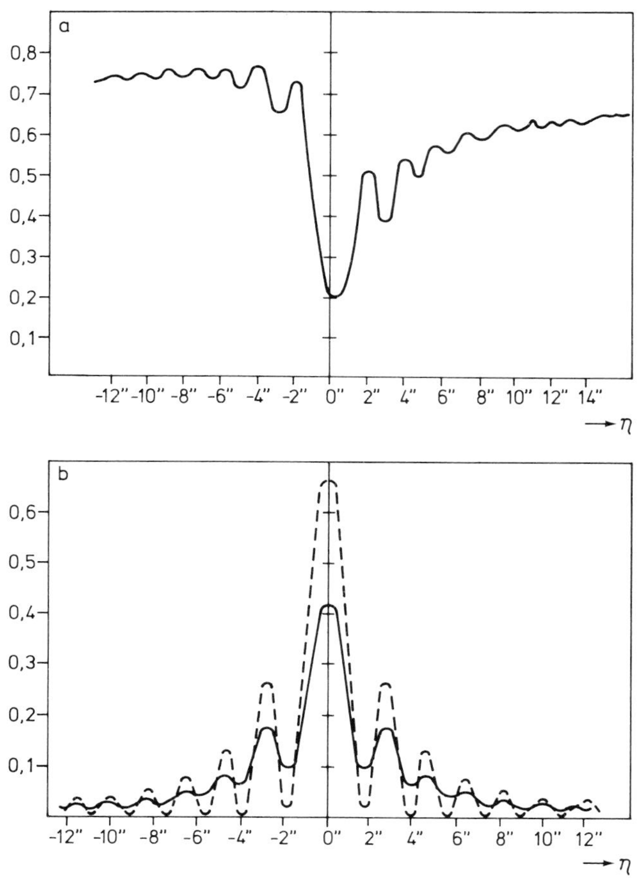

Fig. 9.20 a and b Rocking curves: transmission a) and reflection b) - with subsidiary maxima of Pendellösung (Laue case). Cu $K\alpha_1$ and Ag $K\alpha_1$, 220 Si. Dotted line - theoretical curve (after LEFELD-SOSNOVSKA and MALGRANGE [9.10])

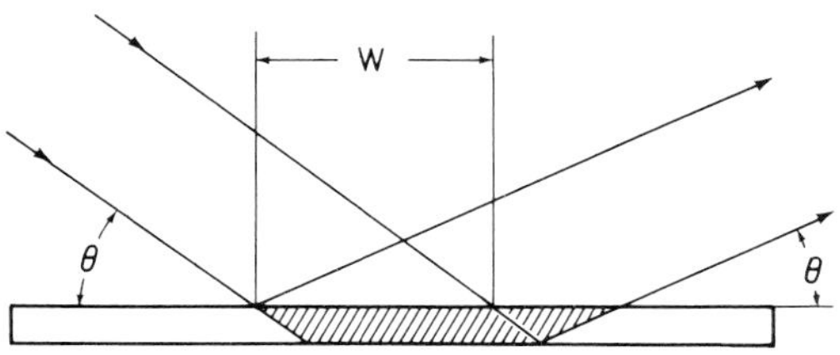

Fig. 9.21 Scheme of Bragg reflection from a plane-parallel slice leading to subsidiary maxima formation (after BATTERMAN and HILDEBRANDT [9.39])

dent wave must be sufficient. The experimental curve (Fig. 9.22) exhibits maxima whose positions are in agreement with theory (see (8.39) and (8.95-97)). However, here too, as in [9.10] for same reasons the contrast is very weak.

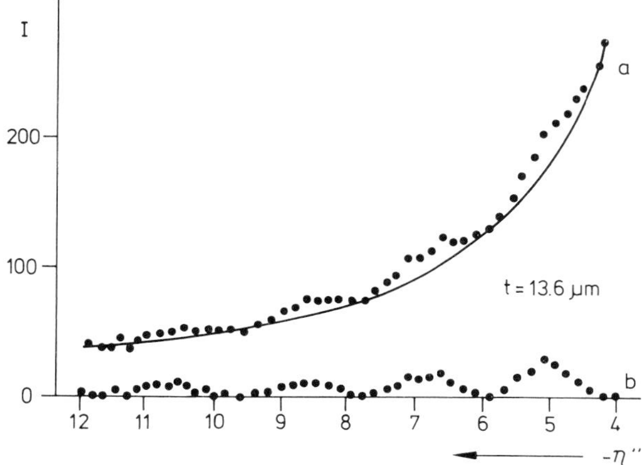

Fig. 9.22 Subsidiary maxima of Bragg reflection rocking-curve obtained by BATTERMAN and HILDEBRANDT [9.39]. The points in a) correspond to experimentally measured ones

10. X-Ray Interferometry. Moiré Patterns in X-Ray Diffraction

In the preceding chapters we described the theoretical and experimental investigations into the effects of the pendulum solution in conditions (or in an approximation) of both the incident plane wave and the spherical wave, to put it more precisely, of a wave packet with a very narrow wave front.

In Chapter 3 we noted that the pendulum solution bands may be regarded as the interference of radiation from two coherent sources. In Chapter 6 we used (as a mathematical approach, of course) an infinite set of plane waves.

In the 1960's and 1970's a number of investigations were been published dealing with the identification and theoretical analysis of diffraction patterns resembling pendulum solution fringe patterns; these patterns are, however, associated with plane defects in crystals [10.1,2], as well as with the successive transmission of an incident beam through two or three crystals separated by nondiffracting lamellae [1.28,51]. This entire group of phenomena and the corresponding theories may be united under the common heading of X-ray interferometry during interaction of coherent beams.

An important contribution to this study was made by BONSE and HART [10.3-6], who have designed some X-ray interferometers. The moiré effect discovered by them and detected, almost simultaneously and independently, by other investigators [10.7,8] has attracted much attention and become a method of X-ray topography; a characteristic feature of it is tremendous sensitivity to deviations in the crystal lattice periods (to values up to $\Delta d/d \sim 10^{-8}$). The most efficient design, developed by BONSE and HART, is an interferometer based on Laue reflection and consisting of three crystal plates. The calculations involved, which were made by the authors themselves, is elementary and based on the plane-incident-wave approximation.

It is worth noting that in their recent paper, BAUSPIESS et al. [10.9] develop a rigorous theory of a three-crystal spectrometer *without absorption* applicable to neutron interference. They use expansion in plane waves with subsequent integration in a way similar to KATO's papers reviewed in Chapter 6. The authors performed detailed numerical calculations of the intensity of the emerging beam $D_h^{(d)}$ (see Fig. 10.1).

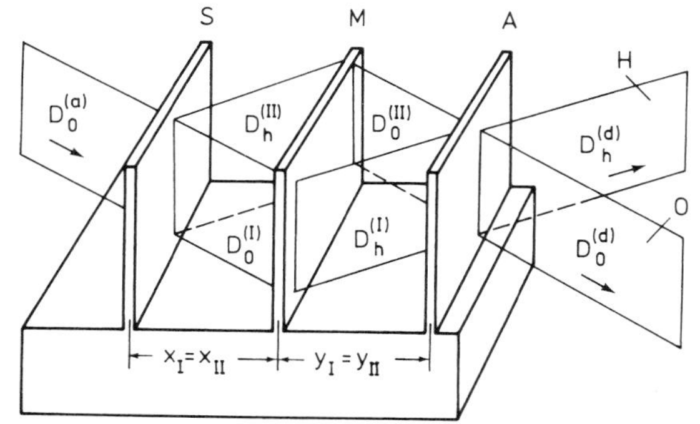

Fig. 10.1 Schematic diagram of an interferometer with Laue reflection (after BONSE and HART [10.3b])

By contrast, the "spherical"-incident-wave approximation (using expansion in plane waves) was applied by KATO et al. [10.1] to analyze the interferometry pattern in a crystal with a plane defect, and especially by AUTHIER et al. [1.51] when investigating an interferometer which consisted of two plates with a nondiffracting lamella. As shown further on in this chapter, and more comprehensively in Chapter 11, along with the plane-wave method, some authors (see Chapter 1) have developed a generalized theory based on Takagi equations; while being totally equivalent to the plane-wave method, this theory is mathematically more compact. The use of this second method will be demonstrated below in the case of the moiré theory developed by SIMON and AUTHIER.

In this chapter we shall also cover some experimental investigations in which the above-mentioned phenomena and equipment have been used in solving more specific problems.

10.1 Three-Crystal Interferometers

In these apparatuses, designed according to the classical scheme, the lenses of visible light optics are replaced by crystal plates which deflect incident beams by sufficiently large angles as a result of diffraction. It is well known that the refraction effect cannot be used because the difference (1-n) is negligibly small for X-rays. As shown in Fig. 10.1, the BONSE and HART interferometer [1.28, 10.3] consists of three plates with a common base and is made of a perfect single-crystal silicon block. Since for all the plates the reflection indices are the same, the geometrical parameters of the scheme are strictly adhered to (see below); the beams interferring in analyzer A must be coherent, with a sufficient degree of accuracy. It is obvious that plate S plays the part of a separator, plates M^{I} and M^{II} (in Fig. 10.1 one common plate M), which are called mirrors, "refract" beams D_o^{I} and D_h^{II}, and in analyzer A we have overlapping of beams D_h^{I} and D_o^{II} and interference of the fields induced in the plate.

In their cross section the emerging beams $D_o^{(d)}$ and $D_h^{(d)}$ will have a certain intensity distribution, depending on the experimental conditions. The author's main purpose was to achieve complete coincidence of the phases of the waves in crystal A forming the emerging beams $D_o^{(d)}$ and $D_h^{(d)}$. Despite the use of Cu $K\alpha_1$ radiation, the authors neglected absorption in their calculations; the wave vectors $\underline{k}_{o,h}^{(i)}$ in the crystal and the accommodation values $\delta^{(i)}$ for the refracted waves in the crystal and $\delta^{(i)'}$ for the diffracted waves were assumed to be real. At the same time, where the total thickness of ≈ 1.5 mm, which corresponds to $\mu t \approx 22$, the authors took no account of the strongly absorbed first fields in plates S and M. They used the conditions for the wave vectors (3.1) and for the amplitudes (3.29) and (3.50) consecutively on the entrance and exit surfaces of the separator and the mirrors, with the indicated reservations regarding the first field, and then the corresponding conditions for the analyzer with an allowance for both fields. They obtained the following expressions for the functions of the refracted and reflected waves inside the crystal of analyzer A:

$$D_{o,h2}^{AI} = -D_o^{(a)} p_{o,h} \exp\left\{-2\pi i K\left[\delta^{(2)'} y_I + \delta^{(2)} x_I\right]\right\} \quad (10.1)$$

$$D_{o,h1}^{AI} = D_o^{(a)} q_{o,h}^{I} \exp\left\{-2\pi i K\left[\delta^{(1)'} - \delta^{(2)'}\right]L + \delta^{(2)} x_I + \delta^{(2)'} y_I\right\} \quad (10.2)$$

$$D_{o,h2}^{AII} = -D_o^{(a)} p_{o,h} \exp\left\{-2\pi i K\left[\delta^{(2)} y_{II} + \delta^{(2)'} x_{II}\right]\right\} \quad (10.3)$$

$$D_{o,h1}^{AII} = D_o^{(a)} q_{o,h}^{II} \exp\left\{-2\pi i K\left[\delta^{(1)} - \delta^{(2)}\right]L + \delta^{(2)'} x_{II} + {}^{(2)} y_{II}\right\} \ . \quad (10.4)$$

The Arabic figures are the numbers of the wave fields, and the Roman ones, the numbers of the mirrors and, respectively, the trajectories in Fig. 9.21; the segments $x_{I,II}$ and $y_{I,II}$ are also shown there.

In this chapter the presence of a comma in the superscript or subscript means that the equation must be considered twice with the respective suffixes.

$$p_h = p_o \cdot c^{(2)} = \frac{c^{(1)^2} c^{(2)}}{[c^{(1)} - c^{(2)}]^3} c^{(2)}$$

$$q_h^I = q_o^I \cdot c^{(1)} = \frac{c^{(1)^2} c^{(2)}}{[c^{(1)} - c^{(2)}]^3} c^1 \quad ;$$

$$q_h^{II} = q_o^{II} c^{(1)} = \frac{c^{(1)} c^{(2)^2}}{[c^{(1)} - c^{(2)}]^3} c^{(1)} \quad . \tag{10.5}$$

It is easy to demonstrate that, even if the anomalous absorption in plate A is neglected, the wave functions of the first field due to the incident beams D_h^I and D_o^{II} with the same phases will have amplitudes whose ratios for refracted and diffracted waves will be pairwise close to -1. As a result, the first field in A will vanish completely.

For the wave functions which are components of the emerging waves $D_o^{(d)}$ and $D_h^{(d)}$ and are associated with transmission through mirrors M^I and M^{II}, respectively, we get the expressions

$$D_o^{i(d)} = -D_o^{(a)} p_o \exp\left\{-2\pi iK\left[\delta^{(2)'} x^i + \delta^{(2)} y^i - \delta^{(2)}(L + t_A)\right]\right\} \tag{10.6}$$

$$D_h^{i(d)} = -D_o^{(a)} p_o \exp\left\{-2\pi iK\left[\delta^{(2)'} y^i + \delta^{(2)} x^i - \delta^{(2)'}(L + t_A)\right]\right\} \tag{10.7}$$

$$i = I, II \quad ; \quad L = t_s + x_I + t_{MI} + y_I = t_s + x_{II} + t_{MII} + y_{II} \quad . \tag{10.8}$$

It is obvious that, if the amplitudes of the wave functions $D_o^{i(d)}$ are equal, the maximum intensity in the beam cross section can be obtained only when their phase factors coincide. The same is true of a pair of waves $D_h^{i(d)}$. Since the ratios of these pairs

$$\left[D_o^{I(d)}/D_o^{II(d)}\right] = \left[D_h^{I(d)}/D_h^{II(d)}\right] =$$

$$= \exp\left\{-2\pi iK\left[\delta^{(2)'}(y_I - x_{II})+\delta^{(2)}(x_I - y_{II})\right]\right\} \quad (10.9)$$

the phase coincidence conditions will be

$$y_I = x_{II} \quad , \quad y_{II} = x_I \quad . \quad (10.10)$$

These conditions may be called the *ideal geometry*.

It can be seen from the above relations that the emerging waves $D_h^{(d)}$ and $D_o^{(d)}$ are formed only by the second field inside the crystal analyzer.

The authors of [10.3] further consider two questions which are essential in interpreting the experiment correctly. First, is "optical" polishing of the surface of the interferometer plates sufficient? It is indicated that such polishing is sufficient because the values $\delta^{(2)}$ and $\delta^{(2)'}$ are of the order of extinction lengths.

The other question refers to the validity of the plane-incident-wave approximation, because plate S receives a beam with a wave front restricted in the direction parallel to the common base of the interferometer. It should be noted that plates S, M, and A act as Authier monochro-collimator (see Fig. 9.4b). The numerical estimates given in the paper under review point to about 1" angular divergence of the radiation on the exit surface of plate A, which corresponds to a coherent section of the wave front B $\sim$ 130μ. This ensures the focussing of pencils $D_o^{I(d)}$, $D_o^{II(d)}$, $D_h^{I(d)}$, $D_h^{II(d)}$ even with small deviations from the ideal geometry. The corresponding equations (10.10) are also extremely important for taking into account the finite angular area of beam focussing in crystal A associated with the width of the source, or its size, i.e., the width of the ribbon-like beam. Since the different points of the source radiate noncoherently, the conditions (10.10) imply that the transverse shift of the beam ensures the equality of the phases on exit from A for all the points of the effective part B_o of the source:

$$B_e = 2T|\tan\varepsilon_m|\cos\vartheta \quad ; \quad T = t_s + t_M + t_A \quad . \quad (10.11)$$

Finally we shall add that the interferometer described is sufficiently achromatic, because variations in λ cause only slight shifts of the irradiated areas in mirrors M.

Papers [10.4] and [10.5] describe both interferometers using Bragg reflection and "mixed" type devices with Laue and Bragg reflection.

BONSE and TE KAAT [10.6] analyze the operation of a misaligned three-crystal interferometer. They consider two effects: deviations from the ideal geometry conditions (10.10) and the use of a source with a width $B > B_e$, where B_e produces beams beyond the angle ε_m. This investigation was carried out on a specially designed device, with plate A slightly inclined to the vertical plane. It was shown that the interference patterns due to the divergence of the waves emerging from the source at an angle $\Delta\psi_o$ are equivalent to the effects due to the deviations from the ideal geometry conditions. The above-mentioned interference patterns are formed by hyperbolas with an interval $\Lambda = \lambda/\Delta\psi_o$ and are described by the equation

$$\ell_1\ell_2 = \lambda Tm\ \tan\vartheta/\alpha|\chi_h| \tag{10.12}$$

where m is a natural set of numbers, $\alpha = \mathrm{arc}\ \tan\alpha'$, α' is the angle of inclination of plate A, and ℓ_1 and ℓ_2 are the coordinates corresponding to the axes perpendicular and parallel to base of the device. To ensure the contrast of the interference patterns, it is important that the difference between the real and effective size of the source be minimized.

Also of interest is the superposition of the above-mentioned interference patterns and the moiré patterns usually observed when photographing beams emerging from the interferometer. According to the authors, an undistorted moiré pattern can be obtained only by complete elimination of the angle of device misalignment.

10.2 Two-Crystal Interferometer

As we have already mentioned in the introductory part of this chapter, the most comprehensive study of interference effects during the transmission of an incident wave with a narrow wave front through two perfect crystals with an air (or generally nondiffracting) interlayer was carried out by AUTHIER et al. [1.51]. The two crystals have an identical structure and orientation with respect to the incident beam (Figs. 10.2,3).

The method of calculating interference patterns with different ratios between the thickness of the crystals and that of the air interlayer is the same as in KATO's investigations described in Chapter 6, i.e., expansion of the incident wave packet in plane waves, followed by integration of the cor-

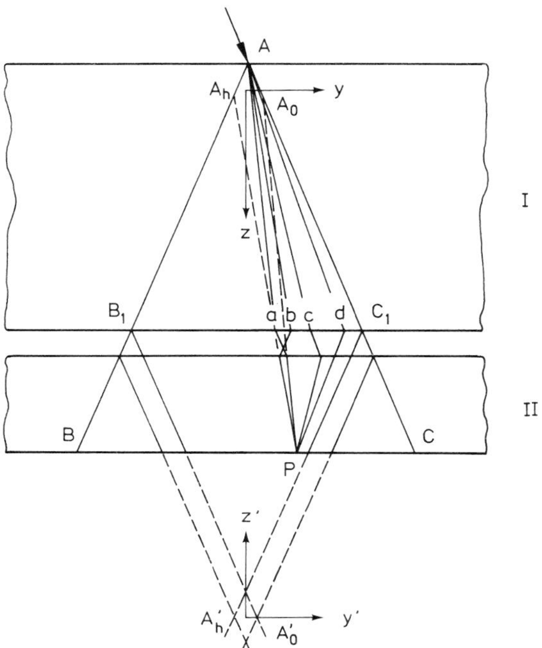

Fig. 10.2 Scheme of beam paths in two-crystal interferometer. Focussing paths from A to P. Path a): waves $D_{R1(T1)}$ and $D_{T1(R1)}$. Path b): waves $D_{R2(T2)}$ and $D_{T2(R2)}$. Path c): waves $D_{R2(T1)}$ and $D_{T2(R1)}$. Path d): waves $D_{R1(T2)}$ and $D_{T1(R2)}$ (after AUTHIER et al. [1.51])

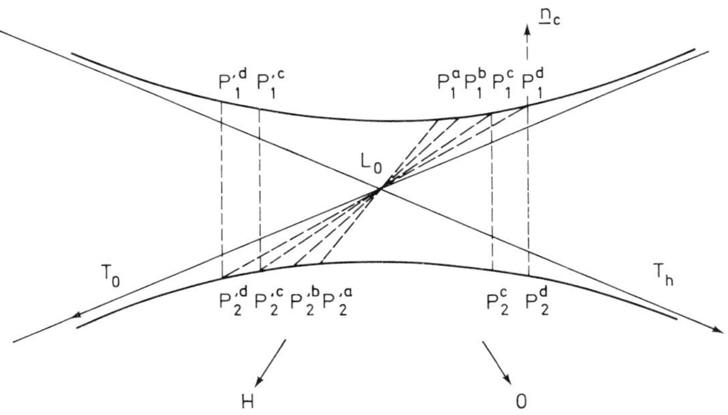

Fig. 10.3 Reciprocal space diagram corresponding to Fig. 10.2 (after AUTHIER et al. [1.51])

responding amplitudes with respect to the angular coordinate. The essential difference lies in the fact that because of the complexity of the task, integration was performed by the approximate "saddle-point", or stationary-phase, method. Therefore the values of the wave functions and intensities in the region of the central focus, as well as where the two plates are of equal thickness, were not determined and are not considered in this paper.

Taking into account the linear nature of the diffraction problem, it was possible to "trace" the path of propagation (or trajectories) of separate plane harmonics in both crystals. These trajectories correspond to the vector $\underline{\underline{OP}}$ in Fig. 6.2, i.e., to the propagation of Poynting's vector for a beam $\underline{\nu}$. The paper under review considers the trajectories of $\underline{\nu}$ referring the pair of waves $\bar{D}_T$ and D_R (see (5.17)). The positions of the trajectories are indicated by the values $Y_{o,h}$ and $Y'_{o,h}$ according to the diagram in Fig. 5.5 (see also (5.53) and (10.15)).

As distinct from Fig. 6.2, according to which complete information on the wave field inside the crystal and on the intensity distribution on the exit surface could be obtained by means of a single variable vector OP, in the case of transmission through two crystals, four typical trajectories were considered for each state of polarization. These trajectories- a,b,c,d - differ as follows: the segments of the trajectories for a and b in the two crystals are parallel, and for c and d, nonparallel. Accordingly, the initial number of the wave field in the second crystal is retained for the first pair (a,b) and changes for the second pair (c,d), for which interbranch scattering takes place.

The paper discusses only those waves diffracted upon transmission through both crystals.

As a result of integration, the following expressions are obtained for the spherical-wave functions on the exit surface corresponding to the standard trajectories indicated (D for pair a,b; D' for c,d,):

$$D = D_{R1(T1)} + D_{T1(R1)} + D_{R2(T2)} + D_{T2(R2)} = \frac{A}{\sqrt{z_I + z_{II}}}\left(\frac{1 + Y_o}{(1 - Y_o^2)^{\frac{1}{4}}} \cdot \left[\exp(-i(\varphi_o - \tfrac{\pi}{2})) + \exp(i\varphi_o)\right] + \frac{1 - Y_h}{(1 - Y_h^2)^{\frac{1}{4}}}\left\{\exp\left[-i(\varphi_h - \tfrac{\pi}{2})\right] + \exp(i\varphi_h)\right\}\right) \quad (10.13)$$

$$D' = D_{R2(T1)} + D_{T2(R1)} + D_{R1(T2)} + D_{T1(R2)} = \frac{A}{\sqrt{|z_I - z_{II}|}} \frac{1 + Y'_o}{(1 - Y'^2_o)^{\frac{1}{4}}}$$

$$\exp\left[-i(\varphi'_o + \frac{\pi}{4} \pm \frac{\pi}{4})\right] - \frac{A}{\sqrt{|z_I - z_2|}} \frac{1 + Y'_h}{(1 - Y'^2_h)^{\frac{1}{4}}} \exp\left[-i(\varphi'_h + \frac{\pi}{4} \pm \frac{\pi}{4})\right]$$

$$- \frac{A}{\sqrt{|z_I - z_{II}|}} \frac{1 + Y'_o}{(1 - Y'^2_o)^{\frac{1}{4}}} \exp\left[i(\varphi'_o - \frac{\pi}{4} \pm \frac{\pi}{4})\right] + \frac{A}{\sqrt{|z_I - z_{II}|}} \frac{1 + Y'_h}{(1 - Y'^2_h)^{\frac{1}{4}}}$$

$$\exp\left[i(\varphi'_h - \frac{\pi}{4} \pm \frac{\pi}{4})\right] \quad . \tag{10.14}$$

In $D_{RM(TN)}$ and $D_{TM(RN)}$ the parenthesized subscripts at D indicate the wave in crystal I, those without parentheses, in crystal II: M,N, the numbers of branches where the tie point lies; z_I and z_{II}, the thickness of crystals; Y_o, Y_n, Y'_o, the deviation parameters. According to (5.53) and the respective values of α, we get

$$\left.\begin{matrix} Y_{o,h} \\ Y'_{o,h} \end{matrix}\right\} = \frac{\tan\alpha}{\tan\vartheta} = \pm \frac{y^2}{1 + y^2} \tag{10.15}$$

for the trajectories of the first wave field in crystal I. The phases φ appearing in the expression for the wave functions (10.13,14) have the following values:

$$\varphi_{o,h} = \frac{\pi}{\tau}(z_I + z_{II})\sqrt{1 - Y^2_{o,h}} \quad ; \quad \varphi'_{o,h} = \frac{\pi}{\tau}\sqrt{1 - Y'^2_{o,h}}$$

$$\tau = \frac{\lambda\cos\psi_o}{C\sqrt{\chi_h\chi_{\bar{h}}}} \quad . \tag{10.16}$$

In the absorbing crystal the phases become complex, and

$$\tau \equiv \tau_r \quad ; \quad \varphi_i = \frac{\pi}{\tau_i}(z_I + z_{II})\sqrt{1 - Y^2} \tag{10.17}$$

where τ_i is the absorption length.
In (10.14) the upper sign in the values of the phases refers to the condition $z_I > z_{II}$.
On transition from wave functions to intensities

$$I = (D + D')(D + D')^* \qquad (10.18)$$

the following phases arise:

$$\Delta_{MP} = \mathrm{Re}[\varphi_{Nj(Mi)} - \varphi_{Qk(B\ell)}] \qquad (10.19)$$

which determine the geometry of the diverse diffraction patterns. Some typical examples are given below.

Interference Between D-Waves

Of the four possible phase differences, two are expressed by the values

$$\Delta_{TT,RR} = \frac{2\pi}{\tau}\sqrt{1 - Y_i^2}\,(z_I + z_{II}) - \frac{\pi}{2} \quad ; \qquad Y_i = Y_o, Y_h \qquad (10.20)$$

and the third, by

$$\Delta'_{TR} \approx \frac{2\pi}{\tau}\sqrt{1 - Y^2}\,(z_I + z_{II}) - \frac{\pi}{2}$$

$$Y = (Y_o + Y_h)/2 \quad . \qquad (10.21)$$

These expressions are typical for the interference of waves with excitation points on the two branches of the dispersion hyperbola symmetrical about point L_o (conjugate tie points) (Figs. 10.2,3). They describe families of hyperbolas with an interval of $1/2\sqrt{A_oA_h}$ similar to the hyperbolas described in Chapter 6. Note that points A_o and A_h are imaginary sources from which the waves converge along straight paths on the focal point P on the exit surface.

The fourth phase difference in the group under study

$$\Delta''_{TR} \equiv \varphi_{DR2(T2)} - \varphi_{DT2(R2)} \approx \frac{\pi}{\tau}(YdY)/(z_I + z_{II})\sqrt{1 - Y^2} \quad ;$$

$$dY = 2g/(z_I + z_{II}) \tag{10.22}$$

leads to a system of planes of equal phases described by the equation

$$y = \pm\left(k\tau \tan\vartheta/g\sqrt{1 + \frac{k^2t^2}{g^2}}\right) z \; ; \quad k = \text{const.} \tag{10.23}$$

where y,z are a rectangular system of coordinates in the plane of reflection with an axis z directed into plate I, perpendicular to its entrance surface. A bundle of these planes, intersecting along the axis perpendicular to the plane of reflection, forms a system of parallel bands on the entrance surface of the second plate (at points P). If $\tau \ll g$ (g is the gap thickness), then for the first orders, i.e., in the central part of this pattern, the period will be constant and will further decrease as k increases. It is extremely important that this interference results from interaction of waves corresponding to different excitation points on one and the same branch of the dispersion hyperbola, and therefore, in distinction to the pendulum solution effects, it does not disappear when one of the fields in the crystal is strongly absorbed. In the investigations of HART and MILNE [10.10,11] as well as of BAKER et al. [10.12] (comp. also [1.52]), such interference patterns were used for precision determination of the absolute values of the structure amplitudes.

At the same time, since the waves forming the interference patterns Δ'_{TR} and Δ''_{TR} pass the gap by nonparallel paths, the band periods depend on the value of g. They can be formed only at a sufficiently small value of g, as a condition for the overlapping of the corresponding waves.

Interference Between D'-Waves

As has already been indicated, on entering crystal II, these waves are identified with a different initial wave field. They propagate in the two crystals along trajectories symmetrical about the reflecting plane.

The four possible phase differences belonging to this group are described by expressions which differ from (10.20-22) in that they include

$z_I - z_{II}$ instead of $\sum z$, and therefore each of the four expressions has plus-minus signs; the plus sign refers to the case $z_I > z_{II}$. Moreover, the value dY' appearing in them, which is similar to dY, is equal to

$$dY' = 2g/(z_I - z_{II}) \qquad (10.24)$$

and the expressions for the two last phase differences (corresponding to the value $\pm \Delta''_{T'R'}$) are applicable only to small values of g and a considerable value of $\pm (z_I - z_{II})$.

The families of hyperbolas corresponding to the first three pairs of these phase differences have been considered previously [10.1,2]. The pattern of the bands appearing for the last pair of values of the phase differences is quite similar to the one described by (10.23).

Interference Between D and D'-Waves

This refers to two groups of phase differences (four in each group). One of them includes interference patterns independent of the values of g. They are related pairwise by expressions which establish a difference by π, so that the conditions for a maximum for one correspond to the conditions for a minimum for the other.
For instance:

$$\Delta'_{TT'} = \Delta'_{RR'} + \pi \ ; \quad \Delta''_{TT'} = \Delta''_{RR'} - \pi \ . \qquad (10.25)$$

The mutual shifts in the extrema on the exit surface should also be mentioned. These extrema are antisymmetrical about a certain, precisely defined straight line on the exit surface.

An analysis of the behaviour of the hyperbola families for one phase diference, which can easily be applied to other differences in this group, was carried out in more detail. If we displace the gap towards the entrance surface, the hyperbolas become flatter, and at $z_I = 0$ the interference pattern disappears, and the hyperbolas merge with its edges, which are dynamical diffraction images of the surface. The situation is reversed when the gap is displaced towards the exit surface.

A decrease in thickness z_I or z_{II} (with the corresponding increase in the gap) leads to interesting limiting expressions. Thus, a decrease results (in the limit) in the merging of the excitation points on one of the branches of the dispersion hyperbola which correspond to the waves a and c,b and d; here, $Y_o \rightarrow Y'_o$ and $Y_h \rightarrow Y'_h$. As a result, for the thin second plate the corresponding phase shifts will be approximately as follows:

$$\Delta_{TT'} \longrightarrow \frac{2\pi}{\tau}\frac{z_{II}}{\sqrt{1-Y_o^2}} \quad ; \quad \Delta''_{RR'} \longrightarrow \frac{2\pi}{\tau}\frac{z_{II}}{\sqrt{1-Y_h^2}} \quad . \tag{10.26}$$

$$\Delta'_{TT'} \longrightarrow \frac{2\pi}{\tau} z_I\sqrt{1-Y_o^2} \quad ; \quad \Delta'_{RR'} \longrightarrow \frac{2\pi}{\tau} z_1\sqrt{1-Y_h} \quad . \tag{10.27}$$

Similar expressions are obtained where there is a decrease in the thickness of the first plate ($Y_o \rightarrow Y'_o$; $Y_h \rightarrow Y'_h$).

Expressions of the type of (10.26) describe the phase differences of two waves induced in a crystal by an incident *plane* wave, while expressions of the type of (10.27) describe the phase differences in waves during incidence on a "spherical" wave. This result was confirmed by experimental observations in the investigations of MALGRANGE and AUTHIER [9.38], and HART and MILNE [10.13]. The ratio between the straight bands and the incident-plane-wave approximation has previously been established by MILNE [10.10].

Another group of phase difference values, although described by expressions similar to the first group, is characterized by nonparallel trajectories in the gap region.

Finally, it is interesting to compare the periods of the interference bands along the lines corresponding to the incidence of the incident beam component at the values $\sqrt{1-Y_o^2} = \sqrt{1-Y_h^2} = 0$. Here, the periods attain extreme values.

Turning to D-waves, for instance $\Delta_{TT,RR}$ we get

$$\Delta_{TT} = \Delta_{RR} = \frac{2\pi}{\tau}(z_I + z_{II}) - \frac{\pi}{2} \tag{10.28}$$

which corresponds to the period of the pendulum solution bands for a crystal of thickness $(z_I + z_{II})$.

For D'-waves we have:

$$\Delta_{T'T'} = \Delta_{R'R'} = \pm\left[\frac{2\pi}{\tau}\,|2z_I-(z_I + z_{II}) \quad - \frac{\pi}{2}\right] \quad . \tag{10.29}$$

This expression, at a constant value of $(z_I + z_{II})$, yields half the value of the pendulum solution bands.

For interference of D- and D'-waves, the extreme values of the periods in the directions indicated on the interference pattern may be as follows:

$$\Delta'_{TT'} = \Delta'_{RR'} + \pi = \frac{2\pi}{\tau} z_I + \frac{\pi}{4} \mp \frac{\pi}{4} \tag{10.30}$$

which corresponds to the period of the pendulum solution bands of a crystal of thickness z_I, and

$$\Delta''_{TT'} + \pi = \Delta''_{RR'} = \frac{2\pi}{\tau} + \frac{\pi}{4} \pm \frac{\pi}{4} \tag{10.31}$$

the latter corresponds to a similar value for the crystal of z_{II}.

The above equations may be useful in interpreting complex patterns, in particular, on projection topographs.

Interference Pattern Intensities

According to (10.18):

$$I = I_1 + I_2 + I_3 \tag{10.32}$$

$$I_1 = |D|^2 \quad ; \quad I_2 = |D'|^2 \quad ; \quad I_3 = 2\mathrm{Re}(DD'^*) \quad . \tag{10.33}$$

As regards the values I_1 and I_2, it should be noted that only those values yield appreciable contrast in conditions of absorption which correspond to the phase difference Δ'_{TR} and Δ''_{TR} (the D-waves) and $\Delta'_{T'R'}$, $\Delta''_{T'R'}$ (the D'-waves). Thus, for the intensity of an interference pattern corresponding to (10.22) we get

$$I''_{TR} = \frac{B^2}{z_1 + z_{II}}\left\{\exp\left[-\frac{\mu}{\cos\vartheta}(z_I + z_{II})\right]\right\}\frac{(1 + Y_o)(1 - Y'_h)}{(1 - Y_o^2)^{\frac{1}{4}}(1 - Y'^2_h)^{\frac{1}{4}}} \quad . \tag{10.34}$$

Here, an essential element is the hyperbolic cosine factor cosh. The distribution of intensity I''_{TR} is symmetrical about the centre line of the section pattern.

When considering the intensities I_3 the attention was focussed on interference patterns independent of the gap thickness.

The group of four expressions takes the form:

$$I'_{TT'}{}'' = \frac{B^2}{\sqrt{|z_I^2 - z_{II}^2|}} \left\{ \exp\left[- \frac{\mu}{\cos\vartheta} (z_I + z_{II}) \right] \right\} \frac{(1 + Y_o)(1 + Y'_o)}{(1 - Y_o^2)^{\frac{1}{4}} (1 - Y'^2_o)^{\frac{1}{4}}} \cdot$$

$$\cdot \cosh(\varphi_{oi} \mp \varphi'_{oi}) \cos\Delta'_{TT'}{}'' \tag{10.35}$$

$$I'_{RR'}{}'' = \frac{B^2}{\sqrt{|z_I^2 - z_{II}^2|}} \left\{ \exp\left[- \frac{\mu}{\cos\vartheta} (z_I + z_{II}) \right] \right\} \frac{(1 - Y_h)(1 + Y'_h)}{(1 - Y_h^2)^{\frac{1}{4}} (1 - Y'^2_h)^{\frac{1}{4}}} \cdot$$

$$\cdot \cosh(\varphi_{hi} - \varphi'_{hi}) \cos\Delta'_{RR'}{}'' \quad . \tag{10.36}$$

The phases φ'_o and φ'_h are negative or positive, depending on the relation $z_{II} \gtrless z_I$; this results in the predominance of the intensities $I'_{TT'}$ and $I'_{RR'}$ or $I''_{TT'}$ and $I''_{RR'}$. The "flatter" hyperbolas are predominant. Besides, the symmetry or asymmetry of the hyperbola intensity on the section patterns relative to the centre line must also depend on the thickness relationship indicated.

The graph given in [1.51], which was obtained by calculating the intensity distribution on the exit surface of a double wedge, illustrates the qualitative analysis of the expressions (10.35) and (10.36)

It can also be shown that the total value of intensity (10.32) leads to the expression for the perfect crystal for g = 0. Here, I_2 and I_3 vanish to zero, and

$$I_1 = \frac{2B^2}{z_I + z_{II}} \left[\exp\left(- \mu \frac{z_I + z_{II}}{\cos\vartheta} \right) \right] \left[\sin^2\varphi_i + \cos^2(\varphi_r - \frac{\pi}{4}) \right] \Big/ \tag{10.37}$$

$$(1 - Y^2)^{\frac{1}{4}} \quad . \tag{10.37}$$

Using the generalized dynamical theory, on the basis of the TAKAGI equations for the perfect crystal, INDENBOM et al. [10.14] considered in detail the wave field distribution in the case of focussing at $z_I = z_{II}$. The authors believe that the principle of obtaining the main intensity maximum (focus) investigated by them may prove useful in obtaining patterns of defects in crystals.

The above-listed papers neglected the effect of moiré, which appears in two-crystal, as well as three-crystal, interferometers under certain conditions. The moiré theory is outlined in Section 10.3.

10.3 Formation and Utilization of X-Ray Moiré Patterns

As is well known, optical (or geometrical) and diffraction moiré differ from one another. According to SHUBNIKOV [10.15], optical moiré is observed on successive transmission of light through images of two lattices: A with repeated image and B with a grating whose periods coincide with (the lattices are nonparallel) or are close to (the lattices are parallel) the periods of A. The images in A, divided into the elements, are synthesized with increased intervals. The intensities of the light beams are added together. Diffraction moiré appears on successive transmission (the incidence of beams in the maximum region) of radiation through two crystals with ratios of the periods and orientations similar to lattices in conditions of optical moiré; here, the wave functions of scattering on separate crystals are added together. It is obvious that in this case the lattice periods are close to the radiation wavelengths.

Moiré in electron diffraction was discovered a little earlier than X-ray moiré. Detailed consideration of electron moiré is given in [1.27, 10.16]. The relevant theory has been developed by HASHIMOTO et al. [10.17] and JEVERS [10.18]. JEVERS' theory can also be applied in describing X-ray moiré in the incident-plane-wave approximation (perfect crystal!), which is equivalent to the use of the generalized dynamical theory based on Takagi-type equations [1.45]. The fundamentals of this theory are outlined in Chapter 11. Here, we will apply it to the formation of moiré in the incident-plane-wave approximation, regarding both crystal-scatterers as perfect. In this case the equations for the wave field in the first crystal will be (see (II.36))

$$i\,\frac{\pi}{\lambda}\,\gamma_o\,\frac{dD_o}{dz} = \chi_o D_o + C\chi_h D_h \quad ,$$

$$i\,\frac{\pi}{\lambda}\,\gamma_h\,\frac{dD_h}{dz} = \chi_o D_h - aD_h + C\chi_h D_o \quad . \tag{10.38}$$

We shall now pass on to symmetrical reflection ($\gamma_o = \gamma_h = \cos\vartheta$) and assign the following values of the wave functions D_o and D_h:

$$D_o = D_o \exp[-i\pi\chi_o(\underline{k}_o\underline{r})] = D_o \exp\left(-i\,\frac{\pi\chi_o}{\lambda\,\cos\vartheta}\right) \quad ,$$

$$D_h = D_h \exp\{-\,i\pi\chi_o[(\underline{k}_h\underline{r})+(\underline{s}\underline{r})]\} \quad , \tag{10.39}$$

$$|\underline{s}| = K\alpha = 2K\eta\,\sin 2\vartheta \quad . \tag{10.40}$$

As a result of substituting (10.39) and (10.40) in (10.38) these last equations take the form

$$i\,\frac{\lambda\cos\vartheta}{\pi}\,\frac{dD_o}{dz}\,\exp\left(-i\,\frac{\pi\chi_o}{\lambda\cos\vartheta}\,z\right) = C\chi_h D_h\,\exp\{-i\pi\chi_o[(\underline{k}_h\underline{r})+(\underline{s}\underline{r})]\}$$

$$i\,\frac{\lambda\cos\vartheta}{\pi}\,\frac{dD_h}{dz}\,\exp\{-i\pi\chi_o[(\underline{k}_h\underline{r})+(\underline{s}\underline{r})]\}$$

$$= C\chi_h D_o\,\exp\left(-i\,\frac{\pi\chi_o}{\lambda\cos\vartheta}\,z\right) \quad ,$$

or, replacing

$$\underline{k}_h = \underline{k}_o + \underline{h} \quad , \qquad |\underline{s}| = K\alpha \quad , \qquad \tau_o = \frac{\lambda\cos\vartheta}{C|\chi_h|}$$

we obtain

$$\frac{dD_o}{dz} = i\,\frac{\pi}{\tau_o}\,D_h\,\exp[-i\pi\chi_o(\underline{h} + \underline{s},\underline{r})] \quad ,$$

$$\frac{dD_h}{dz} = i\,\frac{\pi}{\tau_o}\,D_o\,\exp[i\pi\chi_o(\underline{h} + \underline{s},\underline{r})] \quad . \tag{10.41}$$

Note that the incident-plane-wave approximation finds its expression in the fact that the amplitudes D_o and D_h are functions of the z-coordinate alone. The absence of defects in the crystal-scatterer means that the angular coordinate inside the maximum for the incident wave $\underline{s}$ is independent of the coordinate z.

A model of formation of moiré under conditions of the generalized theory, in particular (10.41), is depicted in Fig. 10.4. Eqs. (10.41) refer to the wave field inside the first crystal. In order to describe the field inside the second crystal one must first introduce some constant increment $\Delta\underline{r}$ to the radius vector $\underline{r}$, which corresponds either to a displacement or to the rotation of the second crystal, and, secondly (for the purpose of keeping χ_h constant) use the well-known approximation

$$\chi_h^{II} = \chi_h^{I}\,\exp[-2\pi i(\Delta\underline{h}\underline{r})] \quad ,$$

and also replace

$$(\underline{h} + \Delta\underline{h},\ \underline{r} + \Delta\underline{r}) \ \text{ by } \ (\underline{h}\Delta\underline{r}) + (\underline{r}\Delta\underline{h}) \quad ; \quad (\underline{h}\Delta\underline{r}) = (-\underline{r}\Delta\underline{h}) \quad .$$

In this case (10.41) are rewritten:

$$\frac{dD_o^{II}}{dz^{II}} = i\,\frac{\pi}{\tau_o}\,D_h[1-(1 - \exp i\alpha)]\,\exp[-i\pi\chi_o(\underline{s}\underline{r})] \quad , \tag{10.42}$$

$$\frac{dD_h^{II}}{dz^{II}} = i\,\frac{\pi}{\tau_o}\,D_o[1-(1 - \exp i\alpha)]\,\exp[i\pi\chi_o(\underline{s}\underline{r})] \quad ,$$

$$i\alpha \equiv 2\pi i\chi_o x|\Delta\underline{h}| \quad . \tag{10.43}$$

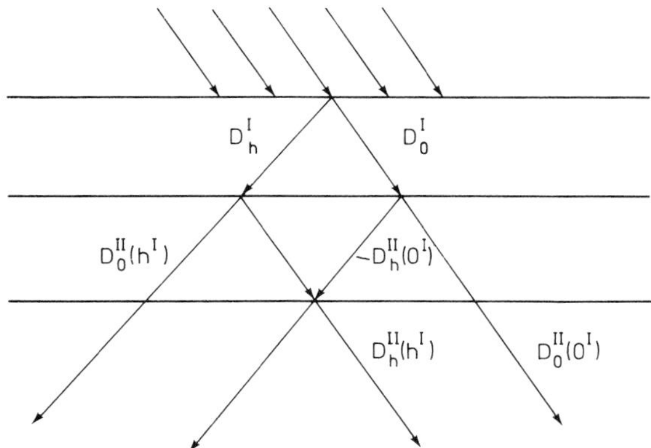

Fig. 10.4 Scheme of diffraction moiré formation (after AUTHIER and SIMON [1.61])

Here, x is a coordinate in the direction of h, i.e., perpendicular to the reflecting planes.

Since the values of the derivatives in (10.42) depend on x only via α, the loci of the points of equal or constant values of the amplitude and scattering intensities will satisfy the conditons

$$x = \frac{\text{const}}{|\Delta \underline{h}|} = \Lambda \text{ const} \quad , \qquad \Lambda = |\Delta \underline{h}|^{-1} \quad . \tag{10.44}$$

Consequently the moiré period Λ is determined by the change in reflection vector $\Delta \underline{h}$. Hence follow the equations for the periods of parallel (dilatation), rotation, and mixed moiré:

$$\Lambda_{par} = d_h/\ell \quad , \quad \ell = \Delta d_h/d_h \quad , \quad \Lambda_{rot} = d_h/\varphi$$

$$\Lambda_{mixed} \approx \left(\sqrt{\Lambda_{par}^{-2} + \Lambda_{rot}^{-2}}\right)^{-1} \quad . \tag{10.45}$$

These formulas coincide with the relation for geometrical moiré, which were derived by SHUBNIKOV.

If the incident wave experiences in crystals A and B reflections from different systems of planes with vectors $\underline{h}$ ang $\underline{g}$, respectively, the value of the moiré period is determined from the equation

$$\Lambda = |\underline{h} - \underline{g}|^{-1} \quad . \tag{10.46}$$

It follows from (10.44) that the bands of parallel moiré are parallel to the reflecting planes, and the bands of rotary moiré are parallel to the reflection vector $\underline{h}$.

Further, following JEVERS [10.18], we can obtain expressions for the moiré pattern intensities. According to Fig. 10.4, moiré appears on interference of a pair of waves either in the direction of the diffracted wave or the wave refracted in the first crystal.

Using the theory in the plane-incident-wave approximation successively for scattering in the first and second crystal, it can be shown that the intensity on the exit surface of the second crystal can be represented as follows (for the direction of reflection from the first crystal):

$$I_h = I_h^I I_o^{II}(h^I) + I_o^I I_h^{II}(0^I) + 2(I_h^I I_o^{II} I_o^I I_h^{II} \cos 2\pi(x - x_o)\lambda^{-1} \quad , \quad (10.47)$$

where x_o is a function of the thickness of both crystals and of the deviation from the Bragg angle.

If we consider the formation of moiré upon incidence of a "spherical" wave, we must use the more general form of Takagi equations, in which the amplitudes of wave functions in the crystal depend on two arguments. These arguments may be related either with the rectangular axes x and z or with the oblique axes $x_o\underline{s}_o$ and $x_h\underline{s}_h$ directed along the wave vectors of the refracted and diffracted wave in the crystal (see (11.36)):

$$i\frac{\lambda}{\pi}\frac{\partial D_o}{\partial x_o} = \chi_o D_o + C\chi_h D_h \quad ,$$

$$i\frac{\lambda}{\pi}\frac{\partial D_h}{\partial x_h} = (\chi_o - \alpha)D_h + C\chi_h D_o \quad . \quad (10.48)$$

AUTHIER and SIMON [1.61] pass on from equations of the type of (10.42) to second-order equations (see, for instance, (11.54)) and then use a solution similar to (11.49,50) for a spherical incident wave. If the front of the incident wave is of arbitrary length:

$$D_h(P) = -\frac{i\pi K\chi_h\gamma_o C}{\sin 2\vartheta}\int_{AB} D_o^{(a)}(x)J_o(B\sqrt{b^2 - x^2})dx \quad , \quad (10.49)$$

$$D_o(P) = D_o^{(a)}(A) - \frac{B}{2} \int_{BA} D_o^{(a)}(x) \sqrt{\frac{b + x}{b - x}} J_1(B\, b^2 - x^2)dx \quad . \tag{10.50}$$

The above values of $D_h(P)$ and $D_o(P)$ apply to the arbitrary point P of the wave field on the exit surface of the crystal plate. Integration is performed over some perimeter $\overline{AB}$ on the entrance surface, which is the cross section of an incident wave with a finite front. This perimeter is bisected by the point of origin, 0; $b = \overline{OA} = \overline{OB}$, x is a variable coordinate on the line $\overline{AB}$ (see Fig. 10.6):

$$B = \frac{2\pi K\sqrt{\chi_h\chi_{\bar{h}}}|C|\sqrt{\gamma_o\gamma_h}}{\sin 2\vartheta} \quad . \tag{10.51}$$

On the other hand, when the length of $\overline{AB}$ shrinks almost to a point, i.e., in the case of a spherical incident wave, the values of the amplitudes D_h and D_o at the arbitrary point p_o on the exit surface are expressed as follows:

$$D_h(p_o) = -iD_o^{(a)}(4r_o)^{-1}\sqrt{Kr_o}\, \frac{C\chi_h}{\sin 2\vartheta} \exp\left(-i\,\frac{\pi}{4}\right)J_o(B\sqrt{b^2 - p_o^2}) \quad , \tag{10.52}$$

$$D_o(p_o) = \frac{D_o^{(a)}}{4\pi r_o} \left\{\exp\left[-\, i\pi\,\frac{K\gamma_o^2}{r_o}(b - p_o)^2\right] - \exp\left(-\, i\,\frac{\pi}{4}\right) \cdot \right.$$

$$\left. \cdot\, \frac{\pi\sqrt{Kr_o}\sqrt{\chi_h\chi_{\bar{h}}}}{\sin 2\vartheta} \sqrt{\frac{\gamma_h}{\gamma_o}} \sqrt{\frac{b + p_o}{b - p_o}}\, J_1(B\sqrt{b^2 - p_o^2})\right\} \quad . \tag{10.53}$$

It is obvious that in this case the region of the wave field in the crystal is bounded by a triangle formed by the directions of $\underline{s}_o$ and $\underline{s}_h$ and the exit surface of the crystal (see Fig. 6.2); $\underline{r}_o$ corresponds to $\underline{P}$ in the figure indicated; further, b and p_o in (10.52) and (10.53) are similar to b and x in (10.49) and (10.50), but refer to the exit surface.

Considering a more general case as compared with the diagram of Fig. 10.4, namely the presence of a small interlayer of air (vacuum) between the crystals, we can write the following expressions for the waves transmitted through this lamella:

$$D_h^I = D_h^I(p^I)\ \exp\left(-2\pi i\,\frac{K\chi_o}{2\gamma_h}\,z^I\right)\ \exp(-2\pi i K_h^I r^I) \quad . \tag{10.54}$$

$$D_o^I = D_o^I(p^I)\ \exp\left(-2\pi i\,\frac{K\chi_o}{2\gamma_o}\,z^I\right)\ \exp(-2\pi i K_o^I r^I) \quad , \tag{10.55}$$

where $D_h^I(p^I)$ and $D_o^I(p^I)$ are given in (10.52) and (10.53).

A characteristic feature of the method used in the paper under review [1.61] is the arbitrary choice of the wave vector in a given crystal. Two conditions are adopted for this vector, viz.: 1) tangential components equal to that of the wave vector of the incident vacuum wave; 2) a modulus equal to Kn, where n is the refractive index.

Now we revert to the formation of moiré (see SIMON and AUTHIER [10.18a]). Assume that a second crystal, which is made of the same material as the first, becomes slightly turned relative to it, and the waves described by (10.54) and (10.55) experience in it reflection from the same plane as in the first crystal. As the wave vectors inside these crystals, we choose the vectors joining points 0 and H with point L_o in reciprocal space (Fig. 10.5).

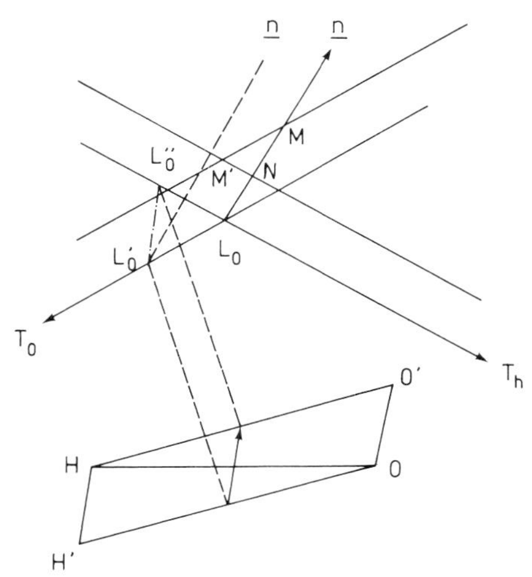

Fig. 10.5 Reciprocal space corresponding to moiré formation (after SIMON and AUTHIER [1.64])

Further, we put

$$|\underline{k}_o^{II}| = |\underline{k}_h^{II}| = |\underline{k}_o'^{II}| = |\underline{k}_h'^{II}| \tag{10.56}$$

where the primed vectors correspond to the reflected waves $D_o^{II}(h^I)$ and $D_h^{II}(h^I)$.

In Fig. 10.5, the vacuum waves transmitted through the first crystal are represented by the wave vectors $\underline{K}_o^I = \overline{OM}, \underline{K}_h^I = \overline{HN}$. If the diffraction vector h for reflection in the first crystal is represented by the straight line $\overline{OH}$, in the second crystal, for the wave transmitted through the first crystal, the diffraction vector is represented by the straight line $\overline{H'O} = \underline{h} - \underline{dh}$. For the wave vectors in the second crystal we have $\underline{k}_h^{II} = \overline{H'L'_o}$, $\underline{k}_o^{II} = \overline{OL'_o}$. Similarly, for a wave reflected from the first crystal, the diffraction vector and the wave vectors in the second crystal will be represented as

$$\overline{O'H} = -\underline{h} + \underline{dh} \quad ,$$

$$\underline{k}'^{II} = \overline{HL''_o} \quad ,$$

$$\underline{k}'^{II}_o = \overline{O'L''_o} \quad .$$

Let us now express the waves incident on the second crystal, with the aid of the following expressions (cf. (10.54) and (10.55))

$$D_o^I = D_o^I(x^I)\left\{\exp\left(-2\pi i\,\frac{K\chi_o z^I}{2\gamma_o}\right)\exp[-2\pi i(\underline{K}_o^I - \underline{K}_o^{II})\underline{r}^I]\right\}\cdot$$

$$\cdot\ \exp(-2\pi i\underline{K}_o^{II}\underline{r}^I) \quad , \qquad (10.57)$$

$$D_h^I = D_h^I(x^I)\left\{\exp\left(-2\pi i\,\frac{K\chi_o z^I}{2\gamma_h}\right)\exp[-2\pi i(\underline{K}_h^I - \underline{K}_h^{II})\underline{r}^I]\right\}\cdot$$

$$\cdot\ \exp(-2\pi i\underline{K}_h^{II}\underline{r}^I) \quad . \qquad (10.58)$$

The vector differences in the phase factors in the braces are determined directly from Fig. 10.5. We expand the scalar products of these values by $\underline{r}^I$ into sums of the products of the vector difference modulus by the projections of $\underline{r}^I$ normal and tangential to the plane of separation. As a result, for the "pseudo-amplitudes" we obtain

$$D'^{I}_{o}(x^I, z^I) = D^I_o(x^I)\exp(-2\pi iK\eta_o\gamma_o x^I)\cdot$$

$$\cdot\exp(2\pi iK\eta_o\alpha_o z^I)\exp\left(-2\pi iK\frac{\chi_o}{2\gamma_o}z^I\right)$$

$$= D'_o(x^I)\exp(2\pi iK\eta_o\alpha_o z^I)\exp\left(-2\pi iK\frac{\chi_o}{2\gamma_o}z^I\right) \quad , \tag{10.59}$$

$$D'_h(x^I z^I) = D'^{I}_{h}(x^I)\exp(2\pi iK\eta_h\alpha_h z^I)\exp\left(-2\pi i\frac{K\chi_o}{2\gamma_h}z^I\right) \quad . \tag{10.60}$$

Upon transmission through both crystals, the summary reflected amplitude is equal to

$$D_h = D^{II}_h(0^I) + D^{II}_o(h^I) \quad . \tag{10.61}$$

In order to calculate each of these waves, we can use equations for a wave with a finite front width (10.49) and (10.50). Here, the waves (10.59) and (10.60) are taken as $D^{(a)}_o$, or the pseudo-amplitude. We obtain

$$D^{II}_h(0^I) = -\frac{i\pi K\chi^{II}_h\gamma_o}{\sin 2\vartheta}\int_{A_2B_2} D'^{I}_{o}J_o\left(B'\sqrt{b^{II2} - x^{II2}}\right)dx^{II} \quad , \tag{10.62}$$

$$D^{II}_o(h^I) = D'^{I}_{h}(x^I = X + b^{II}) - \frac{B'}{2}\int_{A_2B_2} D'^{I}_{h}(x^I)\,\cdot$$

$$\cdot\sqrt{\frac{b^{II} + z^{II}}{b^{II} - z^{II}}}\,J_1\left(B'\sqrt{b^{II2} - x^{II2}}\right)dx^{II} \quad . \tag{10.63}$$

Finally, we introduce the values

$$D^{II}_h(x^{II}) = -\frac{i\pi K\chi^{II}_h\gamma_o}{\sin 2\vartheta}J_o\left(B'\sqrt{b^{II2} - x^{II2}}\right) \quad , \tag{10.64}$$

$$D_o^{II}(x^{II}) = \gamma_o\sqrt{K/r_o}\,\exp\left[-i\pi\,\frac{K\gamma_o^2}{r_o}\,(b^{II} - x^{II})^2 + i\,\frac{\pi}{4}\right]$$

$$- \frac{B'}{2}\sqrt{\frac{b^{II} + x^{II}}{b^{II} - x^{II}}}\,J_1\left(B'\sqrt{b^{II^2} - x^{II^2}}\right) \quad , \qquad (10.65)$$

where, because $\sqrt{\chi_h^{II}\,\chi_{\bar{h}}^{II}} = \sqrt{\chi_h\chi_{\bar{h}}}$, $B' = B$.

The values $D_h^{II}(0^I)$ and $D_o^{II}(h^I)$ can be represented as a function of the amplitudes $D_h^i(x^i)$:

$$D_h^i(x^i) = \begin{cases} D_h^i(x^i) \,, & |x^i| \le b^i \,, \\ 0 \,, & |x^i| > b^i \,, \end{cases} \qquad (10.66)$$

$$D_h^{II}(0^I) = D_o^{'I}(X)*D_h^{II}(X)\,\exp\left[-2\pi i\left(\frac{K\chi_o}{2\gamma_o} - K\eta_o a_o\right)z^1\right] \quad , \qquad (10.67)$$

$$D_o^{II}(h^I) = D_o^{'I}(X)*D_o^{II}(-X)*\exp\left[-2\pi i\left(\frac{K\chi_o}{2\gamma_h} - K\eta_h a_h\right)z^I\right] \,. \qquad (10.68)$$

The final expression for the pseudo-amplitude of the reflected wave upon transmission through both crystals has the form

$$D_h(X) \approx D_o^{'I}(X)*D_h^{II}(X) + D_h^{'I}(X)*D_o^{II}(-X)\,\cdot$$

$$\cdot\,\exp[(-2\pi iK\chi_o/2)z^I(\gamma_h^{-1} - \gamma_o^{-1})]\,\exp(-2\pi i dh_\perp z^I) \quad , \qquad (10.69)$$

where $dh_\perp$ denotes the component of the vector $\underline{dh}$ directed along the normal to the plane of separation.

Figure 10.6 illustrates the transmission of an incident wave packet through both crystals. It also shows the reference axes for the coordinates x^I and x^{II} on the plane of separation and X on the exit surface of the second crystal. According to the notation of this figure, $b^I = \overline{O_1A_1} = \overline{O_1B_1}$,

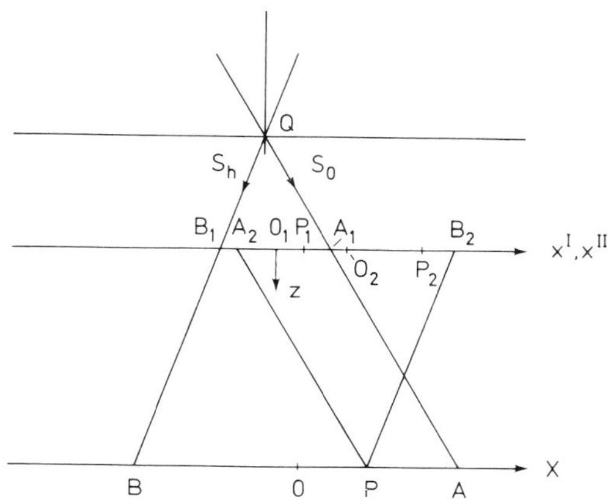

Fig. 10.6 Illustration to moiré theory in case of the wave-packet incidence (after SIMON and AUTHIER [1.64])

$b^{II} = \overline{O_2A_2} = \overline{O_2B_2}$, and $b = \overline{OA} = \overline{OB}$. The segment $\overline{OP} = X$ is a variable coordinate, for which the value of $D_h(X)$ is given in (10.69).

The rather complicated expression (10.69) is simplified in the particular case of a small thickness of the first crystal. With this assumption, the intensity of the moiré pattern can be represented as

$$I_h = \frac{\pi^2 K r_o}{\sin^2 2\vartheta} \chi_h^2 \left[I_o^I J_o^2 (B\sqrt{b^{II^2} - X^2}) \right.$$

$$+ I_h^I \frac{b^{II} - X}{b^{II} + X} J_I^2 (B\sqrt{b^{II^2} - X^2}) + 2\sqrt{I_o^I I_h^I} \tag{10.70}$$

$$\left. \cdot J_o (B\sqrt{b^{II^2} - X^2}) \frac{b^{II} - X}{b^{II} + X} J_1 (B\sqrt{b^{II^2} - X^2}) \cos\Phi \right] ,$$

where I_o^I and I_h^I and the argument Φ have the following values

$$I_o^I = I_o \left(1 - \frac{\sin^2 Bb^I \sqrt{1 + y_o^2}}{1 + y_o^2} \right), \quad I_h^I = I_o \frac{\sin^2 Bb^I \sqrt{1 + y_h^2}}{1 + y_h^2} \quad , \tag{10.71}$$

$$y_o = \eta_o \sin 2\vartheta / C\sqrt{\chi_h\chi_{\bar{h}}}\sqrt{\gamma_h/\gamma_o}$$

$$\Phi = 2\pi(\underline{\underline{dh}}\underline{\underline{r}}) - \pi K\chi_o\left(\frac{1}{\gamma_h} - \frac{1}{\gamma_o}\right)z^I - 2\pi\underline{\underline{dh}}_\perp z^I + \varphi \; . \quad y_h = \eta_h \sin 2\vartheta / C\sqrt{\chi_h\chi_{\bar{h}}}\sqrt{\gamma_h/\gamma_o}$$

$$(10.72)$$

$$\tan\varphi = \frac{y_{o,h}}{\sqrt{1 + y_{o,h}^2}} \tan Bb^I\sqrt{1 + y_{o,h}^2} \quad . \tag{10.73}$$

Using the diagram in Fig. 6.2 to describe the wave field in the first crystal, we notice that (see (3.66))

$$2b^I = d(\tan\psi_o + \tan\psi_h) = \frac{\sin 2\vartheta}{\gamma_o\gamma_h}$$

$$Bb^I = \frac{\pi K|C|t\chi_h\chi_{\bar{h}}}{\sqrt{\gamma_o\gamma_h}} = A \quad . \tag{10.74}$$

As a result, (10.71) for the intensities I_o and I_h are similar to (3.62) for the intensities in the case of the transparent crystal in the plane-incident-wave approximation (assuming $\chi_h = \chi_{\bar{h}}$).

At the same time, the factors at the value I_o and I_h in (10.70) are close to the squares of the moduli of the integrals U_o and U_h in KATO's theory (see (6.29,32)).

Thus, (10.70-74) permit us to make a clear, qualitative interpretation.

On the other hand, (10.64) can be compared with JEVERS' (10.47) only if the intensities I_o and I_h in both the first and the second crystal correspond to the plane-incident-wave approximation. It should be recalled that JEVERS' theory, developed for electron diffraction, is applicable to X-ray scattering only in symmetrical reflection. In this case the expression for the phase Φ given in (10.66) is simplified to

$$\Phi = 2\pi[(\underline{\underline{dh}}, \underline{r}) + \varphi] \quad , \tag{10.75}$$

which is quite similar to the value $2\pi(x - x_0)/\Lambda$ in (10.47), if we take into account the last one of (10.44).

10.4 Experimental Investigations. Three-Crystal Interferometer

When passing on to the experiments in X-ray interferometry, it should primarily be noted that the authors of published papers who used a three-crystal interferometer did not in many cases achieve uniform intensity resulting from realization of ideal geometry conditions (10.10). Usually, a moiré pattern is observed on the cross sections of the emerging beams $D_o^{i(d)}$ and $D_h^{i(d)}$. With the above-indicated high sensitivity of moiré to negligible disturbances in periodicity, the source of these patterns may evidently be either the insufficiently perfect structure of the initial single-crystal Si or quartz block or the distortions appearing during the fabrication of the interferometer. In fact, three-crystal interferometers are used most often for obtaining moiré patterns. At the same time, the presence of parasitic moiré bands on the photographs of beams emerging from plate A of the interferometer does not exclude the possibility of utilizing the instrument in various dynamic scattering investigations.

We noted in Chapter 9 (see (9.68)) the formation of the interference pattern resulting in the introduction of a wedge-shaped sample in the path of one of the interferometer beams. The spacing of these patterns is evidently directly related to the refracting index n of the radiation used. Such measurements were made in BONSE and HELLKÖTTER's work [10.6b], Fig. 10.7.

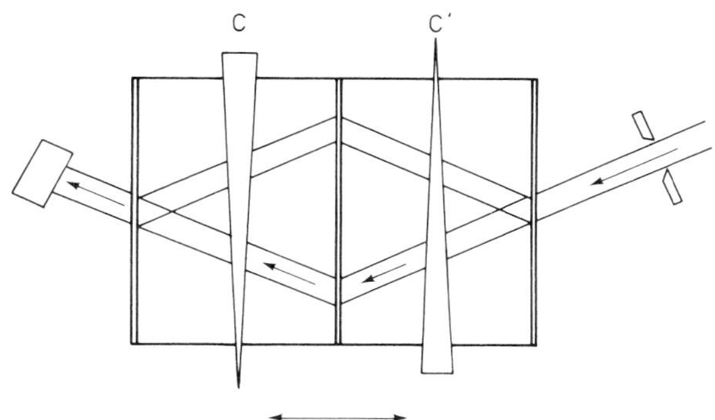

Fig. 10.7 Scheme of experiments allowing measurement of values of $|\chi_{or}|$ and refracting index n on the introduction samples in the path of one of interferometer beams (after BONSE and HELLKÖTTER [10.6b])

The scheme used with two rigidly connected wedges c, c' oscillating as indicated by arrows gives an oscillating curve across the section of the emerging beam $D_h^{(d)}$. The curve is recorded by the counter. The use of only one wedge halves the number of the maxima as a result of the absorption difference on the two trajectories in the interferometer.
Calculation formula given by the authors permitted obtaining the values of the 1-n from experiments carried out for a number of substances. Disagreements with the theory due to noninclusion of the dispersion correction make up ~ 0.3 % for plexiglas and Be, and ~ 1.5 % for LiF and NaF.

Actually the same effect was observed at the formation of interference patterns from plane-concave lens interposed in one of the beam paths in the three-crystal interferometer in the first works of BONSE and HART [1.28b], Fig. 10.8.

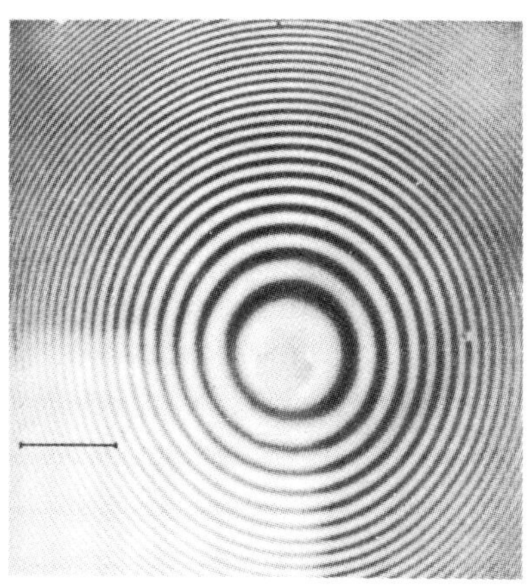

Fig. 10.8 Interference pattern from a plane-concave epoxy plate (scale mark is I μm) (after BONSE and HART [1.28b])

In BONSE and HART's paper [10.3b], part of the silicon single-crystal block adjoining the plate-analyzer A is connected to the other part by a narrow band (Fig. 10.9) and is provided with a contraption (lever, pulley, and weight) which makes it possible to turn the analyzer by a small angle with respect to the other part of the instrument. A turn of plate A by an angle of 0.01" will produce moiré with a band spacing of ~ 4 mm. Fig. 10.10a exhibits a series of patterns obtained at different values of the moment of the couple turning plate A. From the maximum angle of inclination of the bands observed, the ratio $\Delta d/d \approx 8 \cdot 10^{-8}$ can be determined. Fig. 10.10b depicts the dislocation images in moiré patterns. As already noted above, the moiré patterns not only directly reflect the distortions of the lattice

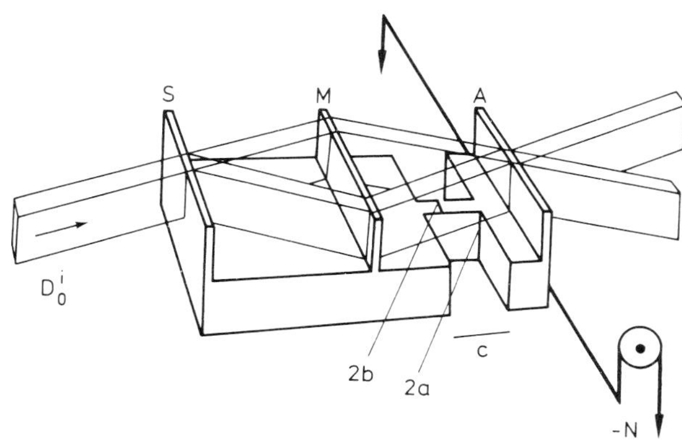

Fig. 10.9 Arrangement for deformation measurements by means of interferometer (see text) (after BONSE and HART [10.3b])

periods or the mutual turns, but also serve to obtain pictorial diffraction patterns of the dislocations. Very much like the excess plane in a crystal, moiré patterns show excess half-bands, whose number N obeys the rule

$$N = |\underline{b} \cdot \underline{h}| \tag{10.76}$$

where $\underline{b}$ is Burger's dislocation vector. BONSE and HART [10.3b] obtained patterns showing dislocations on the bands of the (predominant) dilatational (with vertical bands) and, respectively, rotational moiré. HART [10.19] also used the patterns obtained to determine the indices of Burger's vectors according to (10.76). These indices are determined with an accuracy to just one sign. As for the vector modulus, it was 1/2 110, or 3.84 Å. This value corresponds to the minimum interplanar spacing in Si crystals of the diamond-structure type. The corresponding reflection is forbidden on ordinary diffraction patterns.

In the case of dilatational moiré (the bands are perpendicular to vector $\underline{h}$), at $\Delta d > 0$ the excess bands are situated above the dislocation pattern, and at $\Delta d < 0$, below it. In the case of rotational moiré (the bands are parallel to vector $\underline{h}$), the relationship between the position of the dislocation in the crystal and its pattern depends on the sign of rotation of the plate under investigation with respect to the standard. If our line of vision is directed towards the source, on counterclockwide rotation the excess band on the pattern appears to the right of the dislocation in the crystal; on clockwise rotation, it appears to the left of it.

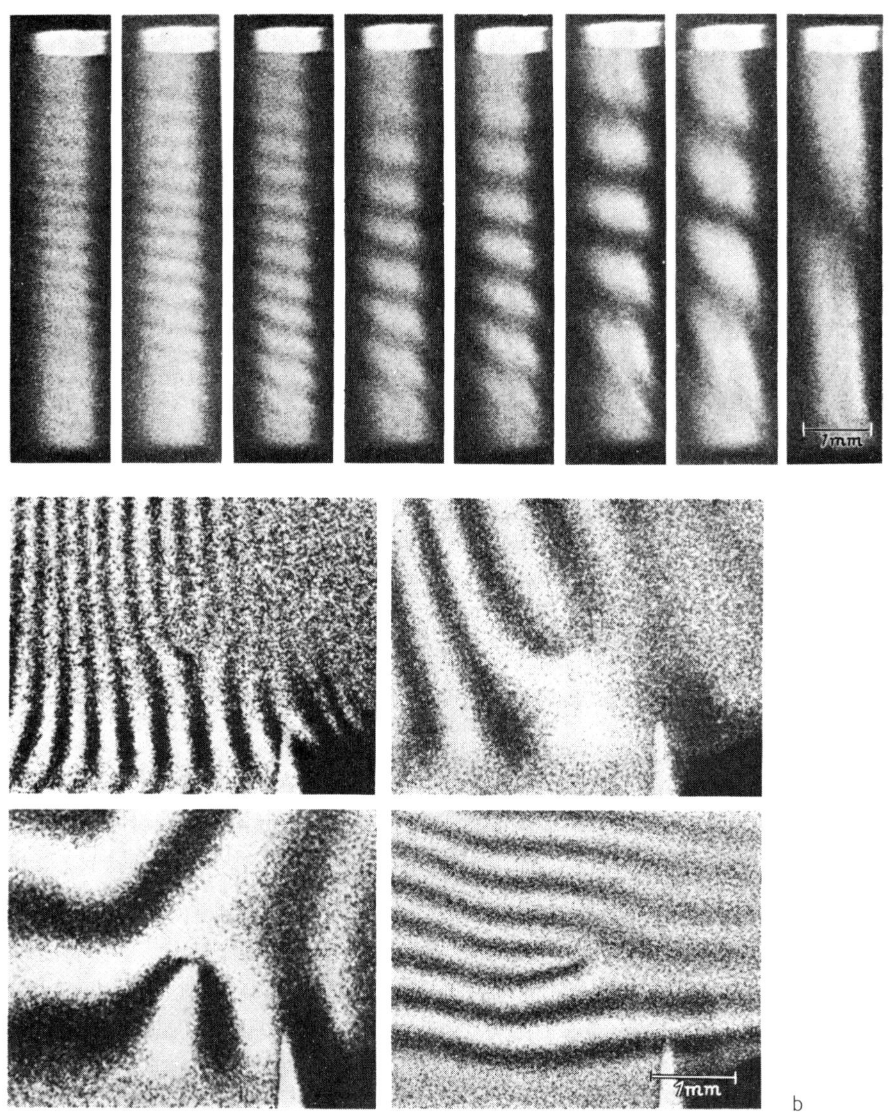

Fig. 10.10 Moiré patterns obtained on three-crystal interferometer at different values of moment of couple turning plate A. a) Patterns resulting from twisting (turn of the analyser A); Moment of couple decreases from left to right. b) Dislocation images in moiré pattern (after BONSE and HART [10.3b])

Among the highly promising investigations are studies of translational moiré appearing when one of two entirely identical-structure crystals *moves* relative to the other, and also when interferometer plate A moves with reference to S and M. As noted above, the three-crystal interferometer theory outlined in Section 10.1 is not quite rigorous, but nevertheless it makes it possible to formulate the ideal geometry conditions (10.10), provided that there is no defect in the structure of the plates forming the instrument. It is more difficult to develop a more precise theory of the moiré which appears when using BONSE and HART's interferometer, and particularly of the translational moiré, or to construct the appropriate model. It should be remembered that HART [10.20] and DESLATTES [10.21] investigated translational moiré in conditions of a nearly transparent crystal. In any case, for a plane incident wave (see (10.43)), in the more general theory as well (see (10.72)), the vector $\underline{\underline{dh}}$, which causes the appearance of moiré bands, appears where there is both common and translation moiré.

The appropriate recording of the intensity of any one of the two waves emerging of the interferometer (Fig. 10.1) makes it possible to determine the period d_h, regardless of the X-radiation wavelength used, by counting the number of bands (maxima) per unit length. A number of authors have solved this problem by simultaneous measurement of the total displacement produced by an optical interferometer. If we measure the interval corresponding to a single band $1/2\ \lambda_o$ with the aid of a radiation source with a wavelength equal to the International Standard of λ_o, then

$$d_h = 1/2\,\lambda/(p + \varepsilon) \tag{10.77}$$

where p is a whole number of bands of X-ray moiré, and ε is a fractional number.

Solving this problem sufficiently accurately essentially depends on the degree to which X-ray and optical measurements refer to one and the same segment of length. The instrument used by HART [10.20] to obtain translation moiré is shown in Fig. 10.11. Using the appropriate pressure, it is possible to deform springs S_1 and S_2 elastically and displace the analyzer with respect to plates S and M. A 20 μm displacement of the micrometric screw deforming the springs results in relative displacement of the interferometer parts by $d_{220} = 1.92$ Å. The translational moiré pattern is given in Fig. 10.12.

DESLATTES [10.21] used a similar device, which was supplemented by an optical interferometer and assembled together with the X-ray interferometer on a common brass base. This combination-type instrument with simultaneous

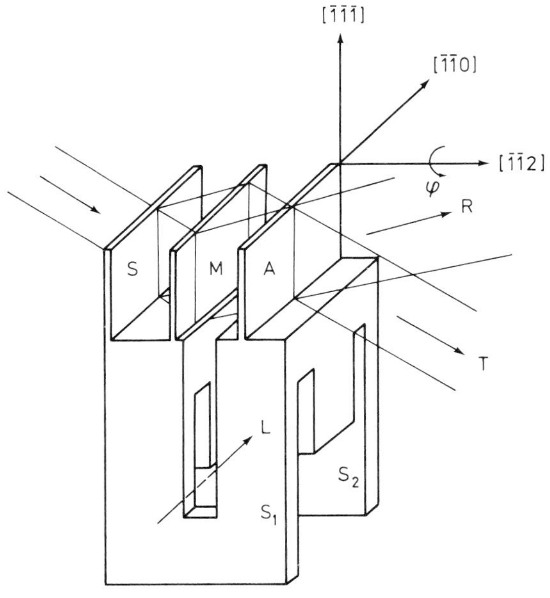

Fig. 10.11 Interferometer for obtaining translation moiré fringes (after HART [10.20])

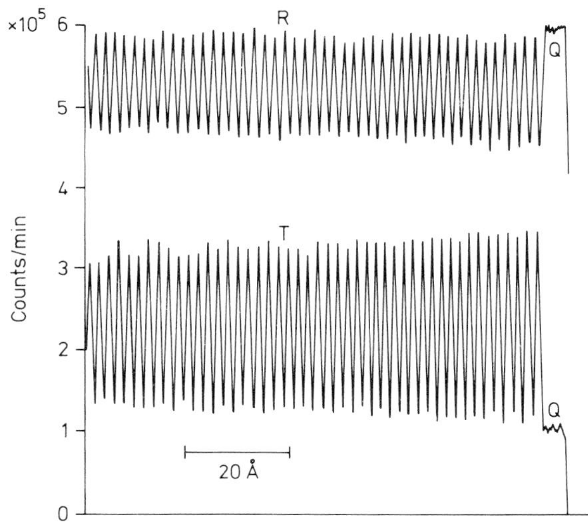

Fig. 10.12 Record of translation moiré fringes (after HART [10.20])

recording on two channels, optical and X-ray, enables one to make absolute measurements of segments up to 50 μm long with an error of the order of 1 Å.

An analysis of the accuracy attained by such parallel measurements in these author's schemes, and also in [10.22], is given in HART's survey lecture [10.23]. It would appear that measurements of d_h with an accuracy to

$\sim 1 \cdot 10^{-7}$ may soon be made. An increase in the indicated accuracy by one order makes these methods similar to the diffractometric ones (see Chapter 9). The possibility is not excluded that in the future the atomic mass standard will be used in place of the present artefact, the kilogram.

Reverting to the estimation of the role of the moiré method in X-ray topography, we note that, apart from the above analysis of dislocation patterns, interpretation of the experiment largely amounts to the use of simple geometrical relations of the type of (10.45). Moiré patterns were observed without a three-crystal interferometer in the investigations of CHIKAWA [10.7], LANG and MIUSKOV [10.8], and BRADLER and LANG [10.24]. These latter authors obtained a clear-cut pattern of rotational moiré (Fig. 10.13) from two single-crystal silicon plates oriented with a slight turn (2.5") during the experiment.

In MIUSKOV's work [10.25], moiré effects were used systematically in the X-ray topography of real crystals (Fig. 10.14a). The pattern was obtained from high-perfect, dislocation-free Si crystals with small oxygen contamination ($\sim 10^{-16}$ cm^{-3}). The top of the pattern shows the dilatation moiré fringes with $\Delta d/d \approx 1 \cdot 10^{-6}$. In the right and left bottom corners there are rotation moiré which correspond to the $\Delta\varphi = 3 \cdot 10^{-6}$ (< I"). The centre of the Si ingot (the central field of pattern) shows mixed type moiré fringes. There are no fringes in the very centre of the pattern which evidences that the central part of the Si ingot has a most perfect structure [1.63, 10.25].

In Fig. 10.14b, obtained by BEZIRGANIAN and EIRAMDZHIAN [10.26] in a three-crystal interferometer, one can observe the modulation of the moiré pattern (on the right) by interference from a wedge (on the left). The wedge was interposed in one of the interferometer beams (see Fig. 10.1).

10.4.1 Double-Crystal Interferometer

A number of experiments described in published papers correspond to various special conditions of wave diffraction upon transmission through two crystals, which is considered in the paper of AUTHIER et al. [1.51]. We have already mentioned the realization of total absorption of one of the two fields induced in crystal I, with a sufficient thickness of crystal II, in [10.11, 12]. The determinations of the absolute values of the structure and atomic amplitudes of Si and Ge performed in them are among the most accurate measurements of these parameters. The authors based their calculation of the interference pattern on (10.34) transformed with an allowance for the two polarization terms. The determinations were carried out on specimens whose

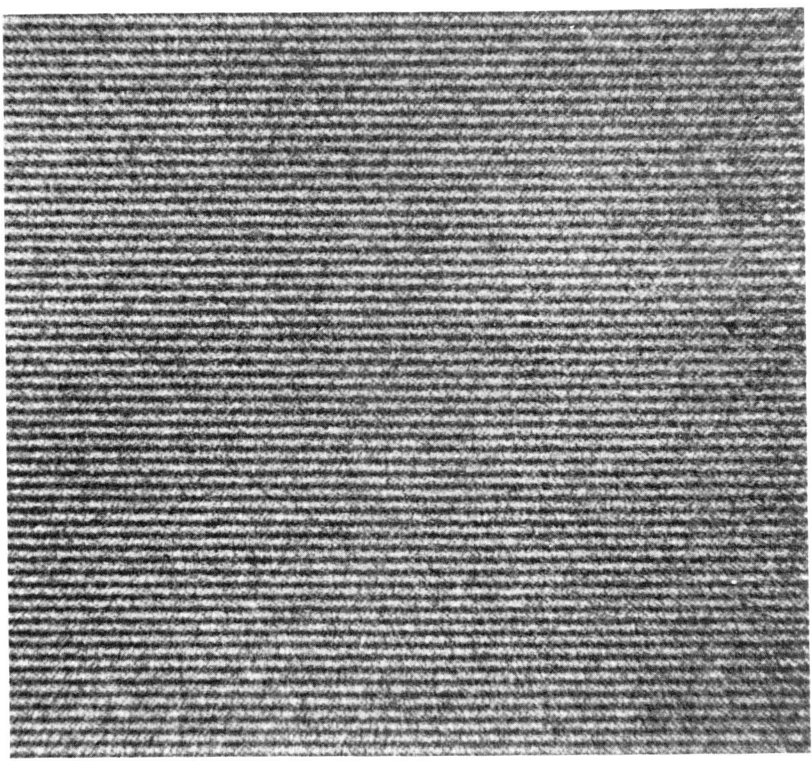

Fig. 10.13 Rotational moiré from two single-crystal Si-plates (relative turn 2.5") (after BRADLER and LANG [10.24])

thicknesses corresponded to $\mu t \approx 10$ for Si in [10.11] and $\mu t \approx 33$ for Ge in [10.12]. Nevertheless, account was taken of small displacements of the maxima due to the π-term of polarization. In both investigations, use was made of ladder-shaped specimens, which enabled the authors to obtain, in a single topograph, several interference patterns corresponding to different values of the thickness $(z_I + z_{II})$. A detailed analysis of the measuring procedure and the sources of errors is given in the papers indicated.

It is significant that this method of determining atomic amplitudes is applicable to medium- and heavy-atomic-weight elements and their compounds, since the topographs obtained correspond to an anomalously transmitted field

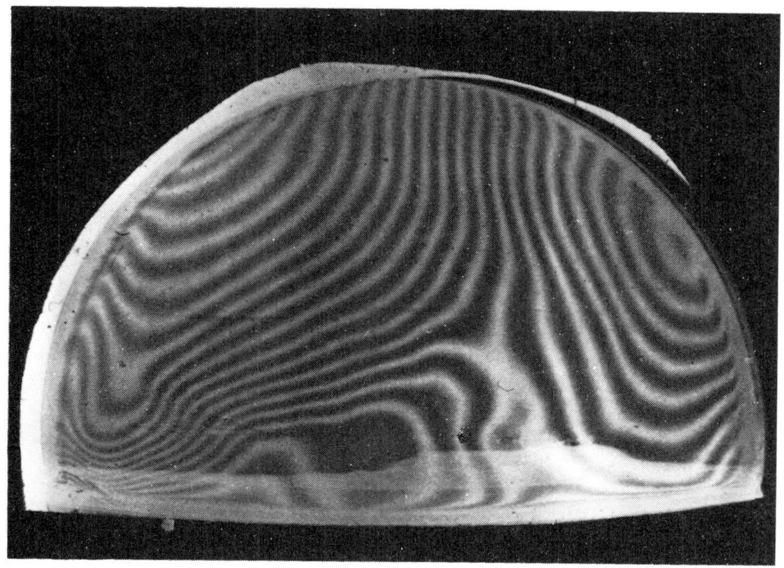

a

b

Fig. 10.14 Moiré pattern obtained on three-crystal interferometer. a) Moiré patterns from silicon crystal. b) Modulation of moiré image resulting from the wedge-shaped crystal plate interference [1.63, 10.26])

in the crystal. At the same time, it is worth mentioning the paper by TANEMURA and LANG [10.27] in which a repeat theoretical analysis of the scheme used by the authors of [10.11,12] was performed (by the same method as in [1.51]) with an allowance for the possible defects in the structures of crystals I and II in the course of filing through the gap. A moiré pattern could have appeared then. Moreover, the authors of 10.27 considered the contribution of a small deviation of the "asymptotes" of the dispersion hyperbolas from linearity, which could have resulted in an error of the order of ~ 1 percent. Of great interest is experimental realization of the specific cases discussed in [1.51], viz.: crystal I or II is wedge-shaped, and $\underline{g}$ is large enough to exclude overlapping of the wave fields in crystal II. Diffraction patterns in such schemes are described by the above relations for the phase differences (10.26) and (10.27). The authors of [9.38] achieved conditions corresponding to (10.26).

A direct comparison of interference hyperbolas corresponding to a plane or spherical incident wave was made by HART and MILNE in [10.13]. The experiment scheme (Fig. 10.15) corresponds to the conditions $z_I < z_{II}$. A beam

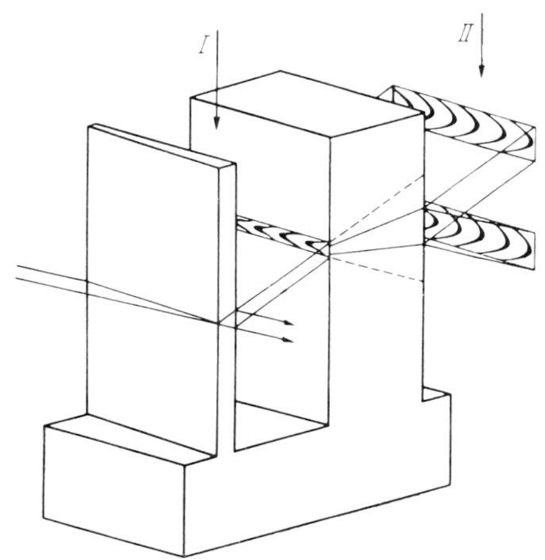

Fig. 10.15 Scheme of arrangement for obtaining hyperbolas corresponding to the case of plane and spherical incident waves (after HART and MILNE [10.13])

transmitted through a system of slits was directed at the first thin wedge-shaped crystal. A photoplate interposed immediately after the first plate shows an interference pattern in the shape of hyperbolas, which is discussed in detail in Chapter 6. The intensity distribution of the beam transmitted through both plates, which is depicted in Fig. 10.16a, was investigated thoroughly. The differences between the two interference patterns largely amount to the following.

1) The absolute position of the bands of the plane wave are displaced by 1/4 period relative to the bands of the spherical wave, in the direction of the exit surface.

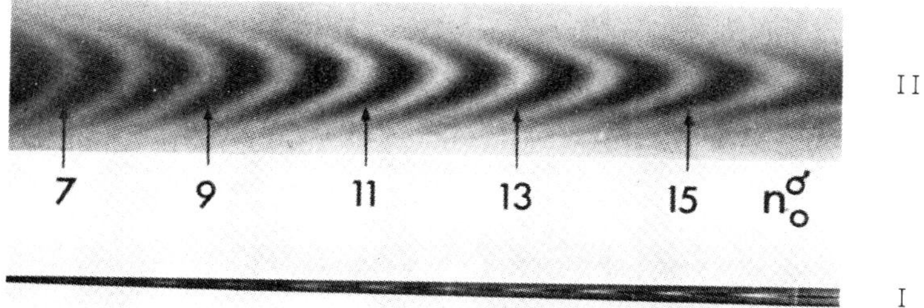

Fig. 10.16 Images of interference hyperbolae for plane (II) and spherical (I) incident waves (see text) (after HART and MILNE [10.13])

2) At the hyperbola midpoint, i.e., in a direction parallel to the reflecting plane, the periods of the two patterns are almost identical at a sufficient depth.

3) While for the plane wave the periods diminish with increasing $|y_r|$ inside the maximum, for the spherical wave the periods increase; this explains the reversal of the hyperbolae on the plane wave pattern.

4) The periods of the two patterns (for each given value of y_r) are slightly different, since the effect of the modulation of the spherical wave patterns due to the interference of the σ- and π-oscillations disappears because of the absorption of the π-component by the thick plate (Fig. 10.15).

Thus, the pattern of Fig. 10.16a unquestionably corresponds to the interference bands on incidence of a plane wave, which however, differ from the straight pendulum solution bands of constant thickness. Namely, they are not formed at a single fixed value of y_r, but represent the entire maximum region. Similar to (10.26), in this experiment we obtain for the phase difference at $z_I \ll z_{II}$

$$\Delta'_{TT'} \approx \frac{2\pi}{\tau} \frac{z_I}{\sqrt{1 - Y_o^2}} \quad ; \quad \Delta'_{RR'} \approx \frac{2\pi}{\tau} \frac{z_I}{\sqrt{1 - Y_h^2}} \quad . \tag{10.78}$$

The value of τ, corresponding to τ_0 in (3.48), reduces to zero with increasing Y (i.e., with increasing y, according to (5.53), which is characteristic of the case of incidence of a plane wave. It should be emphasized that with the condition indicated $z_I \ll z_{II}$, bands corresponding to the difference between the phases of the A- and A'-waves appear: $\Delta''_{TT'}$, $\Delta''_{RR'}$, which correspond to the "spherical"-incident-wave approximation (see, for instance, (10.27)). The problem is solved by relating the intensities of these two types of pattern:

$$I'_{TT'} \sim \cosh(\varphi_{oi} - \varphi'_{oi}) \cos\Delta'_{TT'} \quad ;$$

$$I''_{TT'} \sim \cosh(\varphi_{oi} + \varphi'_{oi}) \cos\Delta''_{TT'} \quad . \tag{10.79}$$

Since φ'_o (and φ'_h) becomes negative at $z_I < z_{II}$, the contrast will be intensified precisely for the bands $I'_{TT'}$.

With conditions: z_I and g are constant, z_I is sufficiently thin for transmission of both wave fields, and z_{II} is variable, in paper [1.51] there is a photograph of a wave field which shows, in the centre, a more intensive area corresponding to $z_I = z_{II}$. As already noted, no analytical form for the intensity was obtained in that work for the latter condition. The relevant problem is discussed in [10.14].

Consideration of the problem of transmitting an incident wave packet through two crystals separated by a nondiffracting lamella, using the generalized theory method, is also presented in [1.61], which has partly been discussed previously. Finally, the authors of [1.51] briefly review more complex topographs obtained from a specially prepared specimen in the form of a thin crystal with a filed-through slit, which is not parallel to the entrance and exit surfaces. This experiment scheme corresponds to more complex interference effects at variable values of z_I and z_{II} and constant g and $(z_I + z_{II})$.

11. Generalized Dynamical Theory of X-Ray Scattering in Perfect and Deformed Crystals

In Chapter 1, which presents a historical survey of the development of conceptions of X-ray propagation in perfect (or nearly perfect) crystals, we note that recent years have seen a kind of revival of DARWIN's dynamical theory. The advantage of the original version of this theory lies in the extreme simplicity of the derivations, with the aid of which its author, as well as PRINCE [7.1] and others, obtained results confirmed by the more rigorous Ewald-Laue-Zachariasen theory. When the task of developing a theory of X-ray scattering in deformed crystals came to the fore, PENNING and POLDER [1.57], and KATO [1.58] considered this problem for slightly deformed crystals with the use of a somewhat modified Laue-Zachariasen theory. They met with considerable difficulties caused by the inapplicability to different scattering conditions of such notions as the ordinary Bloch wave and its approximation by superposition of plane waves, the dispersion surface, and exponential complex wave functions.

At the same time, TAGAKI proposed, in a short item published in 1962 [1.45], a different approach to the same problem and derived fundamental equations for describing the waves in a crystal, which were close in their form to equations of the type

$$\begin{cases} i\dfrac{dD_o}{dz} = g_{oo}\underline{D}_o + g_{oh}\underline{D}_h \\ \\ i\dfrac{dD_h}{dz} = g_{ho}\underline{D}_o + (g_{hh} + 2\alpha)\underline{D}_h \end{cases} \qquad (11.1)$$

where g_{hh} are some complex transmission and reflection coefficients, and 2α is the function of deviation from the exact value of the Bragg angle, for instance, of the type of (3.6).

Incidentally, (11.1) are nothing but DARWIN's recurrence relations recorded in differential form [1.5].

It has been shown for electron diffraction [1.27,44], and by a number of authors for X-ray diffraction [1.45-49, 9.24], that a theory based on such equations can be directly generalized for the deformed crystal and for the general case of the incident wave packet.

In this chapter, following TAUPIN [11.1], we first of all give the derivation of Takagi-type equations form Maxwell's formulas. Then we consider, in accordance with [1.60], the application of the theory in the approximation of the incident wave packet to scattering in the perfect crystal both for the Laue and Bragg cases.

The potentialities of the theory for solving problems on Bragg diffraction as applied to the deformed crystal are illustrated in Section 4.3.

11.1 Deriving Fundamental Equations in the General Case of Deformed Crystal

In this theory, the distortion region is divided into two parts: the slightl deformed one, and the strongly deformed one, in which a dynamical equilibrium of the transmitted and reflected waves is practically nonexistent. For the slightly deformed region, no assumption is made as to the kind of deformation, and only one limitation is imposed on the order of increase in its value, namely, the relative change in deformation is taken to be small compared with unity.

$$\left|\frac{\partial u_i}{\partial x_j}\right| << 1 \quad . \tag{11.2}$$

($u_j(\underline{r})$ is the displacement vector, $\underline{r}$ is the radius vector in the real crystal; i, j = 1,2,3).

The electric displacement vector in a vacuum is described by the expression

$$\underline{D} = \underline{D}_0(\underline{r}) \exp\{i[\omega_0 t - 2\pi\Phi_0(\underline{r})]\} \tag{11.3}$$

where the cyclical frequency is introduced:

$$\omega_0 = 2\pi\nu_0 \quad .$$

The expression (11.3) is a generalization of (2.19) and describes an arbitrary (in the sense of the shape of the equal-phase surface) wave. In particular, for the plane wave: $\Phi_o(\underline{r}) = (\underline{K}_o \underline{r})$, $|\underline{K}_o| = 1/\lambda$. In the general case, $\underline{D}$ satisfies the wave equation for a vacuum:

$$\Delta \underline{D} + \frac{\omega_o^2}{c^2} \underline{D} = 0 \quad . \tag{11.4}$$

Substituting (11.3) in (11.4), we obtain

$$\Delta \underline{D}_o - i4\pi \sum_i \frac{\partial \Phi_o}{\partial x_i} \frac{\partial \underline{D}_o}{\partial x_i} - i2\pi \underline{D}_o \Delta \Phi_o - 4\pi^2 \sum_j \left(\frac{\partial \Phi_o}{\partial x_j}\right)^2 \underline{D}_o + \frac{4\pi^2}{\lambda^2} \underline{D}_o = 0 \quad . \tag{11.5}$$

Since $\underline{D}_o$ and Φ_o are real functions of the coordinates, the division of the real and imaginary parts of (11.5) is carried out in a straightforward manner.

We will consider waves for which the radii of curvature of the equal-phase surfaces are much larger than the wavelength λ; for instance, for a spherical wave of radius R we will assume $\lambda/R \ll 1$. It is easy to see that in this case

$$|\mathrm{grad}\Phi_o| \approx \lambda^{-1}\left[1 + 0\left(\frac{\lambda^2}{R^2}\right)\right] \tag{11.6}$$

where the second term in the brackets of the right-hand side denotes negligibly small terms of higher orders of smallness with respect to λ.

If the incident wave is nearly plane, i.e.,

$$\mathrm{grad}\Phi_o = \underline{K}_o + \Delta \underline{K}_o \tag{11.7}$$

and $\underline{K}_o \Delta \underline{K}_o = 0$ and $|\Delta \underline{K}_o| \ll |\underline{K}_o|$, then it follows from (11.6) that

$$|\underline{K}_o| = \lambda^{-1} \quad , \qquad |\Delta \underline{K}_o| = R^{-1} \tag{11.8}$$

which corresponds to an incident wave packet with an angular width of the order of 1/R.

In describing the wave inside the crystal we will retain the (generalized) function $\underline{D}_o$ for the phase. The amplitude $\underline{D}_o$ becomes a complex function of the coordinates, which is dependent on the difference in the paths of propagation in the crystal and the vacuum. In particular, in the case of the perfect crystal we have a single pseudo-periodic wave corresponding to the right-hand side of (3.34), for each state of polarization. Hence it follows that the boundary condition (3.28) amounts to regarding D_o as a continuous function of the coordinates on transition from the vacuum to the crystal.

The wave field in the crystal will be described with the aid of the Bloch function, but now with variable amplitudes (see (2.36) and (2.38)):

$$\underline{D} = \sum_m \underline{D}_m \exp\{i[\omega t - 2\pi(\underline{k}_m r)]\} \quad . \tag{11.9}$$

In the case of the perfect crystal and plane waves, the following values remain constant (2.37):

$$\underline{k}_m = \underline{k}_o + \underline{h}_m \quad , \quad \underline{k}_o = \mathrm{grad}\Phi_o \quad . \tag{11.10}$$

In the general case, however, $\underline{k}_o$ and $\underline{h}_m$ are functions of the coordinates. We can determine, at each point, the vector $\underline{h}_m$, which is normal to the system of planes m (index triplet), and

$$|\underline{h}_m(\underline{r})| = d_m^{-1}(\underline{r}) \quad . \tag{11.11}$$

Then, on switching from the plane with number n_m to plane $n_m + 1$ of the same system, we have the relation

$$\Delta h_m = 1 = \underline{h}_m \underline{dr} \tag{11.12}$$

where $\underline{dr}$ is the increment of the radius vector $\underline{r}$ in transfer from the given point $\underline{r}$ on the n_m-th plane to the point $r + dr$ on the $(n_m + 1)$-th plane.

Regarding, further, n_m quite formally as a continuous function of the coordinates, which takes integral values on each plane m, we find

$$\underline{h}_m = \mathrm{grad}\; n_m \tag{11.13}$$

$$D = \sum_m D_m \exp[i(\omega_o t - 2\pi\Phi_m)] \qquad (11.14)$$

where

$$\Phi_m(\underline{r}) = \Phi_o(\underline{r}) + n_m(\underline{r}) \quad . \qquad (11.15)$$

Note that the method for introducing the function n_m described here is not rigorous. The continuity of n_m breaks down, in particular, in crystals containing dislocations. Nevertheless it can be shown that the results of the theory outlined are applicable in the general case of arbitrary distribution of n_m if (11.2) is satisfied in the deformed crystal.

We will now make a remark referring to the dielectric constant and polarizability in the general case of the deformed crystal. The dielectric constant here remains a periodic function of the coordinates, but according to condition (11.2), we can approximately assume $\varepsilon(\underline{r})$ to be an exponential functions of n_m in each system m of reflecting planes. For polarizability, (2.24), (2.34), and (2.35) remain valid. In the Fourier expansion (2.30), in accordance with the above remark, we take into account (11.11). Thus we write down

$$\chi(\underline{r}) = \sum_m \chi_m \exp[-2\pi i n_m(\underline{r})] \qquad (11.16a)$$

$$\chi_m \equiv \chi_{h_m} = -\frac{e^2}{mc^2}\frac{\lambda^2}{\pi\Omega}F_{h_m} \quad . \qquad (11.16b)$$

Further, in accordance with the analysis carried out in Sections 4.1 and 4.2 (for instance, (4.19)), we write down for the general case of the absorbing crystal:

$$\chi_o = \chi_{or} + i\chi_{oi} \quad , \quad \chi_h = \chi_{hr} + i\chi_{hi} \quad . \qquad (11.17)$$

One of the characteristic features of scattering in deformed crystals is the dependence of the reflection angle, or deviation η_m, on the coordinates. In this chapter we use, as angular variable, the value

$$\alpha_m = 2\eta_m \sin 2\vartheta_m \qquad (11.18)$$

which corresponds to the doubled value of α from (3.6). It is easy to see that the value of α_m from (11.18) corresponds to the approximate value of the expressions

$$\alpha_m = k_o^{-2}[h_m^2 + 2(\underline{k}_o\underline{h}_m)] = \lambda^2\left(\frac{1}{d_m^2} - \frac{2\sin\vartheta_m}{\lambda d_m}\right) \qquad (11.19)$$

where both d_m and $\sin\vartheta_m$ are coordinate dependent.

We shall now derive the wave equation for the field inside the crystal and the fundamental equations. As in Chapter 2, we shall proceed from the Maxwell equations.

Taking the curl of both sides of (2.2), we get

$$\mathrm{curl\ curl}\ \underline{E} = -\frac{1}{c}\,\mathrm{curl}\,\frac{\partial\underline{H}}{\partial t} = -\frac{1}{c}\frac{\partial}{\partial t}\,\mathrm{curl}\ \underline{H} \quad . \qquad (11.20)$$

Now rewrite (2.2) in the form (see (2.15)).

$$\mathrm{curl}\ \underline{H} = c^{-1}\,\frac{\partial\underline{D}}{\partial t} \quad . \qquad (11.21)$$

Substituting (11.21) in (11.20) and passing over from the vector $\underline{E}$ in the left-hand side of (11.20) to the vector $\underline{D}$, in accordance with the relation

$$\underline{D} = \varepsilon\underline{E} = (1 + \chi)\underline{E} \quad , \quad \underline{E} \approx (1 - \chi)\underline{D} \qquad (11.22)$$

we obtain the wave equation for the displacement waves $\underline{D}$:

$$\mathrm{curl\ curl}\ (1 - \chi)\underline{D} = \frac{4\pi^2}{\lambda^2}\,\underline{D} \quad . \qquad (11.23)$$

Here, it is assumed formally that $\underline{D}$ depends on time in the same way as does each of the plane waves in (11.14).

In order to calculate the left-hand side of (11.23) in explicit form, we first rewrite the value $(1 - \chi)\underline{D}$ using the solution of the wave equation (11.14) and the Fourier expansion of polarizability (11.16a). By analogy with the relation of the reciprocal vectors in the perfect crystal $\underline{h}_k + \underline{h}_l = \underline{h}_{k+l}$, we put $n_k + n_l = n_{k+l}$:

$$\underline{E} = (1 + \chi)\underline{D} = \exp(i\omega_0 t)\left\{ \sum_m \underline{D}_m \exp(-i2\pi\Phi_m) - \sum_m \sum_h \chi_m \underline{D}_h \exp[-i2\pi(n_m + \Phi_h)] \right\}$$

$$= \exp(i\omega_0 t) \sum_m \underline{Q}_m \exp(-i2\pi\Phi_m) \tag{11.24}$$

$$\underline{Q}_m = \underline{D}_m - \sum_h \chi_{m-h} \underline{D}_h \quad . \tag{11.25}$$

Subsequently, the factor $\exp(i\omega_0 t)$ is omitted. Besides, the following relations are used:

$$\underline{k}_0 = \mathrm{grad}\Phi_0 \quad , \quad \underline{h}_m = \mathrm{grad}\, n_m \quad , \quad \underline{k}_m = \mathrm{grad}\Phi_m \quad . \tag{11.26}$$

The components along the axes of the double vector product have the following form according to the well-known vector analysis equations:

$$(\mathrm{curl\ curl}\ \underline{E})_i = -\frac{\partial^2 E_i}{\partial x_k^2} + \frac{\partial^2 E_k}{\partial x_i \partial x_k} \tag{11.27}$$

where the subscripts i and k independently take the values of 1,2,3.

For the term number m in the sum in the right-hand side of (11.24), the following values of the components along the axes will appear after the first differentiation:

$$\frac{\partial}{\partial x_k}(1 - \chi)D_{m_i} = \exp(-i2\pi\Phi_h)\left(\frac{\partial Q_{m_i}}{\partial x_k} - i2\pi k_{mk} Q_{mi}\right) \quad . \tag{11.28}$$

Performing a second differentiation and taking into account that

$$\frac{\partial}{\partial x_i} k_{mk} = \frac{\partial^2 \Phi_m}{\partial x_i \partial x_k} = \frac{\partial}{\partial x_k} k_{mi}$$

we finally get

$$\begin{aligned} \text{curl curl}(1 - \chi)\underline{D} = \sum_m \exp(-i2\pi\Phi_m)\{4\pi^2[k_m^2 \underline{Q}_m - (\underline{k}_m \underline{Q}_m)\underline{k}_m] \\ + i4\pi \underline{k}_m \frac{\partial \underline{Q}_m}{\partial \underline{r}} - i2\pi \text{grad}(\underline{k}_m \underline{Q}_m) - i2\pi \underline{k}_m \text{ div } \underline{Q}_m \\ + i2\pi \underline{Q}_m \Delta\Phi_m - \Delta\underline{Q}_m + \text{grad div } \underline{Q}_m\} = \underline{A} \quad . \end{aligned} \tag{11.29}$$

Substituting (11.29) and the Bloch solution (11.14) in (11.23), we obtain

$$\frac{4\pi^2}{\lambda^2} \sum_m \underline{D}_m \exp(-i2\pi\Phi_m) = \underline{A} \quad . \tag{11.30}$$

Equation (11.30) has an infinite number of unknowns. Assuming that the effective distances at which the preexponential factors with number m in (11.30) change are much larger than the wavelength with the wave vector $\underline{k}_m$ (as will be shown by the subsequent estimates), we can equate the terms referring to different waves m. Setting up scalar products of each of the terms of (11.30) by $\underline{Q}_m$, we get for each wave

$$\begin{aligned} 4\pi^2[k_m^2 Q_m^2 - k_o^2(\underline{D}_m \underline{Q}_m) - (\underline{Q}_m \underline{k}_m)^2] + i2\pi \underline{k}_m \text{ grad } \underline{Q}_m^2 \\ + i2\pi[\underline{Q}_m^2 \Delta\Phi_m - \underline{Q}_m \text{ grad}(\underline{k}_m \underline{Q}_m) - (\underline{k}_m \underline{Q}_m) \text{ div } \underline{Q}_m] \\ + \underline{Q}_m \text{ grad div } \underline{Q}_m - \underline{Q}_m \Delta \underline{Q}_m = 0 \end{aligned} \tag{11.31}$$

Replacing k_m^2 by α_m in (11.31), we obtain, according to (11.19),

$$4\pi^2[k_o^2\alpha_m Q_m^2 - k_o^2 \sum_h \chi_{m-h} \underline{D}_h \underline{Q}_m - (\underline{Q}_m \underline{k}_m)^2]$$

$$+ i2\pi\{\underline{k}_m \text{ grad } Q_m^2 + Q_m^2 \Delta\Phi_m - \underline{Q}_m \text{ grad}(\underline{k}_m \underline{Q}_m)]$$

$$- (\underline{k}_m \underline{Q}_m) \text{ div } \underline{Q}_m + \underline{Q}_m \text{ grad div } \underline{Q}_m - \underline{Q}_m \Delta \underline{Q}_m \quad . \tag{11.32}$$

The system of (11.32) contains terms differing considerably in their order of magnitude. In what follows, we eliminate terms which are small compared to $Q_m^2/\lambda^2 \simeq 1$. Below, we give an estimate of the order of magnitude of all the terms of (11.32). The letter τ here denotes values which are small compared to unity.

1) $k_o^2 \alpha_m Q^2{}_m$: $\alpha_m << 1$, whence $k_o^2 \alpha_m Q_m^2 \sim \tau Q^2 \lambda^{-2}$;

2) $k_o^2 \chi_{m-h} \underline{D}_h \underline{Q}_m$: $|\chi_{m-h}| << 1$, whence

$k_o^2 \chi_{m-h} \underline{D}_h \underline{Q}_m \sim \tau Q_m^2 \lambda^{-2}$;

3) $(\underline{Q}_m \underline{k}_m)^2$; in view of the sufficiently accurate transversality of $\underline{D}_h$ wave: $(\underline{Q}_m \underline{k}_m) \sim \tau \underline{Q}_m/\lambda$, $(\underline{Q}_m \underline{k}_m)^2 \sim \tau^2 Q_m^2/\lambda^2$;

4) $\underline{k}_m$ grad Q_m^2: $\underline{Q}_m$ varies gradually on segments of the order of λ, $\underline{k}_m$ grad $Q_m^2 \sim \lambda Q_m^2/\lambda^2$

5) $Q_m^2 \Delta\Phi_m$: from (11.4-6), and also (11.25), it follows that $\Delta\Phi_m \sim \Phi_m \tau^2/\lambda^2$ hence $Q_m^2 \Delta\Phi_m \sim \tau^2 Q_m^2/\lambda^2$;

6) $\underline{Q}_m$ grad$(\underline{Q}_m \underline{k}_m)$: $\underline{k}_m \underline{Q}_m \sim \tau \underline{Q}_m/\lambda$ vary gradually, hence $\underline{Q}_m$ grad$(\underline{Q}_m \underline{k}_m)$ $\sim \tau^2 Q^2/\lambda^2$;

7) $(\underline{k}_m \underline{Q}_m)$ div $\underline{Q}_m$: $(\underline{k}_m \underline{Q}_m) \sim \tau \underline{Q}_m/\lambda$, since $\underline{Q}_m$ varies gradually on segments of the order of λ, then div $\underline{Q}_m << \underline{Q}_m/\lambda$; $(\underline{k}_m \underline{Q}_m)$ div $\underline{Q}_m \sim \tau^2 Q_m^2/\lambda^2$;

8) $\underline{Q}_m$ grad div $\underline{Q}_m$: div $\underline{Q}_m \sim \tau \underline{Q}_m/\lambda$ varies gradually on segments of the order of λ and $\underline{Q}_m \cdot$ grad div $\underline{Q}_m \sim \tau^2 Q_m^2/\lambda^2$;

9) $\underline{Q}_m \Delta \underline{Q}_m$: $\Delta \underline{Q}_m \sim$ grad div $\underline{Q}_m$, $\underline{Q}_m \Delta \underline{Q}_m \sim \tau^2 Q_m^2/\lambda^2$.

Taking into consideration the above estimates and neglecting, in (11.32), the other terms of the second order of smallness which appear on elimination of $\underline{Q}_m$, in accordance with (11.25) we rewrite (11.32) in the following form (after cancelling by $k_o^2\underline{D}_m$; $\cos X_{mh}$ is the polarization factor):

$$\alpha_m\underline{D}_m - \sum_h \chi_{m-h}\underline{D}_h \cos X_{mh} + \frac{i\lambda^2}{\pi}(\underline{k}_m \operatorname{grad})\underline{D}_m = 0 \quad . \tag{11.33}$$

Besides, it should be emphasized that in (11.33) the values of $\underline{k}_m$ are constant and correspond to the wave vectors in the perfect crystal. Thus, the problem on X-ray propagation in the perfect crystal reduces to the solution of the system of (11.33) in first-order partial derivatives with variable α_m calculated with an allowance for the local deformation field at the given point of the crystal.

In the subsequent text we shall restrict ourselves to consideration of two waves (the two-wave approximation) 0 and h and to scattering in a centrosymmetrical crystal.

Let $\underline{s}_o$ and $\underline{s}_h$ be unit vectors in the directions of propagation of the refracted and reflected wave, respectively:

$$\underline{s}_o = \lambda\underline{k}_o \quad , \quad \underline{s}_h = \lambda\underline{k}_h \quad . \tag{11.34}$$

For any point of the reflection plane in the crystal we have

$$\underline{r} = s_o\underline{s}_o + s_h\underline{s}_h \tag{11.35}$$

and the system of (11.33) in the two-wave approximation is recorded thus

$$-i\,\frac{\lambda}{\pi}\,\frac{\partial D_o}{\partial s_o} = \chi_o D_o + C\chi_{\bar{h}} D_h$$

$$-i\,\frac{\lambda}{\pi}\,\frac{\partial D_h}{\partial s_h} = (\chi_o - \alpha_h)D_h + C\chi_h D_o \quad , \quad C = \cos X_{oh} \text{ (see (2.74))} \quad . \tag{11.36}$$

As in Chapter 2, in deriving the main relations of the Ewald-Laue theory, switching to the two-wave approximation made it possible to obtain the fundamental (11.36) in scalar form.

It is quite evident that in the two-wave case we can include the polarization factor C appearing in (11.33), in the value χ_h by adding the symbols σ (σ-polarization) and π (π-polarization; here,

$$\chi_h^\sigma = \chi_h^\pi(|\cos 2\Theta|)^{-1} \quad . \tag{11.37}$$

The system (11.36) was originally obtained by TAKAGI in 1962 [1.45]. At present it is the principal system of equations used in solving problems on the propagation of space-inhomogeneous X-ray wave packets in a crystal and in analyzing the pattern problem in X-ray diffraction optics (both in perfect and deformed crystal).

11.2 X-Ray Diffraction in Perfect Crystal Under Conditions of Space-Inhomogeneous Dynamical Problem. Influence Functions of Point Source

Let us consider the general case of a wave packet incident on the surface of a perfect crystal in the symmetrical Laue case, introduce, in the reflection plane, a rectangular system of dimensionless coordinates with an axis Ox antiparallel to the reflection vector $\underline{h}$, and convert the oblique system $(s_o s_h)$ in (11.36) to a new rectangular system with the aid of the relations

$$x = \frac{\pi}{\lambda}(C_o - C_h) \quad , \quad z = \frac{\pi}{\lambda}(C_o + C_h) \quad . \tag{11.38}$$

Eqs. (11.36) take the form (see (10.48))

$$\begin{cases} -i\left(\frac{\partial}{\partial z} + \frac{\partial}{\partial x}\right)D_o = \chi_o D_o + \chi_{\bar{h}} D_h \\ \\ -i\left(\frac{\partial}{\partial z} - \frac{\partial}{\partial x}\right)D_h = (\chi_o - \alpha)D_h + C\chi_h D_o \end{cases} \tag{11.39}$$

or, using the substitutions

$$D_o' = D_o \exp(-i\chi_o z) \quad , \quad D_h' = D_h \exp(-i\chi_o z) \tag{11.40}$$

we obtain

$$\begin{cases} -i\left(\dfrac{\partial}{\partial z} + \dfrac{\partial}{\partial x}\right)D_o' = C\chi_{\bar{h}}D_h' \\ \\ -i\left(\dfrac{\partial}{\partial z} - \dfrac{\partial}{\partial x}\right)D_h' = -\alpha D_h' + C\chi_h D_o' \quad . \end{cases} \tag{11.41}$$

In the subsequent text we omit the primes at D_o' and D_h'.

The system (11.41) corresponds to the telegraph equation for the amplitudes D_o and D_h [11.2]:

$$\hat{L}(D_j) = \left[\frac{\partial^2}{\partial z^2} - \frac{\partial^2}{\partial x^2} + i\alpha\left(\frac{\partial}{\partial z} + \frac{\partial}{\partial x}\right) + \chi^2\right]D_j = 0 \tag{11.42}$$

$$(j = 0,h) \quad .$$

In order to construct (11.42) with arbitrary boundary conditions, we introduce into consideration functions $G(\xi - x, \zeta - z)$ (see [11.3])

$$\hat{L}_{\zeta\xi}(G) = \left[\frac{\partial^2}{\partial \xi^2} - \frac{\partial^2}{\partial \xi^2} + i\alpha\left(\frac{\partial}{\partial \xi} + \frac{\partial}{\partial \xi}\right) + \chi^2\right]G = \delta(\xi - x)\delta(\zeta - z) \quad . \tag{11.43}$$

Here, $\delta(\xi)$ is Dirac's delta function. In (11.42,43) we use the notation $\chi^2 = C^2\chi_h\chi_{\bar{h}}$. From the definition of the function $G(\xi - x, \zeta - z)$ and (11.43) also follows

$$\hat{L}_{z,x}(G) = \left[\frac{\partial^2}{\partial z^2} - \frac{\partial^2}{\partial x^2} - i\alpha\left(\frac{\partial}{\partial z} + \frac{\partial}{\partial x}\right) + \chi^2\right]G = \delta(\xi - x)\delta(\zeta - z) \quad . \tag{11.44}$$

The linear differential operators of the function G satisfying equations with the δ-function in the right-hand side are called the Green functions.

We will now set up a bilinear combination - the identity

$$[D_j\hat{L}_{z,x}(G) - G\hat{L}(D_j)] = D_j\delta(\xi - x)\delta(\zeta - z) \quad . \tag{11.45}$$

Integrating (11.45) in the (xz)-plane and transforming the integral over the area to the integral over the contour bounding the given area with the aid of the Green theorem, we find the integral equation relating the wave fields inside the crystal and on an arbitrarily assigned contour C:

$$D_j(\xi,\zeta) = \int_C \left[D_j\left(\frac{\partial G}{\partial z}\,dx + \frac{\partial G}{\partial x}\,dz\right) - G\left(\frac{\partial D_j}{\partial z}\,dx + \frac{\partial D_j}{\partial x}\,dz\right)\right.$$

$$\left. + i\alpha G D_j(dz - dx)\right] \quad . \tag{11.46}$$

If the values of the function D_j and of its normal derivative $\partial D_j/\partial n$ on contour C are known (these values can be assigned arbitrarily only for contours which intersect in straight lines $x + z = \text{const}$ at a single point [11.3]), then (11.46) yields the solution of the problem on the determination of the wave field at any point of the crystal. Here, the function G and its derivatives are influence functions describing the propagation of the local perturbation of the wave field (see Chapter 1).

11.3 Laue Reflection in Perfect Crystal

For simplicity, we will restrict ourselves to the consideration of Laue diffraction (it is possible to show - see (11.7) for more details - that by using the appropriate choice of the coordinate system the general arrangement of nonsymmetrical diffraction can be reduced to a symmetrical one) on a perfect crystal in the shape of a parallel-sided plate of thickness t.

Let us assume that the crystal is of the $z > 0$ type, and its entrance surface coincides with the plane $z = 0$. The distribution of the incident wave has the form

$$D_o = D_o^{(a)}(x,y) \quad , \quad D_h = 0 \quad . \tag{11.47}$$

The wave field in the crystal propagates in the positive direction of the z-axis. From physical considerations it is clear that in this case it is convenient to choose, as function G in (11.45), the retarded Green function $G^{(r)}$

$$G^{(r)}(\zeta - z, \xi - x) = \frac{1}{2} J_o[\chi\sqrt{(\zeta - z)^2 - (\xi - x)^2}] \exp\{-i \frac{\alpha}{2} [(\zeta - z) - (\xi - x)]\} \cdot \Theta(\zeta - z)\Theta(\zeta - z - |\xi - x|) \qquad (11.48)$$

where $J_o(z)$ is a zero-order Bessel function; $\Theta(z)$ is a stepped function: $\Theta(z) = 1$ at $z > 0$; $\Theta(z) = 0$ at $z < 0$, and $\partial\Theta(z)\partial z = \delta(z)$. Direct substitution of (11.48) in (11.43) or (11.44) proves that $G^{(r)}$ turns these equations into equalities. The expression (11.48) for the function G was first obtained, on the basis of the TAKAGI equations, in the work of SLOBODETSKII et al. [1.60] and, independently, by AUTHIER and SIMON [1.61].

From (11.46) we obtain the following expressions for the wave field inside the crystal, using (11.47,48), as well as (11.41), and omitting the intermediate calculations:

$$D_o(x,y,z) = \int_{-\infty}^{\infty} dx' D_o^{(a)}(x',y) G_{oo}(x - x',z) \quad ,$$

$$D_h(x,y,z) = \int_{-\infty}^{\infty} dx' D_o^{(a)}(x',y) G_{ho}(x - x',z) \quad , \qquad (11.49)$$

$$G_{oo}(x,z) = -\exp[-i \frac{\alpha}{2}(z - x)]\Big[\delta(z - x) - \frac{\chi}{2} \sqrt{\frac{z + x}{z - x}} \cdot J_1(\chi\sqrt{z^2 - x^2})\Theta(z)\Theta(z - |x|)\Big] \quad ,$$

$$G_{ho}(x,z) = -\frac{i\chi_h C}{2} \exp[-i \frac{\alpha}{2} (z - x)] \cdot J_o(\chi \sqrt{z^2 - x^2})\Theta(z)\Theta(z - |x|) \quad . \qquad (11.50)$$

Equations (11.49) and (11.50) enable one to plot the wave field inside the crystal with an arbitrary distribution of the incident wave over the entrance surface $z = 0$. Let us consider, as an example of application of the equations obtained, the image of the slit in the field of the diffracted wave when

$D_o^{(a)}(x,y) = [\Theta(x + a) - \Theta(x - a)]$, where a is the slit half-width. In this case we have from (11.49) and (11.50), for the field of the diffracted wave

$$D_h(x,z) = \frac{i\chi_h C}{2} \int_{\max(x-z,-a)}^{\min(x+z,a)} dx' \cdot$$

$$\cdot J_o(\chi\sqrt{z^2 - (x - x')^2}) \exp[-\frac{i\alpha}{2}(z - x + x')] \quad . \tag{11.51}$$

As $a \to \infty$, (11.51) naturally transforms into the classical solution of the dynamical problem (see, for instance, (3.41))

$$D_h(z) = \frac{i \sin[\chi z\sqrt{1 + (\alpha/2\chi)^2}]}{\sqrt{1 + (\alpha/2\chi)^2}} \exp(-\frac{i\alpha z}{2}) \quad . \tag{11.52}$$

If the slit half-width a exceeds the crystal thickness z, (11.52) holds good only for the central, "uniformly illuminated" area $|x| < (a - z)$, while at the pattern edges $(a - z) < |x| < (a + z)$, intensity oscillations must be observed.

If the slit half-width is less than the crystal thickness, the uniformly illuminated area disappears, and the intensity oscillation areas coincide; with a decrease in slit width or an increase in crystal thickness, the intensity maxima increase towards the pattern edges for the transparent crystal. In strongly absorbing crystals, on the contrary, these intensity oscillations smooth out.

Note that, as follows from (11.49) and (11.50), in the limiting case of an infinitely narrow slit its image is given directly by the functions G_{oo} in the transmitted, and G_{ho} in the diffracted wave, which agrees with the result of KATO's theory for a spherical wave incident on a crystal (see Chapter 6).

Also of interest are the derivation and physical interpretation of the influence functions of a point source based on transition from hyperbolical-type equations for the wave field in the crystal (the Cauchy problem) to elliptical-type equations (the Laplace problem) [11.4]. Below, we briefly outline the theory of plotting influence functions as superposition of "generalized" plane waves.

$$\gamma_h(X + iY) = s_o \quad , \quad \gamma_o(X - iY) = s_h \tag{11.53}$$

for the amplitude $D_h(X,Y)$, we find the following equation:

$$\left(\frac{\partial^2}{\partial X^2} + \frac{\partial^2}{\partial Y^2} + 4\pi^2 k^2 C^2 \chi_h \chi_{\bar{h}} \gamma_o \gamma_h\right) D_h = 0 \quad . \tag{11.54}$$

The amplitude $D_h(X,Y)$ can be regarded formally as a complex electromagnetic-wave amplitude satisfying (11.54) and the boundary conditions on the contour $\Gamma(X = 0)$

$$D_h(X,Y)\Big|_\Gamma = 0 \quad , \quad \frac{\partial D_h(X,Y)}{\partial Y}\Big|_\Gamma = 0 \quad ,$$

$$\frac{\partial D_h(X,Y)}{\partial X}\Big|_\Gamma = -2\pi i k C \chi_h \gamma_o D_o(X,Y)\Big|_\Gamma \quad . \tag{11.55}$$

It is clearer from physical considerations that $D_h(X,Y)$ cannot increase at infinity when $(X^2 + Y^2) \to \infty$, i.e., when the asymptotic equality $D_h(X,Y) \to 0$ holds for $X^2 + Y^2 \to \infty$. Eq. (11.54) is conveniently represented as

$$\hat{L} D_h = 0 \quad . \tag{1..56}$$

Here, the differential operator $\hat{L}$ is equal to

$$\hat{L} = \frac{\partial^2}{\partial X^2} + \frac{\partial^2}{\partial Y^2} + \sigma^2 \quad ,$$

$$\sigma^2 = 4\pi^2 k^2 C^2 \chi_h \chi_{\bar{h}} \gamma_o \gamma_h \tag{11.57}$$

and acts on the amplitude $D_h(X,Y)$ in the region $\Omega'(X > 0)$.

Introduce the distribution function $T(X,Y)$ in the following way:

$$T = D_h \Theta(X) \quad , \tag{11.58}$$

where $\Theta(X)$ is a step function.

Acting with the operator $\hat{L}$ on (11.58), we find the equation for the T-function:

$$\hat{L}T = -D_h \frac{d^2}{dX^2}\Theta(X) - 2\frac{\partial D_h}{\partial X}\frac{d\Theta(X)}{dX} \tag{11.59}$$

which is an inhomogeneous (with the right-hand side) two-dimensional wave equation. The solution of this equation can be written in the form of retarded and advanced potentials:

$$T = \frac{i}{4}\iint dX'dY'[H_0^{(1)}(\sigma R) + H_0^{(2)}(\sigma R)]$$

$$\cdot\left[D_h \frac{d^2\Theta(X')}{dX'^2} + 2\frac{\partial D_h}{\partial X'}\frac{d\Theta(X')}{dX'}\right] \quad , \tag{11.60}$$

$$R^2 = (X - X')^2 + (Y - Y')^2$$

where $H_0^{(i)}(\sigma R)$ is a zero-order, first- or second-kind Hankel function.

Taking into account the boundary conditions (11.55) and the relation

$$\frac{d\Theta(X)}{dX} = \delta(X)$$

and also recalling that in the region of Ω' the T-function coincides with the amplitude D_h, we finally find

$$D_h(X,Y) = \frac{1}{2}\int_{-\infty}^{\infty} dY^S \left.\frac{\partial D_h(X,Y^S)}{\partial X}\right|_{X=0} \cdot [H_0^{(1)}(\sigma|\underline{R} - \underline{R}^S|)$$

$$+ H_0^{(2)}(\sigma|\underline{R} - \underline{R}^S|)]\Theta(|\underline{R} - \underline{R}^S|) \quad , \tag{11.61}$$

where $\underline{R}^S$ is the radius vector of the circuit $\Gamma \equiv (0,Y^S)$.

The expression (11.61) for the amplitude D_h is fully equivalent to the result of (11.49) obtained by Riemann's method if we use the functional relation

$$J_o(u) = \frac{1}{2} [H_o^{(1)}(u) + H_o^{(2)}(u)] \quad .$$

On the other hand, the latter relation permits interpreting the influence function (11.48) as superposition of "generalized" plane waves (mathematically described by the Hankel functions) propagating from a point source located on the crystal surface.

Thus, the foregoing interpretation of the process of propagation of X-rays in a crystal in the case of a finite wave front of the incident wave leads to a more general conception according to which the observed intensity distribution can be regarded as an interference pattern. In some specific conditions this pattern corresponding to the presence of two wave fields (AUTHIER's experiment [1.15, 9.38]). In the case of an incident plane wave with an infinite wave front, the interference pattern, which is known as the pendulum solution of the dynamical problem, can be calculated as interference of two wave fields - a physical model, the use of which is optimal. The same result can be obtained by using the influence functions by the Riemann method.

Hence it follows that the Ewald-Laue-Zachariasen theory outlined in Chapters 2 through 5 and 7 and 8 is not the only possible interpretation of dynamical scattering. The multiwave scattering problem, which is covered in Chapter 12 from the standpoint of the Ewald-Laue theory, will require special consideration if it is desired to use the influence functions.

11.4 Bragg Reflection in Perfect Crystal

Taking into account the system of coordinates chosen by us, to the crystal region now corresponds to $x > 0$. The boundary conditions have the form

$$D_o = D_o^{(e)}(z,y) \quad , \quad x = 0$$

$$D_h = 0 \quad , \quad x = t \quad . \tag{11.62}$$

Note that in the case of Bragg diffraction the boundary conditions are assigned for different values of the variable x; therefore, the course of the solution for the wave field is not so obvious as it was in the case of Laue diffraction. Nevertheless, this problem was solved by AFANASIEV and KOHN [1.62] and, independently, by URAGAMI [1.53] using the Riemann method generalized by these authors for the case of assignment of the boundary conditions (11.62).

For completeness sake, we will now give the solution based on the use of Green's functions G.

Let us introduce into consideration the retarded and advanced Green functions:

$$G^{(r)}(\xi - x, \zeta - z) = -\frac{1}{2} J_o\left[\chi\sqrt{(\zeta - z)^2 - (\xi - x)^2}\right] \exp\left[-\frac{i\alpha}{2}(\zeta - z - \xi + x)\right] \cdot \Theta(\xi - x)[\Theta(\zeta - z + \xi - x) - \Theta(\zeta - z - \xi + x)] \tag{11.63a}$$

$$G^{(a)}(\xi - x, \zeta - z) = -\frac{1}{2} J_o\left[\chi\sqrt{(\zeta - z)^2 - (\xi - x)^2}\right] \exp\left[-\frac{i\alpha}{2}(\zeta - z - \xi + x)\right] \cdot \Theta(x - \xi)[\Theta(\zeta - z + \xi - x) - \Theta(\zeta - z - \xi + x)] \tag{11.63b}$$

which satisfies (11.43) and (11.44).

Assuming, in (11.46), $G = G^{(a)}$ and taking into consideration that for $G = G^{(a)}$ the integral over the straight line x = 0 is zero, we have

$$D_h(\xi,\zeta) = \int_{-\infty}^{\infty} dz\left[D_h \frac{\partial G^{(a)}}{\partial x} - G^{(a)} \frac{\partial D_h}{\partial x} + iG^{(a)} D_h\right]\Bigg|_{x = t} . \tag{11.64}$$

Recalling that $D_h|_{x = t} = 0$ and $\partial D_h/\partial x|_{x = t} = -iC\chi_h D_o$ (see (11.46)), (11.64) can be represented as

$$D_h(\xi,\zeta) = -iC\chi_h \int_{-\infty}^{\infty} dz G^{(a)}(\zeta - z, \xi - t) D_o(t,z) \quad . \tag{11.65}$$

To determine $D_o(t,z)$ in (11.65), we again use (11.46) with $G = G^{(r)}$, for which the integral over the line $x = t$ vanishes:

$$D_o(t,z) = \int_{-\infty}^{\infty} dz' \left[D_o \frac{dG^{(r)}}{dx'} - G^{(r)} \frac{\partial D_o}{\partial x'} + i\alpha G^{(r)} D_o \right]_{x = o} \quad . \tag{11.66}$$

Taking into account (11.41) and carrying out identical transformations, we rearrange (11.66):

$$D_o(t,z) = \int_{-\infty}^{\infty} dz' \left\{ D_o^{(a)} \left[\frac{\partial G^{(r)}}{\partial x'} - \frac{\partial G^{(r)}}{\partial z'} \right] \right.$$

$$\left. + i\alpha D_o^{(a)} - iC\chi_h D_h \right\}_{x' = o} \quad . \tag{11.67}$$

Substitute (11.67) in (11.65). Thus we obtain an integral equation of the diffracted wave field for each value of the thickness ξ. Since in the case of Bragg diffraction we are interested in the field of the reflected wave on the entrance surface of the crystal, then, assuming $\xi = 0$ and making allowance for the explicit form of the functions $G^{(a)}$ and $G^{(r)}$ (11.62,63), we get the sought-for integral equation:

$$D_h(0,\xi) + \frac{1}{4} \chi^2 \iint_{-t}^{+t} dz dz' J_o(\chi\sqrt{z^2 - t^2}) J_o(\chi\sqrt{z'^2 - t^2})$$

$$\cdot \exp\left[-\frac{i\alpha}{2}(z + z')\right] D_h(\zeta - z - z') = \frac{1}{4} i\chi^2 \iint_{-t}^{+t} dz dz' \frac{J_1(\chi\sqrt{z'^2 - t^2})}{\sqrt{z'^2 - t^2}}$$

$$(z' + t) \cdot J_o(\chi\sqrt{z^2 - t^2}) \exp\left[-\frac{i\alpha}{2}(z + z')\right] D_o(\zeta - z - z')$$

$$+ \frac{1}{2} i\chi_h C \exp\left(-\frac{i\alpha t}{2}\right) \int_{-t}^{t} dz J_o(\chi\sqrt{z^2 - t^2}) \exp\left(-\frac{i\alpha z}{2}\right) D_o(\zeta - z - t) \quad . \tag{11.68}$$

The integral equation (11.68) is most conveniently solved by the Fourier transformation method:

$$D_h(0,\zeta) = \frac{1}{2\pi}\int_{-\infty}^{\infty} D_h(\alpha)\exp(i\alpha\zeta)d\alpha$$

$$D_h(\alpha) = \int_{-\infty}^{\infty} D_h(0,\zeta)\exp(-i\alpha\zeta)d\zeta \quad . \tag{11.69}$$

As a result of this transformation, after performing direct calculations and using the tabulated integral

$$\int_{-t}^{t} dz \exp(-i\alpha z)J_o(\chi\sqrt{z^2 - t^2}) = \frac{2\sin(t\sqrt{\alpha^2 - \chi^2})}{\sqrt{\alpha^2 - \chi^2}} \tag{11.70}$$

we find

$$D_h(\alpha) = G^{(B)}(\alpha)D_o(\alpha) \tag{11.71}$$

$$G^{(B)}(\alpha) = \frac{\sin(t\sqrt{\alpha^2 - \chi^2})}{\alpha\sin(t\sqrt{\alpha^2 - \chi^2}) - i\sqrt{\alpha^2 - \chi^2}\cos(t\sqrt{\alpha^2 - \chi^2})} \quad . \tag{11.72}$$

The function $G^{(B)}(\alpha)$ sets up a relationship between the Fourier components of the desired function $D_h(0,\zeta)$ and the function $D_o^{(a)}(\zeta)$ given on the crystal surface. In the coordinate space, this relationship has the form

$$D_h(0,\zeta) = \int_{-\infty}^{\infty} G^{(B)}(\zeta - z)D^{(a)}(z)dz \tag{11.73}$$

$$G^{(B)}(\zeta) = \frac{1}{2\pi}\int_{-\infty}^{\infty} G^{(B)}(\alpha)\exp(i\alpha\zeta)d\alpha \quad . \tag{11.74}$$

If a plane wave falls on the crystal, then

$$D_o^{(a)}(z) = D\exp(-i\chi_o z) \quad , \quad D = \text{const} \quad . \tag{11.75}$$

It is easy to check that by substituting (11.74) and (11.75) in (11.73) and integrating, first with respect to z and then with respect to α, we obtain a result which precisely corresponds to the result of the classical dynamical theory for Bragg diffraction.

Note that (11.71) and (11.72) can be obtained directly by decomposing the incident wave into plane waves and solving, for each Fourier component, the dynamical problem in its classical formulation.

Precisely this approach was used effectively by SAKA et al. [1.48] for describing multiple Laue-Bragg diffraction in finite polyhedral crystals.

Thus, (11.72-74) completely solve the posed problem of determining the field of the reflected wave on the entrance surface of the crystal with an arbitrary distribution of the incident wave in the case of Bragg diffraction.

It now remains for us to present the final expression for the integral (11.74). Using the known tabulated integrals, we can obtain for $G^{(B)}(z)$ an expression in the form of a finite sum for any fixed z:

$$G^{(B)}(z) = \frac{i\chi_h C}{2}\exp\left(-i\,\frac{\alpha}{2}\,z\right)\sum_{n=0}^{\infty}\sum_{m=n,\;n+1}(-1)^m\left(\frac{z-2tm}{z+2tm}\right)^n$$
$$\cdot\left\{J_{2n}\left[\chi\sqrt{z^2-(2tm)^2}\right]+\frac{z-2tm}{z+2tm}\,J_{2n+2}\left[\chi\sqrt{z^2-(2tm)^2}\right]\right\}\Theta(z-2tm) \quad . \tag{11.76}$$

In the case of the infinitely thick crystal ($t\to\infty$), which is important for various applications, only the first term of the series with n = 0 remains in (11.76):

$$G^{(B)}(z)\Big|_{t\to\infty} = \frac{i\chi_h C}{2}\exp\left(-\frac{i\alpha}{2}z\right)[J_0(\chi z)+J_2(\chi z)]\Theta(z) \quad . \tag{11.77}$$

11.5 Application of Generalized Theory to Deformed Crystal. Relationship Between Angular Variable α_h and Deformation Field

Let a point with radius vector $\underline{r}^*$ shift to point $\underline{r}$ as a result of deformation. Similarly, let $\underline{h}_m$ be the value of the vector $\underline{h}_m^*$ prior to deformation. Then the equation of the plane with number n_h prior to deformation has the form

$$n_h = \underline{h}_m^* \underline{r}^* \ . \tag{11.78}$$

Assume that on crystal deformation the point 0 is not shifted, i.e.,

$$\underline{h}_m(0) = \underline{h}_m^* \ . \tag{11.79}$$

On the other hand,

$$\underline{r} = \underline{r}^* + \underline{u} \tag{11.80}$$

($\underline{u}$ is the displacement vector). Substituting (11.80) in (11.78) and taking into account (11.79), we find

$$n_h = [\underline{h}_m(0)(\underline{r} - \underline{u})] \ , \tag{11.81}$$

$$\underline{h}_m = \text{grad}\, n_h = \underline{h}_m(0) - \text{grad}(\underline{h}_m(0)\underline{u}) \ . \tag{11.82}$$

Then the value of α is equal to (see (11.18,19))

$$\begin{aligned} \alpha_h &= \lambda^2\{h_m^2(0) + \text{grad}^2[\underline{h}_m(0)\underline{u}] - 2\underline{h}_m(0)\ \text{grad}[\underline{h}_m(0)\underline{u}] \\ &\quad + 2\underline{k}_o\underline{h}_m(0) + 2\underline{k}_o\ \text{grad}[\underline{h}_m(0)\underline{u}]\} \\ &= \alpha_h(0) - 2\lambda^2\underline{k}_h\ \text{grad}[\underline{h}_m(0)\underline{u}] + O(\tau^2) \end{aligned} \tag{11.83}$$

or, using (11.34) and (11.35), we get

$$\alpha_h = \alpha_h(0) - 2\lambda \frac{\partial}{\partial s_h} [\underline{h}_m(0)\underline{u}] \tag{11.84}$$

where α_h (see (11.19)) determines the deviation of the particular region of the crystal structure from the exact Bragg condition. Assuming finally

$$\chi_h = \chi_{hr}(1 + i\varkappa) \quad , \quad \chi_{hr} < 0 \qquad \varkappa = -\left|\frac{\chi_{hi}}{\chi_{hr}}\right| \tag{11.85}$$

(χ_{hr} is a real value), we obtain

$$s_h = -\frac{\lambda A_h}{\pi \chi_{hr} C} \tag{11.86}$$

$$s_o = -\frac{\lambda A_o}{\pi \chi_{hr} C} . \tag{11.87}$$

Represent the system of (11.36) as

$$\begin{aligned} -i \frac{\partial D_h}{\partial A_h} &= \left[\frac{\alpha_h(0)}{\chi_{hr} C} + 2\frac{\partial \Phi}{\partial A_h} - \frac{\chi_o}{\chi_{hr} C}\right] D_h - (1 + i\varkappa) D_o \\ -i \frac{\partial D_o}{\partial A_o} &= \left(-\frac{\chi_o}{\chi_{hr} C}\right) D_o - (1 + i\varkappa) D_h \end{aligned} \tag{11.88}$$

$$\Phi = \pi[\underline{h}_m(0)\underline{u}] \quad . \tag{11.89}$$

11.6 Fundamental Equations of Geometrical Optics of X-Rays

When plotting the X-ray field in slightly deformed crystals, where the characteristic length of change in the local parameter of deviation from the exact Bragg condition ℓ_{eff} greatly exceeds the extinction length Λ, one can use asymptotic methods of geometrical optics in inhomogeneous media.

Acting on analogy with geometrical optics of the visible light, we can single out, in the amplitudes D_o and D_h, the rapidly changing phase factors

$$\hat{D} \equiv \begin{pmatrix} D_o \\ D_h \end{pmatrix} = e^{i\tilde{\Phi}^{(1)}}\hat{D}^{(1)} + e^{i\tilde{\Phi}^{(2)}}\hat{D}^{(2)} \quad . \tag{11.90}$$

Here, in contrast to conventional light optics, two Eikonals, $\tilde{\Phi}^{(1)}$ and $\tilde{\Phi}^{(2)}$, appear, which correspond to the two sheets of the dispersion surface. If we take into account the imaginary part $\chi_{hi} > 0$ in the parameter $(\chi_h\chi_{\bar{h}})^{\frac{1}{2}}$, the first sheet $[\tilde{\Phi}^{(1)}]$ corresponds to strongly absorbed, and the second $[\tilde{\Phi}^{(2)}]$, to weakly absorbed Bloch waves. Substituting (11.90) in the Takagi equations (11.88) gives the equation for determining $\hat{D}^{(1)}$ and $\hat{D}^{(2)}$

$$(\hat{d}_{\tilde{\Phi}} + \hat{d})\hat{D} = 0 \tag{11.91}$$

where

$$\hat{d}_{\tilde{\Phi}} = \begin{bmatrix} \frac{\partial\Phi}{\partial x} + \frac{\partial\Phi}{\partial z} & \chi_{\bar{h}}C \\ \chi_h C & \frac{\partial\Phi}{\partial x} - \frac{\partial\Phi}{\partial z} - 2\alpha(\underline{r}) \end{bmatrix} \tag{11.92}$$

$$\hat{d} = \begin{bmatrix} i\left(\frac{\partial}{\partial z} + \frac{\partial}{\partial x}\right) & 0 \\ 0 & i\left(\frac{\partial}{\partial z} - \frac{\partial}{\partial x}\right) \end{bmatrix} \quad . \tag{11.93}$$

(When deriving (11.91-93) we restrict ourselves, for simplicity, to the case of symmetrical Laue diffraction).

For a nontrivial zero approximation satisfying the equation $\hat{d}_{\tilde{\Phi}}[{}_{(0)}\hat{D}] = 0$ to exist, the determinant of the matrix of $\hat{d}_{\tilde{\Phi}}$ must vanish to zero. In expanded form, this condition yields, for the Eikonals $\tilde{\Phi}^{(1)}$ and $\tilde{\Phi}^{(2)}$, the following equation in first-order partial derivatives:

$$\left[\frac{\partial\tilde{\Phi}}{\partial z} + \;(\underline{r})\right]^2 - \left[\frac{\partial\tilde{\Phi}}{\partial x} - \alpha(\underline{r})\right]^2 = \tilde{\chi}^2$$

$$\tilde{\chi} = C(\chi_h\chi_{\bar{h}})^{\frac{1}{2}} \quad . \tag{11.94}$$

The Eikonal equation (11.94) is similar to the one-dimensional relativistic equation of Hamilton-Jacobi for a particle with a rest mass $\pm\tilde{\chi}$ in some variable external field. It should, however, be borne in mind that (11.94) includes the complex coefficient $\tilde{\chi} = \chi(1 + i\varkappa)$ and, hence, the trajectories and the Eikonal are complex. Physically, this is associated with dynamical attenuation of the X-ray wave field in the absorbing crystal. In the case of an arbitrary relationship between 1 and $|\varkappa|$, (11.94) represents a system of two nonlinear equations with respect to the real and imaginary parts of the Eikonal $\tilde{\Phi} = \mathrm{Re}\,\tilde{\Phi} + i\mathrm{Im}\,\tilde{\Phi}$. For X-rays, usually $|\varkappa| \ll 1$. This permits one to assume the imaginary part of the Eikonal small as compared with its real part. Putting $|\mathrm{Im}\,\tilde{\Phi}| \ll |\mathrm{Re}\,\tilde{\Phi}|$ and introducing the notation

$$H + i\Gamma = -\frac{\partial}{\partial z}\tilde{\Phi} \quad , \qquad P + iQ = \frac{\partial}{\partial x}\tilde{\Phi} \tag{11.95}$$

we find, from (11.94), with an accuracy to terms of the order of $\varkappa^2 \ll 1$, equations for determining $\mathrm{Re}\,\tilde{\Phi}$ and $\mathrm{Im}\,\tilde{\Phi}$

$$(H - \alpha)^2 - (P - \alpha)^2 = \chi^2 \tag{11.96}$$

$$(H - \alpha)\Gamma - (P - \alpha)Q = \chi^2\varkappa \ . \tag{11.97}$$

It is easy to show that the trajectory equations for the Hamilton-Jacobi equations (11.96) and (11.97) coincide and have the form

$$\frac{d}{dz}\left(\frac{\pm\,\chi\dot{x}}{\sqrt{1 - \dot{x}^2}}\right) = F(\underline{r}) \tag{11.98}$$

where $\dot{x} = dx/dz$, and $F(\underline{r})$ is determined by the derivative of the function $\alpha(\underline{r})$ in the direction of propagation of the transmitted wave D_0

$$F(\underline{r}) = -\left(\frac{\partial}{\partial z} + \frac{\partial}{\partial x}\right)\alpha(\underline{r}) \quad . \tag{11.99}$$

The signs $\mp$ in (11.98) correspond to fields weakly and strongly absorbed in the crystal, respectively. Eqs. (11.98) and (11.99) were first obtained by

KATO [1.58] and KAMBE [11.5] on the basis of the general principles of the theory of geometrical optics as applied directly to the Maxwell equations.

Following the general procedure for constructing the asymptotic solution of the Takagi equations, we put for each field

$$\hat{\underline{D}} = \sum_{n=0}^{\infty} {}_{(n)}\hat{\underline{D}} \tag{11.100}$$

where the zero approximation term ${}_{(o)}\hat{\underline{D}}$ satisfies the equation

$$\hat{d}_{\tilde{\Phi}}[{}_{(o)}\hat{\underline{D}}] = 0 \tag{11.101}$$

and the subsequent terms, the equation

$$\hat{d}_{\tilde{\Phi}}[{}_{(n+1)}\hat{\underline{D}}]+\hat{d}[{}_{(n)}\hat{\underline{D}}] = 0 \quad . \tag{11.102}$$

Introduce the left-hand side, $\hat{\underline{l}} = \begin{pmatrix} \chi_h C \\ P-H \end{pmatrix}$, and the right-hand side, $\hat{\underline{r}} = \begin{pmatrix} \chi_h C \\ P-H \end{pmatrix}$ zero vectors of the matrix $\hat{d}_{\tilde{\Phi}}$. According to (11.11), the vector ${}_{(o)}\hat{\underline{D}}$ is equal to

$${}_{(o)}\hat{\underline{D}} = \sigma(x,z)\hat{\underline{r}}$$

where σ (x,z) is a scalar factor. Multiplying (11.102) for n = 0 from the left-hand side by the vector $\hat{\underline{l}}$, we obtain the "transfer" equation

$$\hat{\underline{l}}\hat{d}(\sigma\hat{\underline{r}}) = 0 \quad , \tag{11.103}$$

which makes it possible to determine unambiguously, to a zero approximation, the variation in field amplitude along the trajectory if one knows the initial values of D_o and D_h on the entrance surface of the crystal. Further, (11.102) permits us, in principle, to determine successively all the terms of the expansion (11.100).

The "transfer" equation for the field amplitude in the zero approximation, recorded in the form of the law of conservation of the energy flow $J = \underline{k}_o D_o^2 + (\underline{k}_o + \underline{h})D_h^2$, was first pointed out by KATO [4.41].

The "transfer" equation (11.103) can be rewritten as

$$\left(\frac{\partial}{\partial x}, \frac{\partial}{\partial z}\right)\hat{\underline{J}} = 0 \quad , \quad \hat{\underline{J}} = \sigma^2 \begin{bmatrix} \tilde{\chi}^2 - \left(\frac{\partial\tilde{\Phi}}{\partial x} + \frac{\partial\tilde{\Phi}}{\partial z}\right)^2 \\ \tilde{\chi}^2 + \left(\frac{\partial\tilde{\Phi}}{\partial x} + \frac{\partial\tilde{\Phi}}{\partial z}\right)^2 \end{bmatrix} \quad . \tag{11.104}$$

According to (11.104), the "transfer" equation is also complex. In the case under consideration, $|x| << 1$, and the imaginary part of (11.104) can also be neglected, because its inclusion results in corrections $|x/\chi| << 1$ in the amplitudes D_o, D_h. Thus, the "transfer" equation takes the form

$$\frac{\partial}{\partial z}\{\sigma^2[\chi^2 + (P - H)^2]\} + \frac{\partial}{\partial x}\{\sigma^2[\chi^2 - (\dot{P} - H)^2]\} = 0 \quad . \tag{11.105}$$

Equation (11.96-98), as well as (11.105), represent a complete system of equations of geometrical optics describing X-ray scattering in deformed crystals in the general case of arbitrary smooth displacement fields $\underline{u}(\underline{r})$.

The partial derivative equations (11.96,97) for determining the complex Eikonal $\tilde{\Phi}(\underline{r})$ can be recorded in the form of differential equations along the trajectories (11.98). Taking into account the relationship between the ordinary momentum and the canonical one,

$$p = P - \alpha(\underline{r}) \quad , \quad p = \frac{\pm\chi\dot{x}}{\sqrt{1 - \dot{x}^2}} \tag{11.106}$$

we find the desired equations for Re $\tilde{\Phi}$ and Im $\tilde{\Phi}$

$$\frac{d}{dz}(\mathrm{Re}\,\tilde{\Phi}) = \pm\chi\sqrt{1 - \dot{x}^2} - \alpha(\underline{r})(1 - \dot{x}) \tag{11.107}$$

$$\frac{d}{dz}(\mathrm{Im}\,\tilde{\Phi}) = \pm x\sqrt{1 - \dot{x}^2} \quad . \tag{11.108}$$

Making an allowance for the equality $H - \alpha(\underline{r}) = \mp\chi/\sqrt{1 - \dot{x}^2}$, we arrive at the "transfer" equation

$$\frac{\partial}{\partial x}\left(\dot{x}\,\frac{\sigma^2}{1+\dot{x}}\right)+\frac{\partial}{\partial z}\left(\frac{\sigma^2}{1+\dot{x}}\right)=0 \tag{11.109}$$

whose solution has the form

$$\frac{\sigma^2(x,z)}{1+\dot{x}(x,z)}\,\frac{dx}{dx_o}=\frac{\sigma^2(x_o,z_o)}{1+\dot{x}(x_o,z_o)} \quad . \tag{11.110}$$

We shall now present the complete system of equations of the Eikonal approximation in a form convenient for analytical and numerical calculations. We take into account the factor $\exp(+i\chi_o z)$, $(\chi_o = \chi_{or} + i\chi_{oi})$ in the Eikonal in explicit form. Below, the dot denotes differentiation with respect to z; the upper sign corresponds to weakly absorbed rays and the lower one, to strongly absorbed ones:

$$\dot{P}=F(\underline{r}) \quad , \qquad F(\underline{r})=-\left(\frac{\partial}{\partial z}+\frac{\partial}{\partial x}\right)\alpha(\underline{r}) \quad , \qquad p=\mp\frac{\chi\dot{x}}{\sqrt{1-\dot{x}^2}}$$

$$\dot{x}=\frac{\mp P}{\sqrt{\chi^2+p^2}}\equiv\varphi(p)$$

$$\mathrm{Re}\,\tilde{\Phi}=\chi_o \pm (\chi^2+p^2)^{-1/2}-\alpha(\underline{r})(1-\dot{x})$$

$$\mathrm{Im}\,\tilde{\Phi}=\chi_{oi} \pm \varkappa\chi(\chi^2+p^2)^{-1/2}$$

$$\frac{\sigma^2}{1+\dot{x}}\,\frac{dx}{dx_o}=\mathrm{const} \quad , \qquad \hat{D}=\sigma\left[\frac{1}{\pm\sqrt{\frac{\chi_{hr}}{\chi_{\bar{h}r}}\,\frac{(1+\dot{x})}{(1-\dot{x})}}}\right] \tag{11.111}$$

$$\frac{d\dot{x}}{dx_o}=\frac{dp}{dx_o}\,\frac{\partial\varphi}{\partial p}$$

$$\frac{d\dot{p}}{dx_o}=\frac{dx}{dx_o}\,\frac{\partial F}{\partial x} \quad .$$

Thus, in the Eikonal approximation, the problem on dynamical scattering of X-rays in the deformed crystal reduces to constructing two systems of ray trajectories along which the wave field in the crystal propagates. The system of (11.111) was first used effectively in the papers of CHUKHOVSKII and SHTOLBERG [11.6] for approximate analytical construction of the wave field in a crystal with a single dislocation.

11.7 Approaches Based upon Wave Theory

11.7.1 Rigorous Theory of Laue Diffraction of X-Rays in Crystal with Uniform Strain Gradient

The dynamical theory of X-ray scattering in deformed crystals was first developed with reference to problems on diffraction on an elastically (thermoelastically) bent crystal [1.57-59a, 11.7]. Using the Eikonal approximation of wave optics, the theory explained the diffraction experiments in the case of uniform curvature of the crystals, in particular, the contraction of the bands of the pendulum solution [9.37, 11.8] and the asymmetry of the integral intensity of the diffracted wave with respect to the sign of curvature of the reflecting planes [11.9-11]. Later, the solution of the problem on diffractions of X-rays on a bent crystal, obtained in analytical form, made it possible to give a quantitative description of X-ray topograms in the case of arbitrary distortion for crystal regions in which the strain gradient is not too large [11.12-16].

However, because of the strong restrictions associated with the conditions for the applicability of the Eikonal approximation, the geometrical optics theory does not supply the correct description of X-ray diffraction in strongly distorted crystal regions, as well as in the vicinity of the characteristic rays - the boundaries of the Borrmann triangle. The reason is that it is absolutely impossible to calculate the preexponential factor and the wave field scattering phase while remaining within the framework of the geometrical optics theory.

In this connection, of special interest is the exact solution of the problem on Laue diffraction of X-rays in crystals with a uniform strain gradient (bent crystal) obtained recently in [1.66, 11.17-19] and, independently, in [11.20,21]. Using the Takagi equations [1.45], the authors of [1.66, 11.19] constructed the asymptotics of the Green-Riemann functions of the X-rays in a crystal with a displacement field depending quadratically on the coordinates. The general solution is expressed through confluent hypergeometric functions and is rather complicated. Therefore, it is interesting to analyze the rigorous mathematical solution and obtain the physical results from it.

The purpose of the present consideration is a) to discuss the general properties of the exact solution of the Laue diffraction problem (Section 11.7.2); b) to present a comparative analysis of the quasi-classical asymptotics of the exact solution with an allowance for the preexponential factor and the scattering phase, with the results of KATO's [1.58] and KATAGAWA and KATO's theories [11.20] (Sections 11.7.3 and 11.7.4). Special attention is given to determining the applicability range of the Eikonal theory. Further, the Eikonal theory is generalized for the case of an arbitrary value of the strain gradient.

In Section 11.7.4 we consider the geometro-optical picture of the wave field in the crystal on the basis of the trajectory approach in the cases of an incident plane and spherical wave. The trajectories have been obtained by applying the stationary-phase principle for finding the wave field in accordance with the integral formulation of the Huygens-Fresnel principle. Here, wave field singularities may arise, which can be regarded as focussing points. In the vicinity of the singular points the stationary-phase principle becomes inapplicable, as well as the geometrical optics theory. It is significant that the focussing effect takes place when the component of the strain gradient along the reflection vector is zero and the integrated intensity of the diffracted wave is equal to the corresponding value for the perfect crystal.

For practical calculation of the integrated intensity of the diffracted wave, a simple numerical method has been proposed which is applicable in the general case of asymmetrical Laue diffraction (Section 11.7.5). It is shown that, in the limit of large strain gradients, the Eikonal approximation of the theory yields the correct kinematical limit for the integrated intensity of the diffracted wave, thus eliminating the problem of "divergence", which arises in KATO's geometrical optics theory.

11.7.2 Integrated Formulation of Huygens-Fresnel Principle

The coherent wave field inside a crystal oriented in the vicinity of exact Bragg reflection is described by the system of Takagi's dynamical equations [1.45]

$$i \frac{\partial D_o}{\partial s_o} + \sigma_{\bar{h}} \exp[-i(\underline{\underline{hu}})] D_h = 0$$

$$i \frac{\partial D_h}{\partial s_h} + \sigma_h \exp i(\underline{\underline{hu}}) \; D_o = 0 \tag{11.112}$$

$$\sigma_{\bar{h}} = \frac{1}{\sqrt{|\beta|}} \sqrt{\left|\frac{\chi_{\bar{h}}}{\chi_h}\right|} \; (1 + ix)$$

$$\sigma_h = \frac{\beta}{\sqrt{|\beta|}} \sqrt{\left|\frac{\chi_h}{\chi_{\bar{h}}}\right|} \; (1 + ix) \quad , \quad \beta = \frac{\gamma_o}{\gamma_h}$$

(for more information on the form of the equations and the notation adopted here and henceforward see, for instance, [1.66, 11.17]).

In the general case of a crystal with a constant deformation gradient the function $(\underline{\underline{hu}})$ is quadratic with respect to the coordinates s_o, s_h in the scattering plane

$$(\underline{\underline{hu}}) = 2(As_o^2 + 2Bs_o s_h + Cs_h^2) \quad . \tag{11.113}$$

The coefficients A,B,C for some cases of practical interest are given in [11.22]. In recording (11.113), the terms linear along the coordinates are omitted for simplicity, since their inclusion would only lead to renormalization of the value of the Bragg angle ϑ.

Assume, in (11.112),

$$D_o = \tilde{D}_o \exp(-2iCs_h^2) \quad , \quad D_h = \tilde{D}_h \exp(2iAs_o^2) \quad . \tag{11.114}$$

Then it is obvious that the equations relating the amplitudes D_o and D_h will retain the form of (11.112), with replacement of the displacement field function $(\underline{\underline{hu}})$ by

$$\underline{\underline{h\tilde{u}}} = 4Bs_o s_h \quad . \tag{11.115}$$

The system of (11.112), with an allowance for (11.113-115) leads to the following equations in second-order partial derivatives of the hyperbolical type for the amplitudes of the tramsitted, D_o, and diffracted, D_h, wave, respectively,

$$\begin{cases} \dfrac{\partial^2\tilde{D}_o}{\partial s_o\, s_h} - i\,\dfrac{\partial(\underline{h}\underline{\tilde{u}})}{\partial s_h}\,\dfrac{\partial\tilde{D}_o}{\partial s_o} + \sigma^2\tilde{D}_o = 0 \\[2ex] \dfrac{\partial^2\tilde{D}_h}{\partial s_o\partial s_h} + i\,\dfrac{\partial(\underline{h}\underline{\tilde{u}})}{\partial s_o}\,\dfrac{\partial\tilde{D}_h}{\partial s_h} + \sigma^2\tilde{D}_h = 0 \end{cases} \tag{11.116}$$

($\sigma^2 = \sigma_{\bar{h}}\sigma_h \equiv 1 + 2i\varkappa$, $\varkappa$ is the normalized dynamical absorption coefficient, $\varkappa < 0$).

The solution of (11.116) satisfying the known boundary conditions on the curve RQ in the scattering plane is of the following form [11.17] in accordance with the Riemann method:

$$\tilde{D}_o(P) = \tilde{D}_o(R) + \int_{RQ}\left(\frac{\partial R_o}{\partial s_h} + 4iBs_oR_o\right)\tilde{D}_o ds_h + \int_{RQ} R_o\,\frac{\partial\tilde{D}_o}{\partial s_o}\,ds_o$$

$$\tilde{D}_h(P) = \tilde{D}_h(Q) + \int_{RQ}\left(\frac{\partial R_h}{\partial s_o} - 4iBs_hR_h\right)\tilde{D}_h ds_o + \int_{RQ} R_h\,\frac{\partial\tilde{D}_h}{\partial s_n}\,ds_h \tag{11.117}$$

where the Riemann functions R_o, R_h are determined as solutions of the conjugate homogeneous equations

$$\frac{\partial^2 R_o}{\partial s_o\partial s_h} + 4iB\,\frac{\partial}{\partial s_o}\,(s_oR_o) + \sigma^2R_o = 0$$

$$\frac{\partial^2 R_h}{\partial s_o\partial s_h} - 4iB\,\frac{\partial}{\partial s_h}\,(s_hR_h) + \sigma^2R_h = 0 \tag{11.118}$$

satisfying the following conditions on the characteristics:

$$R_o(s_h = s_{hp}) = 1$$

$$R_o(s_o = s_{op}) = \exp[-4iB(s_h - s_{hp})s_{op}] \tag{11.119a}$$

$$R_h(s_o = s_{op}) = 1$$

$$R_h(s_h = s_{hp}) = \exp[4iB(s_o - s_{op})s_{hp}] \quad . \tag{11.119b}$$

Note that (as follows from (11.118-119)) the functions R_o and R_h transform into one another on replacing the variables $s_o \rightleftarrows s_h$ and changing the sign of the deformation gradient $B \to -B$. Thus, The Riemann method for solving (11.117) results in finding one of the functions, R_o or R_h. The method for finding Riemann functions is described in [11.17,18]. Therefore we will directly give the results for the functions R_o and R_h:

$$R_o = \exp[4iB(s_{hp} - s_h)s_o]\,{}_1F_1\left[i\,\frac{\sigma^2}{4B}\,,\,1\,;\;4iB(s_{op} - s_o)(s_{hp} - s_h)\right]$$

$$R_h = \exp[-4iB(s_{op} - s_o)s_h]\,{}_1F_1\left[-i\,\frac{\sigma^2}{4B}\,,\,1\,;\,-4iB(s_{hp} - s_h)(s_{op} - s_o)\right] \tag{11.120}$$

where ${}_1F_1$ is the degenerate hypergeometrical function.

Substituting (11.120) in (11.117) with an allowance for the boundary conditions on the curve RQ, we obtain the general solution of the problem on the propagation of an arbitrary wave packet in a crystal with a uniform strain gradient. Further on, we will use a special coordinate system (see Fig. 11.1) [1.66]:

$$z = (s_o + s_h) \quad , \quad x = (s_o - s_h) \tag{11.121}$$

in which the characteristic rays bounding the Borrmann triangle, are given by the equations $x = \pm z$ irrespective of the value γ_o/γ_h. This makes the mathematical description of the nonsymmetrical case of Laue diffraction identical with the symmetrical case and simplifies the physical analysis of the

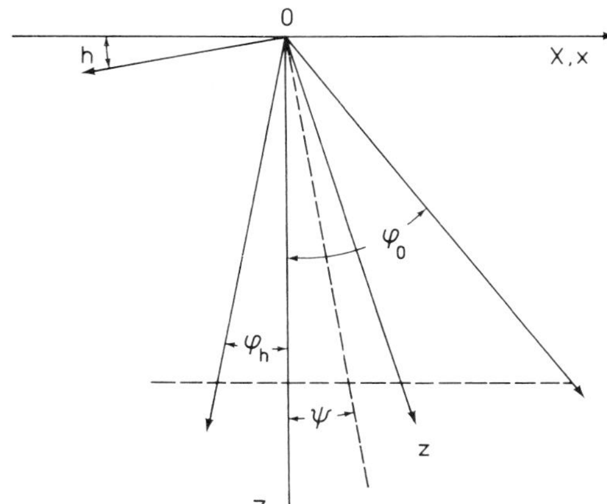

Fig. 11.1 Diffraction geometry and coordinate system used

results. Thus, using (11.117), (11.120), and (11.121), the solution of the Tagaki equations can, after direct calculations, be represented as

$$D_o(P) = \int_{RQ} \mathfrak{G}_{oo}(\underline{r}_p,\underline{r})D_o(\underline{r})(dx - dz) + \int_{RQ} \mathfrak{G}_{oh}(\underline{r}_p,\underline{r})D_h(\underline{r})(dx + dz) \quad ,$$

$$D_h(P) = \int_{RQ} \mathfrak{G}_{ho}(\underline{r}_p,\underline{r})D_o(\underline{r})(dx - dz) + \int_{RQ} \mathfrak{G}_{hh}(\underline{r}_p,\underline{r})(dx + dz) \quad . \tag{11.122}$$

Here, $D_{o,h} = \exp(i\underline{\eta}\underline{r})D_{o,h}$ are the total wave function values of the transmitted and diffracted waves; integration is performed over a segment of the curve RQ intercepted by the characteristics $|x_p - x| = z_p - z$, which are drawn from the observation point $\underline{r}_p$. The influence functions $\mathfrak{G}_{mn}(\underline{r}_p,\underline{r})$ are defined as follows

$$\mathfrak{G}_{mn} = \exp[i\underline{\eta}(\underline{r}_p - \underline{r})]G_{mn}$$

$$G_{oo}(\underline{r}_p,\underline{r}) = \exp(i\Phi_o)\Big[\delta(x - x_p + z_p - z - 0) - \frac{\sigma^2}{4}(z_p - z + x_p - x)$$

$$\cdot \exp\left(-\frac{iB\rho^2}{2}\right){}_1F_1\left(1 + i\frac{\sigma^2}{4B}, 2; -iB\rho^2\right)\Big] \quad ,$$

$$G_{ho}(\underline{r}_p,\underline{r}) = \frac{i\sigma_h}{2}\exp\left[i\Phi_o - i(\underline{h}\underline{u})_p - i\,\frac{B\rho^2}{2}\right]{}_1F_1\left(1 + i\,\frac{\sigma^2}{4B}, 1\,;\, iB\rho^2\right) ,$$

$$(11.123)$$

$$G_{hh}(\underline{r}_p,\underline{r}) = \exp(-i\Phi_h)\left[\delta(x - x_p + z - z_p + 0) - \frac{\sigma^2}{4}(z_p - z - x_p + x)\right.$$

$$\left.\cdot\exp\left(i\,\frac{B\rho^2}{2}\right){}_1F_1\left(1 - i\,\frac{\sigma^2}{4B}, 2\,;\, -iB\rho^2\right)\right]$$

$$G_{oh}(\underline{r}_p,\underline{r}) = \frac{i\sigma_{\bar{h}}}{2}\exp\left[i(\underline{h}\underline{u})_p - i\Phi_h + \frac{iB\rho^2}{2}\right]{}_1F_1\left(1 - i\,\frac{\sigma^2}{4B}, 1\,;\, -iB\rho^2\right)$$

where $\rho^2 = (z_p - z)^2 - (x_p - x)^2$; $\Phi_{o,h}$ are functions depending quadratically on the coordinates

$$\Phi_o(\underline{r}_p,\underline{r}) = \frac{C}{2}\,[(z_p - x_p)^2 - (z - x)^2] + \frac{B}{2}\,[(z_p + x)^2 - (x_p + x)^2]$$

$$(11.124)$$

$$\Phi_h(\underline{r}_p,\underline{r}) = \frac{A}{2}\,[(z_p + x_p)^2 - (z + x)^2] + \frac{B}{2}\,[(z_p - x)^2 - (x_p - z)^2] .$$

The components of the vector $\underline{\eta}$ are related with the deviation from the exact Bragg condition at the origin $\alpha = -2\sin 2\vartheta_B\,\Delta\vartheta_B$ and the zero Fourier component of the crystal polarizability χ_o by the equations

$$\eta_{x,z} = \frac{\chi_o(1 \mp \beta) \pm \alpha\beta}{2C\sqrt{|\beta|\,\mathrm{Re}(\chi_{\bar{h}}\chi_h)}} . \qquad (11.125)$$

The symbols ± 0 in (11.123) mean that the singular points of the δ-functions must be included in the integration area.

The relations (11.122) and (11.123) are the integral formulation of the Huygens-Fresnel principle for a crystal with a uniform strain gradient, because they express the wave field amplitudes at each point $\underline{r}_p$ through their values on the boundary contour. The influence functions have the physical meaning of a wave field induced by a point source of unit intensity placed at point r and observed at point $\underline{r}_p$. With an accuracy to the phase factors, the influence functions depend on the coordinates in translation-invariant form. This is associated with the fact that, from general physical considerations [1.66, 11.19], the intensity distribution (the square of the modulus of the wave field amplitude) of the point source must be invariant to translations $\underline{r} \to \underline{r} + \underline{a}, \underline{r}_p \to \underline{r}_p + \underline{a}$ as a consequence of the cylindrical symmetry of the point source and the displacement field (11.113). Note that the influence function (11.123) transforms into the corresponding expressions for the perfect crystal when the total displacement field (11.113) is identically zero. However, if the field (11.114) vanishes, i.e., B = 0, the influence functions differ from the functions for the perfect crystal only in the phase factors. The latter case is realized, for instance, in symmetrical Laue diffraction on an elastically bent crystal.

In the case of Laue diffraction, the amplitudes $D_{o,h}$ are assigned on the crystal surface z = 0 with the aid of conventional boundary conditions

$$D_o(x,0) = D(x) \quad , \quad D_h(x,0) = 0 \tag{11.126}$$

and from (11.122) we find for the wave field inside the crystal

$$D_o(\underline{r}_p) = \int_{x_p - z_p}^{x_p + z_p} dx \mathfrak{G}_{oo}(x_p, x, z_p, 0) D(x)$$

$$D_h(\underline{r}_p) = \int_{x_p - z_p}^{x_p + z_p} dx \mathfrak{G}_{ho}(x_p, x, z_p, 0) D(x) \quad . \tag{11.127}$$

From the explicit form of (11.118) and the tranlation invariance of the point-source intensity it follows, among other things, that the dependence of the influence functions corresponding to transitions from a transmitted wave to a diffracted one, and vice versa, on the sign of the deformation gradient, sign B, is determined, with an accuracy to the phase factors, only by the parameter $\widetilde{x}_h = x - B$. The transition from x to x_h can be interpreted as renormalization of the dynamical coefficient of absorption in a crystal with a uniform strain gradient. It is significant that such renormalization takes place only for the diffracted wave. Indeed, each influence function is determined by a combination of the form

$$\exp\left(-\frac{iB\rho^2}{2}\right) {}_1F_1\left(1 + i\frac{\sigma^2}{4B}, \quad \nu; \ iB\rho^2\right)$$

$$= \exp\left(-i\varepsilon \frac{|B|\rho^2}{2}\right) {}_1F_1\left(\frac{\nu}{2} + \varepsilon \frac{\tilde{\sigma}^2}{4|B|}, \nu, \varepsilon i|B|\rho^2\right) \tag{11.128a}$$

where

$$\tilde{\sigma}^2 = 1 + 2i\tilde{x} = 1 + 2i(x + \varepsilon)|B|(\nu - 2) \quad , \quad \varepsilon = \text{sign } B \tag{11.128b}$$

($\nu = 1, 2$ for the diffracted and transmitted wave, respectively).

Using Kummer's transformation for the degenerate hypergeometrical function [11.23,24], it is easy to show that the right-hand side of (11.128a) depends on the sign of deformation ε only via $\tilde{\sigma}^2$. Hence it follows at once that since there is no renormalization of the dynamical coefficient of absorption for the transmitted wave, $\tilde{x} = x(\nu = 2)$, the intensity distribution of the transmitted wave induced by a point source is independent of the sign of deformation. For the diffracted wave, the situation depends essentially on whether we deal with a transparent, $|x|z_p < 1$, or an absorbing, $|x|z_p \gtrsim 1$, crystal. In the case of the transparent crystal a change in the sign of deformation is equivalent to transition to a conjugate complex expression for influence functions with $\nu = 1$ and does not affect the intensity distribution of the diffracted wave of the point source and the integrated intensity of the beam of diffracted waves R_h^i, respectively. In an absorbing crystal, the symmetry of the absorption coefficient with respect to the sign of strain gradient is disturbed because of the renormalization (a change in the sign of B may be caused, in particular, by inversion of Bragg reflection, $h \to \bar{h}$), and this, in turn, automatically leads to the violation of the Friedel law for integrated intensity R_h^i.

These features of Laue diffraction of X-rays in a crystal with a constant deformation gradient have been described previously within the framework of the Eikonal approximation [1.58]. The above discussion shows that they are preserved in the exact theory as well.

In accordance with the physical conception of propagation in a crystal, in conditions of dynamical scattering, of two Bloch waves corresponding to the two sheets of the dispersion surface, each of the influence functions can be represented thus [11.24]

$$\exp\left(-i \frac{|B|}{2}\rho^2\right) {}_1F_1\left(\frac{\nu}{2} + i \frac{\tilde{\sigma}^2}{4|B|}, \nu, i|B|\rho^2\right) = \exp\left(-\pi \frac{\tilde{\sigma}^2}{4|B|}\right)\Gamma(\nu) \cdot$$

$$\cdot\left[\frac{\exp\left(-\frac{i|B|}{2}\rho^2 - i\pi\nu/2\right)}{\Gamma\left(\frac{\nu}{2} + i\frac{\tilde{\sigma}^2}{4|B|}\right)}\Psi\left(\frac{\nu}{2} - i\frac{\tilde{\sigma}^2}{4|B|}, \nu; -i|B|\rho^2\right) + \frac{\exp\left(i\frac{|B|}{2}\rho^2 + i\pi\nu/2\right)}{\Gamma\left(\frac{\nu}{2} - i\frac{\tilde{\sigma}^2}{4|B|}\right)}\right.$$

$$\left.\Psi\left(\frac{\nu}{2} + \frac{i\tilde{\sigma}^2}{4|B|}, \nu; i|B|\rho^2\right)\right] \tag{11.129}$$

where Ψ (a,c,x) is a second-order hypergeometrical function according to Tricomi's definition.

Passing over to the limit, $B \to 0$, it is easy to show that at $|B|z_p \ll 1$ the first term in the right-hand side of (11.129) corresponds to a weakly absorbed wave field (2 is a branch of the dispersion surface), whereas the second term corresponds to a strongly absorbed (1-branch) field in the perfect crystal. The behaviour of the wave fields in the deformed crystal is more complicated, however, because of the coordinate dependence of the amplitude of the 2- and 1-Bloch waves. Strictly speaking, the division into weakly and strongly absorbing fields holds good only for a transmitted wave field ($\nu = 2$), and the transmitted component of the (2)-branch is always weakly absorbed. The diffracted component of the 2(1)-branch is weakly (strongly) absorbed at $\tilde{x} < 0$, and vice versa (see (11.128,130) below, and also the asymptotic representations Ψ (a,c,x) at large $|x|$ [11.24]). It is interesting to note that at $x = B$ there is no dynamical absorption, and the diffracted components of the 2- and 1-branches are equal at any ρ. This results from the renormalization of the dynamical coefficient of absorption due to the presence of the strain gradient.

The above discussion shows that the visibility of the pendulum solution bands in a deformed absorbing crystal depends on the sign of deformation. Here, the case $B < 0$ is preferred because of the improved contrast [11.25].

11.7.3 Quasi-classical Wave Field Asymptotes

In a crystal whose thickness greatly exceeds the extinction length, $z_p \gg 1$, investigation of the wave fields (11.123, 129) in the central part (ray zone) of the Borrmann triangle can be conducted with the aid of the asymptotic representation of the confluent hypergeometric function [1.66, 11.19]

$$\exp\left(-i\,\frac{|B|}{2}\,\rho^2\right){}_1F_1\left(\frac{\nu}{2} + i\,\frac{\sigma^2}{4|B|}\,,\nu;\ i|B|\rho^2\right) \sim \left(\frac{\rho}{2}\right)^{1-\nu}\,\frac{1 - \exp(-\pi/2|B|)}{2\pi\rho\sqrt{1 + B^2\rho^2}}^{\frac{1}{2}}$$

$$\cdot\,\{\exp[i\Phi(\rho) - i\varphi(\nu,|B|)] + \exp[-i\Phi(\rho) + i\varphi(\nu,|B|)]\}$$

$$\left[1 + 0\left(\frac{1}{\rho\sqrt{1 + B^2\rho^2}}\right)\right] \qquad (11.130)$$

which can be derived from a more general asymptotic expansion with an allowance for the smallness of the dynamical coefficient of absorption, $|x| \ll 1$ [11.22].

The applicability range of (11.130) depends on the condition

$$\rho\sqrt{1 + B^2\rho^2} \gg 1 \quad . \qquad (11.131)$$

In (11.128) we use the following notation:

$$\Phi(\rho) = \frac{\rho}{2}\,\sqrt{1 + B^2\rho^2} + \frac{1 + 2i\tilde{x}}{2|B|}\,\ln(\sqrt{1 + B^2\rho^2} + |B|\rho) \qquad (11.132)$$

$$\varphi(\nu,|B|) = \frac{\pi}{2}\,(\nu - 1) + \mathrm{Im}\{\ln[\Gamma(1 + \frac{i}{4|B|})]\} + \frac{1}{4|B|}[\ln(4|B|) + 1] \ . \qquad (11.133)$$

The order of the terms in (11.130) is the same as in (11.129); the first term corresponds to the 2-branch of the dispersion surface.

The imaginary part of the Eikonal, $\Phi(\rho)$, can be represented as follows:

$$\mathrm{Im}\,\Phi\,(\rho) = \frac{x}{|B|}\,\mathrm{arc\,sinh}(|B|\rho) + \varepsilon(\nu - 2)\,\ln(\sqrt{1 + B^2\rho^2} + |B|\rho) \quad . \qquad (11.134)$$

The real part, $\mathrm{Re}\{\Phi(\rho)\}$, together with the first term in the right-hand side of (11.134), coincides with KATO's complex Eikonal [1.58]. Conversion to KATO's notation can be achieved with the aid of the relations

$$2Bz = Z \quad , \quad 2Bx = X \quad , \quad \frac{1}{2\,B} = \frac{m_o^2 c}{|f|}$$

where Z, X, m_o, c, f are the parameters used by KATO. The second term in the right-hand side of (11.134) in KATO's theory is included in the amplitude of the diffracted wave.

The wave fields determined by (11.123,127,129,130,132,133) represent a generalized Eikonal approximation of the theory (quasi-classical asymptotes). The generalization results in the appearance in (11.130) of the static factor $[1 - \exp(-\pi/2|B|)]^{\frac{1}{2}}$ and the scattering phase $\varphi(\nu,|B|)$, which depends on the strain value $|B|$. In the case of small, $|B| >> 1$,

$$1 - \exp -(\frac{\pi}{2|B|}) \simeq 1 \quad , \quad \varphi(\nu,|B|) \simeq \frac{\pi}{2}(\nu - 1) + \frac{\pi}{4}$$

and (11.130-133) transform to the corresponding results of the Eikonal approximation. In the case of strong deformation, $|B| >> 1$

$$1 - \exp(-\frac{\pi}{2|B|}) \simeq \frac{\pi}{2|B|} \quad , \quad \varphi(\nu,|B|) \simeq \frac{\pi}{2}(\nu - 1)$$

i.e., the field amplitudes in (11.130) are $(\pi/2|B|)^{\frac{1}{2}}$ times less than in the Eikonal approximation. As is well known [1.58], in the Eikonal approximation the integrated intensity R_h^i for the transparent crystal grows as $|B| >> 1$. Inclusion of the static scattering factor eliminates this nonphysical result and leads to the correct kinematical limit $R_h^{i(kin)}$ at large $|B|$ (see Section 11.7.5). As regards the change in scattering phase at $|B| >> 1$, it is hardly observable because of the great difference between the amplitudes of the two wave fields in (11.130) at large $|B|$.

11.7.4 Ray Trajectories

To wind up the comparison of our results with those of the Eikonal approximation, let us consider the ray trajectories which can be obtained by applying the stationary-phase principle for calculating the integrals in (11.127). In the case of the plane incident wave, the stationary points are determined from the equations

$$\frac{\partial}{\partial x}\,[\mathrm{Re}\{\Phi_{\pm}\}(x_p,x,z_p,0)] = 0 \tag{11.135}$$

where Re $\{\Phi_{\pm}\}$ are the real parts of the Eikonals corresponding to the two wave fields in the crystal

$$\mathrm{Re}\{\Phi_{\pm}\}(\underline{r}_p,\underline{r}) = \underline{\eta}(\underline{r}_p - \underline{r}) + \Phi_o(\underline{r}_p,\underline{r}) \pm [\Phi(\rho) - \varphi] \ .$$

Note that the imaginary $\mathrm{Im}\{\Phi_{\pm}\}$ varies slower than the real one at any $|B|$, provided (11.131) is satisfied. Therefore the stationary-phase principle in the form of (11.135) is applicable at any value of the strain gradient.

Each of (11.135) can be regarded as the equation of a trajectory with an origin $(x,0)$ and a current observation point (x_p,z_p). After some transformations we find the trajectory equations in the form

$$[2B(x_p - x) \mp \xi(x)]^2 - [2Bz_p - \eta(x)]^2 = 1 \tag{11.136}$$

$$\eta(x) = \mathrm{Re}\{\eta_x\} + (C - B)x \quad , \qquad \xi(x) = \sqrt{1 + \eta^2(x)} \ . \tag{11.137}$$

The trajectories (11.136) are similar to those which are obtained in the Eikonal approximation [1.58]. The difference consists in the dependence of the parameters $\eta(x)$ and $\xi(x)$ on the x-coordinate on the entrance surface. This dependence arises due to the phase factor in (11.123), which is common to both wave fields and reflects the variation of the deviation from the exact Bragg condition along the entrance surface due to the displacement field (11.113). Thus, the incident plane wave induces two "fans" of trajectories for the two wave fields in the crystal. Besides, one of the fans will always be divergent and the other, convergent. Hence the possibility of the appearance of points of intersection of rays if the crystal is thick enough. These points can be determined from the condition

$$\frac{\partial^2}{\partial x^2}\,[\mathrm{Re}\{\Phi_{\pm}\}(x_p,x,z_p,0)] = 0 \tag{11.138}$$

together with (11.135).

Eliminating x from (11.135) and (11.138), we obtain the equation of a caustic surface, which consists of two branches in accordance with the two branches of the dispersion surface (or with the two signs in (11.135)). Each point of either branch of the caustic surface is the focussing point for one wave field or the other, depending on the sign of deformation. The lower branch consists of real focussing points corresponding to the convergent trajectory fan, and the upper branch consists of imaginary focussing points corresponding to the divergent fan. In the particular case of symmetrical Laue diffraction the caustic surface is determined by the equation [1.66] $(B = 0)$

$$Cz_p = \pm\,[(Cx_p + \mathrm{Re}\{\eta_x\})^{2/3} + 1]^{3/2} \quad . \tag{11.139}$$

The general case of the geometric-optical image in non-symmetrical Laue reflection is considered in [11.26]. The minimum distance between the entrance surface and the focus is equal to

$$z_{min} = |C|^{-1} \qquad Cx_p + \mathrm{Re}\{\eta_x\} = 0 \quad . \tag{11.140}$$

It can be seen from (11.136) that the point determined by (11.140) lies on the central trajectory x_p = const. Besides, this is a singular point of the caustic surface, since the derivative dz_p/dx_p at it has not been determined. The caustic surface (11.139) and several near central trajectories for the two wave fields are shown in Fig. 11.2.

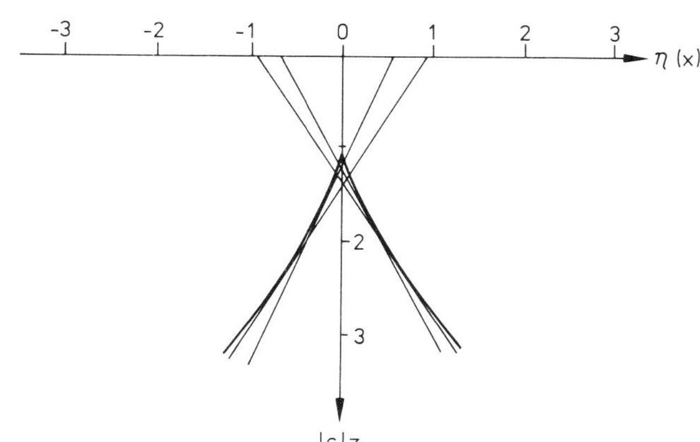

Fig. 11.2 Caustic and ray trajectories in the case of symmetrical Laue diffraction [1.66]

The preceding consideration referred to the case of the incident plane wave, but it can readily be extended to the case of the incident spherical wave. Compared with the plane wave, which has the same direction of incidence (determined by means of $\mathrm{Re}\{\eta_x\}$) at the origin, the spherical wave has an additional phase term of the form $Px^2/2$, which must be added to $\mathrm{Re}\{\Phi_\pm\}$ in (11.135) and (11.138). This leads to the replacement $C \to C - P$ in all the preceding equations. For a point source located at a distance L from the origin on the entrance surface, the coefficient P is equal to

$$P = \frac{\sin^2 2\vartheta}{2\pi\gamma_h^2} \frac{\Lambda}{\lambda L} \quad . \tag{11.141}$$

For the typical values of $\Lambda \sim 10\ \mu$, $\lambda \sim 1$ Å, $L \sim 1$ m, we get $P \gtrsim 1$. This means that in order to obtain diffraction focussing of the incident spherical wave in a thick crystal, $z \widetilde{=} 100$, the coefficients C and P must be of the same order of magnitude, $C \simeq P$ and, hence, $L \sim R$, where R is the curvature radius of the crystal. On the other hand, if $P \gg 1$, $P \gg |C|, |B|$, the wave fields in the crystal can be constructed on the assumption of infinite, P, $P \to \infty$. Physically, the case $P \to \infty$ corresponds to a point source of X-rays placed on the surface of the crystal. In this case the singular points of the two branches of the caustic surface coincide with each other and with the origin, while the caustic equation degenerates into the characteristic-ray equation, $x_p - x = \pm z_p$. The trajectory equation (11.136) now takes different forms at $x = 0$ and $x \neq 0$. At $x \neq 0$ it describes only the characteristic rays $|x_p - x| = z_p$, for which the diffracted intensity vanishes; at $x = 0$ the trajectory equation takes the form

$$\left[2Bx_p \mp \sqrt{1 + (\mathrm{Re}\{\eta_x\})^2}\right]^2 - (2Bz_p - \mathrm{Re}\{\eta_x\})^2 = 1 \tag{11.142}$$

as in the theory of the pendulum solution of the problem on Laue diffraction of X-rays on a bent crystal developed by KATO [1.58]. The angular-deviation parameter $\mathrm{Re}\{\eta_x\}$ is not determined now, and all the trajectories with different $\mathrm{Re}\{\eta_x\}$ are realized simultaneously.

11.7.5 Integrated Intensity of Diffracted Wave

Let us now consider the integrated intensity of the diffracted wave, R_h^i, in the case of a thick crystal. With the aid of the asymptotic representation (11.130) it is easy to obtain, for the nonoscillating part of the integrated intensity,

$$R_h^i(M_o,M_\beta,M,D,|B|) = \exp(-M_o)\,\frac{1-\exp(-\pi/2|B|)}{\pi}$$

$$\cdot\int_{-1}^{1}\frac{d\xi}{\sqrt{1-\xi^2}}\,\frac{\cosh(M_\beta\xi)\cosh\{(\frac{M}{D}+2)\ln[\sqrt{1+D^2(1-\xi^2)}+D\sqrt{1-\xi^2}]\}}{\sqrt{1+D^2(1-\xi^2)}} \tag{11.143}$$

where

$$\xi = \frac{x}{t}\quad,\quad M_o = \frac{\mathrm{Im}\{\chi_o(1+\beta)t\}}{C\sqrt{|\beta|\mathrm{Re}\{(\chi_h\chi_{\bar{h}})\}}}\quad,\quad M_\beta = \frac{1-\beta}{1+\beta}M_o$$

$D = Bt$, $M = 2|x|t$; M_o characterizes the normal absorption, M the Borrmann absorption, M_β the absorption asymmetry because of the nonsymmetrical diffraction geometry, D the effective deformation. It can be shown that the relative error due to the use of (11.130) instead of the exact (11.123) decreases with increasing thickness t and strain gradient B as $(t\sqrt{1+B^2})^{-1}$.

The intensity (11.143) is normalized to unity for the perfect nonabsorbing crystal, $R_h^i\,(0,0,0,0,0) = 1$. For the perfect absorbing crystal

$$R_h^i = (M_o,M_\beta,M,0,0) = \exp(-M_o)I_o(\sqrt{M_\beta^2+M^2}) \quad . \tag{11.144}$$

($I_o(x)$ is the Bessel function of the imaginary argument).

Within the limits of large deformations $|B| \to \infty$, the intensity R_h^i tends to

$$2t\,\exp(-M_o)\,\frac{\sinh(M_\beta)}{M_\beta}\;, \tag{11.145}$$

i.e., it becomes proportional to the crystal thickness, which must, indeed, be the case in kinematical scattering.

With intermediate values of $|B|$ the integral in (11.143) cannot be calculated in explicit form. In the case of symmetrical Laue diffraction, $M_\beta = 0$, the relative integrated intensity

$$E(M,D) = \frac{R_h^i(M_o,0,M,D)}{R_h^i(M_o,M)} \tag{11.146}$$

was tabulated by ANDO and KATO [11.12] for the parameter values $0 \le M \le 25$ and $-5 \le D \le 5$. In the general case the number of parameters determining the relative integrated intensity increases from two (M,D) to four (M_β,M,D, $|B|$), which hinders tabulation.

For approximate numerical calculations of the integrated intensity with an allowance for the particular kind of the integral (11.143), good results can be obtained with the aid of the Gauss-Tschebyshev quadrature equation [11.27]

$$\int_{-1}^{1} \frac{d\xi\eta(\xi)}{\sqrt{1-\xi^2}} \simeq \frac{\pi}{2} \sum_{k=1}^{n} \eta(\xi_k) \tag{11.147}$$

where ξ_k are the roots of the Tschebyshev polynomial of n-th order.

$$\xi_k = \cos\left(\frac{2k-1}{2n}\pi\right) .$$

The practical calculations performed by us (Fig. 11.3 showed that an accuracy of 0.5 percent is achieved at n = 7 [11.22]. The value of the integrated function were calculated only at four points due to its symmetry.

11.7.6 Conclusion

A comparison of the exact solution of the dynamical diffraction problem for a crystal with a constant deformation gradient with the solution obtained in the Eikonal approximation shows that the Eikonal approximation yields a very good physical description of the arising diffraction phenomena. The gener-

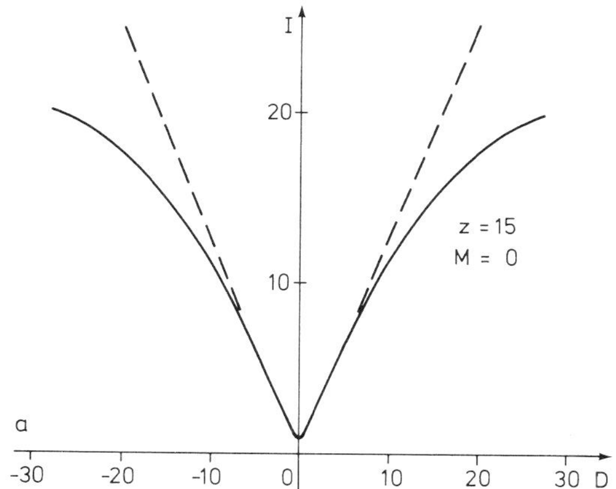

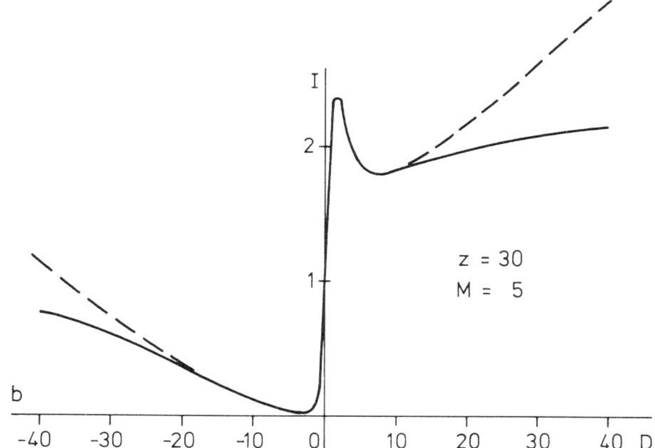

Fig. 11.3 a and b Integrated reflection R_h^i as function of effective deformation. a) M = 0, Z = 15. b) M = 5, Z = 30 [11.22]

alization of the Eikonal approximation of the theory on the basis of the exact solution of the Takagi equations leads to the following results:

1) extension of the applicability range of the theory to the case of an arbitrary (large) strain gradient. This extension results in the appearance of the static scattering factor $[1 - \exp(-\pi/2|B|)]^{\frac{1}{2}}$ in the wave field amplitudes in the crystal, and in the dependence of the additional phase term φ on $|B|$;

2) another difference consists in taking into account the variation in wave field amplitudes as a result of renormalization of the dynamical coefficient of absorption. This conception entails some difficulties in identifying wave fields as strongly or weakly absorbed ones, but still it reflects

their actual behaviour. Besides, this is the most natural way of describing wave fields near the boundaries of the Borrmann triangle in a deformed crystal [1.66, 11,18, 19].

A fact of fundamental importance is that the principal concept of geometrical optics, i.e., the ray trajectory, is preserved even in the case of strong deformation. The trajectories defined in (11.134) tend to the kinematic trajectories $|x_p - x| = z_p$ with increasing $|B|$, and at the same time the limits of applicability of the generalized Eikonal representation of wave fields (11.130) extend so that the trajectories do not lose their meaning. This gives us grounds to believe that the geometrical optics theory can be so modified that the exact (correct) asymptotic solution of the Takagi equations is obtained from it at any deformation.

An analysis of the wave fields in an elastically bent crystal in the case of an incident wave of finite spatial and angular spread involves some difficulties because the ray trajectories intersect. It is obvious that in the vicinity of the caustic, the stationary-phase method (and geometrical optics as a whole) leads to infinite values of intensity and is therefore inapplicable. This case is of some practical interest, for instance, for multi-crystal topography. A special consideration of this problem is given in [11.28].

12. Dynamical Scattering in the Case of Three Strong Waves and More

Choosing the angles of incidence of the primary wave on the crystal makes one easily aware of situations where the Bragg condition will be satisfied, not for one, but for several diffracted waves simultaneously. The first general consideration for such scattering, which included a comprehensive analysis of the properties of the dispersion surface, was carried out by EWALD in 1937 [12.1] on the basis of his dynamical theory. The phenomena associated with multiwave diffraction had been discovered experimentally long before that, in the early 1920s. They included, above all, the "brightening" effect (Aufhellung), discovered by WAGNER in 1923 and thoroughly investigated by BERG, and in particularly by MAYER [12.2]. The Aufhellung phenomenon consists in using a slightly divergent ray beam with a finite spectral interval for obtaining a definite reflection from a single crystal; by rotating the crystal about one of the spectrograph axes, it can be maneuvered into a position which simultaneously corresponds to the second reflection. In this case, the spectral band corresponding to the first reflection shows a bright, narrow line, which is due to the change in the relative intensity of the first (main) reflection on transition to three-wave scattering.

Three-wave scattering, in a more direct sense, was discovered and investigated in 1937 by RENNINGER [1.29], who named it "indirect excitation" (Umweganregung). Studying the reflection of X-rays from a diamond crystal, RENNINGER observed reflection 222 (discovered previously), which is forbidden diamond structure.

This reflection is due to the deviation of the electron shell of the carbon atoms in this structure from spherical symmetry and the concentration of a certain part of the electron density in the atomic "bridges" [4.7]. This explanation, however, may be regarded as a qualitative, rather than a quantitative, confirmation of experiment.

RENNINGER used an ingenious experimental procedure. Setting up a crystal, relative to the incident beam, in a position of second-order (Bragg) reflection from the plane (111), he recorded the intensity of the reflected beam with an ionization chamber, rotating the crystal about an axis normal to the plane (111). He observed a weak reflection 222 and the periodic appearance of more or less strong peaks, which were due to reflections from other planes of the crystal. The period of these maximums of intensity was 60° in accordance with the hexagonal symmetry of the axis 111 in the reciprocal lattice. The formation of the reflections discovered during rotation of a diamond crystal about the axis 111 at an angle of incidence corresponding to reflection 222 is observed directly on the stereographic projection of a cubic crystal along a third-order axis if one traces the corresponding projections of the base circles of the Kossel cones. The circle of the cone of reflections 222 (which is projected without any distortion) alternately intersects the circles of the cones of reflections 111, 133, and 113 at the given intervals of 60°. It is easy to show that, despite the intersection of the cone circles of the stronger reflections 220 and 400, these reflections cannot be formed.

It should be noted that as far back as 1928, MAYER used EWALD's dynamical theory in calculating the brightening effect. MAYER considered the case of three-wave diffraction and, to simplify the analysis, restricted himself to situations where all the three wave vectors $\underline{k}_o$, $\underline{k}_1$, and $\underline{k}_2$ lie on the same plane. In this, so-called coplanar, case the equations describing diffraction scattering are grouped for π- and σ-polarization, respectively.

Thus, the following system of equations covers σ-polarization:

$$(\chi_o - 2\varepsilon_o)\underline{D}_o + \chi_{01}\underline{D}_1 + \chi_{02}\underline{D}_2 = 0$$

$$\chi_{10}\underline{D}_o + (\chi_o - 2\varepsilon_1)\underline{D}_1 + \chi_{12}\underline{D}_2 = 0$$

$$\chi_{20}\underline{D}_o + \chi_{21}\underline{D}_1 + (\chi_o - 2\varepsilon_2)\underline{D}_2 = 0$$

whence one obtains

$$\underline{D}_1 = \frac{\chi_{10}(\chi_o - 2\varepsilon_2) - \chi_{20}\chi_{12}}{(\chi_o - 2\varepsilon_1)(\chi_o - 2\varepsilon_2) - |\chi_{12}|^2}\,\underline{D}_o \quad .$$

This expression reveals two interesting features of multiwave scattering. First, when $\chi_{10} = 0$, i.e., for instance, a given reflection is forbidden, the quantity D_1 and, hence, the reflection intensity, does not necessarily vanish to zero. This result explains the RENNINGER effect, or "nondirect excitation". Conversely, with nonzero values of χ_{01}, χ_{02}, and χ_{12}, the second term in the numerator may have a sign opposite to the first one, with the result that the value of D_1, and hence the intensity of the diffracted wave, may be greatly reduced. This explains the "brightening effect".

More recently, double reflections, including nondirect excitations, have been studied by many authors. A detailed description of these investigations can be found in the survey of TERMINASOV and TUZOV [12.3].

A few years ago, the RENNINGER method, in a slightly modified form [12.4], was used by HOM et al. [12.5] for precision measurement of the parameters of the crystal lattice of germanium and silicon. The accuracy of measurement of the lattice constants achieved in this work is several units per 10^{-6}.

New and detailed investigations of multiwave scattering were stimulated by the work of BORRMANN and HARTWIG carried out in 1965 [1.30]. They investigated simultaneous reflections 111 and $11\bar{1}$, in Laue geometry, of Cu K_α radiation from a germanium crystal plate. The authors found that in the three-wave scattering range, anomalous penetration is strongly enhanced as compared with the two-wave case for reflection 111. The numerical values obtained by them were equal to

$$\mu_0 = 352\ \text{cm}^{-1} \quad , \quad \mu_{\min}^{(2)}\ (111) = 105\ \text{cm}^{-1} \quad , \quad \mu_{\min}^{(3)} = 45\ \text{cm}^{-1} \quad .$$

We also note that the minimum absorption for germanium in the two-wave case is realized on reflection 220 and equals $\mu_{\min}^{(2)}\ 220 = 15\ \text{cm}^{-1}$.

Following this work, a large number of experimental investigations in the Laue-Laue geometry were carried out [12.6-25] which either adjusted the data of Borrmann and Hartwig or studied other reflections. Interest in the theoretical investigations analyzing either the general questions of multiwave scattering [1.31-35] or studying certain situations [12.6-38] in detail has also grown.

It is noteworthy that in all experimental investigations, although a decrease in the absorption coefficient as compared with the two-wave case was observed, the authors failed to obtain a coefficient of absorption less than $\mu_{\min}^{(2)}$ 220 in the case of a germaium crystal. In this respect, particularly interesting is the theoretical investigation of JOKO and FUKUHARA [12.28], who successively considered the three-, four-, and six-wave cases of symmetrical scattering in the Laue geometry. The authors showed that in the symmetrical three-wave case there is no decrease in the absorption coefficient as compared with $\mu_{min}^{(2)}$ 220, but in the four-, and particularly six-wave case, a sharp drop in μ_{min} occurs.

The theoretical analysis of multiwave scattering is much more complicated than is the two-wave case. The dispersion surface does not decompose into lower-order independent surfaces due to lack of standard polarizations, but forms a single surface of the sixth and higher orders. A specific analysis of the dispersion surface, the interference absorption coefficients, and the induction vectors requires the use of numerical methods in this case. Numerical computation methods are considered in Section 12.13. The main part of this chapter discusses the general questions of the theory of multiwave diffraction with qualitative discussion of the effects arising in this case, in particular the enhancement of "anomalous transmission", and also specific cases admitting of analytical solutions.

12.1 Scattering in Nonabsorbing Crystal. Reference Coordinate Systems

For analysis and a description of the wave field in a crystal during multiwave diffraction, as well as in the two-wave approximation, wide use is made of the image of a dispersion surface in a reciprocal space. Various choices of coordinate systems are possible. In the three-wave approximation, the most convenient system is the one related to a triplet of wave vectors drawn from point L_o, the Lorenz point, to the corresponding points of the reciprocal lattice. This vector triplet is shown in Fig. 12.1.

By construction, the Lorenz point is the point of intersection of spheres with a radius

$$k = K(1 - \frac{1}{2}|\chi_o|) \quad , \quad k = Kn \tag{12.1}$$

and with centres at the points of the reciprocal lattice. The respective vectors will be denoted by the letters $\underline{k}^{(o)}$, $\underline{k}^{(1)}$, $\underline{k}^{(2)}$, and the vector $\underline{k}^{(o)} = \overline{\underline{L}_o\,\underline{0}}$ is directed from the Lorenz point to the point 0, while

$$\underline{k}^{(1)} = \underline{k}^{(o)} + \underline{h}_1 \quad , \quad \underline{k}^{(2)} = \underline{k}^{(o)} + \underline{h}_2 \quad . \tag{12.2}$$

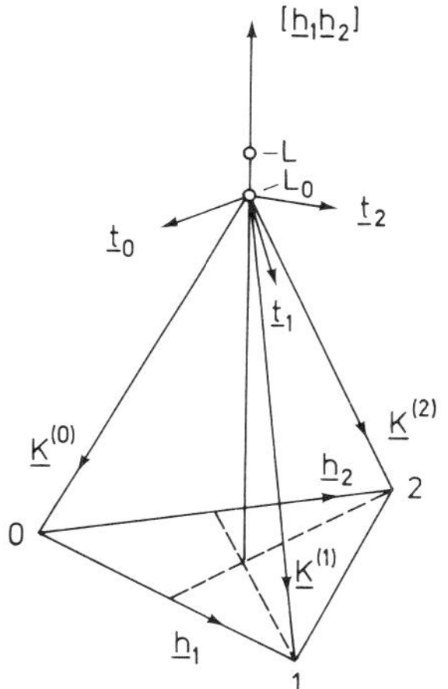

Fig. 12.1 The reference coordinate system in the reciprocal space used in the three-wave scattering theory (after EWALD and HENO [1.31])

It is obvious that the wave vectors of dynamical three-wave scattering $\underline{k}_m$ differ only slightly in direction from the vectors $\underline{k}^{(m)}$. We further introduce the unit vectors

$$\underline{s}_o = \frac{\underline{k}^{(o)}}{k} \ , \quad \underline{s}_1 = \frac{\underline{k}^{(1)}}{k} \ , \quad \underline{s}_2 = \frac{\underline{k}^{(2)}}{k} \ . \tag{12.3}$$

Thus, as well as the standard origin of coordinates in the reciprocal space, point 0, we introduce a new origin, point L_o, which is displaced by the vector $-\underline{k}^{(o)}$ with respect to 0. In order to express the vector $\underline{k}^{(o)}$ in terms of the vectors $\underline{h}_m$, we take advantage of the fact that point L_o lies at the intersection of planes perpendicular to $\underline{h}_1$ and $\underline{h}_2$ and passing through the middle. Let $\underline{R}$ be an arbitrary vector with its origin at the point 0. The equation of these planes take the form

$$(\underline{R}\underline{h}_1) = \frac{1}{2} h_1^2 \ , \quad (\underline{R}\underline{h}_2) = \frac{1}{2} h_2^2 \ , \quad h_m = |\underline{h}_m| \ . \tag{12.4}$$

First, we determine the vector $\underline{R}_o$ lying in the plane of the vectors $\underline{h}_1$ and $\underline{h}_2$ and satisfying (12.4). To do this, we write down $\underline{R}_o = \alpha\underline{h}_1 + \beta\underline{h}_2$ and determine α and β from the system of linear equations (12.4). As a result

$$\underline{R}_o = \frac{1}{2[\underline{h}_1 \times \underline{h}_2]^2} \left\{ h_2^2[h_1^2 - (\underline{h}_1\underline{h}_2)]\underline{h}_1 + h_1^2[h_2^2 - (\underline{h}_1\underline{h}_2)]\underline{h}_2 \right\} . \tag{12.5}$$

The vector $\overline{\underline{OL}}_o = -\underline{k}^{(o)}$ will evidently be equal to the vector sum of the vector $\underline{R}_o$ and some vector perpendicular to the plane of the vectors $\underline{h}_1$ and $\underline{h}_2$,

$$\overline{\underline{OL}}_o = \underline{R}_o + \nu[\underline{h}_1 \times \underline{h}_2] = \underline{R} \quad . \tag{12.6}$$

The unknown ν is found from the condition $R^2 = k^2$. The final expression for the vector $\underline{k}^{(o)}$ has the form

$$\underline{k}^{(o)} = \frac{1}{2[\underline{h}_1 \times \underline{h}_2]^2} \left\{ [h_1^2 - (\underline{h}_1\underline{h}_2)]h_2^2\underline{h}_1 + [h_2^2 - (\underline{h}_1\underline{h}_2)]h_1^2\underline{h}_2 \right.$$
$$\left. + w[\underline{h}_1 \times \underline{h}_2] \right\} , \quad w = \left\{ 4k^2[\underline{h}_1 \times \underline{h}_2]^2 - h_1^2h_2^2(\underline{h}_1 - \underline{h}_2)^2 \right\}^{\frac{1}{2}} \quad . \tag{12.7}$$

By definition the Laue point L corresponds to the point of intersection of spheres of radius K and with centres at the points 0, h_1, and h_2.

Now introduce a vector $\underline{x}_o = \overline{\underline{LO}}$ directed from the Laue point to point 0. It is defined by (12.7), in which k^2 must be replaced by K^2. The distance between the points L_o and L is equal to

$$\overline{\underline{L_oL}} = \underline{k}^{(o)} - \underline{x}_o = \frac{1}{2[\underline{h}_1 \times \underline{h}_2]^2} (w' - w)[\underline{h}_1 \times \underline{h}_2] \quad ,$$
$$w' = \left\{ 4K^2[\underline{h}_1 \times \underline{h}_2]^2 - h_1^2h_2^2(\underline{h}_1 - \underline{h}_2)^2 \right\}^{\frac{1}{2}} \quad . \tag{12.8}$$

Because of the small difference between k and K the difference w' - w can be replaced by (dw/dk^2) $(K^2 - k^2)$. As a result we get

$$\overline{\underline{L_oL}} = \frac{K^2|x_o|}{w} [\underline{h}_1 \times \underline{h}_2] \quad . \tag{12.9}$$

The role of the initial, or real space coordinate system will be played by the triplet of vectors $\underline{s}_m$, according to (12.3), with common origin at point L_o. The volume of the cell supported on the vectors $\underline{s}_m$ is

$$v_s = (\underline{s}_o[\underline{s}_1 \times \underline{s}_2]) = \frac{1}{k^3}(\underline{k}^{(o)}[\underline{h}_1 \times \underline{h}_2]) = \frac{w}{2k^3} \quad . \tag{12.10}$$

In order to investigate the dispersion surface we introduce a coordinate system reciprocal to the system s_m. The vectors of the reciprocal axes t_m are determined with the aid of the relations

$$\underline{t}_o = \frac{[\underline{s}_1 \times \underline{s}_2]}{v_s} \quad , \qquad \underline{t}_1 = \frac{[\underline{s}_2 \times \underline{s}_o]}{v_s} \quad , \qquad \underline{t}_2 = \frac{[\underline{s}_o \times \underline{s}_1]}{v_s} \quad . \tag{12.11}$$

Taking into account (12.2) and (12.3), it is easy to see that the vector sum

$$\underline{t}_o + \underline{t}_1 + \underline{t}_2 = -\frac{[\underline{h}_1 \times \underline{h}_2]}{k^2 v_s} = -\frac{2k}{w}[\underline{h}_1 \times \underline{h}_2] \tag{12.12}$$

coincides in direction with the line $\overline{LL_o}$.

The axis passing through the points L_o and L is called the main axis, and its points of intersection with the dispersion surface, the main points. The position of the excitation point A on the dispersion surface relative to the Lorenz point L_o is defined by the vector $\underline{v}$

$$\overline{\underline{L}_o\underline{A}} = \frac{1}{2}k\underline{v} \quad . \tag{12.13}$$

The Laue point, according to (12.9) and (12.12), is associated with the vector

$$\underline{v}_L = \chi_o(\underline{t}_1 + \underline{t}_2 + \underline{t}_3) \quad . \tag{12.14}$$

In a general case

$$\underline{v} = \sum_m \tau_m \underline{t}_m \quad . \tag{12.15}$$

The factors τ_m are related to the excitation errors ε_m, determined by (2.55), in the following way:

$$\tau_m = \chi_o - 2\varepsilon_m \quad . \tag{12.16}$$

The wave vectors $\underline{k}_m$, which refer to the excitation point A, have the following form according to (12.13):

$$\underline{k}_m = k(\underline{s}_m - \frac{1}{2}\underline{v}) \quad . \tag{12.17}$$

Taking into account (2.55), we obtain for τ_m

$$\tau_m \approx - \frac{(k_m^2 - k^2)}{K^2} \approx - \frac{1}{2} \frac{(|\underline{k}_m| - k)}{K} \quad . \tag{12.18}$$

In order to relate the excitation point A on the dispersion surface to the experimental conditions, we consider a semi-infinite crystal with a plane crystal-vacuum boundary, which will be called the entrance surface. The wave vector of the incident vacuum wave $\underline{K}_o$ lies on a sphere of radius K, its origin being at point O. The part of the sphere in the vicinity of the Laue point L represents a plane perpendicular to the vector $\underline{x}_o = \underline{\overline{LO}}$ and passing through point L. Introduce two mutually perpendicular unit vectors $\underline{a}_1$ and $\underline{a}_2$ lying on this plane. Then the vector $\underline{K}_o$ will be written as

$$\underline{K}_o = \underline{x}_o + \underline{K}(\Theta_1 \underline{a}_1 + \Theta_2 \underline{a}_2) \quad . \tag{12.19}$$

The values $K\Theta_1$ and $K\Theta_2$ define the position of the vector of the incident wave relative to the Laue point and are true independent parameters of the theory. They depend exclusively on the experimental conditions, namely the geometrical position of the crystal relative to the incident beam. The quantities Θ_1 and Θ_2, expressed in seconds of arc, define the angular position of the incident beam (see quantity η, introduced in Chapter 3). Owing to the boundary conditions, the vectors $\underline{K}_o$ and $\underline{k}_o$ are related by the expression (see Chapter 3)

$$\underline{k}_o = \underline{K}_o - K\delta\underline{n}_o \quad . \tag{12.20}$$

where $\underline{n}_o$ is an inward-drawn normal to the entrance surface.

The quantity δ is directly related to ε_o:

$$\varepsilon_o = \frac{|\underline{k}_o| - K}{K} = -\delta(\underline{s}_o\underline{n}_o) = -\delta\gamma_o \quad . \tag{12.21}$$

In deriving (12.21), as well as (12.18), we neglect small additions of the order of $|\chi_o|^2$. We now express the values of ε_m. By definition,

$$\varepsilon_m = \frac{|\underline{k}_m| - K}{K} \approx \frac{1}{2}\alpha_m - \delta(\underline{s}_m\underline{n}_o) \quad . \tag{12.22}$$

The first term in (12.22) has the following form, taking into account (12.19) and keeping in mind that $(\underline{x}_o + \underline{h}_m)^2 = K^2$:

$$\frac{1}{2}\alpha_m = \frac{1}{2}\,\frac{(\underline{K}_o + \underline{h}_m)^2 - K^2}{K^2} = (\underline{a}_1\underline{s}_m)\theta_1 + (\underline{a}_2\underline{s}_m)\theta_2$$

$$= \frac{1}{K}\,[(\underline{a}_1\underline{h}_m)\theta_1 + (\underline{a}_2\underline{h}_m)\theta_2] \quad . \tag{12.23}$$

Denote $(\underline{s}_m\underline{n}_o) = \gamma_m$. As a result we obtain the expression

$$\varepsilon_m = -\delta\gamma_m + \frac{1}{K}\,[(\underline{a}_1\underline{h}_m)\theta_1 + (\underline{a}_2\underline{h}_m)\theta_2] \quad . \tag{12.24}$$

Thus, we have introduced a new system of coordinates with an origin at the Laue point and with vectors $\underline{x}_m = \underline{x}_o + \underline{h}_m$. The position of the excitation point A relative to the Laue point is defined by the vector

$$\underline{\overline{LA}} = K(\delta\underline{n}_o - \theta_1\underline{a}_1 - \theta_2\underline{a}_2) \quad . \tag{12.25}$$

Now consider a multiwave case with N > 3 strong waves. For it to be realized, four or more points of the reciprocal crystal lattice need to lie near the Ewald sphere. Of greatest interest is the particular case where all the vectors of the reciprocal crystal lattice $\underline{h}_m$ lie on the same plane and form a closed polygon inscribed in a circle. This case is the more frequent, the higher is the symmetry of the crystal. Here, all the conceptions developed above for the three-wave case are preserved. By arbitrary choice of two vectors from the system $\underline{h}_m$ we can determine the Lorenz point L_0 from (12.7). The main axis is perpendicular to the plane of the reciprocal lattice vectors, and passes through the centre of the circumscribed circle (Fig. 12.2). The distance between the Laue and Lorenz points is determined by (12.9).

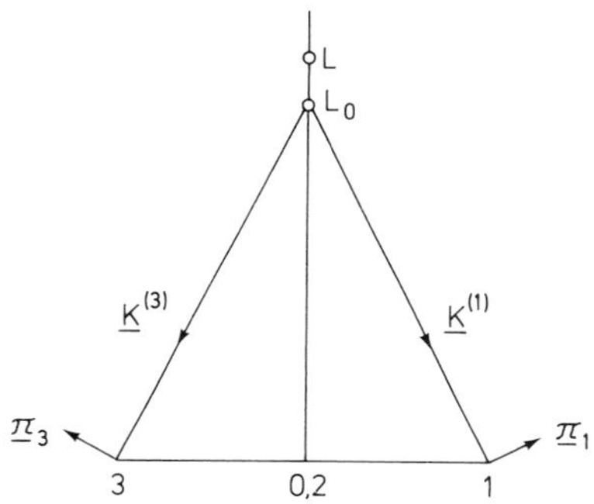

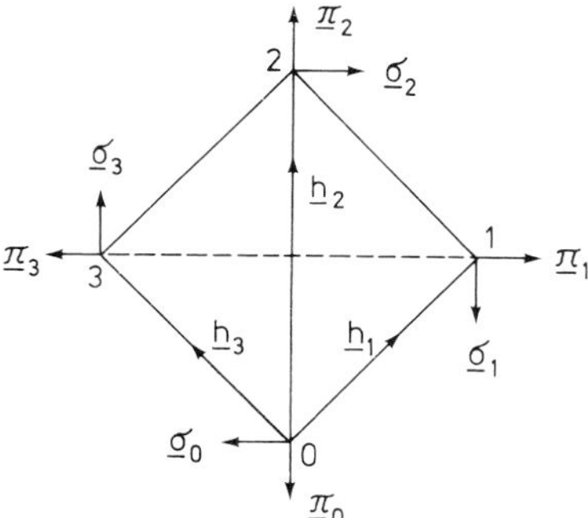

Fig. 12.2 Choice of polarization vectors in the four-wave case. L_0 - Lorenz point, L - Laue point

Because of the three dimensionality of space, however, the dispersion surface can be determined by the dispersion equation only when some additional conditions are satisfied. Let us select three linearly independent vectors $\underline{s}_0$, $\underline{s}_1$, $\underline{s}_2$ out of a system of unit vectors $\underline{s}_n$. The remaining vectors will be denoted by the subscript m, and $\underline{s}_m$ will be expressed in terms of $\underline{s}_n$

$$\underline{s}_m = \sum_{n=o}^{2} c_{mn}\underline{s}_n \quad , \quad m = 3 ,\ldots., N - 1 \quad . \tag{12.26}$$

The wave vectors $\underline{k}_m$ referring to the excitation point A are, as before, defined by (12.15) and (12.17). Not all the values ε_m are independent, however. Taking into account (12.26), we have

$$\varepsilon_m = \frac{1}{2}\,(\chi_o - \underline{s}_m\underline{v}) = \frac{1}{2}\left(\chi_o - \sum_{n=o}^{2} c_{mn}\tau_n\right) \quad , \quad m \geq 3 \quad . \tag{12.27}$$

If we formally determine τ_m from (12.16), from (12.27) we obtain the relation

$$\tau_m = \chi_o - 2\varepsilon_m = \sum_{n=o}^{2} c_{mn}\tau_n \quad , \quad m \geq 3 \quad . \tag{12.28}$$

As regards (12.19-25), they are, of course, generalized for this case.

For complete determination of the wave field in a crystal arising on multi-wave diffraction, one must also know the value and direction of the induction vectors $\underline{D}_m$. Because of the induction, waves are transverse, the vector $\underline{D}_m$ is always perpendicular to $\underline{s}_m$ (with an accuracy to small values of the order of $|\chi_o|$), and hence it has only two independent components. They can be assigned in different coordinate systems, but the most convenient system of vectors is the one where they are directed along the axes of symmetry of a polygon. Introduce unit vectors $\underline{\sigma}_m$, which, by definition, lie on the plane of the reciprocal lattice vectors $\underline{h}_m$ and are perpendicular to $\underline{s}_m$

$$\underline{\sigma}_m = \frac{\underline{\sigma}'_m}{|\underline{\sigma}'_m|} \quad , \quad \underline{\sigma}'_m = k\left[\underline{s}_m \times [\underline{h}_1 \times \underline{h}_2]\right] \quad . \tag{12.29}$$

We express the vector $\underline{\sigma}'_m$ in terms of the reciprocal lattice vectors $\underline{h}_m$. To do this, we substitute (12.2), (12,3), and (12.7) in (12.29). By carrying out certain simplifying transformations we obtain the following expression

$$\underline{\sigma}'_m = \underline{h}_1\left[(\underline{h}_2\underline{h}_m) - \frac{1}{2}\,h_2^2\right] - \underline{h}_2\left[(\underline{h}_1\underline{h}_m) - \frac{1}{2}\,h_1^2\right] \quad . \tag{12.30}$$

The vectors $\underline{\pi}_m$ will be defined in such a way that the system of vectors $\underline{\pi}_m$, $\underline{\sigma}_m$, $\underline{s}_m$ form a right-handed orthogonal system, namely

$$\underline{\pi}_m = [\underline{\sigma}_m \times \underline{s}_m] \quad . \tag{12.31}$$

The vectors $\underline{\pi}_m$, $\underline{\sigma}_m$ are shown in Fig. 12.2. They are called polarization vectors. The induction vector is expressed through them as follows

$$\underline{D}_m = D_{m\pi}\underline{\pi}_m + D_{m\sigma}\underline{\sigma}_m \quad . \tag{12.32}$$

The quantities $D_{m\pi}$, $D_{m\sigma}$ will be called the scalar amplitudes of the multi-wave field.

12.2 System of Fundamental Equations in the Case of Three Strong Waves, and the Dispersion Surface Equation [1.31]

In the case of three strong waves, the general system (2.49), with an allowance for (12.16), can be written

$$\begin{aligned} &\tau_o\underline{D}_o + \chi_{01}\underline{D}_{1[o]} + \chi_{02}\underline{D}_{2[o]} = 0 \\ &\chi_{10}\underline{D}_{o[1]} + \tau_1\underline{D}_1 + \chi_{12}\underline{D}_{2[1]} = 0 \\ &\chi_{20}\underline{D}_{o[2]} + \chi_{21}\underline{D}_{1[2]} + \tau_2\underline{D}_2 = 0 \quad . \end{aligned} \tag{12.33}$$

Represent each of the vectors $\underline{D}_{m[n]}$ as

$$\underline{D}_{m[n]} = \underline{D}_m - \underline{D}_{m||n} \tag{12.34}$$

and rewrite the system (12.33):

$$\tau_o\underline{D}_o + \chi_{01}\underline{D}_1 + \chi_{02}\underline{D}_2 = \lambda_o\underline{s}_o$$

$$\chi_{10}\underline{D}_o + \tau_1\underline{D}_1 + \chi_{12}\underline{D}_2 = \lambda_1\underline{s}_1$$

$$\chi_{20}\underline{D}_o + \chi_{21}\underline{D}_1 + \tau_2\underline{D}_2 = \lambda_2\underline{s}_2 \quad . \tag{12.35}$$

In this section we shall consider the case where the vectors $\underline{s}_m$ do not lie on the same plane. In this case the values λ_m are not identically equal to zero, and therefore the determinant

$$\Delta_3 = \begin{vmatrix} \tau_o & \chi_{01} & \chi_{02} \\ \chi_{10} & \tau_1 & \chi_{12} \\ \chi_{20} & \chi_{21} & \tau_2 \end{vmatrix} \neq 0 \quad . \tag{12.36}$$

The formal solution of (12.35) for the induction vectors has the form

$$\Delta_3\underline{D}_o = \begin{vmatrix} \lambda_o\underline{s}_o & \chi_{01} & \chi_{02} \\ \lambda_1\underline{s}_1 & \tau_1 & \chi_{12} \\ \lambda_2\underline{s}_2 & \chi_{21} & \tau_2 \end{vmatrix} \quad , \quad \Delta_3\underline{D}_1 = \begin{vmatrix} \tau_o & \lambda_o\underline{s}_o & \chi_{02} \\ \chi_{10} & \lambda_1\underline{s}_1 & \chi_{12} \\ \chi_{20} & \lambda_2\underline{s}_2 & \tau_2 \end{vmatrix}$$

$$\Delta_3\underline{D}_2 = \begin{vmatrix} \tau_o & \chi_{01} & \lambda_o\underline{s}_o \\ \chi_{10} & \tau_1 & \lambda_1\underline{s}_1 \\ \chi_{20} & \chi_{21} & \lambda_2\underline{s}_2 \end{vmatrix} \quad . \tag{12.37}$$

In order to obtain the explicit form of the values $\underline{D}_m$ we must exclude the unknown coefficients λ_m. To do this, we multiply each of the expressions (12.37) by the vector $\underline{s}_m$ and keep in mind that $\underline{s}_m\underline{D}_m = 0$. Then we obtain a system of linear homogeneous equations for λ_m. This system has a solution provided the determinant composed of the coefficients at λ_m is equal to zero. If we introduce the notation $s_{mn} = \underline{s}_m\underline{s}_n$, $s_{mm} = 1$, this determinant can be written

$$\Delta = \begin{vmatrix} \begin{vmatrix} \tau_1 & \chi_{12} \\ \chi_{21} & \tau_2 \end{vmatrix} & -s_{10}\begin{vmatrix} \chi_{01} & \chi_{02} \\ \chi_{21} & \tau_2 \end{vmatrix} & s_{20}\begin{vmatrix} \chi_{01} & \chi_{02} \\ \tau_1 & \chi_{12} \end{vmatrix} \\ -s_{01}\begin{vmatrix} \chi_{10} & \chi_{12} \\ \chi_{20} & \tau_2 \end{vmatrix} & \begin{vmatrix} \tau_o & \chi_{02} \\ \chi_{20} & \tau_2 \end{vmatrix} & -s_{12}\begin{vmatrix} \tau_o & \chi_{02} \\ \chi_{10} & \chi_{12} \end{vmatrix} \\ s_{02}\begin{vmatrix} \chi_{10} & \tau_1 \\ \chi_{20} & \chi_{21} \end{vmatrix} & -s_{12}\begin{vmatrix} \tau_o & \chi_{01} \\ \chi_{20} & \chi_{21} \end{vmatrix} & \begin{vmatrix} \tau_o & \chi_{01} \\ \chi_{10} & \tau_1 \end{vmatrix} \end{vmatrix} = 0 \ . \tag{12.38}$$

Equation (12.38) is the dispersion equation for the three-wave case. Taking into consideration (12.16) and (12.24), it is easy to see that it is a sixth degree equation with respect to δ, provided all $\gamma_m \neq 0$. The region of essentially dynamical scattering is determined by the range of the parameters $\theta_m \sim |\chi_o|$. In this case all the values of τ_m in (12.38) are of the same order and hence all the elements of the determinant have approximately equal values.

Introduce the following additional notation:

$$\begin{aligned} &s_{01} = s_1 \ , \quad s_{02} = s_2 \ , \quad s_{12} = s_3 \ , \quad s_1 s_2 s_3 = s^3 \ , \\ &\tau_o \tau_1 \tau_2 = \tau_{012} \ , \quad \tau_{ik} = \tau_i \tau_k \ , \\ &\chi_{01}\chi_{10} = \chi_1^2 \ , \quad \chi_{02}\chi_{20} = \chi_2^2 \ , \quad \chi_{12}\chi_{21} = \chi_3^2 \ , \\ &2c = \chi_{01}\chi_{12}\chi_{20} + \chi_{02}\chi_{21}\chi_{10} \end{aligned} \tag{12.39}$$

and rewrite (12.38) as a series in powers of τ, taking into account (12.39).

$$\begin{aligned} \Delta = \ &\tau_{012}^2 - \tau_{012}[\tau_o \chi_3^2(1 + s_3^2) + \sim + \sim] + \tau_{012} 2c(1 + s^3 - v_s^2) \\ &+ (\tau_o^2 \chi_3^2 s_3^2 + \sim + \sim) + [\tau_{01}\chi_2^2\chi_3^2(1 + 2s^3 - s_1^2) + \sim + \sim] \\ &- 2c[\tau_o \chi_3^2(s_3^2 + s^3) + \sim + \sim] + 4c^2 s^3 - \chi_1^2\chi_2^2\chi_3^2 v_s^2 = 0 \ , \end{aligned} \tag{12.40}$$

$$v_s^2 = 1 + 2s^3 - s_1^2 - s_2^2 - s_3^2 \quad . \tag{12.41}$$

The quantity v_s is the volume of a three-wedge unit cell with periods $a = b = c = 1$ (see also (12.10)). The symbol $\sim$ implies circular permutation of the subscripts $0 \to 1 \to 2 \to 0$ at the quantities τ and simultaneous permutation of the subscripts $1 \to 3 \to 2 \to 1$ at the quantities χ and s. Thus, the sum of the subscripts at τ and χ remains unchanged on permutation. The values χ_m and s_m depend only on the respective reciprocal lattice vectors $\underline{h}_m$, and the role of the vector $\underline{h}_3$ is played by the difference $(\underline{h}_2 - \underline{h}_1)$. In the transparent crystal, the values χ_m are strictly real and equal to the moduli of the respective coupling coefficients with an accuracy to the sign. The relation (12.39) does not determine the sign of the values χ_m. Henceforward we will assume that χ_m have the same sign as χ_o, i.e., they are negative.

12.3 Another Method of Deriving the Dispersion Surface Equation in a Three-Wave Case [1.36]

The method developed in the preceding section enables one to express the sixth-order dispersion determinant Δ as a third-order determinant whose elements are second-order determinants. In this section we shall obtain another form for using the property of its invariance with respect to the transformation of the coordinate system.

Express the induction vectors $\underline{D}_m$ in terms of the scalar amplitude $D_{m\pi}$ and $D_{m\sigma}$ according to (12.32) and substitute these expressions in (12.35). Then multiply the equation with the vector $\underline{s}_m$ in the right-hand side by $\underline{\pi}_m$ and $\underline{\sigma}_m$. The terms with λ_m drop out. As a result, we have a system of six equations

$$\begin{aligned} &\tau_m D_{m\pi} + \sum_{n \neq m} \chi_{mn}[(\underline{\pi}_m \underline{\pi}_n)D_{n\pi} + (\underline{\pi}_m \underline{\sigma}_n)D_{n\sigma}] = 0 \\ &\tau_m D_{m\sigma} + \sum_{n \neq m} \chi_{mn}[(\underline{\sigma}_m \underline{\pi}_n)D_{n\pi} + (\underline{\sigma}_m \underline{\sigma}_n)D_{n\sigma}] = 0 \end{aligned} \tag{12.42}$$

where m,n = 0,1,2. This system has a solution, provided the determinant composed of the coefficients at $D_{n\pi,\sigma}$, i.e., Δ, is equal to zero. Note that although the determinant elements contain polarization vectors, which can be determined in different ways, the value of Δ is independent of the specific choice of polarization vectors. As can easily be seen from (12.38), the determinant Δ depends only on τ_m, χ_{mn}, and s_{mn}.

To determine the explicit form of Δ in the invariant modification, we can do the following. Compose, from the values s_{mn}, all the possible independent invariants of such a type that the vectors $\underline{s}_m$ appear twice in each of them and the total number of vectors does not exceed six. Then represent Δ as a linear combination of these invariants, whose coefficients depend only on the values of τ_m and χ_{mn}. The specific form of these coefficients will be found from the consideration of certain cases in which the determinant Δ can easily be computed directly.

There is, however, a simpler way. First consider the coplanar case where all the vectors $\underline{s}_m$ lie on a single plane. The vectors $\underline{h}_1$, $\underline{h}_2$, and all the $\underline{\sigma}_m$ lie on the very plane, and $(\underline{\sigma}_m\underline{\sigma}_n) = (\underline{s}_m\underline{s}_n)$. The vectors $\underline{\pi}_m$ are equal to each other and perpendicular to this plane, i.e., $(\underline{\pi}_m\underline{\sigma}_n) = 0$. Then, we see at once from (12.42) that the system in this case is greatly simplified, and

$$\Delta = \Delta_3\Delta_{3s} \tag{12.43}$$

where Δ_3 is defined by (12.36), while Δ_{3s} is obtained from Δ_3 by replacing all the χ_{mn} with $\chi_{mn}s_{mn}$.

We further note that the general expansion of Δ in invariants differs from (12.43) in that only such additional invariants are present which vanish to zero in the coplanar case. We can ascertain directly that there is only one such invariant v_s^2. Hence, in a general case

$$\Delta = \Delta_3\Delta_{3s} + v_s^2 B \quad . \tag{12.44}$$

To find the coefficient B we shall consider the case in which the three vectors $\underline{s}_o$, $\underline{s}_1$, $\underline{s}_2$ are mutually perpendicular. In this case we find at once from (12.42), that

$$\Delta = \Delta_{01}\Delta_{02}\Delta_{12} \ , \quad \Delta_{nm} = \begin{vmatrix} \tau_n & \chi_{nm} \\ \chi_{mn} & \tau_m \end{vmatrix} \quad . \tag{12.45}$$

On the other hand, $\Delta_{3s} = \tau_o\tau_1\tau_2$, while Δ remains unchanged. As a result

$$B = \Delta_{01}\Delta_{02}\Delta_{12} - \tau_o\tau_1\tau_2\Delta_3 \quad . \tag{12.46}$$

Thus, we obtain the following expressions for

$$\Delta = \Delta_3\Delta_{3s} + v_s^2(\Delta_{01}\Delta_{02}\Delta_{12} - \tau_o\tau_1\tau_2\Delta_3) \quad . \tag{12.47}$$

It is easy to ascertain, by rearranging (12.47) according to the increasing powers of τ, that it coincides identically with (12.40). The form (12.47) for the dispersion determinant is of interest in its own right, since in near-coplanar cases, when $v_s^2 \ll 1$, the second term in (12.47) is small as compared to the first one and can be taken into account on the basis of the perturbation theory. Eq. (12.47) was first obtained by PENNING [1.33].

12.4 Analysis of the Dispersion Surface Equation in the Case of Nonabsorbing Crystal

1) Dependence on Permutations (Numbers) of Vectors k_n. As is to be expected, the dispersion surface is invariant to circular permutations of the subscripts 0,1,2 of the wave vectors. This follows immediately from the recording of (12.40) for Δ, in which the expressions in the square brackets are formed by the circular permutation rule, while the others are altogether independent of the subscripts. In noncircular permutations the direction of the main axis changes, and the dispersion surface is constructed on the reverse side of the plane of the vectors $\underline{h}_1$ and $\underline{h}_2$.

2) Phase Relations. From (12.40) it follows that the phases of the coupling coefficients $\chi_{mn} = |\chi_{mn}| \exp i\varphi_{mn}$ enter only via the value

$$2c = \chi_{01}\chi_{12}\chi_{20} + \chi_{02}\chi_{21}\chi_{10} = 2\chi_1\chi_2\chi_3 \cos\varphi \tag{12.48}$$

$$\varphi = \varphi_{01} + \varphi_{12} + \varphi_{20} + \pi \quad . \tag{12.49}$$

Thus, the separate phases exert no effect on the dispersion surface and, hence, the dispersion surface geometry can yield only the sum of the phases φ. In a crystal with an inversion centre, $\varphi = 0$.

In this connection we shall make a fundamental remark bearing on the applicability of the Friedel law in three-wave scattering. As has been specially emphasized at the end of Chapter 4, in two-wave scattering, the Friedel law is violated only in the case of an absorbing crystal without an inversion centre. In contrast, in three-wave scattering, the Friedel law is also violated in the case of transparent crystal. Indeed, upon inversion (reorientation) of the polar direction in a noncentro-symmetrical crystal, the sign of φ changes and, naturally, so does the reflection power factor [12.39].

3) Centro-Symmetry of the Dispersion Surface occurs only on condition $c = 0$, since the quantity c appears on both terms with an odd power of τ in (12.40). As follows from (12.48), $c = 0$ either when the equality $\varphi = \pi/2$ holds good or when one of the values χ_m vanishes. The centre of symmetry of the dispersion surface in the cases indicated is obviously point L_o.

4) The Consequence of χ_3 Vanishing. Consider the section of the dispersion surface by the plane of the vectors $\underline{t}_o$ and $\underline{t}_1 + \underline{t}_2$, i.e., put $\tau_1 = \tau_2$. The dispersion equation in this plane takes the form

$$\tau_1^2(\tau_{01}^2 - \tau_{01}2X + Y) = 0 \tag{12.50}$$

$$2X = \chi_1^2(1 + s_1^2) + \chi_2^2(1 + s_2^2)$$

$$Y = \chi_1^4 s_1^2 + \chi_2^4 s_2^2 + \chi_1^2\chi_2^2(1 + 2s^3 - s_3^2) \quad . \tag{12.51}$$

According to (12.50), the section of the dispersion surface in this case represents two coincident straight lines parallel to $\underline{t}_o$ and passing through the Lorenz point L_o, and also two pairs of hyperbolas described by the equations

$$\tau_o\tau_1 = X \pm \sqrt{X^2 - Y} \quad , \qquad \tau_2 = \tau_1 \quad . \tag{12.52}$$

On the main axis $\tau_2 = \tau_1 = \tau_o$ we obtain six points with coordinates

$$\tau_o^{(1,2)} = 0 \quad , \quad \tau_o^{(3,4,5,6)} = \pm\sqrt{X \pm \sqrt{X^2 - Y}} \quad . \tag{12.53}$$

One can see at once that all the values of (12.53) are real, since $X > 0$, $Y > 0$, $X^2 > Y$. In the symmetrical case, when $|\underline{h}_1| = |\underline{h}_2|$, i.e., $\chi_1 = \chi_2$, $s_1 = s_2$, (12.52) will be simplified still further

$$\tau_o\tau_1 = \chi_1^2 \begin{cases} 1 + 2s_1^2 - s_3 \\ 1 + s_3 \end{cases} , \quad \tau_2 = \tau_1 \quad . \tag{12.54}$$

On the main axis, in this case we obtain

$$\tau_o^{(3,4,5,6)} = \pm \chi_1 \begin{cases} \sqrt{1 + 2s_1^2 - s_3} \\ \sqrt{1 + s_3} \end{cases} . \tag{12.55}$$

5) Degenerate Points on the Dispersion Surface in the Case $\varphi = 0$ *(Crystal with Inversion Centre)*. Let us pass on from the variables τ to the new variables z

$$\tau_o = z_o \frac{\chi_1\chi_2}{\chi_3} , \quad \tau_1 = z_1 \frac{\chi_1\chi_3}{\chi_2} , \quad \tau_2 = z_2 \frac{\chi_2\chi_3}{\chi_1} \tag{12.56}$$

and write down the dispersion surface equation in the form of (12.47) with an allowance for (12.48)

$$\begin{aligned} \Delta' = (\chi_1\chi_2\chi_3)^{-2}\Delta = (2 + z_{012} - z_o - z_1 - z_2)(2s^3 + z_{012} \\ - z_o s_3^2 - z_1 s_2^2 - z_2 s_1^2) - v_s^2(1 + 2z_{012} - z_{01} - z_{02} - z_{12}) = 0 \quad . \end{aligned} \tag{12.57}$$

Using direct substitution it is possible to ascertain that on fulfillment of each of the conditions

$$z_1 = z_2 = 1 , \quad z_o = z_2 = 1 , \quad z_o = z_1 = 1 \tag{12.58a}$$

(12.57) is identically satisfied at any values of the third coordinate. Consequently, on the dispersion surface there lie three straight lines parallel to the vectors $\underline{t}_o$, $\underline{t}_1$, $\underline{t}_2$, respectively.

Consider, further, the points of intersection with the line $z_2 = z_1 = z_o$. Eq. (12.57) is transformed on this line to the form

$$\Delta' = (z_o - 1)^3[z_o(z_o + 1)(z_o + 2) + v_s^2(z_o + 1) - 2s^3(z_o + 2)] = 0 \quad . \tag{12.58b}$$

Thus, all the three straight lines (12.58) intersect at a single point, and this is the triple degeneration point. In the symmetrical case, when $|\underline{h}_1| = |\underline{h}_2| = |\underline{h}_3|$ and, hence, all the χ_m and s_m are equal, to the point $z_2 = z_1 = z_o = 1$ there corresponds $\tau_2 = \tau_1 = \tau_o = \chi_1$, i.e., the triple degeneration point lies on the main axis, at a distance of $|\chi_o - \chi_1|$ from the Laue point. Taking into consideration that in this case $v_s^2 = 1 + 2s^3 - 3s^2$, the expression in brackets in (12.58) can be transformed to

$$[\ldots] = (z_o + 1 + s)^2(z_o + 1 - 2s) \tag{12.59}$$

from which it follows that on the main axis there is also a double degeneration point at a distance of $|\chi_o + \chi_1(1 + s)|$ from the Laue point.

Note that the double degeneration point exists in the nonsymmetrical case as well [1.33]. Its coordinates are equal to

$$z_o = -\frac{1}{\omega_3}\left(\omega_1 + \omega_2\frac{s_1 s_3}{s_2}\right) \quad , \quad z_1 = -\frac{1}{\omega_2}\left(\omega_3 + \omega_1\frac{s_2 s_3}{s_1}\right) \quad ,$$

$$z_2 = -\frac{1}{\omega_1}\left(\omega_2 + \omega_3\frac{s_1 s_2}{s_3}\right) \quad , \quad \omega_m = \frac{s_m^2}{s_m^2 - s^3} \quad . \tag{12.60}$$

As proof of this statement is rather complicated, it is not given here.

6) Point Lattice. In some cases it is of interest to consider scattering in a point lattice for which all χ_m have the same value and are equal to χ_o, while $\varphi = 0$. (For instance, scattering of resonance gamma-quanta). This case corresponds, in EWALD's theory, to rigidly fixed point dipoles. According to the results of the preceding section, three sheets of the dispersion surface always pass through the Laue point in this case.

12.5 Deriving the Dispersion Surface Equation in the Case of Four Strong Waves [1.34]

In the four-wave case the system of (2.49) is also transformed (12.35)

$$\tau_m \underline{D}_m + \sum_{n \neq m} \chi_{mn} \underline{D}_n = \lambda_m \underline{s}_m \quad , \qquad m,n = 0,1,2,3 \tag{12.61}$$

which can be written as (12.42) if we use polarization vectors. Assuming, further, that the determinant

$$\Delta_4 = \begin{vmatrix} \tau_0 & \chi_{01} & \chi_{02} & \chi_{03} \\ \chi_{10} & \tau_1 & \chi_{12} & \chi_{13} \\ \chi_{20} & \chi_{21} & \tau_2 & \chi_{23} \\ \chi_{30} & \chi_{31} & \chi_{32} & \tau_3 \end{vmatrix} \neq 0 \tag{12.62}$$

we can write the formal solution of (12.61). Thus

$$\underline{D}_m = \underline{\Delta}_4^{(m)} / \Delta_4 \tag{12.63}$$

where $\underline{\Delta}_4^{(m)}$ differs from Δ_4 by replacement of the m-th column with the product $\lambda_n \underline{s}_n$. Multiplying (12.63) by s_m and equating the result to zero, we obtain a system of linear homogeneous equations with respect to λ_m. The determinant, which is composed of the coefficients at λ_m, must be equal to zero. Note, however, that this condition does not determine the true dispersion equation in the four-wave case.

Indeed, the determinant obtained is a fourth-order determinant composed of third-order determinants. Consequently, upon its expansion the senior power of τ will be equal to twelve, whereas in the true dispersion equation the senior power of τ is equal to eight. This paradox is due to the fact that we ignore the three dimensionality of space, and in particular, of (12.26).

In deriving the dispersion surface equation in the four-wave case, the "direct" method developed in Section 12.3 is more convenient. Expanding the dispersion determinant in the invariants, we can write at once

$$\Delta = \Delta_4 \Delta_{4s} + \sum_m B_m(\tau,\chi) c_m \tag{12.64}$$

where Δ_{4s} differs from Δ_4 by replacement of χ_{mn} with $\chi_{mn} s_{mn}$, while c_m are independent invariants vanishing to zero in the coplanar case. Since the two vectors $\underline{s}_m$ always lie on the same plane, the desired invariants must contain three or four vectors $\underline{s}_m$ simultaneously. There are seven invariants of this type

$$\Omega^2_{012} \quad , \quad \Omega^2_{013} \quad , \quad \Omega^2_{023} \quad , \quad \Omega^2_{123} \quad , \quad \Omega_{kmn} = (\underline{s}_k[\underline{s}_m \times \underline{s}_n]) \quad , \tag{12.65}$$

$$V^2_{01} \quad , \quad V^2_{02} \quad , \quad V^2_{03} \quad , \quad V_{nm} = ([\underline{s}_m \times \underline{s}_n][\underline{s}_k \times \underline{s}_\ell]) \quad ,$$

$$k,m,n,\ell = 0,1,2,3 \quad , \quad k \neq m \neq n \neq \ell \quad . \tag{12.66}$$

Let us now take into account the three dimensionality of space due to which, for instance,

$$\underline{s}_3 = \sum_{n=o}^{2} (\underline{t}_n \underline{s}_3) \underline{s}_n \quad . \tag{12.67}$$

Hence, the value

$$\Omega_{123} = ([\underline{s}_1 \times \underline{s}_2]\underline{s}_3) = (\underline{t}_o \underline{s}_3)([\underline{s}_1 \times \underline{s}_2]\underline{s}_o) = (\underline{t}_o \underline{s}_3)\Omega_{012} \quad . \tag{12.68a}$$

Thus, only three out of the four invariants (12.65) are linearly independent. It is convenient to choose the following six linear combinations of (12.65) and (12.66) as linear-independent invariants

$$C_{01}, C_{02}, C_{03}, C_{12}, C_{13}, C_{23} \quad ,$$

$$C_{mn} = \frac{1}{2}(V^2_{mk} + V^2_{m\ell} - \Omega^2_{mnk} - \Omega^2_{mn\ell}) \quad . \tag{12.68b}$$

The expansion (12.64) with an allowance for (12.68) takes a more specific form

$$\Delta = \Delta_4\Delta_{4s} + \sum_{m<n} C_{mn}B_{mn}(\tau,\chi) \quad . \tag{12.69}$$

In order to find the unknown coefficients B_{mn} in (12.69) we consider six simple cases in which it is easy to calculate the determinant of a system of the type (12.42). Let, for instance,

$$\underline{s}_o \perp \underline{s}_1 \perp \underline{s}_2 \perp \underline{s}_3 \quad . \tag{12.70}$$

In this case we can choose the polarization vectors $\underline{\sigma}_n$ and $\underline{\pi}_n$ in the following way

$$\underline{\sigma}_2 = \underline{\pi}_3 = \underline{s}_o \quad , \quad \underline{\sigma}_3 = \underline{\pi}_o = \underline{\pi}_1 = \underline{s}_2 \quad , \quad \underline{\sigma}_o = \underline{\sigma}_1 = \underline{\pi}_2 = \underline{s}_3 \quad . \tag{12.71}$$

Here, the system of equations for $D_{n\pi,\sigma}$ decomposes into three independent subsystems, and the general determinant is equal to the product of the determinants of these subsystems

$$\Delta = \Delta_{23}\Delta_{012}\Delta_{013} \tag{12.72}$$

where

$$\Delta_{kmn} = \begin{vmatrix} \tau_k & \chi_{km} & \chi_{kn} \\ \chi_{mk} & \tau_m & \chi_{mn} \\ \chi_{nk} & \chi_{nm} & \tau_n \end{vmatrix} \quad , \quad \Delta_{mn} = \begin{vmatrix} \tau_m & \chi_{mn} \\ \chi_{nm} & \tau_n \end{vmatrix} \quad . \tag{12.73}$$

On the other hand, in this case $C_{01} = 1$, while the other $C_{mn} = 0$. Calculating the determinant Δ_{4s}, we find $\Delta_{4s} = \Delta_{01}\tau_2\tau_3$. Thus,

$$B_{01} = \Delta_{012}\Delta_{013}\Delta_{23} - \Delta_{01}\tau_2\tau_3\Delta_4 \quad . \tag{12.74}$$

The other five coefficients are determined similarly from the consideration of particular cases of the type (12.70), in which the two vectors $\underline{s}_m$ are equal, while the others are mutually perpendicular. As a result, the dispersion equation in the case of four strong waves has the form

$$\Delta = \Delta_4\Delta_{4s} + \sum_{m<n} C_{mn}(\Delta_{mnk}\Delta_{mn\ell}\Delta_{k\ell} - \Delta_{mn}\tau_k\tau_\ell\Delta_4) = 0 \quad . \tag{12.75}$$

For the values of C_{mn} we can also obtain the following expressions

$$\begin{aligned} C_{mn} &= \frac{1}{4}(\Omega^2_{mk\ell} + \Omega^2_{nk\ell} - \Omega^2_{mnk} - \Omega^2_{mn\ell} + 2\Omega_{mnk}\Omega_{mn\ell}s_{k\ell} + 2\Omega_{mk\ell}\Omega_{nk\ell}s_{mn}) \\ &= s^2_{mn} - s_{mn}s_{nk}s_{km} - s_{mn}s_{n\ell}s_{\ell m} + s_{mn}s_{nk}s_{k\ell}s_{\ell m} + s_{mn}s_{n\ell}s_{\ell k}s_{km} \\ &\quad - s^2_{mn}s^2_{k\ell} \quad . \end{aligned} \tag{12.76}$$

Note that the coefficient at τ_3^2 in (12.75) is precisely equal to the dispersion determinant for the three-wave case on the vectors $\underline{s}_o, \underline{s}_1, \underline{s}_2$ and can be obtained directly in the form of (12.47). However, because of the existance of (12.28), τ_3 cannot be infinitely large with finite values of the other τ_n. In other words, the transition from the four-wave to the three-wave case is impossible.

12.6 Coefficients of Transmission and Reflection for a Plane-Parallel Plate. Laue Reflection [1.35]

Let a plane wave with a wave vector $\underline{K}_o$

$$\underline{D}^{(a)} = \underline{D}_o^{(a)} \exp[2\pi i(\nu t - \underline{K}_o\underline{r})] \tag{12.77}$$

fall on the vacuum-crystal boundary (entrance surface). In a crystal, the eigensolution of the Maxwell equations is the sum of the Bloch waves (2.36)

$$\underline{D} = \sum_j \exp[2\pi i(\nu t - \underline{k}_o^{(j)}\underline{r})] \sum_m \underline{D}_m^{(j)} \exp(-2\pi i\underline{h}_m\underline{r}) \quad . \tag{12.78}$$

The summation over m extends only to those $\underline{h}_m$ which form strong waves. Let the plate be oriented so that all the parameters γ_m (12.24) are larger than zero (Laue geometry). In this case, at the vacuum-crystal boundary, the solutions (12.77) and (12.78) must coincide. Averaging over the boundary section with dimensions exceeding the interatomic distances, we find at once that this is only possible if

$$\underline{k}_o^{(j)} = \underline{K}_o - K\delta^{(j)}\underline{n}_o \quad ,$$

$$\sum_j \underline{D}_o^{(j)} = \underline{D}_o^{(a)} \quad , \quad \sum_j \underline{D}_m^{(j)} = 0 \quad , \quad m \neq 0 \quad . \tag{12.79}$$

Consequently, the only points on the dispersion surface which can be excited simultaneously in the plate are those of its intersection with a straight line parallel to $\underline{n}_o$, i.e., the vector of a normal to the entrance surface of the crystal. When determining the wave field in the crystal, in this case the most convenient coordinate system is the one with its centre at the Laue point (12.25).

Substituting (12.16) and (12.24) in the system of (12.42), we write it as follows

$$\alpha_n \underline{D}_{ns} - \sum_{ms'} \chi_{nm}^{ss'} \underline{D}_{ms'} = 2\delta\gamma_n \underline{D}_{ns} \tag{12.80}$$

where $\chi_{mn}^{ss'} = \chi_{mn}(\underline{e}_{ms}\underline{e}_{ns'})$, s, s' = π,σ. The vector $\underline{e}_{ms}$ is equal to $\underline{\pi}_m$ for $s = \pi$ and $\underline{\sigma}_m$ for $s = \sigma$. The values D_{ns} are the scalar amplitudes of the field (12.32), while the values α_m (see (12.23))

$$\alpha_m = \frac{2}{K}\left[(\underline{a}_1\underline{h}_m)\theta_1 + (\underline{a}_2\underline{h}_m)\theta_2\right] \tag{12.81}$$

characterize the deviation from Bragg condition in scattering on the vector $\underline{h}_m$.

Multiply (12.80) by the factors $(2\sqrt{\gamma_n})^{-1}$ and pass on to new amplitudes

$$\lambda \underline{B}_{ns} = \sqrt{\gamma_n}\underline{D}_{ns} \quad . \tag{12.82}$$

As a result we get

$$\sum_{ms'} G_{nm}^{ss'} \underline{B}_{ms'} = \delta \underline{B}_{ns} \quad , \tag{12.83}$$

$$G_{nm}^{ss'} = \frac{1}{2} \frac{1}{\sqrt{\gamma_n \gamma_m}} [\alpha_n \delta_{nm}^{ss'} - \chi_{nm}(\underline{e}_{ns}\underline{e}_{ms'})] \qquad (12.84)$$

where $\delta_{nm}^{ss'}$ (Kronecker symbol) is equal to unity at s = s' and n = m, and to zero in the opposite case. Thus, the possible values of $\delta^{(j)}$ are the eigenvalues of the matrix (12.84), and the set of renormalized scalar amplitudes of the multiwave field forms the eigenvector of this matrix. The matrix $\hat{G}$ will be called the scattering matrix.

On emerging from the crystal, the wave field decomposes into a set of noninteracting plane waves

$$\underline{D}_m^{(d)} = \exp[2\pi i(\nu t - \underline{K}_m \underline{r})] \sum_j \underline{D}_m^{(j)} \exp[2\pi i K\delta^{(j)} z] \qquad (12.85)$$

where z is the plate thickness, $\underline{K}_m = \underline{K}_o + \underline{h}_m$. The coefficients of transmission T and reflection R_m are determined in the usual way (see (3.62,63))

$$T = R_o \quad , \quad R_m = \frac{|D_m^{(d)}|^2}{|D_o^{(a)}|^2} \frac{\gamma_m}{\gamma_o} \quad . \qquad (12.86)$$

As can readily be seen from (12.84), in the nonabsorbing crystal the scattering matrix is of the Hermitian type, i.e., $G = G^+$. Hence it follows that all its eigenvalues are real. In other words, the positivity of all the γ_m is a necessary and adequate condition for a straight line parallel to $\underline{n}_o$ intersecting all the sheets of the dispersion surface. The condition for the vector $\underline{n}_o$ to be parallel to the main axis is the equality of all the γ_m, that is to say orientation of the crystal at which the plane of the reciprocal lattice vectors is parallel to the entrance surface of the plate. Another consequence of the Hermitian property of the scattering matrix is the orthogonality of its eigenvectors

$$\left(\underline{B}^{(j)^*} , \underline{B}^{(j')}\right) = \sum_{ns} B_{ns}^{(j)^*} B_{ns}^{(j')} = \delta_{jj'} \quad . \qquad (12.87)$$

In (12.87) we additionally assume that the vectors B_{ns}^j are normalized to unity, i.e., unambiguously defined. Multiplying B_{ns}^j by an arbitrary constant λ_j, we naturally also obtain the solution of (12.83).

To express the values of R_m in terms of the vectors $\underline{B}^{(j)}$, we turn to the boundary conditions (12.79). Passing on to the scalar amplitudes (12.32), we multiply the m-th equation by the factor $\sqrt{\gamma_n}$. Combining all the equalities, we get

$$\sum_j \lambda_j B_{ms}^{(j)} = \sqrt{\gamma_o} D_{os}^{(a)} \delta_{mo} \tag{12.88}$$

where $D_{os}^{(a)}$ are the components of decomposition of the vector $\underline{D}_o^{(a)}$ into vectors $\underline{\pi}_o$ and $\underline{\sigma}_o$. We further multiply (12.88) by $B_{ms}^{(j)*}$, sum up over all the ms, and take into account (12.88). Then the unknown coefficients

$$\lambda_j = \sqrt{\gamma_o} \sum_s D_{os}^{(a)} B_{os}^{(j)*} \quad . \tag{12.89}$$

Thus, we have obtained an important result, namely that the electromagnetic field of the j-th excitation mode depends exclusively on the properties of this mode, while the total field in the crystal is a simple sum of the fields for the separate modes [1.32]. Substituting (12.82) in (12.85), we finally obtain

$$R_m = \left| \sum_{js} \tilde{\lambda}_j B_{ms}^{(j)} \underline{e}_{ms} \exp[2\pi i K \delta^{(j)} z] \right|^2 \tag{12.90}$$

where

$$\tilde{\lambda}_j = \frac{\lambda_j}{\sqrt{\gamma_o} |\underline{D}_o^{(a)}|} = \cos\varphi B_{o\pi}^{(j)*} + \sin\varphi B_{o\sigma}^{(j)*} \quad . \tag{12.91}$$

Here, φ is the angle between the polarization plane of the incident vacuum wave and the vector $\underline{\pi}_o$.

Rewrite (12.90)

$$R_m(\varphi) = \sum_{jj's} B_{ms}^{(j)} B_{ms}^{(j')*} \exp\{2\pi i K[\delta^{(j)} - \delta^{(j')}]z\} c_{jj'}(\varphi) \tag{12.92}$$

where

$$c_{jj'}(\varphi) = \tilde{\lambda}_j\tilde{\lambda}^*_{j'} = \cos^2\varphi\, B^{(j)*}_{o\pi} B^{(j')}_{o\pi} + \sin^2\varphi\, B^{(j)*}_{o\sigma} B^{(j')}_{o\sigma}$$

$$+ \cos\varphi \sin\varphi \left[B^{(j)*}_{o\pi} B^{(j')}_{o\sigma} + B^{(j)*}_{o\sigma} B^{(j')}_{o\pi} \right] . \qquad (12.93)$$

From (12.92) it follows that the terms with $j \neq j'$ contain oscillating factors, and the oscillations of R_m occur both on changes in the plate thickness z and on changes in the distance of the Laue point, i.e., on changes in the parameters θ_1 and θ_2 via the difference $[\delta^{(j)} - \delta^{(j')}]$. Introducing the phases of the complex values $B^{(j)}_{ms} = |B^{(j)}_{ms}| \exp[i\psi^{(j)}_{ms}]$, we can write (12.92) in the form of a set of standing waves in the $\underline{n}_o$-direction; each wave will have its own period and its own dependence on φ. On the whole, the pattern of oscillations of the reflection coefficient will be rather complicated. If we average the crystal thickness over a distance large as compared with the oscillation period, the reflection coefficients will be determined only by the terms with $j = j'$ in the sum (12.92).

In the nonabsorbing crystal, the law of conservation of energy in scattering is naturally obeyed, and in this case it is written thus

$$\sum_m R_m = 1 \quad . \qquad (12.94)$$

To verify this, it will suffice to take the sum of (12.92) over m. Allowing for the orthonormalization of the $\underline{B}^{(j)}$ vectors, we obtain

$$\sum_m R_m = \sum_j c_{jj} = \sum_j |\lambda_j|^2 \frac{1}{\gamma_o |\underline{D}^{(a)}_o|^2} \quad . \qquad (12.95)$$

On the other hand, multiplying (12.90) by the conjugate complex value and using the orthonormalization of the $\underline{B}^{(j)}$ vectors again, we become convinced that the right-hand side of (12.95) is equal to unity.

Most experimenters use a nonpolarized X-ray beam. To compare the results of such experiments, one must average (12.92) over the angle φ

$$\bar{R}_m = \frac{1}{2\pi} \int_0^{2\pi} d\varphi R_m(\varphi) \quad . \qquad (12.96)$$

It is easy to see that here (12.92) holds good, except that $c_{jj'}(\varphi)$ must be replaced by

$$\bar{c}_{jj'} = \frac{1}{2}\left[B_{o\pi}^{(j)*}B_{o\pi}^{(j')} + B_{o\sigma}^{(j)*}B_{o\sigma}^{(j')}\right] . \qquad (12.97)$$

12.7 Scattering in Absorbing Crystal. Introducing Complex Parameters of Scattering and the Coefficient of Absorption [1.35]

As has been shown in Chapter 4, when we consider scattering in the absorbing crystal, the various inelastic scattering channels must be taken into account, the most important of which is the photoeffect. In calculating photoelectric absorption, wide use is made of the expansion in multipoles; in addition to the dipole term, which makes the main contribution, allowance should also be made for the quadrupole term, which shows a more intricate dependence on the polarization of the incident and scattered waves. Since the system (12.35), as well as the more general system (2.49), holds good only in the dipole approximation, when taking into account the quadrupole interaction it is more convenient to use the scattering matrix formalism developed in the preceding section.

In the absorbing crystal the scattering matrix becomes complex and, as a detailed analysis shows, has the following form with an allowance for quadrupole interaction

$$G_{mn}^{ss'} = \frac{1}{2}\frac{1}{\sqrt{\gamma_m\gamma_n}}\Big\{\alpha_n\delta_{mn}^{ss'} - \chi_{mn}^{D}(\underline{e}_{ms}\underline{e}_{ns'}) - \chi_{mn}^{Q}[(\underline{e}_{ms}\underline{e}_{ns'})(\underline{s}_m\underline{s}_n) + (\underline{e}_{ms}\underline{s}_n)(\underline{e}_{ns'}\underline{s}_m)]\Big\} . \qquad (12.98)$$

In our approximation, the real part of χ^D is determined, as before, by (2.34)

$$\chi_{mnr}^{D} = -\frac{e^2}{mc^2}\frac{\lambda^2}{\pi\Omega_o}\sum_j \exp[2\pi i(\underline{h}_m - \underline{h}_n)\underline{r}_j]\exp(-W_{jm-n})f_{jm-n} \qquad (12.99)$$

while the real part of χ^Q is equal to zero. The imaginary part of χ^D and

χ^Q are equal to (see (4.31-44)):

$$\chi^{D,Q}_{mni} = -(2\pi K\Omega_o)^{-1} \sum_j \exp[2\pi i(\underline{h}_m - \underline{h}_n)\underline{r}_j]\mu^{D,Q}_{a,j} \exp(-W_{jm-n}) \tag{12.100}$$

$$\chi_{oi} = -\frac{\mu_o}{2\pi K} = -(2\pi K\Omega_o)^{-1} \sum_j (\mu^D_{aj} + \mu^Q_{aj}) \quad . \tag{12.101}$$

Summation is performed over all the atoms in the unit cell of the crystal: $\underline{r}_j$ is the coordinate of the atom j; Ω_o is the volume of the unit cell; $\mu^{D,Q}_{a,j}$ are the cross sections of photoabsorption on the atom j; f_{jm} is the atomic scattering amplitude at T = 0 K; $\exp(-W_{jm})$ is the Debye-Waller factor for the atom j for scattering on the reciprocal lattice vector $\underline{h}_m$.

As can easily be seen from (12.100), the imaginary parts of the scattering coefficients also form a Hermitian matrix G_i, but the complete scattering matrix (12.98) is no longer Hermitian. The imaginary part of its eigenvalues $\delta^{(j)}_i$ determines, in accordance with (12.79), the imaginary part of the wave vectors

$$\underline{k}^{(j)}_{mi} = \underline{k}^{(j)}_{oi} = -K\delta^{(j)}_i \underline{n}_o \quad . \tag{12.102}$$

Thus, in penetrating the crystal, each excitation mode damps out exponentially with its own coefficient of absorption

$$\mu_j = 4\pi K\delta^{(j)}_i \quad . \tag{12.103}$$

Apart from μ_j, it is also possible to consider the coefficients of absorption in transmission along a different path, which does not coincide in direction with $\underline{n}_o$, for instance along the unit vector $\underline{\ell}$. Then,

$$\mu^{(\ell)}_j = \mu_j(\underline{n}_o\underline{\ell}) \quad . \tag{12.104}$$

The fulfillment of the following inequalities is typical of the interaction between X-rays and a crystal

$$|\chi_{mnr}| \gg |\chi^D_{mni}| \gg |\chi^Q_{mni}| \tag{12.105}$$

which permits us to use the perturbation theory to account for the contributions from χ_{mni}. Assume that the problem (12.87) for the transparent crystal has been solved. The corresponding solutions will be denoted by $\delta_o^{(j)}$ and $\underline{B}_o^{(j)}$. The exact solution of (12.87) will be written in the form of a series, and the formal expansion parameter is the imaginary unit

$$\delta^{(j)} = \delta_o^{(j)} + i\delta_1^{(j)} - \delta_2^{(j)} - i\delta_3^{(j)} + \dots$$

$$\underline{B}^{(j)} = \underline{B}_o^{(j)} + i\underline{B}_1^{(j)} - \underline{B}_2^{(j)} - \dots \quad . \tag{12.106}$$

The standard perturbation theory supplies the following expressions for the separate terms in the series in (12.106):

$$\delta_1^{(j)} = \left[\underline{B}_o^{(j)}, \ \hat{G}_i\underline{B}_o^{(j)}\right] \quad , \tag{12.107}$$

$$\underline{B}_1^{(j)} = \sum_{j' \neq j} \underline{B}_o^{(j')} \frac{1}{\left[\delta_o^{(j)} - \delta_o^{(j')}\right]}\left[\underline{B}_o^{(j')} \ , \ \hat{G}_i\underline{B}_o^{(j)}\right] \quad , \tag{12.108}$$

$$\delta_2^{(j)} = \left[\underline{B}_o^{(j)} \ , \ \hat{G}_i\underline{B}_1^{(j)}\right] \tag{12.109}$$

$$\underline{B}_2^{(j)} = \sum_{j' \neq j} \underline{B}_o^{(j')} \frac{1}{\left[\delta_o^{(j)} - \delta_o^{(j')}\right]}\left\{\left[\underline{B}_o^{(j')} \ , \ \hat{G}_i\underline{B}_1^{(j)}\right] - \delta_1^{(j)}\left[\underline{B}_o^{(j')} \ , \ \underline{B}_1^{(j)}\right]\right\} \tag{12.110}$$

$$\delta_3^{(j)} = \left[\underline{B}_o^{(j)} \ , \ \hat{G}_i\underline{B}_2^{(j)}\right] \quad . \tag{12.111}$$

In most cases it is sufficient to restrict oneself to the first order of the perturbuation theory. Then the eigenvectors are calculated in the transparent-crystal approximation, while the coefficients of absorption are determined by (12.107). Switching to the conventional notation and omitting the superscript j, we obtain

$$\mu = -\frac{2\pi K}{\sum_n (\underline{D}_n^*\underline{D}_n)\gamma_n} \sum_{nm} \left\{ \chi_{nmi}^{D}(\underline{D}_n^*\underline{D}_m) + \chi_{nmi}^{Q}[(\underline{D}_n^*\underline{D}_m)(\underline{s}_n\underline{s}_m) + (\underline{D}_n^*\underline{s}_m)(\underline{s}_n\underline{D}_m)] \right\} . \tag{12.112}$$

Rewrite this equation in a slightly different form which will come in handy later

$$\mu = \mu_o \frac{\sum_n (\underline{D}_n^*\underline{D}_n)}{\sum_n (\underline{D}_n^*\underline{D}_n)\gamma_n} \left(1 + \left[\sum_n (\underline{D}_n^*\underline{D}_n)\right]^{-1} \cdot \sum_{n \neq m} \left\{ \frac{\chi_{nmi}^{D}}{\chi_{oi}} (\underline{D}_n^*\underline{D}_m) \right.\right.$$

$$\left.\left. + \frac{\chi_{nmi}^{Q}}{\chi_{oi}} [(\underline{D}_n^*\underline{D}_m)(\underline{s}_n\underline{s}_m) + (\underline{D}_n^*\underline{s}_m)(\underline{s}_n\underline{D}_m)] \right\} \right) . \tag{12.113}$$

Equation (12.107) is only inapplicable when the scattering matrix for the transparent crystal has degenerate roots, and this degeneration is lifted by perturbation, i.e., it is accidental, not caused by the system's symmetry. In such cases a more concrete analysis is needed. The estimation of $\delta_3^{(j)}$ enables one to determine the accuracy of calculation of μ by (12.113). Concrete numerical calculation shows that in almost all cases, including those for which $\mu_j << \mu_o$,

$$\delta_3^{(j)} \sim \left(\frac{\chi_{oi}}{\chi_{or}}\right)^2 \delta_1^{(j)} . \tag{12.114}$$

The coefficient in front of the square brackets in (12.113) has a simple physical meaning. It is well known that Poynting's vector for the j-th excitation mode, averaged out over the time and the unit cell of the crystal, is determined by the following expression (we consider the separate terms in the sum over j of (5.15))

$$\underline{S}^{(j)} = \frac{c}{8\pi} \left[\sum_n (\underline{D}_n^{(j)*}\underline{D}_n^{(j)})\underline{s}_n\right] \exp[-\mu_j(\underline{n}_o\underline{r})] . \tag{12.115}$$

The average energy density of the j-th mode is equal to

$$W^{(j)} = \frac{1}{8\pi}\left[\sum_n (\underline{D}_n^{(j)*}\underline{D}_n^{(j)})\right] \exp[-\mu_j(\underline{n}_o\underline{r})] \quad . \tag{12.116}$$

The transport velocity is defined as the ratio of $\underline{S}$ to W

$$\underline{v}_t = c\frac{\sum_n (\underline{D}_n^*\underline{D}_n)\underline{s}_n}{\sum_n (\underline{D}_n^*\underline{D}_n)} \quad . \tag{12.117}$$

Thus, the coefficient in front of the square brackets in (12.113) is equal to $c/(\underline{v}_t\underline{n}_o)$. For the coefficient of absorption along the transport velocity we get, in accordance with (12.104),

$$\mu^t = \mu_o \frac{c}{|\underline{v}_t|}\left[1 + \left(\sum_n |\underline{D}_n|^2\right)^{-1} \sum_{n\neq m}\left\{\frac{\chi^D_{nmi}}{\chi_{oi}}(\underline{D}_n^*\underline{D}_m)\right.\right.$$

$$\left.\left. + \frac{\chi^Q_{nmi}}{\chi_{oi}}[(\underline{D}_n^*\underline{D}_m)(\underline{s}_n\underline{s}_m) + (\underline{D}_n^*\underline{s}_m)(\underline{s}_n\underline{D}_m)]\right\}\right] \quad . \tag{12.118}$$

This equation (without the quadrupole term) was first obtained by PENNING and POLDER [1.32].

In conclusion we note that (12.112) for μ becomes exact if the values of $\underline{D}_n$ imply the exact solution of the system of equations in the absorbing crystal. This fact enables us to use (12.113) for qualitative analysis, for instance, for estimating μ_{min}. It can be proved that $\mu_{min}/4\pi K$ is no less than the minimum eigenvalue of the matrix G_i [1.36].

12.8 The Relationship Between the Coefficient of Absorption and the Shape of the Dispersion Surface. EWALD's Criterion [1.31]

If we neglect the quadrupole interaction in photoabsorption, the system of dynamical equations for the absorbing crystal will differ from the system for the transparent crystal (12.61) only by the complex nature of the scattering coefficients. We can consider the complex dispersion equation

$$\Delta_c = \Delta(\tau_{nr} + i\tau_{ni} \quad , \chi_{mnr} + i\chi_{mni}) = 0 \quad . \tag{12.119}$$

Taking into account the inequality (12.105), we can expand Δ_c into a Taylor series in an imaginary unit i, each subsequent term being less than the preceding one in the ratio χ_{oi}/χ_{or}

$$\Delta_c = \Delta(\tau_{nr}, \chi_{mnr}) + i\left(\sum_n \tau_{ni} \frac{\partial\Delta}{\partial\tau_{nr}} + \sum_{m\neq n} \chi_{mni} \frac{\partial\Delta}{\partial\chi_{mnr}}\right) + \ldots \quad . \tag{12.120}$$

Ignoring all the terms except the first two, written out in (12.120), we get

$$\Delta(\tau_{nr}, \chi_{mnr}) = 0 \tag{12.121}$$

$$\sum_n \tau_{ni} \frac{\partial\Delta}{\partial\tau_{nr}} + \sum_{m\neq n} \chi_{mni} \frac{\partial\Delta}{\partial\chi_{mnr}} = 0 \quad . \tag{12.122}$$

Equation (12.122) makes it possible to express the coefficient of absorption in terms of derivatives of the dispersion determinant. Indeed, since, according to (12.16), (12.25), and (12.103),

$$\tau_{ni} = \chi_{oi} - 2\delta_i\gamma_n = \chi_{oi}(1 - \frac{\mu}{\mu_o}\gamma_n) \quad , \tag{12.123}$$

then, substituting (12.123) in (12.122), we obtain for μ

$$\mu = \mu_o \frac{\sum_n \frac{\partial\Delta}{\partial\tau_{nr}}}{\sum_n \frac{\partial\Delta}{\partial\tau_{nr}}\gamma_n}\left[1 + \left(\sum_n \frac{\partial\Delta}{\partial\tau_{nr}}\right)^{-1} \sum_{n\neq m} \frac{\chi_{nmi}}{\chi_{oi}} \frac{\partial\Delta}{\partial\chi_{nmr}}\right] \quad . \tag{12.124}$$

Comparing (12.124) and (12.113) for μ, it can easily be seen that they have similar structures and coincide completely (neglecting the quadrupole effect) if we put

$$\underline{D}_n^*\underline{D}_n = \lambda \frac{\partial\Delta}{\partial\tau_{nr}} \quad , \qquad \underline{D}_n^*\underline{D}_m = \lambda \frac{\partial\Delta}{\partial\chi_{nmr}} \tag{12.125}$$

where λ is an arbitrary constant. Thus, we have demonstrated the connection between the induction vectors and the derivatives of the dispersion determinant.

Turning to (12.124), we notice that the sums appearing in it have a simple geometrical meaning in the three-wave case. The distance from the Lorenz point L_o to the excitation point on the dispersion surface A (see Section 12.1) in units of K/2 is equal to

$$\underline{v} = \overline{\underline{L}_o\underline{A}} = \sum_n \tau_{nr}\underline{t}_n \quad , \quad \overline{\underline{L}_o\underline{L}} = \underline{v}_L = \chi_{or}\sum_n \underline{t}_n \quad , \quad \underline{n}_o = \sum_n \gamma_n\underline{t}_n \quad . \tag{12.126}$$

Regarding Δ as a function of the variables τ_{nr}, we have

$$\mathrm{grad}\Delta(\tau_{nr}) = \sum_n \frac{\partial\Delta}{\partial\tau_{nr}}\underline{s}_n = \underline{c} \quad . \tag{12.127}$$

It is obvious that the vector $\underline{c}$ is directed along the normal to the dispersion surface. Taking into account the above relations, we have

$$\sum_n \frac{\partial\Delta}{\partial\tau_{nr}} = \frac{1}{\chi_{or}}(\underline{c}\underline{v}_L) \quad , \quad \sum_n \frac{\partial\Delta}{\partial\tau_{nr}}\gamma_n = (\underline{c}\underline{n}_o) \quad , \tag{12.128}$$

$$\sum_n \tau_{nr}\frac{\partial\Delta}{\partial\tau_{nr}} = -\sum_{m\neq n}\chi_{mnr}\frac{\partial\Delta}{\partial\chi_{mnr}} = (\underline{c}\underline{v}) \quad . \tag{12.129}$$

The expression (12.124) is transformed as follows

$$\mu = \mu_o\frac{1}{\chi_{or}}\frac{(\underline{c}\underline{v}_L)}{(\underline{c}\underline{n}_o)}\left[1 - \frac{(\underline{c}\underline{v})}{(\underline{c}\underline{v}_L)} + B\right] \quad , \tag{12.130}$$

$$B = \left(\sum_n \frac{\partial\Delta}{\partial\tau_{nr}}\right)^{-1}\sum_{n\neq m}\left(\frac{\chi_{nmi}}{\chi_{oi}} - \frac{\chi_{nmr}}{\chi_{or}}\right)\frac{\partial\Delta}{\partial\chi_{mnr}} \quad . \tag{12.131}$$

Equation (12.130) can be regarded as substantiation of the so-called EWALD criterion according to which the minimum of the absorption coefficient is determined by the areas of the dispersion surface nearest to the Laue point. Indeed, when the dispersion surface passes through the Laue point, the first two terms in the square brackets are cancelled out, and $B \approx 0$ because

$$B' = \frac{\chi_{nmi}}{\chi_{oi}} - \frac{\chi_{nmr}}{\chi_{or}} \approx 0 \quad . \tag{12.132}$$

In crystals with a single type of atom the difference B' is equal to

$$B' = \left(1 - \frac{f_{n-m}}{z}\right) e^{-W_{n-m}} \sum_j e^{2\pi i(\underline{h}_n - \underline{h}_m)\underline{r}_j} \tag{12.133}$$

where $z = f_o$ is the number of electrons in the atom. Consequently, the EWALD criterion is approximately valid for scattering on reciprocal lattice vectors with low indices. For sharply asymmetrical three-wave combinations with large reciprocal lattice vectors, the difference B' is already appreciable, and the criterion becomes less graphic. At the same time it is evidently possible to assert, in a general case, that the minimum of the absorption coefficient is attained in the central area of the dispersion surface, i.e., near the main axis, on the sheet closest to the Laue point.

12.9 Asymptotic Properties of the Dispersion Surface. Transition From the Multiwave to the Two-Wave Region [1.34]

As distinct from the two-wave case, the dispersion surface equation for three or more strong waves has no analytical solution for all the values of the parameters θ_1 and θ_2. In this connection, it is of interest to analyze those areas of the dispersion surface which occupy an intermediate position between the two-wave and multiwave regions, because the problem is considerably simplified in this case. On the other hand, such an analysis enables one to trace certain qualitative differences between a multiwave dispersion surface and a two-wave one, which are of a general nature.

Choose the vector $\underline{a}_2$ in (12.25) so that $\underline{a}_2\underline{h}_1 = 0$. In this case the parameters α_n (12.83) are equal to

$$\alpha_1 = \frac{2}{K}(\underline{a}_1\underline{h}_1)\theta_1 \quad ,$$

$$\alpha_n = \frac{2}{K} [(\underline{a}_1\underline{h}_n)\theta_1 + (\underline{a}_2\underline{h}_n)\theta_2] \quad , \qquad n \neq 1 \quad . \tag{12.134}$$

We will be interested in the region $|\theta_2| >> |\chi_{or}|$, when all the values of α_n, except α_1, and hence all the τ_n, except τ_o, τ_1, become large. The different terms of the multiwave dispersion determinant $\Delta^{(N)}$ will differ in magnitude depending on how many parameters τ_n, $n \neq 0,1$ are contained in each term. In order to single out terms of the same order of magnitude we write $\Delta^{(N)}$ in the following form

$$\Delta^{(N)} = (\tau_2 \ldots \tau_{N-1})^2 \left[\Delta^{(2)}_{o1} + \sum_{m=2}^{N-1} \left(\frac{\partial\Delta^{(3)}_{o1m}}{\partial\tau_m}\right)_{\tau_m=0} \frac{1}{\tau_m}\right.$$

$$\left. + \frac{1}{2} \sum_{m,n=2}^{N-1} \left(\frac{\partial^2\Delta^{(4)}_{o1mn}}{\partial\tau_m\partial\tau_n}\right)_{\tau_m=\tau_n=0} \frac{1}{\tau_m\tau_n} + \ldots\right] \quad , \tag{12.135}$$

where $\Delta^{(2)}_{01}$ is a two-wave dispersion determinant on the vectors $\underline{s}_o$, $\underline{s}_1$, $\Delta^{(3)}_{01m}$ is a three-wave dispersion determinant on the vectors $\underline{s}_o, \underline{s}_1, \underline{s}_m$, and so on. The first term in (12.135) contains all τ_n, $n = 2, \ldots, N-1$ to the highest (second) power. The three-wave determinant $\Delta^{(3)}_{01m}$ can be written as

$$\Delta^{(3)}_{01m} = \tau_m^2\Delta^{(2)}_{01} + \tau_m F_{01m} + H_{01m} \quad . \tag{12.136}$$

The coefficients $\Delta^{(2)}_{01}$, F_{01m}, H_{01m} do not contain any τ_m and therefore have the usual order of magnitude. Differentiating (12.136) with respect to τ_m and equating τ_m to zero, we obtain the coefficient F_{01m}. Thus, each summand in a sum over m in the second term contains all the τ_n, except τ_m, $n, m \neq 0, 1$, to the second power, and τ_m to the first power, and hence the second term is $|\chi_{or}/\theta_2|$ times less than the first. Similarly, the third term is less than the second in the same ratio.

To a zero approximation, we can retain only the first term in (12.135). Then the dispersion equation will have the form

$$\Delta^{(N)} \approx (\tau_2 \ldots \tau_{N-1})^2\Delta^{(2)}_{01} = 0 \quad . \tag{12.137}$$

The dispersion surface has (N-2) twice degenerated planes determined by equations $\tau_2 = 0(\underline{t}_o, \underline{t}_1$ -plane) and (see (12.28))

$$\sum_{n=o}^{2} c_{mn}\tau_n = 0 \quad , \quad m > 3$$

and on all the planes the coefficient of absorption is equal to μ_o and two hyperbolic cylinders (along the $\underline{t}_3$-axis) determined by the equations

$$\tau_o\tau_1 = \chi_1^2 \quad , \quad \tau_o\tau_1 = \chi_1^2 s_1^2 \quad .$$

Here and henceforward we use the notation of (12.39) for m = 2.

We shall further consider the distortion of hyperbolic cylinders due to the second term in (12.135). Let

$$\tilde{\tau}_o^{(j)} = \chi_o - 2\tilde{\delta}^{(j)}\gamma_o \quad , \quad \tilde{\tau}_1^{(j)} = \chi_o - 2\tilde{\delta}^{(j)}\gamma_1 - \alpha_1 \tag{12.138}$$

be two-wave solutions. Put $\delta^{(j)} = \tilde{\delta}^{(j)} - \varepsilon^{(j)}$, where $\varepsilon^{(j)}$ is a small addition of the order of $|\chi_{or}/\theta_2|\tilde{\delta}^{(j)}$. Solving the equation

$$\Delta_{01}^{(2)} + \sum_{m=2}^{N-1} F_{01m}\frac{1}{\tau_m} = 0 \tag{12.139}$$

in the linear approximation with respect to ε, we obtain the following general expression for ε:

$$\varepsilon = \frac{K}{4\theta_2}\left\{\sum_{n=o,1}\gamma_n\left[\frac{\partial\Delta_{01}^{(2)}}{\partial\tau_n}\right]_{\substack{\tau_o=\tilde{\tau}_o\\ \tau_1=\tilde{\tau}_1}}\right\}^{-1}\sum_{m=z}^{N-1}\frac{F_{01m}(\tilde{\tau}_o,\tilde{\tau}_1)}{\underline{a}_2\underline{h}_m} \quad . \tag{12.140}$$

Thus, as $|\theta_2|$ reduces, the dispersion surface approaches the Laue point with one sign of θ_2 and recedes from it with the other. As shown by calculations, with different signs of θ_2 (12.140) characterizes the distortion of different sheets of the multiwave dispersion surface. Fig. 12.3 shows schematically the section of the three-wave dispersion surface by the plane

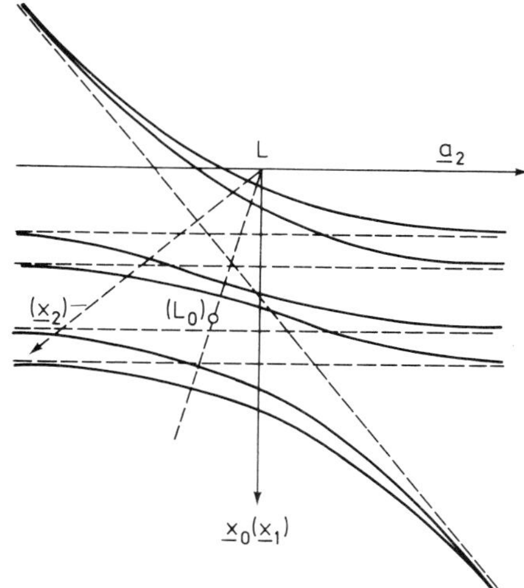

Fig. 12.3 Section of dispersion surface by plane passing through vectors $\underline{x}$ and $\underline{a}_2$

of the vectors $\underline{x}_o$, $\underline{a}_2$ (the vector $\underline{a}_2$ is parallel to the vector $\underline{t}_2$) in the case where $\theta_1 = 0$. The Laue point lies in the plane, while the Lorenz point and the vector $\underline{x}_2$ do not, and the figure shows their projections onto this plane.

The imaginary part of ε determines the difference between the two-wave and multiwave coefficients of absorption at large θ_2

$$\Delta\mu^{(N)} = \mu^{(N)}(\theta_1,\theta_2) - \mu^{(2)}(\theta_1) = -4\pi K\varepsilon_i \quad . \tag{12.141}$$

Equations (12.140) and (12.141) make it possible to analyze qualitatively the equation of enhancement of the Borrmann effect in the multiwave case as compared with the two-wave case in the general form. The two-wave solution corresponding to the Borrmann effect is characterized by the following parameter values

$$\tilde{\tau}_o = \tilde{\tau}_1 = \chi_1 \quad , \quad \gamma_o = \gamma_1 \quad , \quad \theta_1 = 0 \quad . \tag{12.142}$$

In this case

$$\sum_{n=o,1} \left[\frac{\partial \Delta_{o1}^{(2)}}{\partial \tau_n}\right]_{\tau_o=\tau_1=\chi} = 2\chi_1^3(1 - s_1^2) \quad . \tag{12.143}$$

Determining from (12.140) the coefficient at the term τ_2 and assuming in it $\tau_o = \tau_1 = \chi_1$, we get

$$F_{012} = -\chi_1^3(1 - s_1^2 - v_s^2)(\chi_2^2 + \chi_3^2 - 2c/\chi_1) \quad . \tag{12.144}$$

Thus, in crystals without an inversion centre, the value of F_{012} is always different from zero, and hence the enhancement of the effect always takes place. In a crystal with an inversion centre $c = \chi_1\chi_2\chi_3$, that is to say the value of F_{012} may be equal to zero, provided $\chi_2 = \chi_3$. This case has already been considered in Section 12.4. When $\chi_2 = \chi_3$, the line $\tau_o = \tau_1 = \chi_1$ corresponds to the first of the three cases of (12.58), i.e., this two-wave line lies on the three-wave dispersion surface without any changes. As follows from the above analysis, this is the only case where the Borrmann effect is not intensified as compared with the two-wave case.

Substituting (12.143) and (12.144) in (12.140), we obtain the following general expression for $\Delta\mu^{(N)}(\theta_2)$:

$$\Delta\mu^{(N)}(\theta_2) = \frac{1}{\gamma_o}\frac{\pi K^2}{2\theta_2}\sum_{m=2}^{N-1}\frac{[1 - s_1^2 - v_{s(m)}^2]}{(1 - s_1^2)}\frac{1}{(\underline{a}_2\underline{h}_m)} \cdot \mathrm{Im}[\chi_{2(m)} + \chi_{3(m)} - 2c_{(m)}/\chi_1] \quad . \tag{12.145}$$

The subscript (m) at χ, v_s, and c means that in the left-hand side of (12.139) the subscript 2 must be replaced by m. According to (12.145), the decrease in absorption coefficient is the more pronounced, the sharper is the asymmetry of the partial three-wave combinations on the vectors $\underline{s}_o$, $\underline{s}_1$, and $\underline{s}_m$. However, the general multiwave case may have a high symmetry. A characteristic feature of the variations of $\mu^{(N)}$ in the intermediate range is its asymmetry with respect to the change of sign of θ_2. This fact has already been experimentally confirmed in [12.18], whose authors observed intensification and weakening of the two-wave line on the film when approaching the three-wave point from different sides. According to the analysis carried out, this behaviour has a general character in the three-wave case.

In multiwave cases which have a plane of symmetry perpendicular to $\underline{a}_2$, there are, at large $|\theta_2|$, two solutions with one and the same $\mu^{(2)}$. With decreasing $|\theta_2|$, in these cases the dependence $\mu^{(N)}(\theta_2)$ "splits" on both sides into two curves symmetrically diverging upwards and downwards in accordance with (12.145).

12.10 Symmetrical Cases of Multiwave Diffraction. Nonlinear Borrmann Effect [1.37]

Of great importance, from the physical and mathematical points of view, are symmetrical multiwave cases in which the reciprocal crystal lattice vectors form a regular polygon and, besides, the condition $\exp(2\pi \underline{h}_m \underline{r}_j) = 1$ is fulfilled for all vectors and all atoms j in the unit cell. Then, the scattering coefficients χ_{mn} depend only on the modulus of the difference of the vectors $\underline{h}_m$ and $\underline{h}_n$, In cubic-symmetry crystals, they are regular triangles, quadrangles (squares), and hexagons of this type.

The dispersion surface is evidently also symmetrical in these cases, and its extreme points lie on the axis of symmetry, i.e., on the main axis. Let the vector $\underline{n}_o$ lie on the main axis, that is to say all $\gamma_m = \gamma_o$, and let the vector of the incident vacuum wave coincide with $\underline{x}_o$, i.e., all $\alpha_m = 0$. In this case, owing to the symmetry of the scattering matrix $G_{mn}^{ss'}$ (12.86), the general form of its eigenvectors $\underline{B}_{ns}^{(j)} \sim \underline{D}_{ns}^{(j)}$ can easily be guessed at once.

Having in mind the general case, let us turn to the drawing for N = 4 in Fig. 12.2. Regarding m as a discrete variable running through the entire natural series of numbers, it is easy to see that the replacement of m by $m' = m + 1$ is equivalent to rotation of the figure by an angle of $2\pi/N$. Replacement of m by $m' = m + N$ leads to identical transformation of the figure. Thus, for the polarization vectors we obtain the formal condition $\underline{e}_{m+Ns} = \underline{e}_{ms}$. The same condition must be satisfied by the amplitudes B_{ms}, since on crystal rotation by an angle of 2π the induction vector naturally remains unchanged.

Thus, we have a cyclical condition $B_{m+Ns} = B_{ms}$, whose linearly independent solutions have the form

$$B_{ms} \sim \sin\left(\frac{2\pi n}{N} m\right) ; \quad \cos\left(\frac{2\pi n}{N} m\right)$$

where n are integral divisors of N. The subscript s takes on two values, π and σ, and therefore four combinations of sines and cosines are possible. In order to select the necessary ones, we will consider the operation of reflection in a plane x passing through the point 0 and the main axis. Then the projection of the induction onto this plane at the lattice point

$$D^{\times} \sim \sum_m (B_{m\pi}\pi^{\times} + B_{m\sigma}\sigma^{\times})$$

either remains unchanged or changes sign. On the other hand, this operation is equivalent to replacing m by m' = N - m. Since here $\pi^x_{m'} = \pi^x_m$, $\sigma^x_{m'} = -\sigma^x_m$, the amplitudes must behave in the same way. Thus, only combinations of the following type are possible

$$(n\pi):\ B_{m\pi} = \alpha\cos\left(\frac{2\pi n}{N}m\right) \ , \qquad B_{m\sigma} = -\beta\sin\left(\frac{2\pi n}{N}m\right)$$

$$(n\sigma):\ B_{m\pi} = \alpha\sin\left(\frac{2\pi n}{N}m\right) \ , \qquad B_{m\sigma} = \beta\cos\left(\frac{2\pi n}{N}m\right) \qquad (12.146)$$

and the problem reduces to determination of the coefficients α and β. Actually, it is sufficient to determine their ratio, α/β, since the second coefficient is found from the normalization condition.

It is easy to see that in the cases n = N and n = N/2 (for even N) either π - or σ-components are all equal to zero. The solution ($N\pi$) corresponds to the equality of all the π-amplitudes and ($N\sigma$), of all the σ-amplitudes. For N = 4 and 6, there are also antisymmetrical solutions. All these solutions, in symmetry operations, transform only through themselves, i.e., they correspond to nondegenerate eigenvalues. The respective coefficients α and β for them are equal to $1/\sqrt{N}$. Conversely, the solutions ($n\pi$) and ($n\sigma$) for $n \neq N$, N/2 correspond to a degenerate eigenvalue, since they transform through each other in symmetry operations. Substituting either of the solutions (12.146) in any two equations of the system (12.83), we obtain a system of two linear equations for determination of δ and α/β. For each n there are two values for δ, which will be denoted by ($n\pi,\sigma\pm$). For the coefficients α and β of these solutions it is convenient to use the following general representations:

$$\alpha_{\pm} = -\sqrt{\frac{2}{N}}\frac{z_{\pm}}{\sqrt{1 + z_{\pm}^2}} \ , \qquad \beta_{\pm} = \sqrt{\frac{2}{N}}\frac{1}{\sqrt{1 + z_{\pm}^2}} \ ,$$

$$z_{\pm} = \xi \pm \sqrt{1 + \xi^2} \ . \qquad (12.147)$$

We will now pass over to the discussion of the quantities ξ and δ. We define the angle θ by the relation $\cos\theta = \gamma_o = (\underline{n}_o\underline{s}_m)$. The scattering coefficients for reciprocal lattice vectors unequal in modulus will be denoted by χ_m in the order of increasing reciprocal lattice vector moduli. In the three-wave case there are only two coefficients, χ_o and χ_1. Omitting the calculation details, we will give the result

$$\delta' = -2\gamma_o\delta = \begin{cases} \chi_o - \chi_1 + 3\chi_1\sin^2\Theta & (3\pi) \\ \chi_o - \chi_1 & (3\sigma) \\ \left.\begin{array}{l} \chi_o + \frac{1}{2}\chi_1(1 - 3z_\pm\cos\theta) \\ \xi = \dfrac{\sin^2\theta}{2\cos\theta} \end{array}\right\} & (1\pi,\sigma\pm) \end{cases} . \quad (12.148)$$

This case has already been discussed in Section 12.4, where we demonstrated that there is a triply degenerate root here. Indeed, calculating $z_\pm$, we get

$$(1\pi,\sigma+):\quad z_+ = \frac{1}{\cos\theta} \;,\qquad \delta' = \chi_o - \chi_1$$

$$(1\pi,\sigma-):\quad z_- = -\cos\theta \;,\qquad \delta' = \chi_o + 2\chi_1 - \frac{3}{2}\chi_1\sin^2\theta \;. \quad (12.149)$$

Thus, the roots $(1\pi,\sigma+)$ and (3σ) accidentally coincide.

The minimum absorption coefficient along the direction of the incident beam in the case of absorption of a purely dipole nature and crystals with a single type of atoms is equal to

$$\mu_{min}^{(3)} = -2\pi K\delta_i' = \mu_o(1 - e^{-W_1}) \approx W_1\mu_o \quad (12.150)$$

and coincides with the corresponding two-wave value. According to (12.150), the absorption coefficient in this case is proportional to the power index in the Debye-Waller factor $W_1 = Bh^2$, and in this sense one can speak of Borrmann's linear effect.

Consider the four-wave case. Here, there are three unequal scattering coefficients χ_o, χ_1, χ_2, and $h_2^2 = 2h_1^2$. The eigenvalues of the scattering matrix are defined by the following equations

$$\delta' = \begin{cases} \chi_o - \chi_2 + 2(\chi_2 \pm \chi_1)\sin^2\theta & \begin{cases}(4\pi)\\(2\pi)\end{cases} \\ \chi_o - \chi_2 \quad (4\sigma)\ ,\quad (2\sigma) & \\ \chi_o + \chi_2 - 2\chi_1 z_\pm \cos\theta & \\ \qquad \xi = \dfrac{\sin^2\theta}{2\cos\theta}\dfrac{\chi_2}{\chi_1} & \Bigg\} (1\pi,\sigma\pm) \quad . \end{cases} \tag{12.151}$$

According to (12.151), nondegenerate solutions yield a Borrmann effect equal to the two-wave effect on the vector $\underline{h}_2$, and the solutions (4σ) and (2σ) accidentally coincide. Of greatest interest is the solution $(1\pi,\sigma+)$. Consider the limit of small scattering angles when $\sin\theta \sim \lambda/a \ll 1$. Expanding δ' into a Taylor series in θ^2 and using the relation $W_2 = 2W_1$, we get

$$\begin{aligned} \mu_{min}^{(4)} &= -2\pi K\Bigg\{(\chi_o + \chi_2 - 2\chi_1)_i + \theta^2(\chi_1 - \chi_2)_i \\ &\quad + \frac{1}{12}\theta^4\Bigg[\left(4 - 6\frac{\chi_{2r}}{\chi_{1r}}\right)\chi_{2i} - \left(1 - 3\frac{\chi_{2r}^2}{\chi_{1r}^2}\right)\chi_{1i}\Bigg] + \dots\Bigg\} \\ &\approx \mu_o\left[W_1^2 + W_1\theta^2 + \frac{1}{4}\left(1 - \frac{f_2}{f_1}\right)^2\theta^4 + \dots\right] \end{aligned} \tag{12.152}$$

where f_m is the atomic amplitude for the m-th reflection.

Thus, for small scattering angles the absorption coefficient is proportional to W_1^2. Hence, a "quadratic" Borrmann effect arises. A distinguishing property of this effect as compared with the two-wave case is the pronounced dependence on the scattering angle. Thus, at $\theta > \sqrt{W_1}$ the second term in (12.152) becomes larger than the first, and the effect weakens.

Now we pass over to the six-wave case. For the eigenvalues of the scattering matrix we obtain the following equations:

$$
\delta' = \begin{cases}
\chi_0 \pm \chi_1 - \chi_2 \mp \chi_3 + \sin^2\theta(\pm\chi_1 + 3\chi_2 \pm 2\chi_3) & \begin{matrix}(6\pi)\\(3\pi)\end{matrix} \\
\chi_0 \pm \chi_1 - \chi_2 \mp \chi_3 & \begin{matrix}(6\sigma)\\(3\sigma)\end{matrix} \\
\left.\begin{matrix}\chi_0 - \chi_3 - \frac{1}{2}(\chi_1 - \chi_2)(1 + 3z_\pm \cos\theta) \\ \xi = \dfrac{\sin^2\theta}{6\cos\theta} \dfrac{(\chi_1 + 3\chi_2 - 4\chi_3)}{(\chi_1 - \chi_2)}\end{matrix}\right\} & (2\pi,\sigma\pm) \\
\left.\begin{matrix}\chi_0 + \chi_3 + \frac{1}{2}(\chi_1 + \chi_2)(1 - 3z_\pm \cos\theta) \\ \xi = \dfrac{\sin^2\theta}{6\cos\theta} \dfrac{(4\chi_3 + 3\chi_2 - \chi_1)}{(\chi_1 + \chi_2)}\end{matrix}\right\} & (1\pi,\sigma\pm) \quad .
\end{cases}
\tag{12.153}
$$

In this case there are three reciprocal lattice vectors unequal in modulus, and $h_2^2 = 3h_1^2$, $h_3^2 = 4h_1^2$. The same relationships obtain for the Debye-Waller factors $W_2 = 3W_1$, $W_3 = 4W_1$. An analysis of the nondegenerate roots in elementary

$$
\frac{\mu^{(6)}}{\mu_0} = \begin{cases}
6[W_1 + \sin^2\theta(1 - 3W_1)] + \ldots & (6\pi) \\
3W_1^2 \cos^2\theta + \ldots & (3\pi) \\
6W_1 + \ldots & (6\sigma) \\
3W_1^2 + \ldots & (3\sigma) \quad .
\end{cases}
\tag{12.154}
$$

Consequently, the (3σ) solution yields a quadratic Borrmann effect, which is altogether independent of the scattering angle, and for the (3π) field, the coefficient of absorption even reduces with an increase in scattering angle. Among the degenerate solutions, of greatest interest is $(2\pi,\sigma+)$. We will now again consider the limit $\sin\theta \ll 1$. Performing the appropriate calculations, we get

$$\mu_{min}^{(6)} = \mu_o\left(2W_1^3 + \frac{3}{2} W_1^2\theta^2 + \ldots\right) \quad . \tag{12.155}$$

Thus, in the six-wave case, at small scattering angles $\theta < \sqrt{W_1}$, the minimum absorption coefficient is proportional to the third power of the Debye-Waller factor for reflection with minimum indices. What is the physical nature of this effect? It is common knowledge that in the two-wave case the absorption reduces because a standing wave with nodes precisely at the lattice points is formed in the crystal. Due to the finite sizes of the atoms, and mainly owing to the thermal oscillations, there is still some interaction between the atomic electrons and the electric field, but it is much weaker than for the plane wave, when the field energy is evenly distributed throughout the crystal. For point atoms, the interaction is proportional to root-mean-square displacements of the atoms from their equilibrium positions

$$\mu \sim \langle u^2 \rangle \sim W \quad . \tag{12.156}$$

Equation (12.156) presupposes that the electric field increases linearly with distance from the lattice point. It is easy to see that if the electric field gradients at the lattice point also vanish to zero, the dependence of the field energy on the distance to the point will be different. In this case the coefficient of absorption is proportional to the average value of the fourth power of the displacement of the atoms from the equilibrium positions

$$\mu \sim \langle u^4 \rangle \sim W^2 \quad . \tag{12.157}$$

Thus, one can say beforehand that the quadratic dependence of μ on the Debye-Waller factor is ensured by such a structure of the electric field in the crystal with which both the field itself and its derivatives in all directions vanish to zero at the lattice points. This is easily checked by using (12.146) and (12.147). Similarly, in the case of a cubic dependence of μ on W, the tensor of the second derivatives of the field at the lattice point is also equal to zero.

Let us discuss the role of quadrupole absorption. Particular expressions for the quadrupole terms in μ in the case where there is no accidental degeneration can be obtained from (12.113) and from the expressions for the field (12.146). From (12.113) it follows, in particular, that μ^Q is proportional to the product $\underline{s}_n^i \underline{D}_m^k$, where i, k are the vector indices. In a real space, to these terms there correspond derivatives of the field with respect to the coordinate. Hence, generally speaking, the first term of the expansion of the minimum absorption coefficient in W will have a power index one unit less for dipole interaction, as in the two-wave case. However, its coefficients depend on the scattering angle and, in the limit of small scattering angles, the ratios (μ^Q/μ_0^Q) and (μ^D/μ_0^D) almost concide. Note that since the ratio (μ_0^Q/μ_0^D) is less than W in most of the practically important cases, quadrupole absorption does not disturb the Borrmann effect of any power.

It should, however, be remembered that in conditions where photoelectric absorption is strongly suppressed, the role of other inelastic scattering channels, in particular Compton scattering and inelastic scattering on phonons, may increase (see Section 4.1).

Symmetrical cases in germanium were first considered in the paper of JOKO and FUKUHARA [12.28]. The equations for eigenvalues obtained in this work differ, as regards the form of recording, from those given in this section, but the difference is purely formal; both forms of recording can be derived from one another by identical transformations. The authors also estimate the coefficients of absorption for polygons of side (220) and Cu K_α radiation, and point to a sharp decrease in absorption coefficient in the six-wave case.

For observation of multiwave effects, HUANG and POST [12.21] proposed an ingenious scheme of experiment in which divergent and polychromatic radiation of a sharp-focus X-ray tube impinges directly on the crystal, and a photographic film is set up at a large distance (1 to 2 metres) from the source in a direction close to the multiwave one. To reduce the exposure time, the air is pumped out in the path of the X-rays. The pattern obtained on the film has the form of a family of dark lines against a light background, which intersect at a single point. It is easy to see that the dark lines are areas of Kossel cones corresponding to two-wave diffraction, and at their point of intersection we have Borrmann's multiwave effect.

Figure 12.4 shows the photograph (positive) obtained in [12.21]. It clearly reveals a lighter spot at the point of intersection of lines (111) and $(11\bar{1})$, which characterizes the enhancement of "anomalous penetration", but in the four-wave symmetrical case such intensification is not observed at the point of intersection of lines $(0\bar{2}2)$, $(0\bar{2}\bar{2})$, and $(0\bar{4}0)$. HUANG et al. [12.22] specially considered the six-wave symmetrical case, and again no enhancement was detected.

In this connection we must mention the paper of KSHEVETSKY and MIKHAILYUK [12.23], who, in fact, repeated the American experiment, but with some modi-

Fig. 12.4 Forward diffracted divergent-beam photograph (after HUANG and POST [12.21]). A perfect (100) cut germanium crysta was oriented to diffract the Cu $K_{\alpha 1}$ radiation for the (0$\bar{2}$2, 0$\bar{2}\bar{2}$, 0$\bar{4}$0) four-beam case

fications. In particular, their specimen was more than 2 metres distant from the source, while the film was actually placed right behind the specimen, at a distance of 3 cm.

Figure 12.5 exhibits a photograph from [12.23] for the symmetrical six-wave case. A lighter spot is clearly seen at the place of intersection of the two-wave lines (220) and ($\bar{2}$02). The weaker lines (242), ($\bar{2}$24), and

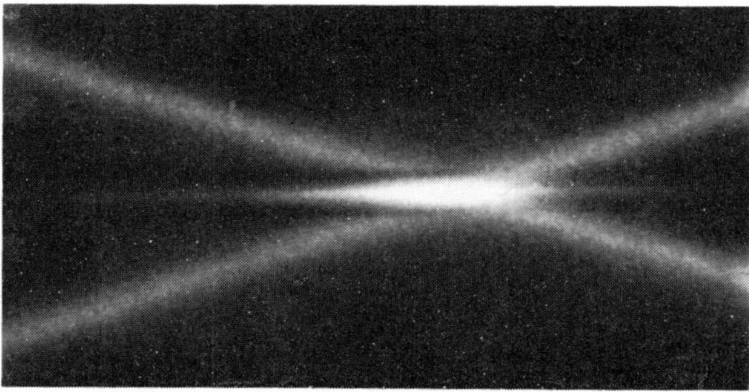

Fig. 12.5 Enhancement of anomalous transmission in six-wave case obtained by KSHEVETSKY and MIKHAILYUK [12.23]. Only (220) - two-wave lines are seen, but the more weaker lines (224) and (044) are not visible, since their development would overexposed the bright spot

(044) are not seen in the figure, since their development leads to overexposure of the light spot. The experimental pattern obtained, however, does not coincide with the calculated pattern of angular dependence of "anomalous penetration" in the case of an incident plane wave. An important feature of the experimental data obtained is the fact that this effect is observed only on not-too-thick crystal plates and disappears with increasing thickness. All this indicates that the incident-plane-monochromatic-wave approximation does not correspond to the scheme of the experiment. A group from the Chernovtsy State University under the guidance of MIKHAILYUK has also considered various four- and three-wave configurations, and the enhancement effect was observed throughout [12.24]. Besides, patterns of distortion of pendulum solution in multiwave diffraction were obtained; in these experiments, good agreement with theory was observed [12.25].

12.11 Scattering in Germanium and Silicon Crystals

The application of the outlines general theory to the currently most important case of scattering in Ge and Si crystals is undoubtedly of great interest. These crystals have a diamond lattice, which represents two face-centred cubic sublattices displaced by one-fourth of the cube diagonal with respect to each other. The inversion centre in the lattice lies at one-eighth of the cube diagonal. The structure amplitudes S_{mn}, which appear in the scattering coefficients χ_{mn}, depend exclusively on the sum of the indices of the vector $(\underline{h}_m - \underline{h}_n)$. Choosing the origin at the inversion centre, we have the following equation for them

$$S_{mn} = \frac{1}{N}\sum_j \exp[2\pi i(\underline{h}_m - \underline{h}_n)(\underline{r}_j - \underline{r}_o)] = \cos[\frac{\pi}{4}(H + K + L)] \quad ,$$

$$(\underline{h}_m - \underline{h}_n) = \frac{1}{a}(H,K,L) \qquad (12.158)$$

where N is the number of atoms in the unit cell and $\underline{r}_o$ is the coordinate of the inversion centre. Equation (12.158) holds good for cases where all the indices H,K,L are either even or odd. In the other cases, S_{mn} is always equal to zero.

According to (12.158), the following three types of reflection can be clearly distinguished:

$$
\begin{aligned}
&\text{A.}\quad H + K + L = 4n \;, \qquad S_{mn} = (-1)^n \\
&\text{B.}\quad H + K + L = 4n + 2 \;, \qquad S_{mn} = 0 \\
&\text{C.}\quad H + K + L = 4n \pm 1 \;, \qquad S_{mn} = \frac{1}{\sqrt{2}}(-1)^n \;.
\end{aligned}
\tag{12.159}
$$

Combinations of these types lead to different multiwave cases. When recording multiwave cases, the vectors $\underline{h}_m$ are usually indicated, which is not always convenient. We will indicate the vectors lying on the polygon sides in the following sequence

$$
(\underline{h}_1/\underline{h}_2 - \underline{h}_1/ \;\dots\; /\underline{h}_{N-1} - \underline{h}_{N-2}/-\underline{h}_{N-1}) \;. \tag{12.160}
$$

In the three-wave case, the vectors $\underline{h}_1$, $\underline{h}_2 - \underline{h}_1$, $-\underline{h}_2$ form a closed triangle. The following connection schemes are possible here:

Scheme	0 → 1	1 → 2	2 → 0
(1)	A	A	A
(2)	A	C	C
(3)	B	C	C
(4)	C	A	C
(5)	C	B	C

(12.161)

When inspecting the diagram we notice that cases 4 and 5 are obtained by circular permutation of the indices in the cases 2 and 3. The dispersion surfaces in these cases are the same, differing exclusively in the angle of rotation relative to the vectors $\underline{s}_m$. But since the wave 0 is always isolated because of the boundary conditions, these cases are still distinguishable.

To ascertain this, we shall consider a case of type 4: (111/$\bar{2}\bar{2}0$/$11\bar{1}$). In this case $\chi_1 = \chi_2$ and, according to (12.56) and (12.58), the dispersion equation has the solution

$$
\tau_1 = \tau_2 = \chi_3 = \chi_{220} \;, \qquad \tau_0 \text{ - any} \;. \tag{12.162}
$$

The absorption coefficient for this solution along the vector $\underline{s}_1$ coincides with the two-wave value for reflection on the vector (220) and is the smallest of all the possible coefficients. It is, however, easy to see directly from (12.42) that to this excitation mode there corresponds such a distribution of the eigen multiwave field in the crystal at which $D_{0\pi} = D_{0\sigma} = 0$. Consequently, according to (12.94), an incident vacuum wave does not excite this mode. Thus, in experiment we will not obtain the expected "anomalous transmission". This fact was experimentally confirmed in [12.18].

Quite a different situation obtains in the case of type 2: $(\bar{2}\bar{2}0/11\bar{1})111)$. Here we have $\chi_2 = \chi_3$. The type (12.162) solution is different now:

$$\tau_0 = \tau_1 = \chi_1 = \chi_{220} \quad , \quad \tau_2 \text{ - any} \quad .$$

Hence this mode is excited by an incident wave, and for it $D_2 = 0$, i.e., anomalous penetration will fail to manifest itself only when the coefficient of reflection R_2 is measured.

Similarly, type 3 and 5 cases also differ from each other. Case 3 corresponds to "nondirect excitation", because the vector $\underline{h}_1$ in this case is forbidden. On switching from a three-wave point to a two-wave line with a vector $\underline{h}_1$ we at once obtain normal absorption, and in the case of Bragg reflection, we have absence of extinction. This case is used in RENNINGER's method [1.29].

Case 5 corresponds to "incomplete excitation" in the sense that waves 1 and 2 do not interact directly. This case was first experimentally investigated by BORRMANN and HARTWIG [1.30], who detected considerable enhancement of "anomalous transmission" at the three-wave point compared with the two-wave line on $\underline{h}_1$. By lucky coincidence this case admits of analytical solution over the entire plane, and therefore it has been considered a number of times [1.31,33, 12.26]. The most detailed consideration is presented in HILDEBRANDT's paper [12.26].

We will now discuss the BORRMANN and HARTWIG case $(111/\bar{2}00/1\bar{1}\bar{1})$, in which $\chi_1 = \chi_2$. The two-wave minimum absorption coefficient (along the vector of the incident wave) in this case equals

$$\mu_{111} = \mu_0\left(1 - \frac{1}{\sqrt{2}}\, e^{-W_{111}}\right) \quad . \tag{12.163}$$

Here and henceforward we shall neglect the quadrupole contribution to absorption for simplicity sake. In the three-wave case, on the plane of symmetry of the three-wave tetrahedron we have the solution (12.54). Switching to the variable δ, we write it down as

$$-2\delta\gamma_o = \chi_o + \frac{1}{2}\{-[\alpha\beta + \chi_o(1-\beta)] \pm \sqrt{[\alpha\beta + \chi_o(1-\beta)]^2 + 4\beta\chi_1^2 C}\}$$

(12.164)

$$\alpha_1 = \alpha_2 = \alpha \ , \quad \gamma_1 = \gamma_2 \ , \quad \beta = \gamma_o/\gamma_1 \ , \quad C = \begin{cases} 1 + 2s_1^2 - s_3 \\ 1 + s_3 \end{cases} \ .$$

In the case $\beta = 1$, $\alpha = 0$ the absorption coefficient is

$$\mu_{min}^{(3)} = -4\pi K\delta_i\gamma_o = \mu_o\left(1 - \sqrt{\frac{C}{2}}\, e^{-W_{111}}\right) \ . \tag{12.165}$$

Thus, the factor $\sqrt{C}$, which appears due to three-wave scattering, almost cancels out the $\sqrt{2}$ in the denominator, which appears from the structure amplitude, and this cancellation is the stronger, the shorter is the wavelength λ, since

$$C_1 = 1 + s_3 = 2(1 - \lambda^2/a^2)$$

$$C_2 = 1 + 2s_1 - s_3 = 2\left[1 - 2(\lambda/a)^2 + \frac{9}{4}(\lambda/a)^4\right] \ . \tag{12.166}$$

The induction vectors in the plane $\gamma_1 = \gamma_2$, $\alpha_1 = \alpha_2$ are obtained most easily from the system (12.42) by introducing the polarization vectors according to (12.29) and (12.31) (see Fig. 12.6).

Here, to the solution with $C = C_1$ there corresponds a field with $D_{o\pi}^{(1)} = 0$

$$-D_{1\pi}^{(1)} = D_{2\pi}^{(1)} = \frac{\chi_1 c}{\tau_1} D_o^{(1)} \ , \qquad D_{1\sigma}^{(1)} = D_{2\sigma}^{(1)} = \frac{\chi_1 a}{\tau_1} D_{o\sigma}^{(1)} \tag{12.167}$$

and to the solution with $C = C_2$ there corresponds a field with $D_o^{(2)} = 0$

$$D_{1\pi}^{(2)} = D_{2\pi}^{(2)} = \frac{\chi_1 b}{\tau_1} D_{o\pi}^{(2)} \ , \qquad D_{1\sigma}^{(2)} = -D_{2\sigma}^{(2)} = \frac{\chi_1 c}{\tau_1} D_{o\sigma}^{(2)} \tag{12.168}$$

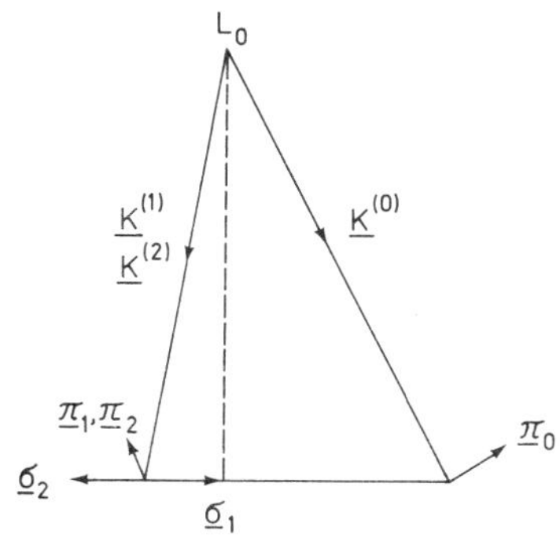

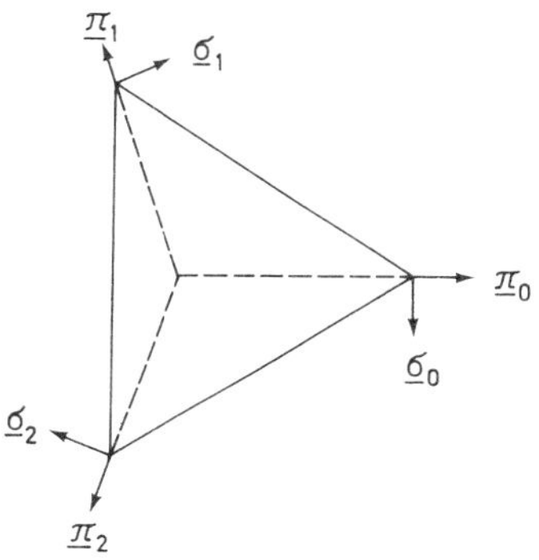

Fig. 12.6 Choice of polarization vectors for Borrmann-Hartwig case

where

$$a = -\underline{\sigma}_o\underline{\sigma}_1 = -\underline{\sigma}_o\underline{\sigma}_2 \quad , \quad b = -\underline{\pi}_o\underline{\pi}_1 = -\underline{\pi}_o\underline{\pi}_2 \quad , \quad c = \underline{\pi}_o\underline{\sigma}_2 = -\underline{\pi}_o\underline{\sigma}_1 \quad .$$

Thus, the two values of C correspond to different directions of the polarization plane of the incident wave, as in the two-wave case.

What is the nature of the enhancement of BORRMANN's effect in this case? To answer this question, we shall consider the electric field energy distribution over the crystal unit cell. We introduce the value

$$F(\underline{r}) = \frac{|\underline{D}(\underline{r})|^2}{\int\limits_{\Omega} dr|\underline{D}(\underline{r})|^2} = \frac{\sum\limits_{mn} \underline{D}_m\underline{D}_n^* \exp[-2\pi i(\underline{h}_m - \underline{h}_n)\underline{r}]}{\sum\limits_{m} \underline{D}_m\underline{D}_m^*} \quad . \tag{12.169}$$

In the two-wave case on the vector $\underline{h}_1$ the value of $F(\underline{r})$ changes only in the direction of the vector $\underline{h}_1$ and has the following form when the Bragg condition (the origin at the inversion centre) is rigorously fulfilled

$$F(x) = 1 + \cos(2\pi x) = 2\cos^2(\pi x) \quad , \quad x = \underline{h}_1(\underline{r} - \underline{r}_o) \quad .$$

To the atomic positions there correspond the values $x = x_n$, where

$$x_n = -\frac{3}{8}, \frac{3}{8}, \frac{5}{8}, \frac{11}{8}, \dots , \quad F(x_n) = 1 - \frac{1}{\sqrt{2}} \quad .$$

Consequently, in the two-wave case, for a reflection of the C type the field nodes do not coincide with atoms.

In the three-wave case, $F(\underline{r})$ changes already in the plane of the reciprocal lattice vectors $\underline{h}_1$ and $\underline{h}_2$. Fig. 12.7 exhibits the distribution of $F(\underline{r})$ and the projections of the atoms onto this plane for solution with

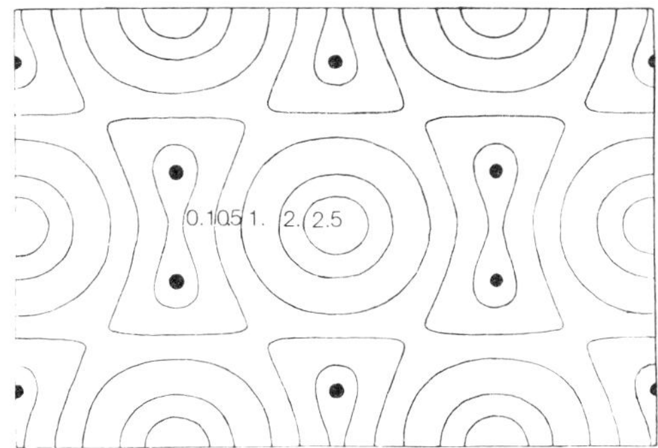

Fig. 12.7 X-ray wave energy density distribution over crystal atoms in Borrmann-Hartwig case

$C = C_1$. The calculation has been performed for a germanium crystal and Ag K_α radiation. Note that strictly at the lattice points the value of F is equal to

$$F(r_n) = \left.\frac{\mu}{\mu_o}\right|_{W=0} = 1 - \frac{1}{\sqrt{2}}\sqrt{C_1} \approx 0.0048 \quad ,$$

that is to say, a practically complete compensation takes place, and the absorption coefficient in the oscillating lattice is determined mainly by the Debye-Waller factor for (111) reflection and may be appreciably smaller than the two-wave value on (220). For the Cu K_α line, we obtain $F(r_n)=0.039$.

Since $W_{220} = 0.035$ for T = 300 K, in this case the three-wave absorption coefficient already exceeds the two-wave one on (220) reflection.

Let us now consider the question of direct intensification of (220)-transmission in the three-wave case. Since a plane perpendicular to (220) is the plane of symmetry of the reciprocal lattice, only isosceles triangles with a base (220) are possible. But, as follows from Section 12.9, in such cases the coefficient of absorption, as measured along the incident beam, is precisely equal to (and, along the normal to the entrance surface, even larger than), the two-wave value, and thus only amplitude intensification is possible.

When switching to four, five and more strong waves, of greatest interest (as already noted) are cases where the reciprocal lattice vectors lie in one and the same plane. This requires that the vectors form a polygon inscribed in a circle. In actual fact, such cases occur rather often because of the high axial symmetry of the cubic lattice. Therefore when proceeding to an analysis of some three-wave case or other, one must always check whether any other points of the reciprocal lattice lie on its circumscribed circle.

This checkup can be made most conveniently in the following manner. We assign vectors $\underline{h}_1, \underline{h}_2$ defining the plane. The vector $\underline{R}_o$, drawn from the centre of the circumscribed circle to the point 0, is defined by (12.5). Let $\underline{e}_1, \underline{e}_2$ be the basis in the plane of the vectors $\underline{h}_1$ and $\underline{h}_2$, i.e., two linearly independent permitted vectors with a minimum modulus lying in this plane. Write down $\underline{h} = \alpha\underline{e}_1 + \beta\underline{e}_2$. The condition for the vector $\underline{h}$ to belong to this circle has the form of (12.4), from which we obtain the following equation for α and β:

$$\alpha^2 - 2\alpha(A - C\beta) - \beta(B - D\beta) = 0 \quad ,$$

$$A = (\underline{R}_o\underline{e}_1)/e_1^2 \quad , \quad B = 2(\underline{R}_o\underline{e}_2)/e_1^2 \quad , \tag{12.170}$$

$$C = (\underline{e}_1\underline{e}_2)/e_1^2 \quad , \quad D = e_2^2/e_1^2 \quad .$$

Note that A and B can be calculated directly from (12.170) for the specific values of α_1, β_1 and α_2, β_2, which correspond to the vectors $\underline{h}_1$ and $\underline{h}_2$ in the selected basis.

Expressing α in terms of β, we have

$$\alpha_{1,2} = (A - C\beta) \pm \sqrt{(A - C\beta)^2 + \beta(B - D\beta)} \quad . \tag{12.171}$$

Since we are interested exclusively in integral values of α, only such integral values of β are possible for which

$$E = (A - C\beta)^2 + \beta(B - D\beta) \geq 0 \quad . \tag{12.172}$$

Such values of β lie between the roots of the equation $E = 0$, because D is always larger than C. Taking all the integral values of β from the indicated interval and calculating for them, we find all the complementary vectors $\underline{h}$ lying on the circle, if any.

A general survey of the possible multiwave combinations in various planes can readily be carried out proceeding from the form of the basis. Let us consider a plane perpendicular to (001). The basis in it is formed by mutually perpendicular vectors (220) and ($\bar{2}$20). Hence, we have a square lattice, in which squares, rectangles, octagons, trapezia, and so on may be formed, and all the cases will contain reflection of type A. In a plane perpendicular to (1$\bar{1}$1), the vectors (220) and (022) form a triangular lattice. Therefore, here too, only combinations of the type 1 of the scheme (12.161) are possible. The probable figures are triangles, hexagons, dodecagons, but there may also be quadrangles. A plane perpendicular to (1$\bar{1}$0) is the richest in possible combinations. The basis in it is formed by the vectors (111) and (11$\bar{1}$) belonging to the type C. By linear combination of these vectors one can obtain reflections of any type. Appendix C lists some of the possible multiwave cases and their radii of the circumscribed circle in units of 1/a.

12.12 Bragg Reflection of X-Rays in Multiwave Diffraction

In contrast to the two-wave case, multiwave diffraction offer different schemes of experiments depending on how many diffracted waves are Bragg-reflected, i.e., emerge from the crystal through the entrance surface. In particular, intermediate cases are possible, where part of the diffracted waves are Bragg-reflected and the other part, Laue-reflected. It is only natural to refer all of them to the Bragg case, because the nature of the physical phenomena occurring in this geometry of experiment differs sharply from pure Laue geometry, when the entire radiation passes through the crystal. The main difference is in the formation of a new effect, extinction, i.e., an abrupt increase in reflection of the radiation from the crystal at a certain orientation of the incident beam. The existence of additional diffracted waves naturally results in new qualitative features of this effect.

Let us consider, as in Section 12.6, reflection and transmission of X-rays by a crystal in the shape of a plane-parallel plate; we assume that the incident vacuum wave is plane (12.79). The vector of induction in the crystal has the form of (12.80) as before, but the boundary conditions for the amplitudes $D_m^{(j)}$ are different now. Having in mind the general case, we assume that the reciprocal lattice vectors $\underline{h}_m$ are ordered in such a way that with increasing m the values of γ_m (12.24) for $m \neq 0$ decrease, the last N_1 values being negative

$$\gamma_o > 0 \quad , \quad \gamma_1 \geq \gamma_2 \geq \cdots \geq \gamma_{N-N_1-1} > 0 \quad , \quad 0 > \gamma_{N-N_1} \geq \cdots \geq \gamma_{N-1} \quad . \tag{12.173}$$

Averaging over a region with dimensions exceeding interatomic distances, both on the entrance and the exit surface of the plate, we obtain the following conditions for the amplitudes

$$\sum_j \underline{D}_o^{(j)} = \underline{D}_o^{(a)} \quad , \quad \sum_j \underline{D}_m^{(j)} = 0 \quad \text{for} \quad m \leq N - N_1 - 1$$

$$\sum_j \underline{D}_m^{(j)} \exp[2\pi i K\delta^{(j)} z] = 0 \quad \text{for } m \geq N - N_1 \tag{12.174}$$

where z is the plate thickness. The incident wave and the diffracted waves for $m = 1, \ldots , N - N_1 - 1$ emerge from the crystal through the exit surface, from the side of which plane waves

$$\underline{D}_m^{(d)} = \exp[2\pi(\nu t - \underline{K}_m \underline{r})] \sum_j \underline{D}_m^{(j)} \exp[2\pi i K\delta^{(j)} z] \tag{12.175}$$

are excited in the vacuum. On the side of the entrance surface, in addition to the incident wave, there are Bragg-reflected diffracted waves

$$\underline{D}_m^{(a)} = \exp[2\pi i(\nu t - \underline{K}_m \underline{r})] \sum_j \underline{D}_m^{(j)} \quad , \quad m \geq N - N_1 \quad . \tag{12.176}$$

The coefficients of transmission and reflection are determined as follows

$$T = R_o \ , \quad R_m = \frac{|\underline{D}_m^{(d)}|^2}{|\underline{D}_o^{(a)}|^2} \, \frac{\gamma_m}{\gamma_o} \ , \qquad m \leq N - N_1 - 1 \ ,$$

$$R_m = \frac{|\underline{D}_m^{(a)}|^2}{|\underline{D}_o^{(a)}|^2} \, \frac{|\gamma_m|}{\gamma_o} \ , \qquad m \geq N - N_1 \ . \tag{12.177}$$

The system of fundamental equations is naturally independent of the geometry of experiment. As in Section 12.6, we can switch to the scalar field amplitudes and then introduce the eigenvector B_{ns} of the scattering matrix $\hat{G}$, which is of the form of (12.102) in the case of the absorbing crystal. However, due to the presence of negative values among γ_m, the scattering matrix $\hat{G}$ is non-Hermitian even in the case of the nonabsorbing crystal because of the factor $(\gamma_n\gamma_m)^{1/2}$. Hence, the eigenvalues of the scattering matrix can have an imaginary part even when no ordinary absorption occurs. This actually leads to extinction.

Another important difference from the Laue case is the fact that the imaginary part of the matrix, $\hat{G}_i$, loses the property of positive determinacy, i.e., the values $\delta^{(j)}$ may include those with a negative imaginary part. From (12.175) we conclude that for the corresponding excitation mode the induction vector increases exponentially with crystal thickness. Note that for the absorbing crystal the number of such solutions is precisely equal to $2N_1$.

Express the coefficients of reflection in terms of the eigenvectors $B_{ns}^{(j)}$ of the scattering matrix; as before, we assume that the latter are normalized to unity, that is to say

$$\sum_{ms} B_{ms}^{(j)} B_{ms}^{(j)^*} = 1 \ . \tag{12.178}$$

The corresponding expressions will be

$$R_m = \sum_{jj's} B_{ms}^{(j)} B_{ms}^{(j')^*} \exp\{2\pi i K[\delta^{(j)} - \delta^{(j')^*}]z\} \tilde{\lambda}_j \tilde{\lambda}_{j'}^* \qquad \text{for } m \leq N - N_1 - 1 \tag{12.179}$$

$$R_m = \sum_{jj's} B_{ms}^{(j)} B_{ms}^{(j')*} \tilde{\lambda}_j \tilde{\lambda}_j^* , \quad \text{for} \quad m \geq N - N_1$$

where

$$\tilde{\lambda}_j = \lambda_j \frac{1}{\sqrt{\gamma_o} |\underline{D}_o^{(a)}|} . \quad (12.180)$$

The excitation coefficients λ_j are found from a system of 2N linear homogeneous equations which is obtained by substituting the relation (12.84) in the boundary conditions (12.174)

$$\sum_j \lambda_j B_{ms}^{(j)} C_{mj} = \sqrt{\gamma_o} D_{os}^{(a)} \delta_{mo} ,$$

$$C_{mj} = \begin{cases} 1 & m \leq N - N_1 - 1 \\ \exp[2\pi i K \delta^{(j)} z] & m \geq N - N_1 \end{cases} . \quad (12.181)$$

Thus, (12.179-181) enable one to calculate the coefficients of reflection in the general case if the problem on the eigenvalues of the scattering matrix has been solved.

We now pass over to the semi-infinite absorbing crystal. In this case no direct solution of the system (12.181) is possible, and additional analysis is necessary. Indeed, the values of C_{mj} for $m \geq N - N_1$ tend either to zero or to infinity for $z \to \infty$, depending on the sign of the imaginary part of the values $\delta^{(j)}$. Let us arrange the values $\delta^{(j)}$ so that with increasing j their imaginary part decreases. Then the last $2N_1$ eigenvalues will have a negative imaginary part, and the others, a positive one. Consequently, the coefficients at λ_j for $j \leq 2(N - N_1)$ in the last $2N_1$ equations of (12.181) will be strictly equal to zero, and the remaining values λ_j must be equated to zero so that the last $2N_1$ equations are satisfied, i.e.,

$$\lambda_j = 0 \text{ for } j \geq 2(N - N_1) + 1 \text{ for which } \delta_i^{(j)} < 0 . \quad (12.182)$$

Hence, for a semi-infinite absorbing crystal, not all the eigensolutions of the scattering matrix are excited in the crystal, but only such for which

$\delta_i^{(j)} > 0$. This result has a simple physical meaning. Since the induction vector for solutions with $\delta_i^{(j)} < 0$ increases exponentially with crystal thickness, for a semi-infinite crystal such solutions have no physical meaning, and therefore must be discarded. For a thin crystal these solutions are preserved, but the excitation coefficients for them strongly depend on the plate thickness and tend to zero with an increase in z.

The values λ_j for j from 1 to $2(N - N_1)$ are determined from the abbreviated system (12.181)

$$\sum_j \lambda_j B_{ms}^{(j)} = \sqrt{\gamma_o} D_{os}^{(a)} \delta_{mo}$$

$$j \leq 2(N - N_1) \quad , \quad m \leq N - N_1 - 1 \quad , \quad s = \pi,\sigma \quad . \tag{12.183}$$

The simplest form of (12.183) refers to the purely Bragg case, where all the diffracted waves are Bragg-reflected. In this case only two solutions with j = 1, 2 are essential. The corresponding values, $\tilde{\lambda}_1$ and $\tilde{\lambda}_2$, are found from the system of two equations

$$\lambda_1 B_{o\pi}^{(1)} + \lambda_2 B_{o\pi}^{(2)} = \cos\varphi$$

$$\lambda_1 B_{o\sigma}^{(1)} + \lambda_2 B_{o\sigma}^{(2)} = \sin\varphi \tag{12.184}$$

where φ is the angle between the plane of polarization of the incident vacuum wave and the vector $\underline{\pi}_o$. Solving the system, we obtain

$$\tilde{\lambda}_1 = \frac{\cos\varphi B_{o\sigma}^{(2)} - \sin\varphi B_{o\pi}^{(2)}}{B_{o\pi}^{(1)} B_{o\sigma}^{(2)} - B_{o\sigma}^{(1)} B_{o\pi}^{(2)}} \quad , \quad \tilde{\lambda}_2 = \frac{\cos\varphi B_{o\sigma}^{(1)} - \sin\varphi B_{o\pi}^{(1)}}{B_{o\pi}^{(2)} B_{o\sigma}^{(1)} - B_{o\sigma}^{(2)} B_{o\pi}^{(1)}} \quad . \tag{12.185}$$

By way of example we will again consider the case $\underline{h}_{1,2} = (111/\bar{1}11)$ in germanium, but with the condition $\gamma_1 = \gamma_2 = -\gamma_o$. The roots of the dispersion equation are in this case obtained from (12.164) at $\beta = -1$

$$-2\delta\gamma_o = \frac{1}{2}\alpha \pm \frac{1}{2}\sqrt{(\alpha - 2\chi_o)^2 - 4\chi_1^2 C} \quad . \tag{12.186}$$

At any value of C there is a root with a negative imaginary part. As for the doubly degenerate solution of the equation

$$\tau_1 = \chi_0 + 2\delta\gamma_1 - \alpha = 0 \quad ,$$

it also has a negative imaginary part at $\gamma_1 < 0$. Thus, in accordance with the general theory, we have two solutions with a positive imaginary part.

The expressions for λ in this case are simplified still further since $B^{(1)}_{o\pi} = 0$, $B^{(2)}_{o\sigma} = 0$. Here,

$$\tilde{\lambda}_1 = \frac{\sin\varphi}{B^{(1)}_{o\sigma}} \quad , \quad \tilde{\lambda}_2 = \frac{\cos\varphi}{B^{(2)}_{o\pi}} \quad . \tag{12.187}$$

Taking into consideration the relations for the amplitudes (12.167) and (12.168) and also the relations $2(a^2 + c^2) = C_1$, $2(b^2 + c^2) = C_2$ (see (12.166)), it is easy to see that the coefficients of reflection for the case where the incident radiation is either σ- or π-polarized, have the form

$$R_1 = R_2 = \frac{2|\chi_1|^2 C}{|2\chi_0 - \alpha) + \sqrt{(2\chi_0 - \alpha)^2 - 4\chi_1^2 C}|^2} \quad . \tag{12.188}$$

Here, the sign of the square root in the denominator must be chosen so that it has a positive imaginary part. This condition automatically singles out, in the vicinity of the maximum, three areas similarly to the consideration of the two-wave case in Chapter 7, and in different areas the real part of the root will have a different sign. The value of C is equal to C_1 for σ-polarization and C_2 for π-polarization.

It is easy to see that in this case (12.188) for the coefficients of reflection differs from the corresponding two-wave expression only by the new polarization factor. The corresponding half-width of the extinction maximum will be $\sqrt{C}$ times larger than in the two-wave case at σ-polarization. Besides, the maximum of the reflection coefficient in the extinction region is equal to 0.5 (for the nonabsorbing crystal), which is in full agreement with the law of conservation of energy, according to which $R_1 + R_2 = 1$. Equation (12.188) is valid only on the line $\alpha_1 = \alpha_2$. With another ratio of α_1 and α_2, the values R_1 and R_2 are no longer èqual to each other. Thus, for instance, at α_1 = const, as we recede from the three-wave point, R_2 tends to zero and R_1, to its two-wave value, which is close to unity, if the value α_1 corresponds to the extinction region.

As in the Laue case, the dependence of the induction vector for diffracted waves on the polarization of the incident wave can be represented as

$$\underline{D}_m(\varphi) = \underline{D}_m(\pi)\cos\varphi + \underline{D}_m(\sigma)\sin\varphi \tag{12.189}$$

where $\underline{D}_m(\pi)$ and $\underline{D}_m(\sigma)$ are the solutions for the cases where the incident wave is either π-, or σ-polarized. Hence, for the intensity of the diffracted waves we get

$$I_m(\varphi) = |\underline{D}_m(\varphi)|^2 = |\underline{D}_m(\pi)|^2\cos^2\varphi + |\underline{D}_m(\sigma)|^2\sin^2\varphi$$

$$+ \mathrm{Re}\{\underline{D}_m(\pi)\underline{D}_m^*(\sigma)\}\sin 2\varphi \quad . \tag{12.190}$$

If the incident radiation is not polarized, the intensities are equal to

$$\bar{I}_m = \frac{1}{2\pi}\int_0^{2\pi} d\varphi I_m(\varphi) = \frac{1}{2}\{|\underline{D}_m(\pi)|^2 + |\underline{D}_m(\sigma)|^2\} \quad . \tag{12.191}$$

Interesting general relations can be obtained by using the reciprocity principle [2.1]. Let a plane wave with a wave vector $\underline{K}_o^{(1)}$ fall on a crystal in a direction close to the multiwave one. The wave vectors of the diffracted waves will be denoted by $\underline{K}_m^{(1)}$. Let a wave with $\underline{K}_1^{(1)}$ correspond to Bragg reflection. Consider a second case, in which the wave vector of the incident wave $\underline{K}_o^{(2)} = -\underline{K}_1^{(1)}$. Because of the uniqueness of the multiwave pyramid, for the diffracted waves of the second case we obtain $\underline{K}_1^{(2)} = -\underline{K}_o^{(1)}$, $\underline{K}_m^{(2)} = -\underline{K}_m^{(1)}$, $m = 2, \ldots, N-1$ (see Fig. 12.8 for $N = 3$). Thus, the reciprocal lattice vectors for both cases are identical, but the point 0 is chosen differently. In particular, in the second case the number of Bragg-reflected waves will be equal to the number of Laue-transmitted waves in the first case. The reciprocity principle makes it possible to establish the relationship between the induction vectors $\underline{D}_1^{(1)}$ and $\underline{D}_1^{(2)}$.

The general formulation of the reciprocity principle stages that if $\underline{j}_s^{(1)}(\underline{k})$ and $\underline{j}_s^{(2)}(\underline{k})$ are the Fourier components of the currents of two radiation sources, and $\overset{(1)}{\underline{E}}(\underline{k})$, $\overset{(2)}{\underline{E}}(\underline{k})$ are the Fourier components of the electric fields, each induced by these sources separately, then

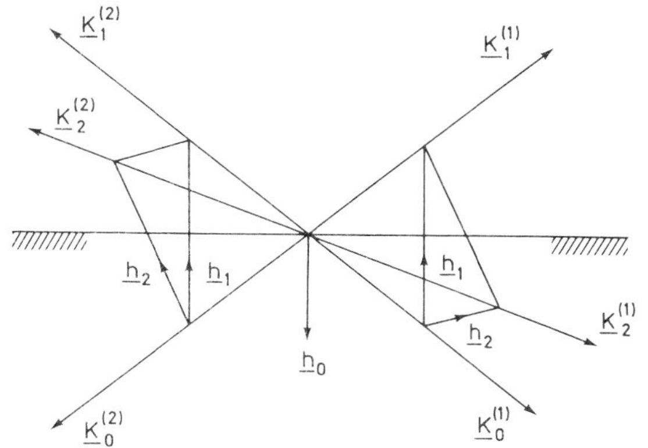

Fig. 12.8 Schemes of reciprocal lattice vectors and wave vectors related by reciprocity principle

$$\sum_k \underline{j}_s^{(1)}(\underline{k})\underline{E}^{(2)}(-\underline{k}) = \sum_k \underline{j}_s^{(2)}(\underline{k})\underline{E}^{(1)}(-\underline{k}) \quad . \tag{12.192}$$

Choosing the sources in the form

$$\underline{j}_s^{(i)}(\underline{k}) = \text{const } \underline{e}_{oa}^{(i)}\delta(\underline{k} - K_o^{(i)})$$

where $\underline{e}_{oa}^{(i)}$ is a unit polarization vector of the incident plane wave, and neglecting the small difference between $\underline{E}$ and $\underline{D}$, we obtain from (12.192)

$$\underline{e}_{oa}^{(1)}\underline{D}_1^{(2)} = \underline{e}_{oa}^{(2)}\underline{D}_1^{(1)} \quad . \tag{12.193}$$

Determining the polarization vectors in each case in the usual way (see Section 12.1), we note that $\underline{\pi}_o^{(1)} = -\underline{\pi}_1^{(2)}$, $\underline{\pi}_o^{(2)} = -\underline{\pi}_1^{(1)}$ and we have similar relations for $\underline{\sigma}$-vectors.

Let now both sources be π-polarized. Then we obtain directly from (12.103)

$$D_{1\pi}^{(2)}(\pi) = D_{1\pi}^{(1)}(\pi) \quad . \tag{12.194a}$$

Assuming similarly that both sources are σ-polarized, or one of these is π-polarized and the other σ-polarized, we find

$$D_{1\sigma}^{(2)}(\sigma) = D_{1\sigma}^{(1)}(\sigma) \tag{12.194b}$$

$$D_{1\pi}^{(2)}(\sigma) = D_{1\sigma}^{(1)}(\pi) \tag{12.194c}$$

$$D_{1\sigma}^{(2)}(\pi) = D_{1\pi}^{(1)}(\sigma) \quad . \tag{12.194d}$$

Taking into account (12.194) and (12.191), we can easily see that if the sources are not polarized, the intensities of the waves directed along $\underline{K}_1^{(i)}$ coincide in the two cases. Note that relations of the type (12.194) also exist in the case of pure Laue geometry.

The first experimental study of multiwave diffraction was carried out by RENNINGER in 1937 [1.29] precisely in the case of Bragg reflection. RENNINGER used the following scheme of experiment. A crystal plate is carved out so that the entrance surface is perpendicular to some reciprocal lattice vector $\underline{h}_1$. The X-ray source is so set up in relation to the crystal that the incident wave diffracts on the vector $\underline{h}_1$, while the detector measures the intensity of the Bragg-reflected radiation (Fig. 12.9). If we rotate the crystal relative to the vector $\underline{h}_1$, the detector readings remain unaffected.

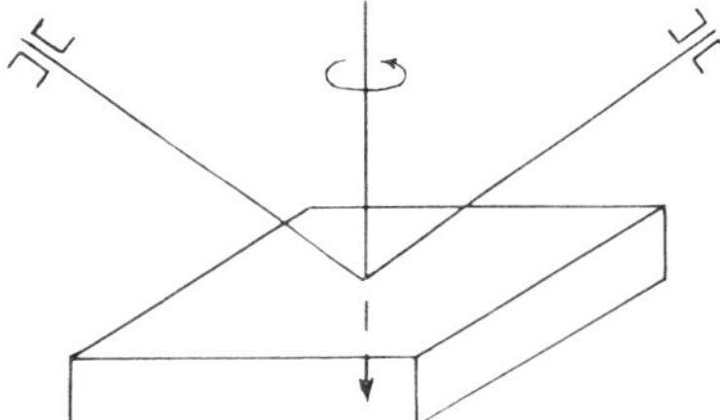

Fig. 12.9 Scheme of experiment in RENNINGER method

However, at a certain angle of rotation the crystal finds itself in a position in which multiwave diffraction occurs, and this position is immediately fixed by the detector. If, for the vector $\underline{h}_1$, we choose a vector of the type B in the scheme (12.159) with a zero structure amplitude, the multiwave positions of the crystal will yield a sharp peak against a weak background on the curve of the diffracted beam intensity via angle of rotation. It is easy to see that the width of this peak will be of the order of several seconds of arc. This is just what RENNINGER observed in his experiments.

The RENNINGER method is still one of the principal methods for studying multiwave diffraction in Bragg reflection [12.4,5,29,30,37,38]. The main advantage of the method is that the radiation source and the detector remain fixed during the measurements, while the rotation of the crystal plate can easily be arranged with sufficient accuracy.

Recently, this method was successfully used for precision measurement of the parameters of the germanium and silicon crystal lattice [12.4,5]. The idea of these measurements is as follows. Since the source is fixed, the centre of the Ewald sphere of radius $(1/\lambda)$ will be assumed fixed in space, while the reciprocal lattice vectors change their positions during crystal rotation. Decompose the vector $\underline{h}_m$ into two components, one parallel and the other perpendicular to $\underline{h}_1$, and denote them by $\underline{h}_{m\parallel}$ and $\underline{h}_{m\perp}$, respectively. During the rotation of the crystal about the vector $\underline{h}_1$, the component $\underline{h}_{m\parallel}$ remains unchanged, while the vector $\underline{h}_{m\perp}$ is also rotating and intersects the Ewald's sphere twice, which is clearly evidenced by Fig. 12.10.

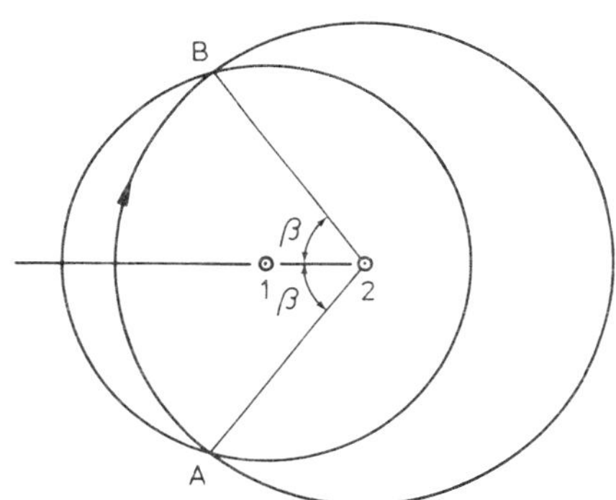

Fig. 12.10 To problems of precision measurement of crystal lattice spacing (after POST [12.4])

The figure shows the section of Ewald's sphere by a plane perpendicular to $\underline{h}_1$ - a circle with centre at point 1, and also a circle of radius $R_2 = |\underline{h}_{m\perp}|$ with centre at point 2, point of intersection of the vector $\underline{h}_1$ with the plane of figure. The radius of the first circle $R_1 = \sqrt{\lambda^{-2} - (0.5\underline{h}_1 - \underline{h}_{m\parallel})^2}$, and the distance between the points 1 and 2 is equal to $x = \sqrt{\lambda^{-2} - 0.25h_1^2}$. It can be seen from the figure that if the crystal is set up in the multiwave position A, then on rotation by an angle of 2β it finds itself in the multiwave position B with the same reciprocal lattice vector $\underline{h}_m$. It is this angle that is measured in experiment with a high accuracy. Here we have

$$\cos\beta = \frac{R_2^2 + x^2 - R_1^2}{2xR_2} = \frac{h_m^2 - \underline{h}_m\underline{h}_1}{2\sqrt{\left[h_m^2 - \frac{(\underline{h}_m\underline{h}_1)^2}{h_1^2}\right]\left(\frac{1}{\lambda^2} - \frac{h_1^2}{4}\right)}} \quad . \tag{12.195}$$

The right-hand side of (12.195) depends only on the wavelength of the radiation and on the crystal lattice spacing. Therefore the accuracy of determination of the lattice spacing depends only on the accuracy with which the values of λ and 2β are known. In particular, it is necessary to know the displacement of the maximum of the reflection coefficient with respect to the direction predetermined by the Bragg conditions, i.e., of the point 0 on the plane of the parameters θ_1 and θ_2.

12.13 Methods for Numerical Determination of Dispersion Surface and Electric Displacement Vectors

As has been shown in Section 12.6, the determination of the wave field in a crystal with a plane entrance surface completely reduces to the problem on the eigenvalues of the scattering matrix

$$\sum_{\beta=1}^{N} g_{\alpha\beta} B_{\beta j} = \delta_j B_{\alpha j} \quad , \quad \hat{g}\underline{B}_j = \delta_j \underline{B}_j \quad . \tag{12.196}$$

Then, the values of δ_j, together with the parameters θ_1, θ_2, determine the shape of the dispersion surface in a coordinate system with origin at the Laue point (12.25), while the vectors $\underline{B}_j$ are related with the induction vectors by (12.84).

In the general case of multiwave diffraction, i.e., at arbitrary values of the parameters θ_1, θ_2, and γ_m, the solution of (12.196) can be obtained only by numerical methods with the use of a computer. In this connection, of great importance is the development of such methods of calculation which require the minimum possible computation expenses, yield sufficiently accurate results and, besides, are easily formalized. At present, many such methods have been developed [12.40,41], all of them having their advantages and shortcomings.

1) The simplest computation program consists in the following. First we compute the coefficients of the characteristic polynomial of the matrix $\hat{g}$ ($\hat{I}$ is the unit matrix)

$$\varphi(\delta) = \det(\delta\hat{I} - \hat{g}) = \delta^N + \sum_{m=1}^{N} A_m \delta^{N-m} = 0 \quad . \tag{12.197}$$

This is the first stage. In the second stage we compute the roots of the characteristic polynomial δ_j. Finally, in the third stage we find the vectors $\underline{B}_j$, using the information on δ_j and A.

A typical method of this type is the Krylov method. To find the coefficients A_m, this method uses the Hamilton-Kelly theorem according to which the matrix is the root of its characteristic polynomial, i.e., $\varphi(\hat{g}) = 0$. Multiplying this identity from the left by the arbitrary vector $\underline{c}$, we obtain a system of linear nonhomogeneous equations for determining A_m

$$\sum_{m=1}^{N} x_{\alpha m} A_m = D_\alpha \quad , \quad \hat{x}\underline{A} = \underline{D} \tag{12.198}$$

where

$$x_{\alpha m} = (\hat{g}^{N-m}\underline{c})_\alpha \quad , \quad \underline{D} = -\hat{g}^N \underline{c} \quad . \tag{12.199}$$

Thus, the first stage amounts to solving (12.198). Solution may be carried out, say, by the Gauss method.

The second stage is not difficult, as a rule. The roots are readily found by iterational procedures, for instance by the method of Newton or Müller, with the elimination of the root after finding it, i.e., by dividing the polynomial by the factor $(\delta - \delta_j)$. The degree of the polynomial decreases all the time as the computations progress and, besides, the method converges on the new root with any choice of initial approximation.

The eigenvectors $\underline{B}_j$ are then computed by the equation

$$\underline{B}_j = \prod_{j' \neq j} \frac{(\delta_{j'}\hat{I} - \hat{g})}{(\delta_{j'} - \delta_j)} \underline{c} = \frac{1}{\psi_j(\delta_j)} \psi_j(\hat{g})\underline{c} \tag{12.200}$$

where

$$\psi_j(\delta) = \varphi(\delta)/(\delta - \delta_j) = \delta^{N-1} + \sum_{m=2}^{N} Q_{mj}\delta^{N-m} \quad . \tag{12.201}$$

Then,

$$[\psi_j(g)c]_\alpha = \chi_{\alpha 1} + \sum_{m=2}^{N} \chi_{\alpha m} Q_{mj} \tag{12.202}$$

and the coefficients Q_{mj} are easily found from the recurrence relations

$$Q_{2j} = A_1 + \delta_i \ , \quad Q_{mj} = A_{m-1} + \delta_j Q_{m-1,j} \ , \quad m \geq 3 \ . \tag{12.203}$$

The total number of the multiplications and additions required in this method is proportional to N^3, and the number of operations needed for computing the matrix x, solving (12.198), and computing the vectors $\underline{B}_j$ is about the same. The Krylov method is one of the fastest methods, but it also has a grave shortcoming in the first stage. To determine the coefficients A_m from (12.198) it is necessary that the matrix $\hat{x}$ be nonsingular, i.e., det $(\hat{x}) \neq 0$. This condition, however, is not always adequately fulfilled. For instance, if in the decomposition of the vector $\underline{c}$ into eigenvectors $\underline{B}_j$

$$\underline{c} = \sum_j \lambda_j \underline{B}_j \ , \tag{12.204}$$

not all the coefficients λ_j are different from zero, the system of vectors

$$\hat{g}^m \underline{c} = \sum_j \lambda_j \delta_j^m \underline{B}_j \tag{12.205}$$

will always be linearly dependent, and the matrix $\hat{x}$ will be a singular matrix. For the same reason the method is inapplicable when the matrix $\hat{g}$ has degenerate roots. For large N the method gives poor results if the eigenvalues of g differ widely in modulus. The first columns of the matrix x will contain too large numbers and, besides, will be almost linearly dependent and proportional to the eigenvector for the eigenvalue maximal in modulus, which can easily be seen from (12.205). In this case (12.198) is poorly justified, and the coefficients A_m are grossly in error.

This shortcoming is easy to avoid in the case of three- and four-wave diffraction if we find the coefficients directly from (12.40) and (12.77). In the Laue case it is convenient to use the vacuum vector $\sqrt{\gamma_0}\, D_{os}^{(a)} \delta_{mo}$ as the vector $\underline{c}$ (12.91). Then, all the $\lambda_j = 1$, and (12.200) yields at once a solution satisfying the boundary conditions. The Krylov method is also useful in the case of Bragg reflection when the scattering matrix is non-Hermitian, even for the nonabsorbing crystal. It should be remembered, however, that in many cases the method results in considerable errors, and should therefore be applied with caution.

2) The most attractive method, in the sense of stability of the computation procedure and the absence of complication, is the iterational method of direct diagonalization of the matrix. We restrict ourselves to the important particular case where the matrix $\hat{g}$ is Hermitian. In this case the method described below is called the Jakobi method, or the rotation method. We rewrite (12.196) in matrix form

$$B^{-1}gB = \delta \tag{12.206}$$

where B is a matrix whose columns are eigenvectors and δ is a diagonal matrix, whose diagonal elements are equal to the eigenvalues. The problem is to find the matrix B, which transforms the scattering matrix g into a similar matrix.

Let us first consider the simplest case of the 2nd-order matrix

$$g = \begin{pmatrix} g_{11} & g_{12} \\ g_{21} & g_{22} \end{pmatrix} , \qquad g_{21} = g_{12}^{*} \quad . \tag{12.207}$$

In this case the problem is solved accurately. The second-order matrix B is conveniently written as

$$B = \begin{pmatrix} \cos\varphi & -e^{i\psi}\sin\varphi \\ e^{-i\psi}\sin\varphi & \cos\varphi \end{pmatrix} , \qquad B^{-1} = B^{+} \tag{12.208}$$

where the angles φ and ψ are determined from the equations

$$\tan 2\varphi = \frac{2|g_{12}|}{(g_{11} - g_{22})} , \qquad e^{i\psi} = \frac{g_{12}}{|g_{12}|} \quad . \tag{12.209}$$

Suppose now that the matrix order exceeds two, and $|g_{ij}|^2$, $i \neq j$ reaches its maximum value for the indices k and m. Compose the matrix B_1 according to the following rule. The elements located at the intersection of two rows and columns with numbers k and m are determined from (12.208) and (12.209) with replacement of the subscripts 1 and 2 by k and m, while the remaining elements are the same as in the unit matrix. Having performed the similarity transformation of the matrix B_1, we obtain a new matrix

$$g^{(1)} = B_1^{-1} g B_1 \quad . \tag{12.210}$$

This completes one iteration.

To estimate the convergence, we can obtain the following relation

$$\sum_{i<j} |g_{ij}^{(1)}|^2 = \sum_{i<j} |g_{ij}|^2 - |g_{km}|^2 \le \left[1 - \frac{2}{N(N-1)}\right] \sum_{i<j} |g_{ij}|^2 \ . \tag{12.211}$$

Repeating the above procedure n times, we arrive at the matrix

$$g^{(n)} = B_n^{-1} \dots B_1^{-1} g B_1 \dots B_n$$

for which the following estimate is valid

$$\sum_{i<j} |g_{ij}^{(n)}|^2 < \left[1 - \frac{2}{N(N-1)}\right]^n \sum_{i<j} |g_{ij}|^2 \approx \exp\left[-\frac{2n}{N(N-1)}\right] \sum_{i<j} |g_{ij}|^2 \quad . \tag{12.212}$$

Thus, in order to decrease the sum of the squares of the moduli of the nondiagonal elements by a factor of exp(-M), not more than $MN^2/2$ iterations are needed. The number of operations required for performing one iteration is actually proportional to N, since the transformation (12.210) affects only two columns and two rows of the matrix. Hence, the total number of operations required in this method is proportional to N^3 as well as in the Krylov method, but the proportionality coefficient is considerable.

Note that for non-Hermitian matrices it is also possible to construct the matrix $\hat{B}$, which reduces to zero two nondiagonal elements of the matrix $\hat{g}$. In this case, however, the relation of the type (12.201) has the form

$$\sum_{i<j} g_{ij}^{(1)} g_{ji}^{(1)} = \sum_{i<j} g_{ij}g_{ji} - g_{km}g_{mk} \tag{12.213}$$

and conveys nothing about the convergence of the method. A correct generalization of this method for the case of non-Hermitian matrices is given in [12.42].

Appendix A

The Values of the Integral $\int_0^{2A} J_0(x)dx$ for $A < 5$

2A	$\int$	2A	$\int$	2A	$\int$
0.00	0.00000	3.40	1.26056	6.80	0.89512
0.10	0.09991	3.50	1.22330	6.90	0.92470
0.20	0.19933	3.60	1.18467	7.00	0.95464
0.30	0.29775	3.70	1.14509	7.10	0.90462
0.40	0.39469	3.80	1.10496	7.20	1.01435
0.50	0.48968	3.90	1.06471	7.30	1.04354
0.60	0.58224	4.00	1.02473	7.40	1.07190
0.70	0.67193	4.10	0.98541	7.50	1.09917
0.80	0.75834	4.20	0.94712	7.60	1.12508
0.90	0.84106	4.30	0.91021	7.70	1.14941
1.00	0.91973	4.40	0.87502	7.80	1.17192
1.10	0.99399	4.50	0.84186	7.90	1.19243
1.20	1.06355	4.60	0.81100	8.00	1.21074
1.30	1.12813	4.70	0.78271	8.10	1.22671
1.40	1.18750	4.80	0.75721	8.20	1.24021
1.50	1.24144	4.90	0.73468	8.30	1.25112
1.60	1.28982	5.00	0.71531	8.40	1.25939
1.70	1.33249	5.10	0.69920	8.50	1.26494
1.80	1.36939	5.20	0.68647	8.60	1.26777
1.90	1.40048	5.30	0.67716	8.70	1.26787
2.00	1.42577	5.40	0.67131	8.80	1.26528
2.10	1.44528	5.50	0.66891	8.90	1.26005
2.20	1.45912	5.60	0.66992	9.00	1.25226
2.30	1.46740	5.70	0.67427	9.10	1.24202
2.40	1.47029	5.80	0.68187	9.20	1.22946
2.50	1.46798	5.90	0.69257	9.30	1.21473
2.60	1.46069	6.00	0.70622	9.40	1.19799
2.70	1.44871	6.10	0.72263	9.50	1.17944
2.80	1.43231	6.20	0.74160	9.60	1.15927
2.90	1.41181	6.30	0.76290	9.70	1.13772
3.00	1.38756	6.40	0.78628	9.80	1.11499
3.10	1.35992	6.50	0.81147	9.90	1.09134
3.20	1.32928	6.60	0.83820	10.00	1.06701
3.30	1.29602	6.70	0.86618		

Appendix B

The Integrated Bragg-Reflection R_i^y Values for Infinite Thick Crystal (A>4)

$0 \le -g \le 3.000$

$0 \le x \le 1.000$

-g	k = 0	0.025	0.50	0.075	0.100	0.125	0.150	0.175	0.200	0.250	0.300	0.400	0.500	0.600	0.700	0.800	0.900	1.000
0.000	2.6667																	
0.005	2.6357																	
0.010	2.6059																	
0.015	2.5769																	
0.020	2.5487																	
0.025	2.5207	2.5278																
0.030	2.4937	2.4998																
0.035	2.4674	2.4730																
0.040	2.4415	2.4468																
0.045	2.4162	2.4212																
0.050	2.3913	2.3959	2.4113															
0.055	2.3669	2.3713	2.3855															
0.060	2.3430	2.3472	2.3606															
0.065	2.3195	2.3236	2.3364															
0.070	2.2964	2.3004	2.3126															
0.075	2.2738	2.2775	2.2893	2.3106														
0.080	2.2515	2.2552	2.2664	2.2864														
0.085	2.2296	2.2332	2.2440	2.2631														
0.090	2.2081	2.2115	2.2220	2.2404														
0.095	2.1870	2.1903	2.2005	2.2182														
0.100	2.1662	2.1694	2.1793	2.1961	2.2230													
0.105	2.1457	2.1488	2.1584	2.1748	2.1994													
0.110	2.1256	2.1286	2.1380	2.1538	2.1774													
0.115	2.1058	2.1088	2.1178	2.1332	2.1557													
0.120	2.0863	2.0892	2.0980	2.1130	2.1347													
0.125	2.0671	2.0700	2.0785	2.0931	2.1142	2.1442												
0.130	2.0483	2.0510	2.0594	2.0736	2.0941	2.1224												
0.135	2.0297	2.0324	2.0405	2.0544	2.0743	2.1016												
0.140	2.0014	2.0140	2.0220	2.0355	2.0549	2.0810												
0.145	1.9934	1.9960	2.0037	2.0169	2.0358	2.0611												
0.150	1.9757	1.9781	1.9858	1.9936	2.0170	2.0416	2.0773											
0.155	1.9582	1.9606	1.9681	1.9806	1.9986	2.0225	2.0540											
0.160	1.9410	1.9434	1.9506	1.9629	1.9804	2.0038	2.0339											
0.165	1.9240	1.9264	1.9335	1.9455	1.9626	1.9854	2.0145											
0.170	1.9073	1.9096	1.9166	1.9283	1.9450	1.9672	1.9956											
0.175	1.8909	1.8931	1.8999	1.9114	1.9278	1.9494	1.9771	2.0129										
0.180	1.8746	1.8768	1.8835	1.8948	1.9107	1.9320	1.9589	1.9930										
0.185	1.8586	1.8608	1.8673	1.8783	1.8940	1.9147	1.9410	1.9740										
0.190	1.8429	1.8450	1.8514	1.8622	1.8776	1.8978	1.8234	1.8555										
0.195	1.8273	1.8294	1.8357	1.8463	1.8614	1.8812	1.9062	1.9373										
0.200	1.8120	1.8140	1.8202	1.8306	1.8454	1.8648	1.8893	1.9196	1.9576									
0.210	1.7820	1.7840	1.7899	1.7999	1.8141	1.8328	1.8562	1.8849	1.9202									
0.220	1.7529	1.7547	1.7605	1.7701	1.7838	1.8017	1.8242	1.8516	1.8851									
0.230	1.7244	1.7263	1.7319	1.7411	1.7544	1.7716	1.7932	1.8195	1.8512									
0.240	1.6969	1.6986	1.7040	1.7130	1.7257	1.7424	1.7632	1.7885	1.8187									
0.250	1.6700	1.6717	1.6769	1.6856	1.6979	1.7140	1.7341	1.7584	1.7874	1.8637								
0.260	1.6438	1.6455	1.6505	1.6589	1.6709	1.6864	1.7058	1.7292	1.7571	1.8290								
0.270	1.6183	1.6199	1.6248	1.6330	1.6446	1.6596	1.6783	1.7010	1.7278	1.7964								
0.280	1.5935	1.5951	1.5998	1.6077	1.6190	1.6335	1.6517	1.6735	1.6994	1.7652								
0.290	1.5693	1.5708	1.5754	1.5831	1.5940	1.6082	1.6257	1.6469	1.6719	1.7350								
0.300	1.5457	1.5472	1.5516	1.5592	1.5697	1.5835	1.6005	1.6210	1.6452	1.7059	1.7886							
0.310	1.5227	1.5242	1.5285	1.5358	1.5461	1.5594	1.5760	1.5959	1.6192	1.6779	1.7561							
0.320	1.5003	1.5017	1.5059	1.5130	1.5230	1.5360	1.5521	1.5714	1.5941	1.6507	1.7256							
0.330	1.4784	1.4798	1.4839	1.4908	1.5006	1.5132	1.5288	1.5476	1.5695	1.6244	1.6963							
0.340	1.4571	1.4584	1.4625	1.4692	1.4787	1.4910	1.5062	1.5244	1.5458	1.5989	1.6681							
0.350	1.4363	1.4376	1.4415	1.4481	1.4573	1.4693	1.4841	1.5019	1.5227	1.5742	1.6409							
0.360	1.4160	1.4172	1.4211	1.4275	1.4365	1.4482	1.4626	1.4799	1.5001	1.5502	1.6147							
0.370	1.3962	1.3974	1.4011	1.4074	1.4162	1.4276	1.4417	1.4585	1.4782	1.5268	1.5893							
0.380	1.3768	1.3780	1.3816	1.3877	1.3964	1.4075	1.4212	1.4376	1.4568	1.5041	1.5647							
0.390	1.3578	1.3591	1.3626	1.3686	1.3770	1.3879	1.4013	1.4173	1.4360	1.4821	1.5409							
0.400	1.3394	1.3406	1.3441	1.3499	1.3581	1.3687	1.3818	1.3975	1.4157	1.4606	1.5178	1.6816						
0.410	1.3213	1.3225	1.3259	1.3316	1.3397	1.3500	1.3628	1.3781	1.3960	1.4397	1.4953	1.6524						
0.420	1.3037	1.3048	1.3082	1.3138	1.3216	1.3318	1.3443	1.3592	1.3767	1.4194	1.4735	1.6252						
0.430	1.2865	1.2876	1.2909	1.2963	1.3040	1.3139	1.3262	1.3408	1.3578	1.3995	1.4523	1.5990						
0.440	1.2697	1.2707	1.2739	1.2793	1.2868	1.2965	1.3085	1.3228	1.3394	1.3802	1.4317	1.5739						
0.450	1.2532	1.2542	1.2574	1.2626	1.2700	1.2795	1.2913	1.3052	1.3215	1.3613	1.4116	1.5496						
0.460	1.2371	1.2381	1.2412	1.2463	1.2536	1.2629	1.2744	1.2881	1.3040	1.3430	1.3920	1.5261						
0.470	1.2213	1.2223	1.2253	1.2304	1.2375	1.2466	1.2579	1.2713	1.2869	1.3250	1.3730	1.5035						
0.480	1.2059	1.2069	1.2098	1.2148	1.2218	1.2307	1.2418	1.2549	1.2702	1.3075	1.3544	1.4815						
0.490	1.1909	1.1918	1.1947	1.1996	1.2064	1.2152	1.2260	1.2389	1.2539	1.2904	1.3363	1.4602						
0.500	1.1761	1.1770	1.1799	1.1847	1.1913	1.2000	1.2106	1.2233	1.2379	1.2737	1.3186	1.4396	1.6178					
0.525	1.1406	1.1415	1.1442	1.1488	1.1551	1.1634	1.1735	1.1856	1.1996	1.2337	1.2763	1.3903	1.5540					
0.550	1.1069	1.1078	1.1103	1.1148	1.1209	1.1287	1.1384	1.1500	1.1634	1.1959	1.2365	1.3444	1.4969					
0.575	1.0749	1.0758	1.0783	1.0824	1.0883	1.0959	1.1052	1.1162	1.1290	1.1601	1.1988	1.3014	1.1446					
0.600	1.0446	1.0453	1.0477	1.0517	1.0574	1.0646	1.0735	1.0841	1.0964	1.1262	1.1633	1.2610	1.3962	1.5852				
0.625	1.0156	1.0164	1.0187	1.0226	1.0280	1.0350	1.0436	1.0537	1.0655	1.0941	1.1296	1.2229	1.3511	1.5257				
0.650	0.9881	0.9889	0.9911	0.9948	1.0000	1.0067	1.0150	1.0248	1.0361	1.0636	1.0988	1.1871	1.3091	1.4726				
0.675	0.9619	0.9626	0.9648	0.9683	0.9734	0.9798	0.9878	0.9972	1.0082	1.0346	1.0674	1.1531	1.2696	1.4238				
0.700	0.9369	0.9376	0.9396	0.9431	0.9479	0.9542	0.9619	0.9710	0.9815	1.0071	1.0386	1.1210	1.2324	1.3786	1.5759			
0.725	0.9130	0.9137	0.9157	0.9191	0.9238	0.9298	0.9372	0.9460	0.9562	0.9808	1.0113	1.0905	1.1974	1.3365	1.5198			
0.750	0.8902	0.8909	0.8928	0.8961	0.9006	0.9065	0.9136	0.9221	0.9320	0.9553	0.9852	1.0616	1.1642	1.2972	1.4695			
0.775	0.8685	0.8691	0.8710	0.8741	0.8785	0.8842	0.8911	0.8993	0.9089	0.9319	0.9603	1.0341	1.1329	1.2620	1.4234			
0.800	0.6476	0.8482	0.8500	0.8531	0.8573	0.8628	0.8696	0.8776	0.8868	0.9091	0.9366	1.0079	1.1032	1.2253	1.3806	1.5844		
0.825	0.8277	0.8282	0.8300	0.8330	0.8371	0.8424	0.8490	0.8567	0.8657	0.8873	0.9140	0.9830	1.0749	1.1924	1.3407	1.5309		
0.850	0.8086	0.8091	0.8108	0.8137	0.8178	0.8229	0.8293	0.8368	0.8455	0.8665	0.8923	0.9591	1.0480	1.1612	1.3032	1.4828		
0.875	0.7903	0.7908	0.7925	0.7953	0.7992	0.8042	0.8104	0.8177	0.8261	0.8465	0.8716	0.9364	1.0224	1.1317	1.2680	1.4387		
0.900	0.7727	0.7732	0.7749	0.7776	0.7814	0.7863	0.7922	0.7993	0.8075	0.8274	0.8517	0.9147	0.9980	1.1036	1.2348	1.3976	1.6069	
0.925	0.7558	0.7564	0.7579	0.7606	0.7643	0.7690	0.7749	0.7818	0.7898	0.8090	0.8327	0.8938	0.9747	1.0769	1.2034	1.3593	1.5553	
0.950	0.7396	0.7401	0.7416	0.7443	0.7478	0.7525	0.7582	0.7649	0.7727	0.7914	0.8145	0.8739	0.9525	1.0514	1.1736	1.3222	1.5090	
0.975	0.7240	0.7245	0.7260	0.7286	0.7321	0.7366	0.7421	0.7487	0.7563	0.7745	0.7970	0.8548	0.9311	1.0271	1.1453	1.2894	1.4663	
1.000	0.7091	0.7096	0.7110	0.7135	0.7169	0.7213	0.7267	0.7331	0.7405	0.7583	0.7802	0.8366	0.9107	1.0040	1.1184	1.2573	1.4266	1.6402
1.050	0.6808	0.6813	0.6826	0.6850	0.6882	0.6924	0.6976	0.7037	0.7107	0.7277	0.7485	0.8021	0.8724	0.9605	1.0682	1.1981	1.3544	1.5452
1.100	0.6546	0.6550	0.6563	0.6585	0.6617	0.6657	0.6706	0.6764	0.6831	0.6993	0.7192	0.7702	0.8371	0.9207	1.0225	1.1446	1.2902	1.4649
1.150	0.6301	0.6306	0.6318	0.6339	0.6369	0.6407	0.6455	0.6510	0.6574	0.6729	0.6919	0.7407	0.8044	0.8840	0.9805	1.0958	1.2324	1.3944
1.200	0.6073	0.6078	0.6090	0.6110	0.6139	0.6176	0.6221	0.6274	0.6335	0.6484	0.6666	0.7133	0.7741	0.8500	0.9419	1.0512	1.1800	1.3314
1.250	0.5861	0.5865	0.5877	0.5896	0.5924	0.5959	0.6003	0.6053	0.6112	0.6254	0.6429	0.6877	0.7460	0.8185	0.9062	1.0101	1.1321	1.2746
1.300	0.5662	0.5666	0.5677	0.5696	0.5722	0.5756	0.5798	0.5847	0.5904	0.6041	0.6209	0.6638	0.7198	0.7892	0.8730	0.9722	1.0881	1.2228
1.350	0.5474	0.5479	0.5490	0.5508	0.5534	0.5566	0.5606	0.5654	0.5708	0.5840	0.6000	0.6415	0.6953	0.7620	0.8422	0.9370	1.0476	1.1755
1.400	0.5300	0.5304	0.5314	0.5332	0.5356	0.5388	0.5426	0.5472	0.5525	0.5651	0.5807	0.6205	0.6723	0.7365	0.8135	0.9044	1.0100	1.1318
1.450	0.5136	0.5139	0.5149	0.5166	0.5190	0.5220	0.5257	0.5302	0.5353	0.5475	0.5624	0.6004	0.6508	0.7126	0.7867	0.8740	0.9751	1.0915
1.500	0.4980	0.4983	0.4993	0.5010	0.5033	0.5062	0.5098	0.5141	0.5190	0.5308	0.5454	0.5824	0.6306	0.6902	0.7616	0.8455	0.9427	1.0541
1.550	0.4834	0.4837	0.4347	0.4862	0.4885	0.4913	0.4948	0.4989	0.5037	0.5151	0.5290	0.5651	0.6116	0.6691	0.7380	0.8189	0.9123	1.0193
1.600	0.4696	0.4698	0.4708	0.4723	0.4744	0.4772	0.4806	0.4846	0.4892	0.5003	0.5139	0.5486	0.5937	0.6493	0.7158	0.7938	0.8839	0.9868
1.700	0.4440	0.4443	0.4452	0.4467	0.4487	0.4513	0.4545	0.4582	0.4625	0.4730	0.4858	0.5184	0.5607	0.6129	0.6752	0.7481	0.8321	0.9277
1.800	0.4211	0.4214	0.4222	0.4235	0.4254	0.4279	0.4309	0.4344	0.4386	0.4484	0.4605	0.4913	0.5311	0.5803	0.6389	0.7074	0.7861	0.8755
1.900	0.4003	0.4006	0.4014	0.4026	0.4045	0.4068	0.4096	0.4130	0.4169	0.4262	0.4376	0.4668	0.5045	0.5510	0.6063	0.6709	0.7449	0.8290
2.000	0.3815	0.3817	0.3824	0.3837	0.3854	0.3876	0.3903	0.3935	0.3972	0.4060	0.4169	0.4446	0.4804	0.5244	0.5767	0.6379	0.7079	0.7872
2.100	0.3642	0.3645	0.3652	0.3664	0.3680	0.3701	0.3727	0.3757	0.3792	0.3877	0.3980	0.4244	0.4584	0.5003	0.5501	0.6081	0.6744	0.7495
2.200	0.3485	0.3487	0.3494	0.3505	0.3521	0.3541	0.3566	0.3594	0.3628	0.3708	0.3807	0.4059	0.4383	0.4782	0.5257	0.5809	0.6440	0.7152
2.300	0.3341	0.3343	0.3349	0.3359	0.3374	0.3394	0.3418	0.3445	0.3477	0.3554	0.3648	0.3889	0.4199	0.4580	0.5034	0.5560	0.6161	0.6840
2.400	0.3207	0.3209	0.3216	0.3225	0.3240	0.3258	0.3281	0.3307	0.3338	0.3412	0.3502	0.3732	0.4029	0.4395	0.4828	0.5331	0.5906	0.6554
2.500	0.3084	0.3086	0.3092	0.3102	0.3115	0.3133	0.3155	0.3180	0.3210	0.3280	0.3367	0.3588	0.3873	0.4223	0.4639	0.5121	0.5671	0.6291
2.600	0.2970	0.2971	0.2977	0.2987	0.3000	0.3017	0.3038	0.3062	0.3091	0.3158	0.3242	0.3454	0.3728	0.4064	0.4464	0.4927	0.5455	0.6049
2.700	0.2863	0.2865	0.2871	0.2880	0.2892	0.2909	0.2929	0.2952	0.2980	0.3045	0.3126	0.3330	0.3594	0.3917	0.4301	0.4746	0.5254	0.5825
2.800	0.2764	0.2766	0.2771	0.2780	0.2792	0.2808	0.2828	0.2850	0.2877	0.2940	0.3017	0.3215	0.3469	0.3780	0.4150	0.4579	0.5067	0.5617
2.900	0.2672	0.2674	0.2679	0.2687	0.2699	0.2714	0.2733	0.2755	0.2780	0.2842	0.2916	0.3106	0.3352	0.3653	0.4009	0.4423	0.4894	0.5423
3.000	0.2586	0.2587	0.2592	0.2600	0.2612	0.2626	0.2644	0.2666	0.2690	0.2749	0.2821	0.3006	0.3243	0.3533	0.3878	0.4277	0.4732	0.5242

$0 \le -g\ x = \le 0.190$					
k = g	R_H^y	k = g	R_H^y	k = g	R_H^y
0.000	2.6667	0.195	1.9681	0.700	1.5759
0.005	2.6366	0.200	1.9576	0.725	1.5765
0.010	2.6098	0.210	1.9371	0.750	1.5782
0.015	2.5801	0.220	1.9175	0.775	1.5809
0.020	2.5535	0.230	1.8988	0.800	1.5844
0.025	2.5278	0.240	1.8809	0.825	1.5889
0.030	2.5030	0.250	1.8637	0.850	1.5941
0.035	2.4790	0.260	1.8474	0.875	1.6001
0.040	2.4558	0.270	1.8317	0.900	1.6069
0.045	2.4332	0.280	1.8167	0.925	1.6143
0.050	2.4113	0.290	1.8023	0.950	1.6223
0.055	2.3901	0.300	1.7886	0.975	1.6310
0.060	2.3694	0.310	1.7755	1.000	1.6402
0.065	2.3493	0.320	1.7630	1.050	1.6603
0.070	2.3297	0.330	1.7511	1.100	1.6824
0.075	2.3106	0.340	1.7397	1.150	1.7062
0.080	2.2920	0.350	1.7288	1.200	1.7316
0.085	2.2739	0.360	1.7184	1.250	1.7585
0.090	2.2563	0.370	1.7085	1.300	1.7867
0.095	2.2391	0.380	1.6991	1.350	1.8161
0.100	2.2230	0.390	1.6902	1.400	1.8466
0.105	2.2059	0.400	1.6816	1.450	1.8781
0.110	2.1899	0.410	1.6736	1.500	1.9105
0.115	2.1743	0.420	1.6659	1.550	1.9438
0.120	2.1590	0.430	1.6586	1.600	1.9779
0.125	2.1442	0.440	1.6517	1.700	2.0481
0.130	2.1297	0.450	1.6452	1.800	2.1208
0.135	2.1154	0.460	1.6390	1.900	2.1957
0.140	2.1016	0.470	1.6332	2.000	2.2724
0.145	2.0880	0.480	1.6277	2.100	2.3507
0.150	2.0773	0.490	1.6226	2.200	2.4305
0.155	2.0618	0.500	1.6178	2.300	2.5115
0.160	2.0492	0.525	1.6071	2.400	2.5937
0.165	2.0368	0.550	1.5981	2.500	2.6768
0.170	2.0247	0.575	1.5909	2.600	2.7608
0.175	2.0129	0.600	1.5852	2.700	2.8457
0.180	2.0013	0.625	1.5809	2.800	2.9313
0.185	1.9900	0.650	1.5780	2.900	3.0175
0.190	1.9789	0.675	1.5763	3.000	3.1044

Appendix C

Table of Multiwave Cases

n	$R(\frac{1}{a})$	N	$(\underline{h}_1/\underline{h}_2 - \underline{h}_1/ \ldots /\underline{h}_{N-1} - \underline{h}_{N-2}/ - \underline{h}_{N-1})$
			Plane (001)
1.	2.000	4	$(220/\bar{2}20/\bar{2}\bar{2}0/2\bar{2}0)$
2.	3.162	4	$(220/\bar{4}40/\bar{2}\bar{2}0/4\bar{4}0)$
3.	4.472	8	$(220/040/\bar{2}20/\bar{4}00/\bar{2}\bar{2}0/0\bar{4}0/2\bar{2}0/400)$
4.	5.375	4	$(220/\bar{4}80/\bar{6}\bar{6}0/8\bar{4}0)$
5.	5.831	4	$(220/\bar{8}80/\bar{2}\bar{2}0/8\bar{8}0)$
6.	6.519	4	$(220/\bar{6},10,0/\bar{6}\bar{6}0/10,\bar{6},0)$
7.	7.211	8	$(220/080/\bar{2}20/\bar{8}00/\bar{2}\bar{2}0/0\bar{8}0/2\bar{2}0/800)$
8.	2.828	4	$(400/040/\bar{4}00/0\bar{4}0)$
9.	5.946	4	$(400/2,10,0/\bar{8}00/2,\overline{10},0)$
10.	6.325	8	$(400/440/040/\bar{4}40/\bar{4}00/\bar{4}\bar{4}0/0\bar{4}0/4\bar{4}0)$
11.	4.000	4	$(440/\bar{4}40/\bar{4}\bar{4}0/4\bar{4}0)$
12.	5.099	4	$(440/\bar{6}60/\bar{4}\bar{4}0/6\bar{6}0)$
13.	5.701	4	$(440/\bar{2}60/\bar{8}\bar{8}0/6\bar{2}0)$
14.	6.146	4	$(620/080/\bar{6}20/0,\overline{12},0)$
			Plane $(1\bar{1}1)$
15.	2.828	6	$(220/022/\bar{2}02/\bar{2}\bar{2}0/0\bar{2}\bar{2}/20\bar{2})$
16.	4.320	6	$(220/044/\bar{2}02/\bar{4}\bar{4}0/0\bar{2}\bar{2}/40\bar{4})$
17.	5.099	4	$(220/\bar{4}48/\bar{2}\bar{2}0/4\bar{4}\bar{8})$
18.	5.888	6	$(220/066/\bar{2}02/\bar{6}\bar{6}0/0\bar{2}\bar{2}/60\bar{6})$
19.	6.524	4	$(220/\bar{4},6,10/\bar{4}\bar{4}0/6,\bar{4},\overline{10})$
20.	7.483	12	$(220/242/022/\bar{2}24/\bar{2}02/\bar{4}\bar{2}2/\bar{2}\bar{2}0/\bar{2}\bar{4}\bar{2}/0\bar{2}\bar{2}/2\bar{2}\bar{4}/20\bar{2}/42\bar{2})$
21.	3.742	4	$(42\bar{2}/044/\bar{4}\bar{2}2/0\bar{4}\bar{4})$
22.	4.899	6	$(42\bar{2}/242/\bar{2}24/\bar{4}\bar{2}2/\bar{2}\bar{4}\bar{2}/2\bar{2}\bar{4})$
23.	5.396	4	$(42\bar{2}/264/\bar{8}44/2\bar{4}\bar{6})$
24.	6.164	4	$(42\bar{2}/088/\bar{4}\bar{2}2/0\bar{8}\bar{8})$
25.	6.350	4	$(42\bar{2}/286/\bar{8}\bar{4}4/2\bar{6}\bar{8})$
26.	5.193	4	$(440/\bar{2}46/\bar{6}\bar{6}0/4\bar{2}\bar{6})$
27.	5.657	6	$(440/044/\bar{4}04/\bar{4}\bar{4}0/0\bar{4}\bar{4}/40\bar{4})$
28.	6.745	4	$(440/\bar{2}68/\bar{8}\bar{8}0/6\bar{2}\bar{8})$
29.	7.118	6	$(440/066/\bar{4}0\bar{4}/\bar{6}6\bar{0}/0\bar{4}\bar{4}/60\bar{6})$

Appendix C (continued)

n	$R(\frac{1}{a})$	N	$(\underline{h}_1/\underline{h}_2 - \underline{h}_1/ \dots / \underline{h}_{N-1} - \underline{h}_{N-2}/ - \underline{h}_{N-1})$
			Plane $(1\bar{1}0)$
30.	2.031	4	$(111/002/\bar{1}\bar{1}1/00\bar{4})$
31.	2.598	4	$(111/\bar{2}\bar{2}4/\bar{1}\bar{1}\bar{1}/22\bar{4})$
32.	3.182	5	$(111/004/\bar{1}\bar{1}1/\bar{3}\bar{3}\bar{3}/33\bar{3})$
33.	3.775	4	$(111/\bar{4}\bar{4}0/11\bar{1}/220)$
34.	4.373	6	$(111/006/\bar{1}\bar{1}1/\bar{4}\bar{4}\bar{2}/00\bar{4}/44\bar{2})$
35.	4.975	6	$(111/113/\bar{5}\bar{5}6/\bar{1}\bar{1}\bar{1}/\bar{1}\bar{1}\bar{3}/55\bar{5})$
36.	5.277	5	$(111/115/\bar{4}\bar{4}4/\bar{3}\bar{3}\bar{3}/55\bar{7})$
37.	5.579	4	$(111/008/\bar{1}\bar{1}1/0,0,\overline{10})$
38.	5.882	4	$(111/\bar{2},\bar{2},10/\bar{5}\bar{5}\bar{5}/66\bar{6})$
39.	6.185	6	$(111/117/\bar{8}\bar{8}0/11\bar{7}/11\bar{1}/440)$
40.	6.640	4	$(111/\bar{5},\bar{5},11/\bar{2},\bar{2},\overline{10}/66\bar{2})$
41.	6.792	6	$(111/0,0,10/\bar{1}\bar{1}1/\bar{7}\bar{7}\bar{5}/00\bar{2}/77\bar{5})$
42.	7.399	4	$(111/\bar{6},\bar{6},12/\bar{1}\bar{1}\bar{1}/6,6,\overline{12})$
43.	2.669	4	$(002/\bar{3}\bar{3}1/00\bar{4}/331)$
44.	3.878	4	$(002/\bar{5}\bar{5}1/00\bar{4}/551)$
45.	4.016	4	$(002/\bar{3}\bar{3}3/00\bar{8}/333)$
46.	4.704	4	$(002/\bar{5}\bar{5}3/00\bar{8}/553)$
47.	5.198	4	$(002/\bar{7}\bar{7}1/00\bar{4}/771)$
48.	5.794	4	$(002/\bar{7}\bar{7}3/00\bar{8}/773)$
49.	6.093	8	$(002/\bar{1}\bar{1}3/\bar{4}\bar{4}2/\bar{3}\bar{3}\bar{3}/00\bar{6}/33\bar{3}/442/113)$
50.	6.326	4	$(002/\bar{3}\bar{3}5/0,0,\overline{12}/335)$
51.	6.558	4	$(002/\bar{9}\bar{9}1/00\bar{4}/991)$
52.	7.025	4	$(002/\bar{9}\bar{9}3/00\bar{8}/993)$
53.	2.450	4	$(220/004/\bar{2}\bar{2}0/00\bar{4})$
54.	2.872	4	$(220/113/\bar{4}\bar{4}0/11\bar{3})$
55.	3.407	4	$(220/115/\bar{4}\bar{4}0/11\bar{5})$
56.	4.175	4	$(220/117/\bar{4}\bar{4}0/11\bar{7})$
57.	4.243	6	$(220/224/\bar{2}\bar{2}4/\bar{2}\bar{2}0/\bar{2}\bar{2}\bar{4}/22\bar{4})$
58.	5.036	4	$(220/119/\bar{4}\bar{4}0/11\bar{9})$
59.	5.196	4	$(220/228/\bar{6}\bar{6}0/22\bar{8})$

Appendix C (continued)

n	$R(\frac{1}{a})$	N	$(\underline{h}_1/\underline{h}_2 - \underline{h}_1/ \ldots /\underline{h}_{N-1} - \underline{h}_{N-2}/ - \underline{h}_{N-1})$
			Plane $(1\bar{1}0)$
60.	5.679	4	$(220/335/\bar{8}\bar{8}0/33\bar{5})$
61.	5.817	4	$(220/337/\bar{8}\bar{8}0/33\bar{7})$
62.	5.943	4	$(220/1,1,11/\bar{4}\bar{4}0/1,1,\overline{11})$
63.	6.164	4	$(220/0,9,12/\bar{2}\bar{2}0/0,0,\overline{12})$
64.	6.327	4	$(220/339/\bar{8}\bar{8}0/33\bar{9})$
65.	6.652	4	$(220/333/\bar{8}\bar{8}0/33\bar{3})$
66.	4.246	4	$(113/\bar{6}\bar{6}0/11\bar{3}/440)$
67.	4.426	4	$(113/\bar{6}\bar{6}2/22\bar{6}/331)$
68.	5.718	4	$(113/\bar{8}\bar{8}0/11\bar{3}/660)$
69.	6.659	4	$(113/\bar{7}\bar{7}\bar{7}/4,4,\overline{12}/222)$
70.	3.787	4	$(222/\bar{1}\bar{1}5/\bar{4}\bar{4}\bar{4}/333)$
71.	5.631	4	$(222/\bar{3}\bar{3}9/\bar{4}\bar{4}\bar{4}/557)$
72.	7.252	6	$(222/117/\bar{5}\bar{5}5/\bar{5}\bar{5}\bar{7}/22\bar{6}/55\bar{1})$
73.	3.369	4	$(004/\bar{3}\bar{3}1/00\bar{6}/331)$
74.	3.464	4	$(004/\bar{4}\bar{4}0/00\bar{4}/440)$
75.	4.690	4	$(004/\bar{6}\bar{6}0/00\bar{4}/660)$
76.	5.012	4	$(004/\bar{3}\bar{3}3/0,0,\overline{10},333)$
77.	5.404	4	$(004/\bar{5}\bar{5}3/0,0,\overline{10}/553)$
78.	5.574	4	$(004/\bar{7}\bar{7}1/00\bar{6}/771)$
79.	6.000	6	$(004/\bar{4}\bar{4}4/\bar{4}\bar{4}\bar{4}/00\bar{4}/44\bar{4}/444)$
80.	6.334	4	$(004/\bar{7}\bar{7}3/0,0,\overline{10}/773)$
81.	6.446	4	$(004/\bar{6}\bar{6}4/0,0,\overline{12}/664)$
82.	6.858	4	$(004/\bar{9}\bar{9}1/00\bar{6}/991)$
83.	4.205	4	$(331/006/\bar{3}\bar{3}1/00\bar{8})$
84.	5.111	4	$(331/008/\bar{3}\bar{3}1/0,0,\overline{10})$
85.	6.056	4	$(331/0,0,10/\bar{3}\bar{3}1/0,0,\overline{12})$
86.	6.538	6	$(331/224/\bar{4}\bar{4}8/\bar{3}\bar{3}\bar{1}/\bar{2}\bar{2}\bar{4}/44\bar{8})$
87.	3.571	4	$(224/\bar{3}\bar{3}3/\bar{2}\bar{2}\bar{4}/33\bar{3})$

Appendix C (continued)

n	$R(\frac{1}{a})$	N	$(\underline{h}_1/\underline{h}_2 - \underline{h}_1/ \dots /\underline{h}_{N-1} - \underline{h}_{N-2}/ - \underline{h}_{N-1})$
			Plane $(1\bar{1}0)$
88.	4.500	5	$(115/\bar{3}\bar{3}3/\bar{3}\bar{3}\bar{3}/11\bar{5}/440)$
89.	5.763	4	$(115/\bar{8}\bar{8}0/11\bar{5}/660)$
90.	6.935	5	$(115/\bar{4}\bar{4}6/\bar{3}\bar{3}\bar{3}/2,2,\overline{10}/662)$
91.	5.545	4	$(333/\bar{4}\bar{4}8/\bar{3}\bar{3}\bar{3}/44\bar{8})$
92.	4.899	4	$(440/008/\bar{4}\bar{4}0/00\bar{8})$
93.	5.076	4	$(440/117/\bar{6}\bar{6}0/11\bar{7})$
94.	5.793	4	$(440/119/\bar{6}\bar{6}0/11\bar{9})$
			Plane $(1\bar{1}2)$
95.	1.658	4	$(\bar{1}11/\bar{2}\bar{2}0/1\bar{1}\bar{1}/220)$
96.	2.622	4	$(\bar{1}11/\bar{3}\bar{1}1/3\bar{3}\bar{3}/131)$
97.	2.958	4	$(\bar{1}11/\bar{4}\bar{4}0/1\bar{1}\bar{1}/440)$
98.	3.469	4	$(\bar{1}11/\bar{5}\bar{3}1/3\bar{3}\bar{3}/351)$
99.	4.330	4	$(\bar{1}11/\bar{6}\bar{6}0/1\bar{1}\bar{1}/660)$
100.	4.504	4	$(\bar{1}11/\bar{6}\bar{2}2/5\bar{5}\bar{5}/262)$
101.	4.677	6	$(\bar{1}11/\bar{7}\bar{5}1/351/0\bar{8}\bar{4}/53\bar{1}/042)$
102.	5.374	4	$(\bar{1}11/\bar{8}\bar{4}2/5\bar{5}\bar{5}/482)$
103.	5.723	4	$(\bar{1}11/\bar{8}\bar{8}0/1\bar{1}\bar{1}/880)$
104.	5.985	4	$(\bar{1}11/\bar{9}\bar{7}1/3\bar{3}\bar{3}/791)$
105.	6.072	4	$(\bar{1}11/\bar{7}\bar{1}3/7\bar{7}\bar{7}/173)$
106.	2.958	4	$(220/\bar{3}33/\bar{2}\bar{2}0/3\bar{3}\bar{3})$
107.	3.741	4	$(220/\bar{4}44/\bar{2}\bar{2}0/4\bar{4}\bar{4})$
108.	4.373	4	$(220/\bar{1}53/\bar{6}\bar{6}0/5\bar{1}\bar{3})$
109.	4.555	4	$(220/\bar{5}55/\bar{2}\bar{2}0/5\bar{5}\bar{5})$
110.	4.830	4	$(220/\bar{2}64/\bar{6}\bar{6}0/6\bar{2}\bar{4})$
111.	5.441	4	$(220/\bar{3}75/\bar{6}\bar{6}0/7\bar{3}\bar{5})$
112.	5.664	4	$(220/131/\bar{6}\bar{6}0/31\bar{1})$
113.	3.469	4	$(131/\bar{2}22/\bar{3}\bar{1}1/4\bar{4}\bar{4})$
114.	4.388	4	$(131/\bar{3}33/\bar{3}\bar{1}1/5\bar{5}\bar{5})$
115.	5.342	4	$(131/\bar{4}44/\bar{3}\bar{1}1/6\bar{6}\bar{6})$
116.	6.315	4	$(131/\bar{5}55/\bar{3}\bar{1}1/7\bar{7}\bar{7})$

n	$R(\frac{1}{a})$	N	$(\underline{h}_1/\underline{h}_2 - \underline{h}_1/ \dots /\underline{h}_{N-1} - \underline{h}_{N-2}/ - \underline{h}_{N-1})$
			Plane $(1\bar{1}2)$
117.	4.017	4	$(\bar{2}22/\bar{5}\bar{3}1/4\bar{4}\bar{4}/351)$
118.	5.078	4	$(\bar{2}22/\bar{7}\bar{5}1/4\bar{4}\bar{4}/571)$
119.	6.298	4	$(\bar{2}22/\bar{9}\bar{7}1/4\bar{4}\bar{4}/791)$
120.	4.472	4	$(440/\bar{4}44/\bar{4}\bar{4}0/4\bar{4}\bar{4})$
121.	5.172	4	$(440/\bar{5}55/\bar{4}\bar{4}0/5\bar{5}\bar{5})$
122.	5.664	4	$(440/\bar{1}53/\bar{8}\bar{8}0/5\bar{1}\bar{3})$
123.	6.377	4	$(440/\bar{3}75/\bar{8}\bar{8}0/7\bar{3}\bar{5})$
124.	5.409	4	$(351/\bar{4}44/\bar{5}\bar{3}1/6\bar{6}\bar{6})$
125.	6.187	4	$(351/\bar{5}55/\bar{5}\bar{3}1/7\bar{7}\bar{7})$
			Plane $(1\bar{1}3)$
126.	3.303	4	$(220/\bar{2}42/\bar{4}\bar{4}0/4\bar{2}\bar{2})$
127.	4.899	4	$(220/\bar{6}64/\bar{2}\bar{2}0/6\bar{6}\bar{4})$
128.	5.721	6	$(220/062/\bar{4}22/\bar{6}\bar{6}0/\bar{2}\bar{4}\bar{2}/60\bar{2})$
129.	4.431	4	$(4\bar{2}\bar{2}/440/\bar{2}42/\bar{6}\bar{6}0)$
130.	5.477	4	$(440/\bar{6}64/\bar{4}\bar{4}0/6\bar{6}\bar{4})$
131.	6.606	4	$(440/\bar{4}84/\bar{8}\bar{8}0/8\bar{4}\bar{4})$
			Plane $(1\bar{1}4)$
132.	2.872	4	$(220/\bar{1}31/\bar{4}\bar{4}0/3\bar{1}\bar{1})$
133.	5.036	4	$(220/\bar{5}73/\bar{4}\bar{4}0/7\bar{5}\bar{3})$
134.	6.164	4	$(220/\bar{8}84/\bar{2}\bar{2}0/8\bar{8}\bar{4})$
135.	6.327	4	$(220/\bar{3}93/\bar{8}\bar{8}0/9\bar{3}\bar{3})$
136.	6.652	4	$(220/151/\bar{8}\bar{8}0/51\bar{1})$
137.	4.246	4	$(\bar{3}11/\bar{4}\bar{4}0/1\bar{3}\bar{1}/660)$
138.	5.718	4	$(\bar{3}11/\bar{6}\bar{6}0/1\bar{3}\bar{1}/880)$
139.	6.093	4	$(\bar{5}\bar{1}1/\bar{4}42/151/8\bar{8}\bar{4})$
140.	5.793	4	$(440/\bar{5}73/\bar{6}\bar{6}0/7\bar{5}\bar{3})$
141.	6.633	4	$(440/\bar{8}84/\bar{4}\bar{4}0/8\bar{8}4)$

References

1.1 M.v. Laue, W. Fridrich, P. Knipping: Münchener Sitzungsberichte 303 (1912); -Ann. Physik 41, 971 (1913)
1.2 E.S. Fedorov: Transactions (Zapiski) S. Petersbourgs Mineralogical Society, 2-d series 27, 448 (1891); - 28, 1
1.3 A. Schoenflies: *Kristallsysteme und Kristallstruktur* (Leipzig 1891)
1.4 P.P. Ewald: *Fifty Years of X-ray Diffraction* (N.V.A. Oosthock's Nitgevers Mij, Utrecht 1962) p. 248
1.5 C.G. Darwin: Phil. Mag. (1914) 27, 315, 675
1.6 A.H. Compton, S.K. Allison: *X-rays in Theory and Experiment*, 2nd ed. (D. Van Nostrand Co., New York 1935)
1.7 P.P. Ewald: Phys. Zschr. 14, 465 (1913) - Fortschr. Chem. Phys. u. phys. Chemie, Ser. B 18, 492 (1925); - Physica 4, 234 (1924)
1.8 P.P. Ewald: Ann. Physik 54, 519 (1917); - Z. Physik 2, 332 (1920); 30, 1 (1924); - Phys. Zschr. 26, 29 (1925); - Handbuch der Physik, Vol. XXIII/2, 2nd ed. (Springer, Berlin 1933) pp. 207-476
1.9 R.W. James: *The Optical Principle of the Diffraction of X-rays* (Bell, London 1950)
1.10 H.A. Bethe: Ann. Physik 87, 55 (1928)
1.11 M.v.Laue: Ergeb. Exakt. Naturw. 10, 133 (1931)
1.12 M. Kohler: Ann. Physik 18, 265 (1933); Sitzungsberichte Preuss. Akad. Wissensch. - Math.-Phys. Kl. 1935, 334
1.13 M.v.Laue: Röntgenstrahlinterferenzen, 1. Aufl., 1940; 2. Aufl., 1941; 3. Aufl., 1960 (Akademische Verlagsgesellschaft, Frankfurt a.M.)
1.14 W.H. Zachariasen: *Theory of X-rays Diffraction in Crystals* (J. Wiley and Sons, New York 1945)
1.15 A. Authier: Bull. Soc. Franc. Mineral. 84, 51 (1961); - *Advances Str. Res. Diffr. Math.* Vol. 3, ed. by L. Braunschweig et al. (1970)
1.16 R.W. James: Solid State Phys. 15, 53 (1963)
1.17 B.W. Batterman, H. Cole: Rev. Mod. Phys. 35, 681 (1964)
1.18 A.H. Compton: Phys. Rev. 9, 29; -10, 95 (1917)
1.19 K. Nakayama, H. Hashisume, A. Miyoshi, S. Kikuta, K. Kohra: Z. Naturforsch. 28a, 632 (1973)
1.20 G. Borrmann: Phys. Zschr. 42, 157 (1941); Zschr. Phys. 127, 297 (1950)
1.21 M.v.Laue: Acta Cryst. 2, 106 (1949); H.N. Campbell: Acta Cryst. 4, 180 (1951); W.H. Zachariasen: Proc. Nat. Acad. Sci. 38, 378 (1952)
1.22 G. Borrmann: Z. Kristall., Mineral., Petrogr. Abt. A 106, 109 (1955)
1.23 N. Kato, A.R. Lang: Acta Cryst. 12, 787 (1959)
1.24 N. Kato: a) Acta Cryst. 14, 526,627 (1961); b) J. Appl. Phys. 39, 2225, 2231 (1968)
1.25 N. Kato: a) J. Phys. Soc. Japan 21, 1160 (1966); b) Acta geologica et geographica Universitatis Comenianae. Geologica 1968, no. 4, 43; c) Acta Cryst. A25, 115 (1969)
1.26 Z.G. Pinsker: *Electron Diffraction* (Butterworths, London 1953)
1.27 P.B. Hirsch, A. Howie, R.B. Nicholson, D.W. Pashley, M.J. Whelan: *Electron Microscopy of Thin Crystals* (Butterworths, London 1965)
1.28 U. Bonse, M. Hart: Appl. Phys. Letters a) 6, 155; b) 7, 99; c) 7, 239 (1965)

1.29 M. Renninger: Naturwissenschaften 25, 43 (1937); - Zschr. Phys. 106, 141 (1937); - Z. Kristall., Mineral., Petrogr., Abt. A 97, 107 (1937)
1.30 G. Borrmann, W. Hartwig: Z. Kristall 121, 401 (1965)
1.31 P.P. Ewald, Y. Hêno: Acta Crystall. A24, 5 (1968); Y. Hêno, P.P. Ewald: Acta Crystall. A24, 16 (1968)
1.32 P. Penning, D. Polder: Philips Res. Rep. 23, 1 (1968)
1.33 P. Penning: Philips Res. Rep. 23, 12 (1968)
1.34 A.M. Afanasiev, V.G. Kohn: Acta Cryst. A32, 2, 308 (1976)
1.35 V.G. Kohn: Sov. Fizika Tverd. Tela 18, 9, 2538 (1976)
1.36 A.M. Afanasiev, V.G. Kohn: Phys. Status Solidi 28, 61 (1975a)
1.37 A.M. Afanasiev, V.G. Kohn: Acta Cryst. A433, 178 (1977)
1.38 G. Hildebrandt, J.D. Stephenson, H. Wagenfeld: Z. Naturforsch. 28a, 588 (1973)
1.39 A.M. Afanasiev, Yu. Kagan: Acta Cryst. A24, 163 (1968)
1.40 P.H. Dederichs: Phys. Kondens. Materie 5, 347 (1966)
1.41 A.M. Afanasiev, Yu. Kagan, F.N. Chukhovskii: Phys. Status Solidi 28, 287 (1968); F.N. Chukhovskii: Sov. Kristallografiya 13, 960 (1968)
1.42 Y.N. Ohtsuki: J. Phys. Soc. Japan 19, 2285 (1964)
1.43 M.D. Giardina, A. Merlini: Z. Naturforsch. 28a, 1360 (1973)
1.44 A. Howie, M.J. Whelan: Proc. Roy. Soc. A263, 217 (1961)
1.45 S. Takagi: Acta Cryst. 15, 1311 (1962); - J. Phys. Soc. Japan 26, 1239 (1968)
1.46 G. Borrmann, K. Lehmann: *Crystallography and Crystal Perfection* (Academic Press, London 1963) p. 101
1.47 K. Lehmann, G. Borrmann: Z. Kristall. 125, 234 (1967)
1.48 T. Saka, T. Katagawa, N. Kato: Acta Cryst. A28, 102, 112 (1972)
1.49 N. Kato, K. Usami, T. Katagawa: Advances in the X-ray Analysis 10, 46 (1967)
1.50 A. Authier: Phys. Status Solidi 27, 77 (1968)
1.51 A. Authier, A.D. Milne, M. Sauvage: Phys. Status Solidi 26, 469 (1968)
1.52 M. Hart, A.D. Milne: Acta Cryst. A25, 134 (1969)
1.53 T.S. Uragami: J. Phys. Soc. Japan 27, 147 (1969)
1.54 T.S. Uragami: J. Phys. Soc. Japan 28, 1508 (1970)
1.55 T.S. Uragami: J. Phys. Soc. Japan 29, 1141 (1971)
1.56 Z.G. Pinsker, F.N. Chukhovskii: Sov. Kristallografiya 20, 501 (1975)
1.57 P. Penning, D. Polder: Philips Res. Rep. 16, 419 (1961)
1.58 N. Kato: J. Phys. Soc. Japan 18, 1785 (1963); - 19, 67, 971 (1964)
1.59 a) U. Bonse: Z. Phys. 177, 385 (1964); b) U. Bonse, W. Graeff: Z. Naturforsch. 28a, 558 (1973)
1.60 I.Ch. Slobodetskii, F.N. Chukhovskii, V.L. Indembom: Sov. Phys. JETP Letters 8, 90 (1968)
1.61 A. Authier, D. Simon: Acta Cryst. A24, 517 (1968)
1.62 A.M. Afanasiev, V.G. Kohn: *Dynamical Theory of X-rays in Nonperfect Crystals* - Reprint (I.V. Kurchatov Institute,Moscow 1969);- Acta Cryst. A27, 421 (1971)
1.63 See: Z.G. Pinsker: *Dinamicheskoye rasseyaniye rentgenovskikh luchei* (Nauka, Moscow 1974) Chap. 11
1.64 D. Simon, A. Authier: Acta Cryst. A24, 527 (1968)
1.65 V.L. Indenbom, F.N.Chukhovskii: Sov. Uspekhi fizicheskikh nauk 107, 229 (1972)
1.66 P.V. Petrashen, F.N. Chukhovskii: Sov. JETP 69, 477 (1975)
2.1 L.D. Landaw, E.M. Lifshits: *Elektrodinamika sploshnykh sred (Electrodynamics of Continuous Media)* (Nauka, Moscow 1959, G.I.T.T.L.)
2.2 H. Wagenfeld: Acta Cryst. A24, 170 (1968)
3.1 S. Koshino, K. Kohra: Japan. J. Appl. Phys. 10, 551 (1971)
3.2 S. Koshino, A. Noda, K. Kohra: J. Phys. Soc. Japan 33, 158 (1972)
3.3 M. Renninger: Acta Cryst. a) A24, 143 (1968); b) A25, 3, 214 (1969)
3.4 J.J. De Marco, R.J. Weiss: Acta Cryst. 19, 68 (1965); R.J.Weiss: Acta Cryst. 18, 814 (1965)

4.1 Atomic Scattering Factors. In: *International Tables for X-Ray Crystallography*, Vol. III (Kynoch Press, Birmingham 1962)pp. 201-207
4.2 E. Clementi: J. Phys. Developm. (I.B.M.) 9, 2 (1965)
4.3 B. Dowson: Proc. Roy. Soc. A298, 379 (1967)
4.4 B. Dowson: Proc. Roy. Soc. A298, 395 (1967)
4.5 D.T. Cromer: Acta Cryst. 19, 224 (1965); D.T. Cromer, J.T. Waber: Acta Cryst. 18, 104 (1965)
4.6 P.A. Doyle, P.S. Turner: Acta Cryst. A24, 390 (1968)
4.7 B. Dowson: Proc. Roy. Soc. A298, 255, 264 (1967); B. Dowson, A.C. Hurley, V.W. Maslen: Proc. Roy. Soc. A298, 289 (1967)
4.8 S.T. Konobeyevskii: Acta Cryst. 8, 606 (1957)
4.9 B. Dowson, B.T.M. Willis: Proc. Roy. Soc. A298, 307 (1967); B. Dowson: Acta Cryst. A25, 12 (1969)
4.10 D.T. Cromer: Acta Cryst. 18, 17 (1965)
4.11 C.H. Dauben, D. Templton: Acta Cryst. 8, 84 (1955)
4.12 U. Bonse, G. Materlik: Z. Physik 253, 232 (1972)
4.13 H. Wagenfeld: Phys. Rev. 144, 216 (1966)
4.14 H. Sano, K. Ohtaka, H. Ohtsuki: J. Phys. Soc. Japan 27, 1254 (1969)
4.15 A.M. Elistratov, O.N. Efimov: Sov. Fizika Tverdogo Tela 4, 9, 2397 (1962); O.N. Efimov: Phys. Status Solidi 22, 297 (1967)
4.16 V.G. Kohn: Sov. Kristallografiya 15, 20 (1970)
4.17 C. Chezzi, A. Merlini, S. Pace: Phys. Rev. B4, 1833 (1971)
4.18 G. Grimvall, E. Persson: Acta Cryst. A25, 417 (1969)
4.19 E. Persson, O.N. Efimov: Phys. Status Solidi 2, 757 (1970a)
4.20 Z.G. Pinsker: Sov. Kristallografiya 15, 658 (1970)
4.21 P.J.E. Aldred, M. Hart: Proc. Roy. Soc. A332, 223, 239 (1973)
4.22 T. Matsushita, K. Kohra: Phys. Status Solidi 24, 531 (1974)
4.23 K. Nakayama, S. Kikuta, K. Kohra: Phys. Letters A37, 29 (1971)
4.24 J.J. De Marco, R.J. Weiss: Phys. Rev. 137A, 1869 (1965)
4.25 L.D. Jennings: J. Appl. Phys. 40, 5038 (1969)
4.26 J.F.C. Baker, M. Hart, J. Hellier: Z. Naturforsch. 28a, 553 (1973)
4.27 E. Persson, E. Zielinska-Rohozinska, L. Gerward: Acta Cryst. A26, 514 (1970)
4.28 B.W. Battermann, D.R. Chipman: Phys. Rev. 127, 690 (1962)
4.29 J. Ludewig: Z. Naturforsch. 28a, 1204 (1973)
4.30 L.S. Salter: Adv. Phys. 14, 1 (1965)
4.31 E. Zielinska-Rohosinska: Acta Phys. Polon. 27, 4, 587 (1965)
4.32 M. Lefeld-Sosnovska, E. Zielinska-Rohosinzka: Acta Phys. Polon. 21, 329 (1962)
4.33 P.P. Ewald: Acta Cryst. 11, 888 (1958)
4.34 Z.G. Pinsker: Sov. Kristallografiya 16, 1117 (1971)
4.35 P.B. Hirsch: Acta Cryst. 5, 176 (1952)
4.36 G.N. Ramachandran: Proc. Indian Acad. Sci. A39, 65 (1954)
4.37 N. Kato: J. Phys. Soc. Japan 10, 46 (1955)
4.38 E.K. Koviev, O.N. Efimov, L.I. Korovin: Phys. Status Solidi 35, 455 (1969)
4.39 A. Gray, G.B. Mathews: *A Treatise on Bessel Functions, A. Their Applications to Physics* (1922)
4.40 N.W. McLachlan: *Bessel Functions for Engineers* (1934)
5.1 N. Kato: Acta Cryst. 11, 885 (1958)
6.1 a) R. Courant, D. Hilbert: *Methods of Mathematical Physics*, Vol. 1 (Interscience Publ., New York 1953); b) I.N. Sneddon: *Fourier Transforms* (McGraw-Hill, New York 1951) Appendix; c) A. Sommerfeld: *Optics* (Academic Press, New York 1954)
6.2 A. Authier: Acta geologica et geographica. Universitatis Comenianae. Geologica no. 14, 37 (1968)
6.3 a) H. Hattori, late H. Kuriyama, N. Kato: J. Phys. Soc. Japan 20, 1047 (1965); b) M. Hart, A.R. Lang: Acta Cryst. 19, 73 (1965)

6.4 H. Hattori, late H. Kuriyama, T. Katagawa, N. Kato: J. Phys. Soc. Japan 20, 988 (1965)
6.5 S. Tanemura, N. Kato: Acta Cryst. A28, 69 (1972) (see the references to the previous papers)
7.1 J.A. Prince: Z. Physik 63, 477 (1930)
7.2 H. Wagner: Z. Physik 146, 127 (1956)
7.3 U. Bonse: Z. Physik 161, 310 (1961)
7.4 R. Bubakova: Czech. J. Phys. B12, 776 (1962)
8.1 a) S.K. Allison: Phys. Rev. 41, 1 (1932); b) M. Renninger: Z. Kristall. 89, 344 (1934)
8.2 See [1.9], Chap. 3, paragraph 8
8.3 G.N. Ramachandran, C. Kartha: Proc. Indian Acad. Sci. 35, 145 (1952)
8.4 H. Cole, N.R. Stemple: J. Appl. Phys. 33, 7, 2227 (1962)
8.5 P.B. Hirsch, G.N. Ramachandran: Acta Cryst. 3, 187 (1950)
8.6 M. Renninger: Acta Cryst. 8, 597 (1955)
8.7 R. Bucksch, J. Otto, M. Renninger: Acta Cryst. 23, 507 (1967)
8.8 A.M. Afanasiev, J.P. Perstnev: Acta Cryst. A25, 520 (1969)
8.9 A. Erdelyi: *Higher Transcendental Functions*, Vol. 2 (McGraw-Hill, New York 1953)
8.10 R.J. Weiss: Acta Cryst. 18, 814 (1965)
9.1 a) M.A. Blokhin: *Physics of X-Ray* (G.T.I., Moscow 1953) Chap. 3; - b) W. Bambynek et al.: Rev. Mod. Phys. 44, 716 (1972)
9.2 G. Brogren: Arkiv Fys. 8, 391 (1954)
9.3 G. Brogren, Ö. Adell: Arkiv Fys. 8, 97 (1954)
9.4 R. Bubakova, J. Drahokoupil, A. Fingerland: Czech. J. Phys. B11, 205, 199 (1961); - B12, 776 (1962)
9.5 M. Renninger: Z. Naturforsch. 16a, 1110 (1961)
9.6 M. Renninger: Advances in X-Ray Analysis 10, 32 (1967)
9.7 K. Kohra, S. Kikuta, S. Nakano: J. Phys. Soc. Japan 21, 1565 (1966)
9.8 K. Kohra, S. Kikuta: Acta Cryst. A24, 200 (1968)
9.9. K. Kohra: *Lecture on the I.S. School* (Limoges, France, August 18-26, 1975)
9.10 M. Lefeld-Sosnovska, C. Malgrange: Phys. Status Solidi 34, 635 (1969)
9.11 M.v. Laue: Z. Physik 72, 472 (1931)
9.12 J. Du Mond: Phys. Rev. 52, 872 (1937)
9.13 K. Kohra: J. Phys. Soc. Japan 17, 589 (1962)
9.14 a) T. Matsushita, S. Kikuta, K. Kohra: J. Phys. Soc. Japan 30, 1136 (1971); b) S. Kikuta, K. Kohra: J. Phys. Soc. Japan 29, 1322 (1970)
9.15 H. Hashizume, K. Nakayama, T. Matsushita, K. Kohra: J. Phys. Soc. Japan 29, 806 (1970)
9.16 S. Kikuta: Phys. Status Solidi 45, 333 (1971b)
9.17 E.K. Koviev, V.E. Baturin: Sov. Kristallografiya 20, 17 (1975)
9.18 U. Bonse, M. Hart: Appl. Phys. Letters 7, 238 (1965)
9.19 G. Brogren: Arkiv Fys. 8, 371 (1954); - 22, 87 (1962)
9.20 G. Brogren, Ö. Adell: Arkiv Fys. 8, 401 (1954); - G. Brogren, E. Hörnström: Arkiv Fys. 23, 81 (1963); - 24, 81 (1964)
9.21 B.T. Zakharov: Sov. Kristallografiya 10, 442 (1965); - 11, 227 (1966)
9.22 E.K. Koviev, L.P. Novikova, O.N. Efimov: Phys. Status Solidi 45, 385 (1971b)
9.23 E.K. Koviev, V.E. Baturin: Sov. Kristallografiya 21, no. 4, 683 (1976)
9.24 W.L. Bond: Acta Cryst. 13, 814 (1960)
9.25 M. Hart: Proc. Roy. Soc. A309, 281 (1969)
9.26 M.V. Koval'chuk, E.K. Koviev, Z.G. Pinsker: Sov. Kristallografiya 20, 142 (1975)
9.27 J.F. Baker, M. Hart: Acta Cryst. A31, 364 (1975)
9.28 S.K. Allison, In. Williams: Phys. Rev. 36, 1702 (1930)
9.29 P. Ross: Phys. Rev. 3, 253 (1932)
9.30 G. Brogren: Arkiv Fys. 8, 507 (1951)
9.31 R. Bubakova: Czech. J. Phys. B12, 695 (1962)

9.32 M.V. Koval'chuk, E.K. Koviev, Z.G. Pinsker: Sov. Kristallografiya 19, 1062 (1974)
9.33 U. Bonse, G. Materlik, W. Schröder: J. Appl. Cryst. 9, 3223 (1976)
9.34 J.H. Beaumont, M. Hart: J. Physics E, Sci. Instrum. 7, 823 (1974)
9.35 M. Hart: J. Appl. Cryst. 8, 436 (1975)
9.36 N. Kato: J. Phys. Soc. Japan 19, 971 (1964)
9.37 M. Hart: Z. Physics 189, 269 (1966)
9.38 C. Malgrange, A. Authier: C.R. Acad. Sci. Paris 261, 3774 (1965)
9.39 B.M. Battermann, G. Hildebrandt: Acta Cryst. A24, 150 (1968)
10.1 N. Kato, K. Usami, T. Katagawa: Advances in X-Ray Analysis 10, 46 (1967)
10.2 A. Authier, M. Sauvage: J. de Phys. (Paris) 27, Suppl. 7/8, C-3-137 (1966)
10.3 U. Bonse, M. Hart: Z. Phys. a) 188, 154 (1965); b) 190, 455 (1966)
10.4 U. Bonse, M. Hart: Z. Physik 194, 1 (1966)
10.5 U. Bonse, M. Hart: Acta Cryst. A24, 240 (1968)
10.6 U. Bonse, E. te-Kaat: a) Z. Physik 214, 16 (1968); b) U. Bonse, H. Hellkötter: Z. Physik 223, 347 (1969)
10.7 J. Chikawa: Appl. Phys. Letters 7, 193 (1965)
10.8 A.R. Lang, V.F. Miuskov: Appl. Phys. Letters 7, 214 (1965)
10.9 W. Bauspiess, U. Bonse, W. Graeff: J. Appl. Cryst. 9, 68 (1976)
10.10 A.D. Milne: M. Sci. Thesis (Bristol, 1966)
10.11 M. Hart, A.D. Milne: Acta Cryst. A26, 223 (1970)
10.12 J.F.C. Baker, M. Hart, J. Helliar: Z. Naturforsch. 28a, 553 (1973)
10.13 M. Hart, A.D. Milne: Phys. Status Solidi 26, 185 (1968)
10.14 V.L. Indenbom, I.Sh. Slobodetsky, K.G. Truni: Sov. Phys. JETP 66, 1110 (1974); I.Sh. Slobodetsky, F.N. Chukhovskii: Sov. Kristallografiya 15, 1101 (1970)
10.15 A.V. Shubnikov: Nature (Sov. Priroda) no. 2, 83 (1927); - no. 1, 20 (1953); - no. 11, 61 (1965)
10.16 S. Amelinckx: *The Direct Observation of Dislocations* (Academic Press, New York 1964)
10.17 H. Hashimoto, M. Mannami, T. Naiki: Phil. Trans. Roy. Soc. A253, 459 (1961)
10.18 R. Jevers: Phil Mag. 7, 1681 (1962)
10.18a D. Simon, A. Authier: Acta Cryst. A24, 527 (1968)
10.19 M. Hart: Phil Mag. 26, 821 (1972)
10.20 M. Hart: Brit. J. Appl. Phys. 1, 2, 1405 (1968)
10.21 R.D. Deslattes: Appl. Phys. Letters 15, 386 (1969)
10.22 U. Bonse, E. te-Kaat, P. Spieker, I. Curtis, I.G. Morgan, M. Hart, A.D. Milne: *Conference on Precision Measurement, A. Fundamental Constants* (Gaithersburg, Md. 1971) (N.B.S. Special Publication) p. 291
10.23 M. Hart: Proc. Roy. Soc. A346, 1 (1975)
10.24 J. Bradler, A.R. Lang: Acta Cryst. A24, 246 (1968)
10.25 V.F. Miuskov: see [1.63], p. 186; - *Sov. Problems of Modern Crystallography, A.V. Shubnikov Memory* (Nauka, Moscow 1975) p. 186
10.26 a) See [1.63], p. 293; - b) see also: F.O. Eiramdzhian, P.A. Bezirganian: Isvestia Armenian Acad. Sci. 6, 280 (1971)
10.27 S. Tanemura, A.R. Lang: Z. Naturforsch. 28a, 668 (1973)
11.1 D. Taupin: Bull. Soc. Fr. Mineral. 84, 51 (1961)
11.2 A.N. Tikhonov, A.A. Samarskii: *The Equations of the Mathematical Physics* (Nauka, Moscow 1966)
11.3 R. Courant: *Partial Differential Equations*, Vol. II (New York, London 1962)
11.4 F. Balibar: *Thesis* (Paris 1969)
11.5 K. Kambe: Z. Naturforsch. 20a, 770 (1965)
11.6 F.N. Chukhovskii, A.A. Shtolberg: Phys. Status Solidi 41, 815 (1970); - Sov. JETP 64, 957 (1973)

11.7 P. Penning: *Thesis* (Eindhoven 1966)
11.8 N. Kato, Y. Ando: J. Phys. Soc. Japan 21, 964 (1966)
11.9 E. Fukushima, K. Nayakawa, H. Nimura: J. Phys. Soc. Japan 18, 11, 348 (1963)
11.10 E.S. Meieraw, A. Blech: J. Appl. Phys. 36, 3162 (1965)
11.11 G.H. Schwuttke, I.R. Howard: J. Appl. Phys. 39, 1581 (1968)
11.12 Y. Ando, N. Kato: J. Appl. Cryst. 3, 74 (1970)
11.13 Y. Ando, I.R. Patel, N. Kato: J. Appl. Phys. 44, 4405 (1973)
11.14 Yu.M. Fishman, V.G. Lutzau: Phys. Status Solidi 18, 443 (1973)
11.15 N. Kato, I.R. Patel: J. Appl. Phys. 44, 965 (1973)
11.16 I.R. Patel, N. Kato: J. Appl. Phys. 44, 971 (1973)
11.17 P.V. Petrashen: Sov. Fizika Tverdogo Tela 15, 3131 (1973)
11.18 F.N. Chukhovskii: Sov. Kristallografiya 19, 482 (1974); - *Collected Abstracts II European Crystallogr. Meeting* (Keszthely 1974) p. 61
11.19 F.N. Chukhovskii, P.V. Petrashen: C.R. Dan SSRF 222, 599 (1975)
11.20 T. Katagawa, N. Kato: Acta Cryst. A30, 830 (1974)
11.21 O. Litzman, Z. Janecek: Phys. Status Solidi 25, 663 (1974a)
11.22 F.N. Chukhovskii, V.P. Petrashen: Acta Cryst. A33, 311 (1977)
11.23 A. Erdelyi: *Higher Transcendental Functions*, Vol. I (McGraw-Hill Book Co. New York, London 1958)
11.24 L.J. Slater: *Confluent Hypergeometric Functions* (Cambridge University Press, Cambridge 1962)
11.25 H. Hashisume, K. Kohra: Phys. Status Solidi 13, K9 (1972a)
11.26 F.N. Chukhovskii, V.P. Petrashen: C.R. DAN SSSR 228, 1087 (1976)
11.27 Gr.A. Korn, Th.M. Korn: *Mathematical Handbook* (McGraw-Hill Book Co., New York, San Francisco, Toronto, London, Sydney 1968)
12.1 P.P. Ewald: Z. Kristall. A97, 1 (1937)
12.2 G. Mayer: Z. Kristall. 66, 585 (1928)
12.3 Yu.S. Terminasov, L.V. Tuzov: Uspekhi fizicheskikh nauk 83, 223 (1964)
12.4 B. Post: J. Appl. Cryst. 8, 452 (1975)
12.5 T. Hom, W. Kiszenick, B. Post: J. Appl. Cryst. 8, 457 (1975)
12.6 E.J. Saccosio, A. Zajac: Phys. Rev. A139, 255 (1965)
12.7 W. Uebach, G. Hildebrandt: Z. Kristall 129, 1 (1968)
12.8 G. Hildebrandt: Acta Cryst. A25, 209 (1969)
12.9 G. Hildebrandt: Phys. Status Solidi 15, K83 (1973a)
12.10 W. Uebach: Z. Naturforsch. 28a, 1214 (1973)
12.11 A.A. Katznelson, V.I. Kissin, N.A. Polyakova: Sov. Kristallografiya 14, 965 (1969)
12.12 T.I. Borodina, A.A. Katznelson, V.I. Kissin: Phys. Status Solidi 3, 105 (1970a)
12.13 V.I. Iveronova, A.A. Katznelson, T.I. Borodina, I.G. Sapkova: Phys. Status Solidi 11, 39 (1972)
12.14 T.I. Borodina, V.I. Iveronova, A.A. Katznelson: Sov. Kristallografiya 19, 1140 (1974)
12.15 T.I. Borodina, V.I. Iveronova, A.A. Katznelson, T.K. Runova: Phys. Status Solidi 28, 365 (1975a)
12.16 T.I. Borodina, V.I. Iveronova, A.A. Katznelson, T.K. Runova: Sov. Kristallografiya 20, 490 (1975)
12.17 M.A. Umeno: Phys. Status Solidi 2, K203 (1970a)
12.18 S. Balter, R. Feldman, B. Post: Phys. Rev. Letters 27, 307 (1971)
12.19 R. Feldman, B. Post: Phys. Status Solidi 12, 273 (1972a)
12.20 T.C. Huang: Phys. Status Solidi 10, K149 (1972a)
12.21 T.C. Huang, B. Post: Acta Cryst. A29, 35 (1973)
12.22 T.C. Huang, M.H. Tillinger, B. Post: Z. Naturforsch. 28a, 600 (1973)
12.23 S.A. Kshevetsky, I.P. Mikhailyuk: Sov. Kristallografiya 21, 381 (1976)
12.24 V.D. Kos'mik, S.A. Kshevetsky, M.L. Kshevetskaya, I.P. Mikhailyuk, M.V. Ostapovich: Sov. Kristallografiya 21, 899 (1976)

12.25 V.D. Kos'mik, S.A. Kshevetsky, M.L. Kshevetskaya, I.P. Mikhailyuk, M.V. Ostapovichi: Sov. Kristallografiya 21, 38 (1976)
12.25a B. Post, S.L. Chang, T.C. Huang: Acta Cryst. A33, 90 (1977)
12.25b M. Umeno: Phys. Status Solidi 37, 561 (1976); 38, 701 (1976)
12.26 G. Hildebrandt: Phys. Status Solidi 24, 245 (1967)
12.27 A.L. Dalisa, A. Zajac, C.H. Ng: Phys. Rev. 168, 859 (1968)
12.28 T. Joko, A. Fukuhara: J. Phys. Soc. Japan 22, 507 (1967)
12.29 L. Lafourcade, J.J. Couderc, P. Larroque: C.R. Acad. Sc., Paris 260, 5752 (1965)
12.30 L. Lafourcade, P. Larroque, M. Baux, J.J. Couderc: J. Microscopie 1966, 537
12.31 V.G. Kohn: Sov. Kristallografiya 20, 6, 1152 (1975)
12.32 M.L. Kshevetskaya, S.A. Kshevetsky, I.P. Mikhailyuk, N.R. Raransky: Ukrainskii fizicheskii zhurnal 18, 578, 1168, 2052 (1973)
12.33 V.D. Kos'mik, M.L. Kshevetskaya, S.A. Kshevetsky, I.P. Mikhailyuk, N.D. Raransky: Ukrainskii fizicheskii zhurnal 19, 1640 (1974)
12.34 A.D. Fofanov, A.V. Kuznetsov, G.A. Karacheva: Sov. Kristallografiya 18, 1126 (1973)
12.35 A.D. Fofanov, A.V. Kuznetsov, B.G. Razgulyaev: Sov. Kristallografiya 21, 30 (1976)
12.36 B.J. Isherwood, C.A. Wallace: Acta Cryst. A27, 119 (1971)
12.37 H. Cole, F.W. Chambers, H.M. Dunn: Acta Cryst. 15, 138 (1962)
12.38 R. Colella: Acta Cryst. A30, 413 (1974)
12.39 K. Kambe: J. Phys. Soc. Japan 12, 25 (1957)
12.40 D.K. Faddeev, V.N. Faddeeva: *Numerical Methods of Linear Algebra* (Fizmatgiz, Moscow 1963)
12.41 J.H. Wilkinson: *The Algebraic Eigenvalue Problem* (Oxford University Press, London 1965)
12.42 P.J. Eberlein: Numer. Math. 14, 232 (1970)

Subject Index

Zeitschrift für Physik B

Condensed Matter and Quanta

EPS Europhysics Journal
ISSN 0340-224X

Zeitschrift für Physik appears in two parts.
Section A (Atoms and Nuclei) started with Vol. 272.
Section B (Condensed Matter and Quanta) with Vol. 20.
Each part can be ordered separately.

Section B covers the physics of condensed matter and general physics. In this section papers on the physical properties of crystalline, disordered, and amorphous solids, and on classical and quantum liquids will be published. Examples would be papers on superconductivity, phase transitions, surface effects, and studies of dynamic processes performed with the help of photon, electron, or neutron scattering. Emphasis is also put on quantum optics and statistical physics, especially in the area of nonequilibrium processes and cooperative phenomena. Papers on molecular physics that relate to problems of condensed matter are also invited.

Springer-Verlag
Berlin
Heidelberg
New York

For subscription information or sample copies write to:
Springer-Verlag Berlin Heidelberg New York
P. O. Box 105280
D-6900 Heidelberg 1
or to your bookseller

X-Ray Optics

Applications to Solids
Editor: H. J. Queisser
1977. 133 figures, 14 tables. XI, 227 pages
(Topics in Applied Physics, Volume 22)
ISBN 3-540-08462-2
Contents:
H. J. Queisser: Introduction: Structure and Structuring of Solids. – *M. Yoshimatsu, S. Kozaki:* High Brilliance X-ray Sources. – *E. Spiller, R. Feder:* X-ray Lithography. – *U. Bonse, W. Graeff:* X-ray and Neutron Interferometry. – *A. Authier:* Section Topography. – *W. Hartmann:* Live Topography.

G. Leibfried, N. Breuer

Point Defects in Metals I

Introduction to the Theory
1978. 138 figures, 29 tables.
XIV, 342 pages
(Springer Tracts in Modern Physics, Volume 81)
ISBN 3-540-08375-8
Contents:
Introduction and survey. – Harmonic approximation and linear response (Green's function) of an arbitrary system. – Lattice theory. – Continuum theory. – Transition from lattice to continuum theory. – Statics and dynamics of simple single point defects. – Scattering of neutrons and X-rays by crystals. – Probability, distributions and statistics. – Properties of crystals with defects in small concentration. – Appendix.

Optical Data Processing

Applications
Editor: D. Casasent
1978. 170 figures, 2 tables.
XIII, 286 pages
(Topics in Applied Physics, Volume 23)
ISBN 3-540-08453-3
Contents:
D. Casasent, H. J. Caulfield: Basic Concepts. – *B. J. Thompson:* Optical Transforms and Coherent Processing Systems with Insights from Crystallography. – *P. S. Considine, R. A. Gonsalves:* Optical Image Enhancement and Image Restoration. – *E. N. Leith:* Synthetic Aperture Radar. – *N. Balasubramanian:* Optical Processing in Photogrammetry. – *N. Abramson:* Nondestructive Testing and Metrology. – *H. J. Caulfield:* Biomedical Applications of Coherent Optics. – *D. Casasent:* Optical Signal Processing.

Dynamics of Solids and Liquids by Neutron Scattering

Editors: S. Lovesey, T. Springer
1977. 156 figures, 15 tables. XI, 379 pages
(Topics in Current Physics, Volume 3)
ISBN 3-540-08156-9
Contents:
Introduction. – Phonons. – Phonons and Structural Phase Transformations. – Dynamics of Molecular Crystals, Polymers, and Adsorbed Species. – Diffusion in Solids, in Particular Hydrogen in Metals. – Collective Modes in Classical Monoatomic Liquids. – Magnetic Scattering.